Twenty-Ninth Hanford Symposium
on Health and the Environment

Indoor Radon and Lung Cancer: Reality or Myth?

October 15–19, 1990

PART 1

Edited by

Fredrick T. Cross

Sponsored by
the United States Department of Energy
and Battelle, Pacific Northwest Laboratories

BATTELLE PRESS
Columbus • Richland

DISCLAIMER

This report was prepared as an account of work sponsored by an agency of the United States Government. Neither the United States Government nor any agency thereof, nor Battelle Memorial Institute, nor any of their employees, makes **any warranty, expressed or implied, or assumes any legal liability or responsibility for the accuracy, completeness, or usefulness of any information, apparatus, product, or process disclosed, or represents that its use would not infringe privately owned rights.** Reference herein to any specific commercial product, process, or service by trade name, trademark, manufacturer, or otherwise does not necessarily constitute or imply its endorsement, recommendation, or favoring by the United States Government or any agency thereof, or Battelle Memorial Institute. The views and opinions of authors expressed herein do not necessarily state or reflect those of the United States Government or any agency thereof.

PACIFIC NORTHWEST LABORATORY
operated by
BATTELLE MEMORIAL INSTITUTE
for the
UNITED STATES DEPARTMENT OF ENERGY
under Contract DE-AC06-76RLO 1830

Library of Congress Cataloging-in-Publication Data

Hanford Symposium on Health and the Environment (29th: 1990: Richland, Wash.)
Indoor radon and lung cancer, reality or myth?: Twenty-ninth Hanford Symposium on Health and the Environment, October 15–19, 1990 / edited by Fredrick T. Cross; sponsored by the United States Department of Energy and Battelle, Pacific Northwest Laboratories.
p. cm.
Includes bibliographical references and index.
ISBN 0-935470-69-7: $89.95
1. Lungs—Cancer—Epidemiology—Congresses. 2. Radon-Carcinogenicity—Congresses. I. Cross, Fredrick T., 1935– . II. United States. Dept. of Energy. III. Pacific Northwest Laboratory. IV. Title.
RC280.L8H34 1990 pt. 1 92-37686
616.99'424071—dc20 CIP

Available to DOE and DOE contractors from the Office of Scientific and Technical Information, P.O. Box 62, Oak Ridge, TN 37831; prices available from (615) 576-8401. FTS 626-8401.

Available to the public from the National Technical Information Service, U.S. Department of Commerce, 5285 Port Royal Rd., Springfield, VA 22161, or through Battelle Press, 505 King Avenue, Columbus, Ohio 43201-2693. 614/424-6393. Toll free 1-800-451-3543.

CONTENTS

Part 1

RADON TRANSPORT IN SOILS AND INTO STRUCTURES

RADON AND RADON PROGENY SOURCES

BIOLOGICAL AND STATISTICAL MODELING STUDIES

EPIDEMIOLOGICAL STUDIES

FOREWORD

The 29th Hanford Symposium on Health and the Environment was a landmark event for the disparate, multidisciplinary, radon research community. While there have been other radon conferences for other audiences, none prior to this one has addressed solely the radon health issues in such a prestigious and comprehensive way. As a result, this meeting brought together people investigating radon and those who had industrial and policy-making concerns. The symposium presented current research results and their applications to risk and mitigation issues. International concerns were also discussed by international scientists and policy-makers from a broad range of backgrounds.

The Department of Energy Office of Health and Environmental Research (OHER), long a sponsor of environment and health research, cosponsored this symposium with Battelle, Pacific Northwest Laboratories. For OHER, the topic of this symposium was especially timely and useful because the DOE Radon Research Program (initiated in 1987) has begun to produce significant results. Participation in the symposium was an opportunity for researchers in the DOE program, as well as other scientists, to focus on their recent accomplishments, for policy-makers to integrate newly available knowledge, and for funding agencies to identify areas that are amenable to further productive research efforts.

The meeting was rewarding and noteworthy for all who attended and a gratifying and positive reflection on the Pacific Northwest Laboratory staff. It received significant national and international prominence and praise. These proceedings will be extremely interesting to anyone involved in the radon issue and, especially, to those attempting to resolve the uncertainties that surround it.

Susan L. Rose, Ph.D.
Manager, Radon Research Program
Office of Health and Environmental Research
U.S. Department of Energy

PREFACE

The Twenty-Ninth Hanford Symposium on Health and the Environment, "Indoor Radon and Lung Cancer: Reality or Myth?," was held in Richland, Washington, on October 15-19, 1990. Sponsors were the United States Department of Energy and Battelle, Pacific Northwest Laboratories. The symposium addressed the most important public health issue in radiation today: Is indoor radon causing lung cancer? The meeting explored diverse methods and approaches, including statistical, biological, and biophysical modeling, to provide perspective on this currently important public issue. Despite a budget crisis in the United States at the time, approximately 250 people, including participants from 15 nations, attended the symposium; 71 papers were presented, including the provocatively titled banquet address, "Radon is Out," by Dr. John Harley, former director, Environmental Measurements Laboratory, New York, New York.

At the time of the Symposium, significant results were beginning to emerge from the recently initiated, multidisciplinary Department of Energy (DOE) Radon Program and from the Commission of European Communities (CEC) radon-related studies. Therefore, it was the intent of the organizers to broaden the base of topics on the radon issue that would be discussed at the symposium while, at the same time, emphasizing the health-effect studies.

Sessions of the symposium included:

- Radon and Progeny Exposure Assessment
- Dosimetry Modeling
- Radon Transport in Soils and Into Structures
- Radon and Radon Progeny Sources
- Methods to Control Radon and Radon Progeny Exposure
- Molecular/Cellular-Level Studies
- Animal Studies and Exposure Systems
- Biological and Statistical Modeling Studies
- Epidemiologic Studies
- Public Strategy, Information, and Risk Communication
- Scientific Activities and Programs to Understand and Control Exposure to Radon (Panel).

The organization of the Proceedings follows this general order. Paper and panel discussions are also included in the Proceedings because they provide greater understanding of the topics discussed and additionally reflect the mood of the participants specific to a radon issue.

A brief description of the evidence bearing on the main question and title of the symposium, "Indoor Radon and Lung Cancer: Reality or Myth?," is provided.

The evidence supporting the opinion that the connection between indoor radon and lung cancer is a myth came from ecologic-type epidemiology studies. Experts in epidemiological and statistical evaluation cast doubts on the utility of such studies for assessment of risk, pointing particularly to the possibility of false findings due to inadequate data and study design. Indeed, the findings of an ecological analysis of the New Jersey data was contradictory to the results of the case-control analysis. In this respect, the consensus at the Hanford Symposium echoed that of the study design group of a recent international workshop on residential radon epidemiology [U.S. Department of Energy and Commission of European Communities (1989), International Workshop on Residential Radon Epidemiology: Workshop Proceedings, July 1989. National Technical Information Service, Springfield, VA, CONF-8907178, pp. 7-12], which recommended against ecologic studies and endorsed case-control studies for risk assessment of indoor radon exposure.

The evidence (not all formally presented, but discussed) supporting the opinion that indoor radon exposure's connection with lung cancer induction represents reality was threefold. First, there are molecular/cellular-level data on mutagenic and cytogenetic damage from radon exposure in combination with current carcinogenesis theory. It was stated that we may never be fully convinced that exposure to indoor radon causes lung cancer until we understand the mechanisms of radon carcinogenesis. A large carcinoma data base from animal radon studies was well fitted by the two-mutation recessive oncogenesis model. This model postulates that two hereditary (at the level of the cell) genetic changes (mutations) are needed to convert normal cells to cancer cells. We know from experience in animal and underground miner studies that radon exposure causes lung cancer; the evidence is particularly clear in animals, even in the absence of other associated exposures; for example, to cigarette smoke. In order for a lung cancer to be radon-related, however, only one mutation is necessary from radon exposure. Other mutations could come from additional exposure to radon or from other mutagenic agents, even from spontaneous mutational events, such as errors in replication of dividing cells. Thus, at the risk of gross oversimplification, it was believed to be theoretically and mechanistically possible that a single alpha-particle hit to the nucleus (the lowest exposure and exposure rate possible) could produce a viable mutation, and, under the right conditions, ultimately a cancer.

Second, there is the evidence from animal studies. A large data base in rats from both the United States and France supports the evidence from underground miners; namely, lung-cancer risk in rats is similar to that in humans at occupational exposure levels (Cross, F. T. 1988. Radon Inhalation Studies in Animals, DOE/ER-0396, Report prepared for the U.S. Department of Energy, Office of Health and Environmental Research, Washington, DC, September 1988, NTIS, Springfield, VA; and Symposium discussions). Except for minor differences, tumor pathology is also comparable, as is the cancer-related dose of radon exposures. It is, of course, a leap of faith to say that risks are also similar at the lower exposure levels typical of lifetime cumulative exposures in houses, where rats still show excess carcinoma development.

Third (evidence that even skeptics might accept) is that from carefully conducted case-control studies of indoor radon exposure. The earlier evidence of radon and lung cancer in New Jersey women was strengthened at this symposium with follow-up data. It was also supported by data from Sweden and by reference to risk data from a forthcoming case-control study report on Missouri women. Where possible, eventual data-pooling with other case-control studies might not only strengthen the positive correlation of increasing risk with increasing indoor radon concentration but might also allow the development of reasonable risk coefficients for residential exposure.

Future meetings will undoubtedly expand on the data presented at the 1990 Hanford Symposium. The residential epidemiology studies were, for the most part, at their genesis in 1990. And the very important experimental carcinogenesis studies, including radon and cigarette-smoke inhalation exposures, were also just beginning to produce data useful for incorporation in biologically based models of risk.

The 1990 Hanford Symposium on Health and the Environment was an exciting and rewarding event for those privileged to attend it. It is the hope of the organizers that the Proceedings will provide a sense of the spirit and accomplishments of this meeting.

The organizers of the symposium (F. T. Cross, chairman; L. A. Braby, A. L. Brooks, G. E. Dagle, E. S. Gilbert, A. C. James, J. R. Johnson, R. F. Jostes, J. A. Mahaffey, and P. C. Owczarski) are grateful to the authors, session chairpersons, and all other participants who contributed to its success. Special thanks are due T. A. Zinn and P. J. Parker, symposium secretaries; R. W. Baalman and D. Felton, technical editors; V. G. Horstman, local arrangements; and M. Cross and H. B. Crow, word processors.

F. T. Cross, Editor

H. M. Parker
Lecture

H. M. Parker Lecturer, 1990

J. C. Villforth

INTRODUCTION

W. J. Bair

The H. M. Parker Lectures are presented each year on the occasion of the Hanford Symposium on Health and the Environment. These lectures are sponsored by the Department of Energy; Battelle, Pacific Northwest Laboratories; and the Herbert M. Parker Foundation for Education in the Radiological Sciences.

The objectives of the H. M. Parker Lecture series are:

1. to enhance the public's understanding of radiation health issues
2. to honor contemporary scientists who evoke H. M. Parker's high technical standards and concern for protection of the health of workers and the public
3. to memorialize H. M. Parker and his outstanding contributions to radiological sciences and radiation protection.

The modern-day practices of medical physics and of health physics were shaped to a large extent by Herbert M. Parker, whose career extended from the early 1930s to the mid 1980s. As a medical physicist, Herb was a pioneer in the use of radiation to treat cancer. In the 1930s, in England, he was codeveloper of the Paterson-Parker dosimetry system, which revolutionized radium therapy for cancer patients. This system is still in use today. His work as a radiological physicist at Swedish Hospital in Seattle in the early 1940s advanced cancer treatment with supervoltage radiation.

Asked to join the atomic bomb project in 1942, Herb Parker turned his attention to radiation protection. He was the principal architect of the health physics program at the Clinton Laboratories, which later became the Oak Ridge National Laboratory. Herb came to Hanford in 1944 to establish a radiation protection program, which was the first to emphasize the environment as well as protection of workers. Since Herb Parker believed strongly in the need to provide information about scientific matters to the public, and about radiation in particular, and because he was himself an outstanding communicator, it is appropriate to memorialize him with this public lecture series.

I am pleased that his wife, Margaret, is present for the lecture. You will have an opportunity to greet her at the reception for the speaker immediately following the lecture.

For the fifth lecture of the series, our speaker is John Villforth. Until quite recently, John was Assistant Surgeon General of the U.S. Public Health Service and Director of the Center for Devices and Radiological Health of the Food and Drug Administration. The Center is responsible for carrying out a nationwide program designed to assure the safety and effectiveness of medical devices and to control unnecessary radiation exposures to humans from medical, industrial, and consumer electronic products. Herb Parker knew John and applauded his efforts to reduce radiation exposures of the general public.

John received Bachelor of Science and Master of Science degrees in sanitary engineering from Pennsylvania State University and later a Master of Science in physics from Vanderbilt. He served with the U.S. Air Force from 1954 to 1961. He was sent to Oak Ridge National Laboratory to gain practical experience in radiation protection and was then assigned as Commander of the USAF Radiological Health Laboratory at Wright Patterson Air Force Base in Ohio, which provided specialized radiation laboratory support for the entire Air Force.

After leaving the Air Force, John joined the U.S. Public Health Service, beginning an association that lasted nearly 30 years. Commissioned into the Service in 1961, he served first as Chief of the Radiation Surveillance Network, which was responsible for monitoring radioactive fallout and evaluating the public health consequences of fallout. From 1963 to 1967 he was Chief of the Radioactive Material Section, where he was responsible for developing programs to protect people from the medical uses of radium and accelerator-produced radioactive materials. In 1967 he was made Chief of Medical and Occupational Radiation Programs and was responsible for developing means to minimize x-ray exposure in medicine and industry by modifying x-ray machines, enhancing operator techniques, and improving the judgment of physicians who order x-ray examinations. The next year, he was appointed Chief of the Division of Medical Radiation Exposure, where he developed programs to reduce unnecessary radiation exposure from the use of medical and dental x-rays and the medical uses of radioisotopes.

From 1969 to 1982 he was Director of the Bureau of Radiological Health, the Food and Drug Administration's organization responsible for implementing the provisions of the Radiation Control for Health and Safety Act. The Bureau developed and enforced performance standards to assure that ionizing and nonionizing radiation emissions from electronic products used in medicine and industry and by consumers are safe. In 1982 he assumed the position of Director of the Food and Drug Administration's Center for Devices and Radiological Health, from which he retired in August. Last month he assumed the position of Executive Director of the Food and Drug Law Institute in Washington, D.C.

In 1972 John was promoted to Assistant Surgeon General (Rear Admiral) in the U.S. Public Health Service.

John also serves as Head of three of the World Health Organization's Collaborating Centers for radiation. These Centers provide advice, consultation, and training for users of medical devices, including those that generate nonimaging radiations such as microwaves, lasers, and ultrasound.

John has represented the Public Health Service in many capacities: he coordinated activities after the accident at Three Mile Island, he is the senior policy official on the Committee on Interagency Radiation Research and Policy Coordination, and he is a participant in the Health and Human Services Program to assure the continuation of government operations in the event of a national emergency.

John is the recipient of numerous awards and honors. The Public Health Service awarded him the Distinguished Service Medal, twice; the Meritorious Service Medal; and the Outstanding Service Medal. He was also awarded the US Air Force Commendation Medal, and the Health Physics Society honored him with the Elda Anderson Award (in 1970). He received the Pennsylvania State University, College of Engineer's Outstanding Engineering Alumnus Award. In 1982 he was the American Academy of Dental Radiology's H. Cline Fixott Memorial Lecturer.

John is a member of a number of professional societies and is a past president of the Health Physics Society. He is a Diplomat of the American Board of Health Physics.

He has been a member of the National Council on Radiation Protection and Measurements, and of the International Commission on Radiological Protection's committee that addresses protection in medicine.

It is a great pleasure to introduce the fifth H. M. Parker Lecturer, John C. Villforth. He will speak on the topic, "Radiation With an On/Off Switch" ("Radiation Protection and the Public Health Service").

H. M. PARKER LECTURE

RADIATION PROTECTION AND THE PUBLIC HEALTH SERVICE

J. C. Villforth*

President, Food and Drug Law Institute, Washington, D.C.

Key words: *History, radium, x rays, electronic products, public health, radiological health*

INTRODUCTION

As the radiation community prepares to celebrate the centennial of Roentgen's 1895 discovery of X rays, there is an opportunity for us to reflect on the leaders who have contributed to making radiation one of the most important diagnostic and therapeutic tools in medicine. It is a tool that is, perhaps, one of the best understood by the scientific community and least understood by the general public.

Dr. Herbert M. Parker is one of these contributing leaders, and I am honored to present a lecture in his name. His leadership started in the field of medical physics, where his system of determining gamma-ray doses from radium tubes and needles used in interstitial therapy has been the system accepted for over 50 y by radiation therapists. His leadership in dosimetry, using Sievert ionization chambers or film badges for personnel monitoring (in 1936), put him in demand for the radiation protection activities of the Manhattan Project. He was invited to join the Health Division in Chicago (1942), went to Oak Ridge (1943) to head the health physics program, then to Hanford (1944). Besides being a prolific writer in the open scientific literature, he also contributed to the National Council on Radiation Protection (NCRP) in the area of "permissible doses" and safe handling of isotopes, which set the stage for radiation protection, or health physics (a term that Herb did not appreciate), for years to come.

---

*Former Director, Center for Devices and Radiological Health, Food and Drug Administration, U.S. Public Health Service.

Dr. Parker was also a consultant and advisory committee chairman to the U.S. Public Health Service (PHS) radiological health program, giving us his wisdom and scientific advice over the years. Because the PHS program did not have as much budgetary support as the Atomic Energy Commission's program in health physics, it was greatly overshadowed. But the PHS made a significant contribution in reducing the risks from unnecessary exposure and by monitoring the effects of environmental radiation. Dr. Parker's counsel was often sought by the leaders of the PHS program during evaluation.

It is appropriate to examine some of the milestones in PHS history to see what it contributed to the overall history of radiation protection.

THE EARLY YEARS

One of the earliest milestones is the work, in 1923, of R. C. Williams, of the Office of Industrial Hygiene and Sanitation. Dr. Williams assisted the National Bureau of Standards (NBS; now the National Institute for Science and Technology) by assessing the exposure of staff while calibrating radium sources used by medical professionals. He was one of the earliest to use dental films as "film badges." He established that NBS employees were being routinely exposed to radiation and reported a correlation with film density and apparent decreases in their red and white blood cell counts. This work resulted in recommendations (Williams, 1923) that are summarized as follows:

- Blood examinations and blood pressure readings should be made at regular intervals on all employees of the radium section.
- All new employees of the section, before beginning work, should be given complete physical examinations, including examinations of the blood.
- In handling radium, employees must utilize, to the greatest extent possible, all practical protective devices, such as screens, lead-lined carrier boxes, and handling forceps.
- Employees should minimize unavoidable unprotected exposure to radiation and should not remain in the vicinity of radium longer than necessary.

At about the same time, the International X-Ray and Radium Protection Committee (predecessor to the International Committee for Radiation

Protection) published similar guidance dealing with protective measures for x-ray procedures and radium calibration. These recommendations focused on working-hour limitations, protective procedures, and electrical precautions for using x-ray equipment (ICRP, 1928).

Dr. Williams set the stage for many important considerations but, ironically, the PHS required some prompting to become fully involved in radiation protection. A 1928 letter from Public Health Service Surgeon General Cumming to the New Jersey Health Commissioner underscores PHS's reluctance to participate in the investigation of risks from the use of radium dial paint in industry. "Our function is to prevent the spread of diseases, more especially communicable disease," according to Cumming, and "...the use of radium dial paint is an industrial hazard and there is nothing catching about it." (New York Times, 1928) Notwithstanding this initial reaction, the PHS reversed its decision only a year later and announced that its Industrial Hygiene and Sanitation Division would investigate health hazards in radium dial painting. The investigators were charged with submitting recommendations for a state law to prevent additional deaths from that occupation (New York Times, 1919).

In later years, against the backdrop of the upheaval created by World War II, awareness of the importance of radiation protection for medical personnel continued. A report prepared by the PHS National Institutes of Health in 1941 noted that the therapeutic use of radiation was increasing and that methods to ensure the protection of health professionals should not be ignored. The National Institutes of Health studied radiation protection practices in 45 hospitals and noted that none passed muster. Skin changes were observed in one-quarter of the radiologists surveyed.

The report also cited problems associated with x-ray equipment operators entering the treatment room when the x-ray tube was in operation, and that medical personnel were routinely holding the patient in position during treatment. The report recommended that more attention be paid to hazards associated with these practices. According to the report, fluoroscopy was "one of the worst offenders in causing overexposure," and additional study was advised to determine the adequacy of lead barriers and lead-lined control rooms for x-ray equipment operators. Also mentioned was the need for methods more reliable than dental films to monitor radiation exposure. In this early report, two important areas that are still stressed today emerge as critical: educating personnel about protective techniques, and improvements in equipment that would afford greater radiation protection.

The report concludes that

> "...no hospital visited had provided optimal protection in every regard. It was observed, however, that most of the radiologists were becoming increasingly aware of their problems and were taking steps to reduce exposures."

The report also suggested that the skin damage seen indicated that much remained to be accomplished in preventing overexposure, and that it would be interesting to see the results of a careful nationwide survey of injuries among radiation workers. Furthermore, educating radiation personnel about protection practices should be stressed as much as providing protective devices.

The report also noted that the new and increasing demands of national defense activities, along with new radiation methods, would create the need for additional equipment and personnel. The report therefore urged that protection surveys and educational programs include people working in industry and defense activities (Cowie and Scheele, 1941).

AFTER WORLD WAR II

With its recommendations for additional study, increased protection measures, and the need to include defense activities, the National Institutes of Health report foreshadowed the next period in radiation protection. A number of PHS activities spanning the period of the 1950s through the early 60s focused on the effects of low levels of ionizing radiation. Much of the federal and state effort at that time was devoted to measuring levels of radioactivity in the environment, searching for potential biologic consequences of low-level contamination, and seeking ways to minimize it. This was because, during this time, a substantial proportion of the world's population was receiving low levels of exposure to ionizing radiation as a result of the aboveground testing of nuclear weapons. During this period, the PHS promoted research, information, and training in radiological health. Environmental radiation responsibilities for conducting surveillance and monitoring programs created the need for new and/or improved measuring instruments. In addition to developing these instruments, the PHS also provided consultative services to government and nongovernment agencies.

In 1949 the first cooperative agreement between the Atomic Energy Commission (AEC) and the PHS was signed. Its purpose was to conduct a study of stream characteristics of the Columbia River Basin to determine the effect of radioactive wastes in the vicinity of weapons production sites. Other cooperative agreements with the AEC were subsequently established. For example, joint AEC-PHS research on decontamination of radioactive waters was initiated in 1950 at Oak Ridge National Laboratory.

In 1951, the AEC and the PHS studied stream characteristics of the Savannah River, as they related to AEC activities there. Also during 1951 the AEC first requested that the PHS provide personnel to assist in offsite radiological monitoring at the Nevada Weapons Testing Site. In 1953, the PHS measured fallout in an offsite area that covered a radius of 300 mi from the test site.

Perhaps the most significant undertaking by the AEC and the PHS was the joint agreement of February 1, 1954. In it, the PHS agreed to support the weapons-testing activities of the AEC by conducting community education and public relations programs; performing radiological monitoring, operating air sampling and background recording equipment at designated stations, and collecting other environmental samples (water, food, vegetation); operating chemical and counting laboratories, and assisting in the analysis of data; and preparing required reports.

A satellite office of the PHS radiological health program was established in Las Vegas, Nevada, to provide coverage for Test Site operations. Between test series, the permanent PHS staff at Las Vegas performed environmental sampling (milk, water, food) and laboratory analysis, maintained public contacts, and provided offsite coverage during safety shots intended to verify the stability and safety of nuclear weapons.

Another facet of the PHS support and cooperation with other government agencies was an agreement, in 1956, with the Department of Defense to provide offsite radiation safety support for inhabited areas during nuclear weapons testing in the Pacific Ocean. In addition, along with the AEC's Division of Biology and Medicine, the PHS established a National Radiation Surveillance network to make immediately available to the public factual data about environmental monitoring during weapons testing at Pacific Proving Grounds test sites.

The major direction of this period was environmental radiation monitoring, but the PHS was also managing activities in the area of radiation exposure prevention. In 1953, an x-ray equipment inspection program was carried out in PHS hospitals, and a guide for inspection of medical and dental installations was developed. Also in that year, the Radiological Health Branch of the PHS issued its first report on x-ray exposure in the healing arts

DIVISION OF RADIOLOGICAL HEALTH

In 1958, the Surgeon General created a National Advisory Committee on Radiation (NACOR) to address the nation's concerns about the effects of low-level radiation and the increasing need for protecting the public against its adverse effects. This committee provided the Surgeon General with guidance on matters pertaining to the control of radiation hazards in the United States.

Among the principal points made in NACOR's first report (NACOR, 1959) were: "Most of the ionizing radiation received by the population today, other than that received from natural sources, has been received from the x-ray machines employed by the health professions." And, "the absence of a comprehensive program through which the health hazards of all sources of ionizing radiation may be brought under supervision appears to this Committee to be an important weakness in this nation's efforts to control radiation safety."

The report made the following proposals and recommendations:

- Primary responsibility for the nation's protection from radiation hazards should be established in a single agency of the federal government; logically, this should be the U.S. Public Health Service.
- The agency should be granted authority for broad planning in the field of radiation control, including coordination of state and local regulatory programs with safety operations of federal and private groups in a manner which would provide a unified attack on problems associated with control of radiation hazards.
- The agency should be given the authority to develop a comprehensive program for all sources of radiation.

Early reports of the United Nations Scientific Committee on the Effects of Atomic Radiation (UNSCEAR) also recognized that the

"...highest genetically significant doses were caused by diagnostic x-ray exposures." In 1957, UNSCEAR issued the following statement:

> "It appears most important ... that medical irradiations of any form should be restricted to those which are of value and importance, either in investigation or treatment, so that irradiation of the population may be minimized without any impairment of the efficient medical use of radiation." (UNSCEAR, 1958)

Based largely on NACOR's recommendations, the Division of Radiological Health was created, on July 1, 1958, within the PHS Bureau of State Services. The Division, headquartered in Washington, D.C., was allocated 51 positions and an annual budget of $393,000.

A Presidential Executive Order in 1959 directed the Department of Health, Education, and Welfare to "...intensify its radiological health efforts and have primary responsibility for the collation, analysis, and interpretation of data on environmental levels ... so that the Secretary of Health, Education, and Welfare may advise the President and the general public." As a result of this directive, the Division of Radiological Health established three regional laboratories to handle the required measurement and surveillance work.

The first of these, the Southwestern Regional Radiological Health Laboratory, was established on December 4, 1959, in Las Vegas, Nevada. On February 10, 1960, the Southeastern Laboratory in Montgomery, Alabama, was established, and on October 12, 1961, responsibility for the AEC's facility in Winchester, Massachusetts, was transferred to the PHS, creating the third in this network of Regional Radiological Health Laboratories. In April, 1960, the PHS published the first issue of "Radiological Health Data," a monthly publication to give the public access to nationwide data on radiation levels in the environment (Terrill, 1982).

In 1961, the Division of Radiological Health was authorized to create a training grant program to meet the national need for radiological health specialists and technicians. Training grant funds were used to strengthen and extend programs of basic instruction and to encourage greater enrollment in the field by providing financial assistance to qualified students preparing for careers in radiological health. Training included study in radiobiology, atomic and nuclear physics, hazards evaluation, epidemiology, biostatistics, and other areas of radiation science and public health. Over the years, there were as many as

35 university-based specialist grants throughout the country, in which as much as $2,500,000 a year was invested before the program ended in 1975.

Today, a program of extramural research grants and contracts, begun by the Division in 1962, continues. It utilizes the scientific competence of the academic community to solve defined and immediate problems in the fields of radiological health. A foremost example of the type of research supported by this program is the collaborative research project initiated with the Colorado State University in Fort Collins, Colorado, in 1962. This study of the long-term effects of low-level gamma-radiation in dogs has cost over 20 million dollars. It was designed to address the consistently observed radiation effect of lifespan shortening, but its specific aim was to examine differential age sensitivity. Dogs were irradiated at three prenatal and three postnatal periods with whole-body doses of 0, 16, or 83 cGy (OST, 1989). All the dogs have died; the Colorado State University is expected to compile the final contractor's report in July, 1991. This study may be the first of the extensive, large-animal lifespan studies to complete its original goal.

This comprehensive study has examined lifespan shortening, effects on reproduction, developmental abnormalities, fatal diseases, and diseases that appear to heighten radiosensitivity. In the females, lifespan shortening appears to be due entirely to excess cancer mortality. In males, the reason is not clear. Overall, mortality from cancer accounted for more than 30% of deaths. About 25% of the deaths in the study were from heritable diseases, particularly chronic renal disease and thyroid insufficiency. About 20% died from problems involving immune and/or hormone functions, and almost 10% were from central nervous system disturbances.

The study has shown a number of developmental abnormalities: birth defects, ocular damage, growth retardation, defects of dentition. Kidney damage consistent with the pathology of chronic renal disease has been statistically significant among dogs irradiated in utero with 16 cGy. Also, a consistent reduction of brain weight has been observed among the dogs irradiated in utero. Incidence of fatal cancers during the first 4 y of life increased in dogs irradiated near birth. Relatively high and statistically significant increases in fatal cancers were observed among dogs irradiated near birth. There is a statistically significant increase in fatal thyroid carcinoma among dogs irradiated as juveniles. So far, no consistent relationship between irradiation and appearance of mammary carcinoma has been detected. To date, possible relationships of noncancer fatalities and irradiation have not been tested, although such analyses are expected in the final report.

RADIATION CONTROL FOR HEALTH AND SAFETY ACT

In January, 1967, the Division of Radiological Health became the National Center for Radiological Health, with a budget of some 20 million dollars. On December 20, 1968, the name was changed to the Bureau of Radiological Health, a component of the now-defunct Environmental Control Administration, Consumer Protection and Environmental Health Service. The total staff in Washington and the field comprised some 800 budgeted positions as well as about 250 reimbursable positions from the AEC. These reimbursable positions were to support offsite monitoring for the AEC's weapons testing program.

As the environmental programs (air, water, solid waste, and occupational waste) matured, concerns intensified. Consumer and Congressional anxiety over unnecessary radiation exposure (e.g., diagnostic x rays, lasers, microwaves) led to Congressional hearings on the x-ray emissions from color television sets and other electronic products. These hearings ultimately resulted in the passage of the Radiation Control for Health and Safety Act of 1968 (US Congress, 1968), signed by the President on October 18, 1968. Through this Act, Congress declared that the public health and safety must be protected from the dangers of "electronic product radiation." The term electronic product radiation means any ionizing or nonionizing electromagnetic or particulate radiation or any sonic, intrasonic, or ultrasonic wave which is emitted from an electronic product as the result of the operation of an electronic circuit in such product. The purpose of the Act was:

> "to provide for the establishment of an electronic product control program which shall include the development and administration of performance standards to control the emissions of ...radiation from electronic products and the undertaking by public and private organizations of research and investigations into the effects and control of such radiation emissions." (U.S. Congress, 1968)

The primary means for accomplishing the purpose of the Act was by promulgating federal mandatory performance standards. To assure that these standards were scientifically based and reflected the state of the art of the industry and the concerns of the consumers or users of the products, the Act required that the Bureau consult with a 15-person Technical Electronic Product Radiation Safety Standards Committee (TEPRSSC). The committee consisted of five scientists from industry, five from government, and five from the consumer or private sector. Their task was to provide advice to the Bureau after reviewing draft proposals for standards prepared by the Agency and after hearing the comments of concerned parties.

Herb Parker was a member of TEPRSSC and served as its first chairman, from 1969 through 1972. During his chairmanship two of the most significant standards were promulgated. One, for television receivers, limited x-ray emission. Another, for microwave ovens, not only limited emissions but also established a testing method and proposed interlock safety features. The importance of Dr. Parker's skillful leadership of this Committee cannot be overemphasized: This was the first time that the federal government had regulated ionizing and nonionizing radiation from electronic products. The consequences of this regulation were especially important because color television sets and microwave ovens were becoming widely available to consumers.

During Dr. Parker's tenure, the Committee considered the efficacy and the specific details of performance standards that would be promulgated after he left. As a result, the Bureau promulgated nine federal mandatory performance standards for a variety of electronic products:

- television receivers, January 16, 1970
- cold-cathode gas discharge tubes, May 19, 1970
- microwave ovens, October 7, 1971
- diagnostic x-ray systems and their major components, August 2, 1974
- cabinet x-ray systems (to include x-ray baggage inspection systems), April 10, 1975
- laser products, August 2, 1976
- ultrasonic therapy products, February 17, 1979
- mercury vapor lamps, March 8, 1980
- sunlamps, May 7, 1980.

On January 20, 1971, the President established an entirely new government agency, the U.S. Environmental Protection Agency (EPA), as a result of public concerns over inadequate federal attention to environmental protection. Responsibilities for radiation surveillance and monitoring in the environment were reassigned to the EPA from the PHS, along with 318 Bureau employees and over $7,000,000. Programs related to machine-produced electronic product radiation and the Radiation Control for Health and Safety Act remained in the PHS.

MEDICAL DEVICE AND RADIOLOGICAL HEALTH PROGRAMS COMBINE

The current radiation protection program of the PHS is administered by the Center for Devices and Radiological Health, one of the Food and Drug Administration (FDA) Centers. The Center was formed in 1982 when the FDA consolidated the programs of the Bureau of Radiological Health and the Bureau of Medical Devices to capitalize on the common elements of these programs and the engineering and scientific expertise in both organizations.

The radiological health program in the PHS began in the late fifties and was initially directed to minimizing exposure of the general population to environmental sources of ionizing radiation. Many of the basic concepts identified in the infancy of the program continue to be viable today. This is especially true for the issue of unnecessary medical x-ray exposure, which is the most significant source of exposure to man-made radiation.

CENTER PROGRAMS IN THE HEALING ARTS

One aspect of the PHS's radiological health program is improving radiation equipment. Equipment safety is prescribed at the manufacturing stage by the Standard for Diagnostic X-Ray Machines. This standard, which became effective in 1974, was developed under the authority provided by the Radiation Control for Health and Safety Act. It remains in force today, addressing such issues as collimation, filtration, and maximum tube-housing leakage.

In cooperation with the PHS, the states conduct inspections of x-ray equipment, including machines used for dental examinations, with a focus on major equipment problems and enforcing the performance standard. As a result of these efforts and technological improvements, equipment problems have largely been eliminated.

Attention has now shifted to techniques. The Center, state radiation control programs, and professional groups have concentrated on educational programs that promote techniques to produce optimal diagnostic images with minimal radiation exposure. The cornerstone of the Center's program in this area has been quality assurance; that is, using all appropriate tools and procedures that are available to optimize the diagnostic image and lessen the patient's exposure. To a great extent as a result of the Center's collaborative work, quality assurance is a byword in radiology today.

To help determine facilities that required assistance in perfecting radiation techniques, the PHS and the states cosponsored the Dental Exposure Normalization Technique (DENT) program. This effort was directed toward measuring the exposure delivered during a typical dental exam and comparing it with normal values. When a facility delivered an atypical exposure, a followup visit by state personnel was scheduled with the intent of evaluating the technique used in the facility and recommending corrections. Coincident with this effort was the Nationwide Evaluation of X-Ray Trends (NEXT) program, designed to monitor typical exposures delivered from various projections. Partly as a result of Center programs and also because of technological improvements, a typical exposure from a dental bitewing radiogram in the United States dropped from over 1200 mR in 1964 to less than 300 mR in recent years (Figure 1). Similar reductions in x-ray exposure can be seen in other x-ray procedures (Figures 2 and 3). There is also a need to limit the size of the x-ray beam to the area of the film or to the area of clinical interest. X rays that extend beyond this area are "wasted" (Figures 4 and 5). This concern, which was greatest when the medical radiation program was started in the early 1960s, has now been essentially eliminated for most common diagnostic procedures.

Evaluating clinical judgment is also included in the Center's medical radiation protection program. For years, physicians and medical organizations believed that, although the Center had a part to play in improving equipment and technique, it had no analogous role in determining when a particular radiologic examination was appropriate. In recent years, in cooperation with the American College of Radiology, the Center has convened panels of radiologists and representatives of other specialty groups to develop what are known as "referral" or selection criteria for radiographic procedures. This voluntary guidance can help the clinician decide on the usefulness of the radiographic examination under various clinical circumstances. We have published such criteria for x-ray pelvimetry, routine chest radiographic screening, presurgical chest radiography, skull radiography after trauma, and dental radiography.

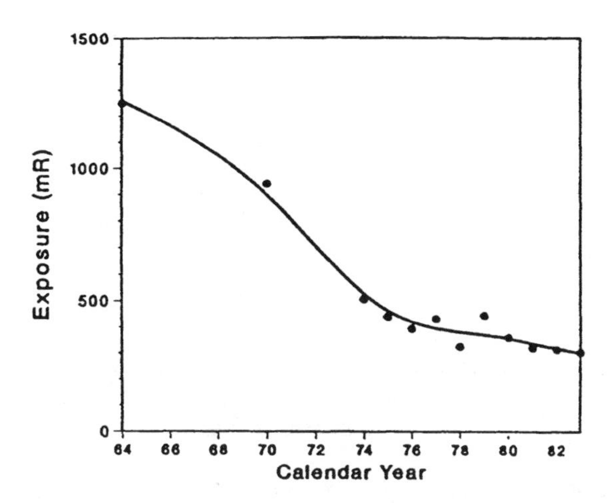

Figure 1. Mean exposure at cone tip during dental bitewing x ray, 1964-1983.

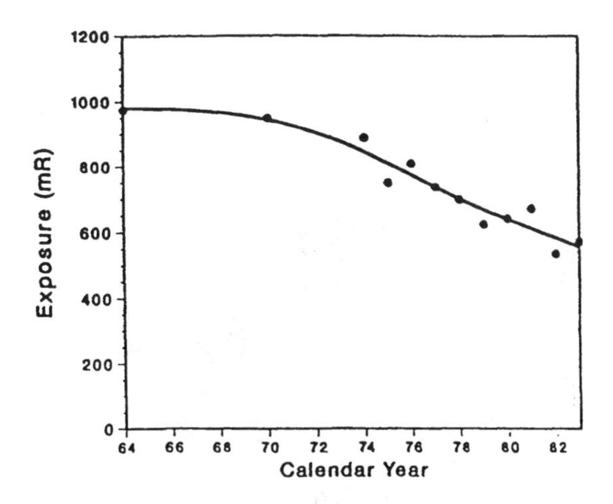

Figure 2. Mean exposure at the point of entrance of the primary beam on the skin during anterior/posterior lumbosacral spine x-ray examinations, 1964-1983.

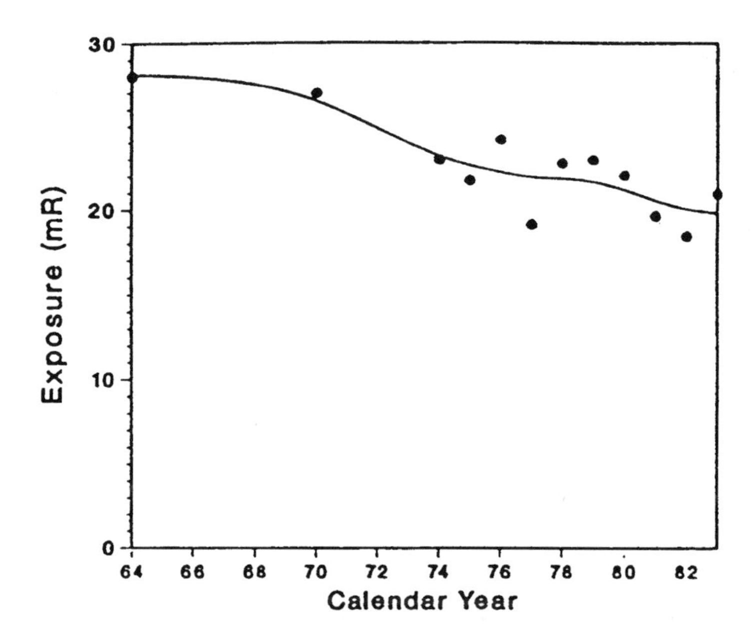

Figure 3. Mean exposure at the point of extrance of the primary beam on the skin during posterior/anterior chest x-ray examinations, 1964-1983.

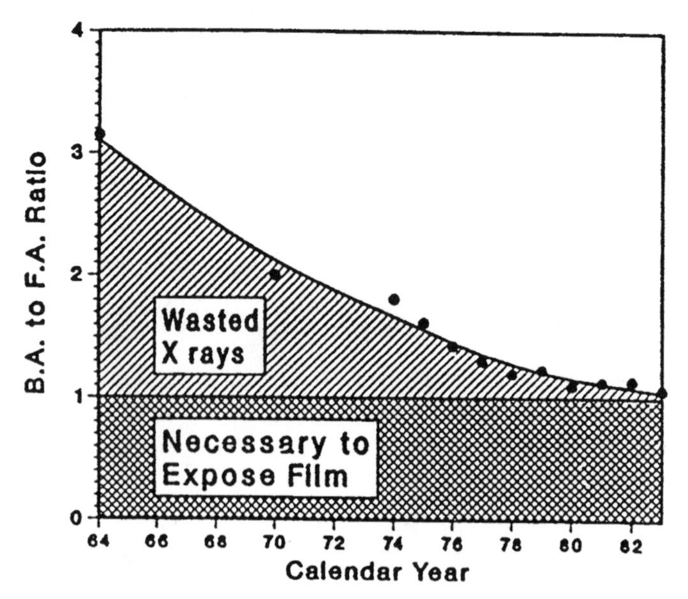

Figure 4. Mean ratio of the area of x-ray beam to the area of x-ray film during posterior/anterior chest x-ray examinations, 1964-1983.

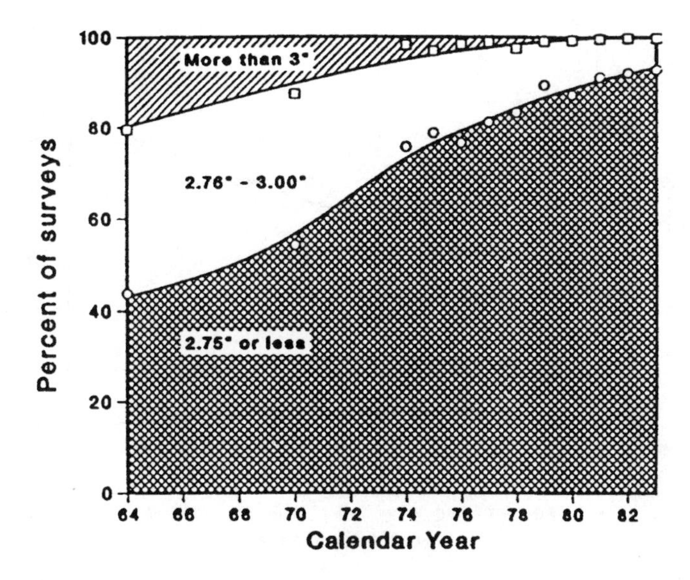

Figure 5. Dental x-ray beam diameter (in.), 1964-1983.

In spite of the efforts of the clinical community to limit the number and frequency of diagnostic x-ray examinations, the number continues to increase in the United States. This increase (Johnson, 1988), from an estimated 110 million examinations in 1964 to an estimated more than 200 million in 1985, is much greater than the proportionate increase in our population during the same period (Figure 6). In part, the increase may reflect the use of the new x-ray modalities that provide the clinician with better diagnostic information. It may also be due to the greater availability of diagnostic x-ray procedures to formerly underserved portions of the U.S. population.

Acceptance by the medical profession of the Center's role in evaluating clinical judgment is relatively new. Criteria recently published for dental radiography are the result of many years of work in this area and provide a good indication of how the radiation protection program has evolved in the modern age of "cost effectiveness" and "patient activism." The recent success of these recommendations is the result of public and professional concern about potential risks from radiation exposure and the increasing costs of radiographic examinations.

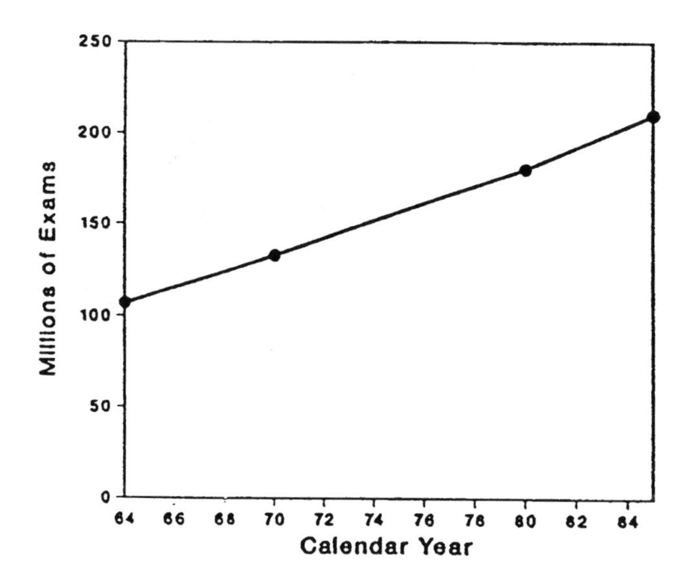

Figure 6. Estimated number of annual medical x-ray examinations in United States, 1964-1985.

The dental radiography guidelines mark a departure from the customary practice of taking bitewing x rays every 6 mo or yearly, and full-mouth radiographs every 3 y. Using these guidelines, a dentist takes an x ray based on clinical observation and the patient's health history. The guidelines also serve as recommendations as to the type of radiograph needed, how frequently, and under what conditions it should be taken (DHHS, 1987).

THE FUTURE

More challenges remain. The opportunity to begin a program to minimize fluoroscopy exposure has eluded us for some 30 y; however, the Center has recently established the methodology to assess the radiation dose delivered during an upper gastrointestinal tract examination. This is the first step in developing a comprehensive program in fluoroscopy.

Mammography is now a major screening program in the United States. Because the imaging task in mammography is very demanding, the

Center is working with the American College of Radiology to develop a certification program for facilities that perform mammography screening. This will ensure that these facilities conduct examinations which result in good clinical radiographs. Because increases in use of the procedure for diagnosis are anticipated, the challenge will be even greater to ensure that the use of radiation is optimized so that the patient is exposed only to the amount of radiation which is clinically necessary.

New technologies will continue to develop, and new challenges will continue to confront the radiation protection community. We must be ever-vigilant in developing the appropriate guidance and regulations to ensure that these technologies are used wisely. Dr. Parker contributed significantly to the progress of radiation protection in general and to the activities of the PHS specifically. It is therefore particularly appropriate that the public health principles he believed in should be considered in confronting these new challenges.

REFERENCES

Cowie, DB and L Scheele. **1941** A survey of radiation protection in hospitals. J Natl Cancer Inst 1:767-787.

DHHS. 1987. *The Selection of Patients for X-Ray Examinations: Dental Radiographic Examinations*, DHHS Publication No. FDA 88-8273. Department of Health and Human Services, Food and Drug Administration, Rockville, MD.

ICRP. 1928 Protocol for the General Assembly of the Second International Congress of Radiology Held at the House of Parliament in Stockholm, 27th July, 1928, Attachment to Report No. ICRP/78/S:G 4. International Commission on Radiological Protection, Oxford, England.

Johnson, DV. **1988** *Population Doses from X-Rays*, CRCPD Publication 88-6. Conference of Radiation Control Program Directors, Frankfort, KY.

NACOR. (March **1959**). *Report to the Surgeon General U.S. Public Health Service on The Control of Radiation Hazards in the United States*. National Advisory Committee on Radiation, Washington, DC.

New York Times. July 16, 1928. "Radium Injury Denied."

New York Times. February 26, 1919. "To Fight Radium Hazards."

OST. 1989. *Annual Report of the Office of Science and Technology, Fiscal Year 1988*. Center for Devices and Radiological Health, Food and Drug Administration, Rockville, MD.

Terrill, JG, Jr. **1982**. *The Role of the U.S. Public Health Service in Radiological Health: 1946-1969*, DHHS Publication FDA 82-8198. Department of Health and Human Services, Rockville, MD.

UNSCEAR. *Sources, Effects and Risks of Ionizing Radiation.* 1958 Report to the General Assembly, with annexes. United Nations Scientific Committee on the Effects of Atomic Radiation, New York, NY.

US Congress. Radiation Control for Health and Safety Act of 1968, Public Law 90-602, 90th Congress, HR 10790. Government Printing Office, Washington, DC.

Williams, RC. **1923**. *Preliminary Note on Observations Made on Physical Condition of Persons Engaged in Measuring Radium Prepartion*, Public Health Reports, Vol. 38, No. 51, pp. 3007-3028. US Public Health Service, Washington, DC.

RADON AND PROGENY EXPOSURE ASSESSMENT

Measurement and Modeling of Exposure Levels

CONTROL OF OCCUPATIONAL EXPOSURE TO RADON IN THE WORKPLACE

L. D. Brown

Saskatchewan Human Resources, Labor and Employment, Regina, Saskatchewan, Canada

Key words: *Radon, occupational exposure, workplace, schools*

ABSTRACT

Mining high-grade uranium ore is an important industry in Saskatchewan; inhalation of radon progeny is therefore a major concern. Strenuous efforts by the Provincial Radiation Safety Unit have kept the exposures received by the majority of miners to less than one working level month (WLM) in a year. Many workers in non-radiation-related employment may be receiving comparable, or occasionally higher, exposures from radon levels in buildings where they work. Because this cannot easily be justified, a draft regulation has been introduced to limit to 1 WLM the annual radon-progeny exposure received by any worker as a result of high natural radon levels in his place of employment.

The Occupational Health Branch is anxious to assess the potential impact of this proposed regulation before it is formally adopted by measuring radon levels in a representative group of buildings selected at random from all parts of the province. Schools have been chosen for this purpose because they provide a convenient sample size, are architecturally very similar to many typical workplaces, represent buildings of all ages, are uniformly distributed in all habitable areas, and, in any given community, have a fairly constant numerical relationship to the number of residential buildings. Initially, single readings are being made in each school, using the charcoal canister procedure. Wherever a nontrivial initial reading is obtained, follow-up samples are taken as necessary. This paper reviews the progress of the study and reports the measurements so far completed.

INTRODUCTION

Saskatchewan has a long history of active involvement not only with radon but also with other radiation-related issues. For example, in 1941 Saskatchewan was the first jurisdiction in North America to introduce mass miniature x-ray screening for tuberculosis, using a photofluorographic x-ray unit which was developed in the province. In 1951, another major pioneer development was the design and use of a cobalt-60 irradiation unit for cancer therapy. At that time, radon was primarily of interest in the form of seeds used as a radiation source for therapy. These were originally supplied from a laboratory in New

York, but by 1931 the first radon plant in western Canada had been established in Saskatoon. It remained active until 1962, when more suitable isotopes became available. Provincial regulations for working with radiation sources date back to November, 1931, and in 1962 Saskatchewan became the first province in Canada to enact a Radiological Health and Safety Act.

Mining radium for medical purposes commenced at Port Radium on the eastern shore of Great Bear Lake in the North West Territories by the Eldorado Mining and Refining Company in 1930. Milling uranium from this ore began in 1933 as an adjunct to refining radium. The same company discovered uranium deposits near Lake Athabasca in northern Saskatchewan in 1932. The development of uranium mining for nuclear power in this area commenced in 1949, and in 1952 the township of Uranium City was laid out close by. Within a few years 19 other small mining companies had commenced operations nearby. This new township was in a very remote area, where all supplies had to be flown or barged in over long distances. This contributed to the extensive use of waste rock from the mines in construction and to eventual concerns about the resulting high radon levels that developed. A joint federal-provincial study in 1976-1977 examined 544 sites in the city and found working levels above 0.01 associated with about half of them. The highest value recorded was 0.8 WL.

An extensive program of remedial work was initiated jointly by the federal and provincial governments in 1978 with the objective of reducing all levels to below 0.02 WL. This pioneer project generated a great deal of information about control and remedial procedures for residential radon problems. Subsequently, a joint federal-provincial meeting established a limit for residential radon levels, 800 Bq m^{-3}, above which remedial work was urgently recommended. This level has frequently been compared unfavorably with the 4 pCi L^{-1} limit recommended by the Environmental Protection Agency in the United States; however, this is not a fair comparison. The establishment of limits in Canada is essentially a provincial responsibility, and this national guideline was always intended to be a ceiling on what is tolerable, rather than a standard for what should be generally acceptable.

In the late 1970s a second uranium mine was in production by Gulf Minerals in Saskatchewan, and a proposal to mine another very high-grade uranium deposit at Cluff Lake was under consideration. By this time the hazards associated with mining uranium ores were beginning to be more fully understood both in other countries and in Canada, where a Royal Commission on the Health and Safety of Workers in mines had been set up in Ontario to report on current conditions in the uranium mines at Elliot Lake. Concerned about the possible implications of mining at Cluff Lake, Saskatchewan appointed its own Board of Enquiry to advise on whether current technology would permit safe mining of these ores. The report, released in 1978, recommended that mining should be allowed to proceed but only after much more rigorous occupational health controls had been put in place. A recommendation of particular interest dealt with health physics control of the mining operation. This involved using what we now call the ALARA principal and also a proposal that external and internal radiation doses (from gamma radiation and the inhalation of radon daughters, respectively) should be monitored for each individual worker and should then be summed by expressing each dose as a fraction of the permissible limit and adding the two fractions together. One result of these recommendations is that Cluff Lake is still the only producing uranium mine where all the workers at risk are issued personal radon dosimeters. The reading from these dosimeters provides the definitive radon-daughter dose subsequently entered in the Canadian National Dose Register.

GEOLOGY

Mining of uranium in Canada today is primarily associated with heavily mineralized deposits formed around the perimeter of the Athabasca basin formation of sandstone and sedimentary rocks situated in the Canadian shield just south of Lake Athabasca (Figure 1). The only exception to this general rule is the Cluff Lake mine, which is situated in the middle of the basin at a point where the underlying structures break through to reach the surface of the basin. This is known as the Carswell dome structure. The other major deposits of high-grade ore where mining is already in progress or is currently planned all lie in the southeast corner of the basin between the east shore of Cree Lake and the west shore of Wollaston Lake. The Beaverlodge deposits near Uranium City are geologically much older and were not associated with the same type of crystallization, so that the ore grades in these formations were never comparable with those subsequently discovered in the Athabasca basin. This, combined with the greater remoteness of the site, which prevented an all-season road being constructed, led to termination of uranium mining in the Beaverlodge area in 1982.

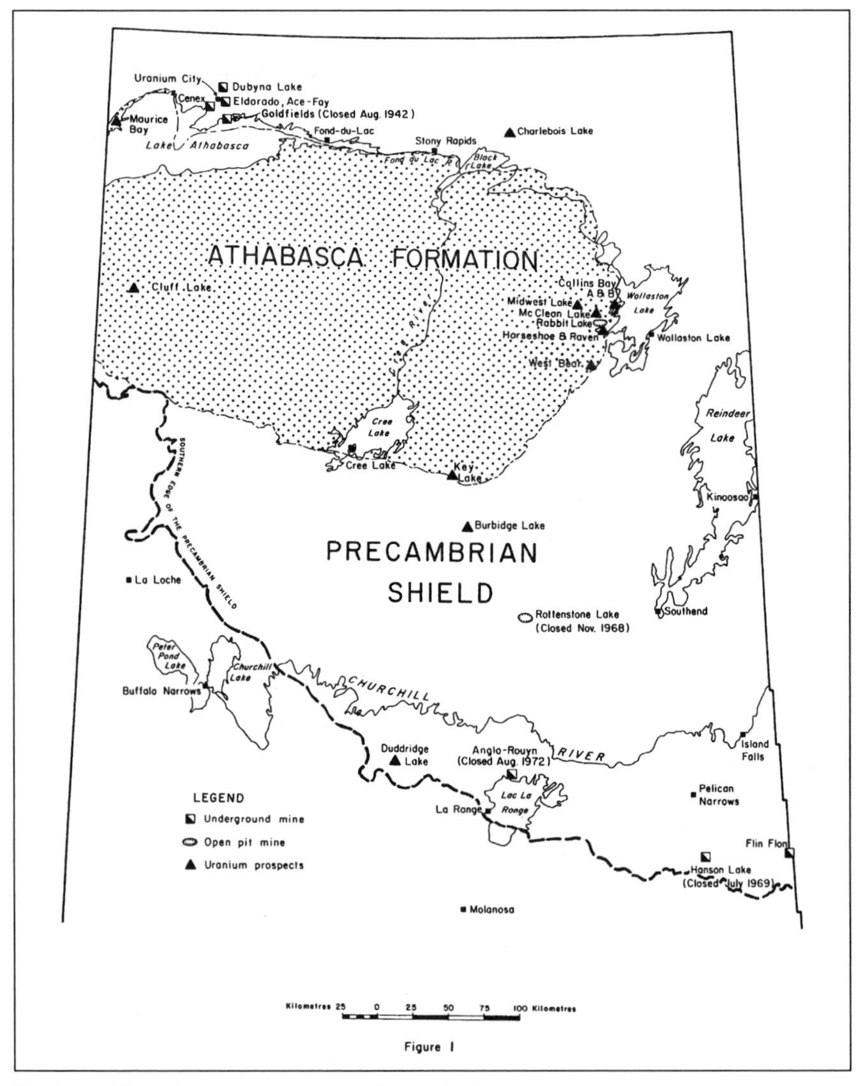

Figure 1. Mineral deposits of northern Saskatchewan.

Following the qualified approval given in the 1978 report, the Cluff Lake mine came into production in 1980. The exceptionally high-grade, open-pit D ore body is now mined out, but production continues from nearby underground ore bodies also associated with the Carswell dome structure. The deposits at Key Lake (the largest uranium mine in the world) were discovered in 1975. Production commenced in 1983, and estimated reserves approach 100 million kg. After terminating their mining operation at Beaverlodge, Eldorado bought out the Gulf Minerals mine at Rabbit Lake. The original open pit is worked out, but production continues from deposits at Collins Bay and will eventually extend to recently discovered rich reserves at Eagle Point. Total reserves are estimated at more than 60 million kg. In addition to these three producing mines, development work to evaluate three new mines at Cigar Lake, MidWest Lake, and McLean Lake is being actively pursued; however, these mines still have few employees and very limited dose records. Histograms of the radiation doses received at the three producing mines show that, despite the high grades, doses received by workers are generally below 0.1 of the maximum permissible dose (Figure 2).

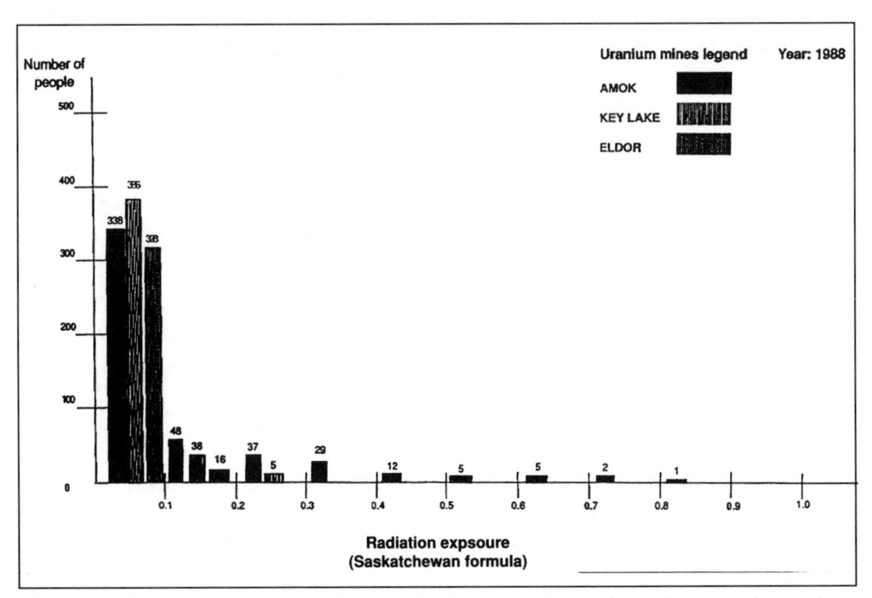

Figure 2. Radiation exposure doses of miners at three Canadian uranium mines (Saskatchewan formula).

HEALTH PHYSICS ASPECTS OF URANIUM PRODUCTION

Unlike the situation with many low-grade uranium deposits, where the risks from gamma-ray exposures are frequently regarded as negligible in comparison with those from radon-daughter inhalation, gamma-ray dose rates associated with high-grade ore bodies containing 5% or more of uranium are considerable. Readings of 0.1 mSv h^{-1} are typical of much of the working area, and readings up to 0.5 mSv h^{-1} have been found in localized areas where the ore grade is exceptionally rich. Radon-daughter concentrations underground vary dramatically with the effectiveness of the ventilation system. Improvements in ventilation are the main reason why worker radiation exposures are much lower today than they were in the past, even where the ore bodies being worked are much richer and more hazardous.

The limit of 4 WLM per annum for exposure to radon daughters was introduced in the United States in 1971 and adopted in Canada soon afterward. However, the gamma dose received by uranium miners is usually considered separately from this. Miners are subject to exposure to up to 50 mSv of gamma radiation in addition to inhalation of up to 4 WLM of radon daughters. This practice is followed despite clear indications that the International Commission on Radiological Protection intended internal and external doses to be combined when assessing compliance with limits. In many mines, workers also experience appreciable internal doses from inhalation or ingestion of dust containing long-lived radioactive elements from the uranium decay series. In some cases, the ore bodies are also high in silica and arsenic, and their dust constitutes an additional source of long-term lung damage to the work force.

Following the publication of the 1978 report, the Saskatchewan government introduced a combined dose limit such that

$$E1/P1 + E2/P2 + ... + Ex/Px = <1, \tag{1}$$

where E1, E2, ... Ex are actual exposures to each source of hazard, and P1, P2, ... Px are the corresponding permissible limits. This requirement is enforced by the Saskatchewan Mine inspectors on an ongoing basis. The present annual permissible limits for gamma-ray and radon-daughter exposures are 50 mSv and 4 WLM, respectively, but these limits are being reviewed in the light of recent increases in the risk estimates associated with radiation exposure. A third term in this summation to cover the inhalation of long-lived radioactive dust is currently under consideration.

RADON IN BUILDINGS

Following early recognition that a significant part of the radon-daughter exposure of miners in Uranium City may have taken place in their homes, Saskatchewan has remained very conscious of the problems associated with high radon levels in dwellings. A number of limited surveys indicate that high levels are not general but occur in significant numbers in limited areas. These include communities located on a ridge known geologically as the Ravenscrag spur formation in the south, many communities in the north, and a few elsewhere in the province. An early federal survey showed that the capital cities in the neighboring provinces of Manitoba and Saskatchewan were the two Canadian cities with the highest radon levels. Both are located on thick clay beds on the sites of ancient lakes. It appears that these high radon levels are associated with the take-up of radium from water on the outside of grains of clay as a result of ion exchange processes, not with abnormally high concentrations of radium in the soil. From the information currently available, a few Saskatchewan residents are exposed to radon-daughter levels in their homes that are higher than those encountered by most provincial uranium miners. However, provincial policy is that remedial work in homes is the responsibility of the owner. Information leaflets are freely available and are accompanied by a list of agencies capable of conducting radon monitoring. A booklet on carrying out remedial work is made available when needed, and it is emphasized that advice can be obtained on request. However, unless the measured radon level exceeds 800 Bq m^{-3}, no pressure is applied to do remedial work.

Nevertheless, it is clear that exposure to radon in other places should be just as carefully regulated as in uranium mines. The province has identified a number of underground vaults, associated with different city services, where some workers spend considerable amounts of time in very high radon levels. Furthermore, office buildings or other workplaces, like some homes, may have high radon levels, which can lead to considerable exposure during the course of a year. It is impossible to justify the stringent regulation of uranium mines if this problem is ignored. Consequently, the province introduced a draft regulation for public comment a year ago which will require the exposure to radon progeny of any nonradiation worker in any workplace to be limited to a maximum of 1 WLM in a year. This draft regulation has been well received and will probably be formally introduced later this year.

The Provincial Radiation Safety Unit thought it would be desirable to determine the potential impact of this new regulation before it is introduced. Therefore, we decided to carry out a comprehensive monitoring survey of a representative selection of buildings in all parts of the province. Primarily because they are readily reused and afford a relatively inexpensive survey, the charcoal canister technique was used. A radon chamber was constructed and calibrated to provide reference levels for the study, and a part-time worker was hired to carry out the measurements. It was decided that the initial measurements would be made in hospitals throughout the province. Initial measurements were completed on 120 such hospitals last spring. One canister was issued to each hospital, with instructions to place it in such a way that a near "worst-case" reading would be obtained. Analysis of the returned canisters followed EPA protocols. In all cases where the measured radon level exceeded 110 Bq m^{-3}, follow-up measurements were initiated. Usually, these consisted of 4 to 6 additional canisters (depending on the size of the hospital) to be issued every 3 mo over a 1-y period. As expected, this protocol required follow-up measurements in about 20% of the centers originally surveyed. Fortunately, in most cases the follow-up measurements appear to indicate a much lower level of hazard than the original worst-case reading. In two cases where the initial survey indicated very high levels, an immediate follow-up survey was made by a member of the Unit, using a continuously recording, instant, working-level meter.

After this survey of hospitals was completed, a similar study in schools began. It has a number of advantages: (1) Schools represent a convenient size sample; there are about 1300 public schools in Saskatchewan. (2) The objective of a sampling program is to take measurements in a group of buildings that are fully representative of the general stock in age, distribution, architectural style, and maintenance. Schools fit these requirements very well, particularly with regard to location throughout the province, where the basic requirement for an optimum sample is that the number of buildings to be tested is proportional to the population density and not to the geographical area involved. (3) The proportion of schools relative to homes is adequate to enable statistically meaningful conclusions to be drawn. (4) When measurements are made in a random selection of private properties, there is likely to be considerable pressure to carry out similar "free" measurements in neighboring properties. A survey of schools avoids this problem.

The worst-case measurements required to complete this survey have now been completed. Results show that the incidence of nontrivial readings where follow up is required is a little lower than among the hospitals. Only one of these exceeded 400 Bq m^{-3} and occasioned any surprise. The high readings have all been restricted to those areas where they were expected.

As we anticipated, additional testing where nontrivial radon levels were recorded proved necessary in about 10% of the schools. This follow-up survey was accomplished by exposing several charcoal canisters once a month for 5 mo. This measurement protocol has reduced the number of schools where major remedial work is contemplated to two or three. Continuously recording working-level monitors have been placed in these schools as a preliminary step to advising on what corrective actions will be required in them.

In the few places where concerns arose about the radon levels in the schools, a more general survey of radon in other public and private buildings was initiated. This showed that in one community, the township of Eston, a large proportion of the buildings had unacceptable radon levels which have necessitated urgent remedial work; some of the readings were as high as 5000 Bq m^{-3}. It is interesting to note that Eston, like the cities of Regina and Winnipeg referred to earlier, is located on heavy clay, part of an old lake bed.

CONCLUSION

These developments have confirmed the value of a survey of radon levels in schools for locating areas where radon may be a widespread problem.

DOSE ASSESSMENT OF POPULATION GROUPS EXPOSED TO ELEVATED RADON LEVELS IN RADIOACTIVE ITALIAN SPAS

G. Sciocchetti, S. Tosti, P. G. Baldassini, R. Sarao, and E. Soldano

Energia Nucleare e delle Energie Alternative -C.R.E., Rome, Italy

Key words: *Radon spa, thermal springs, occupational exposure, natural radioactivity*

ABSTRACT

The natural spring waters on the Isle of Ischia are among the most radioactive in the world. Therapeutic application of these waters, which contain very high radon concentrations, increases the radon exposure of people treated with them. People who live and work at radioactive spas may be good subjects for testing to evaluate detectable biological effects, especially because their exposures will be less influenced by synergistic factors than those of underground miners. The aim of our investigation was to characterize radon exposure for population groups exposed to high radon levels.

Our approach takes into account some peculiar requirements of our epidemiological investigations. To obtain representative dose values, workers were classified into groups to obtain significant results suitable for epidemiological pilot studies. Investigations were carried out on the geological aspects of radon sources, environmental parameters, physical and dosimetric factors which influence radon levels, and related exposures in therapeutic facilities in order to model patterns of radon exposures for the various population groups.

We inventoried hyper-radioactive springs on the island. We identified workers in radon spas who were exposed to radiation from inhaled radon daughters and retrospectively assessed their radon exposures. Results showed that, under some conditions, spa employees may have been exposed to much higher than usual levels of radon, which produced up to about 60 mSv y^{-1} effective dose equivalent.

INTRODUCTION

In waters of thermal spas the radioactive gas radon is often dissolved at very high levels. Radioactive gas, released in the air, irradiates personnel who work inside the confined environment of the thermal baths. In some thermal spas the public is also exposed to this radiation.

In certain countries, i.e., Austria (Steinhäusler, 1988), Bulgaria (Uzunov et al., 1981), and the USSR (Bogolyubov et al., 1983), patients are exposed to elevated radon levels for therapeutic purposes. Such practices are classified by the United Nations Scientific Committee on the Effects of Atomic Radiation (UNSCEAR) as "deliberate exposures to radon" (UNSCEAR,1982). Also, in some Italian spas treatments are inadvertently practiced in the presence of high levels of radon even though radon is not part of the therapy. In this paper we present characteristics of such radioactive sources, and the monitoring and dosimetric procedures used to assessoccupational and public exposure related to thermal bath treatment. Results of investigations in Italian radioactive spas at Ischia and Merano are described, and data from other countries are compared.

GEOLOGICAL CHARACTERISTICS OF ITALIAN THERMAL SPRINGS

The peculiar geological characteristics of Italy give rise to many radioactive springs. The radioactivity of their waters originates both from the lixiviation of rocks and from exhalative phenomena in deep fractures. Radioactive springs on the Isle of Ischia, in southern Italy (near Naples), and at Merano, in northern Italy, are near deep fractures. From them rise hypercritical fluids and subterranean "gases," including radiogenic isotopes. These fluids and "gases" are involved in transporting the radon (and thoron) to underground waters.

Some of the radioactivity in the spa waters also originates in the uranium content of the rocks, which is high in the acid igneous and metamorphic rocks at Merano and higher still in potassic rocks on the Isle of Ischia. The uranium content of the rocks at Ischia is on the order of 10 g t^{-1}, an order of magnitude higher than in rocks (gneiss) in the Merano area (Nascimben, 1989).

RADIOACTIVE ENVIRONMENT ON ISLE OF ISCHIA

At Ischia are, perhaps, the most important thermal springs in Italy, with the highest radon levels. The "Greca" is one of the most radioactive springs in the world.

The presence of solfataras (vents that emit sulfurous vapors) and hot springs demonstrate the "recent" volcanic origin of Ischia. The volcanic outcroppings consist mostly of tuff that is permeable and uraniferous. The origin of thermal mineral radioactivity of hydrothermal manifestations of the Isle of Ischia, is due to the interference between exhalative phenomena and ground waters from meteoric origin.

We found the highest temperatures and radioactivity levels in the springs in the western half of the island, where the depth of the fractures and the permeability and radioactivity of the rocks most favored their occurrence.

METHODS OF MEASURING AND MONITORING PROCEDURES

Instrumentation and experimental apparatus were developed to characterize the atmosphere inhaled in thermal radon spas.

Radon and its progeny were measured from representative air samples in the various spa environments. Grab samples were collected in different areas inside each spa, simulating treatment practices. Concentrations were also monitored using experimental apparatus designed and developed at the Ente per le Nuove Technologie, l'Energia e l'Ambiente radon laboratory.

An automatic, lightweight field instrument based on alpha spectroscopy was used. During operation, filtered air samples are collected, particulates are counted after collection on a silicon barrier detector, and statistics are automatically computed and data stored. The operation of the instrument is fully automatic: timing of sample collection, counting alpha-activity on the filter, computation of results, and data storage.

Performance tests of time-integrating passive devices were also carried out. Work is now in progress to assess cumulative exposures of personnel in the various treatment facilities.

RADON CONTENT OF SPRINGS

Table 1 shows radon concentrations of the water supply at selected Italian spas. At Ischia, the concentration of radon in the water varies from about 50 to 19,000 kBq m^{-3}, with the highest value found in Santa Restituta. The variation depends on the procedure used to transfer water from the source to the treatment facilities.

Table 1. Radon concentrations of water supply at selected Italian spas.

Spas	Radon concentration (kBq m^{-3})
Terme di Lurisia	23,500
Ischia	
Santa Restituta	19,400
Regina Isabella	3500
Santa Montano	762
Felix	444
Antiche Terme Comunali	111
Parco Edera	52
Merano	
Centro Termale	2035[a]
Montecatini Terme	
Tettuccio	984
Bagni di Lucca	
Bernabo'	104
Grotte di Sotto	80

[a] Sampled in bath area

Table 2 shows radon concentrations of the water supply and baths in some spas on the Isle of Ischia. The coefficient of radon transfer varies from 0.13 to about 0.60 for different spas. Data indicate that during transport, the water loses a large amount of radon into the atmosphere.

Table 2. Radon concentrations of water supply and baths in selected spas on the isle of Ischia.

	Radon concentration (kBq m^{-3})		
Location	At source	In bath area	Ratio
Santa Restituta	19,400	2590	0.13
Terme Militari	165	30	0.18
Antiche Terme Comunali	111	65	0.59

CHARACTERIZATION OF RADON IN BATHS

The buildup of radon concentration in the air of the baths is influenced by the radon content of the water, the ventilation rate in the room, the duration of treatment, etc.

We performed a survey of radon concentration in various spas throughout the country. Table 3 compares radon concentration in air with the concentration in water, measured in wells near the spas. Generally, the two levels were not correlated because of several factors. Table 4 shows data from spas in countries other than Italy.

Table 3. Comparison of radon concentrations in air and in water supply at selected Italian spas

Location	Radon concentration in water supply (kBq m^{-3})	Radon concentration in air (kBq m^{-3})
Ischia (17 spas)	111–19,400	0.44–514
Merano	2035[a]	3.9
Bagni di Lucca	80–104	1.1–2.9

[a] Sample taken in bath area.

Table 4. Comparison of radon concentration in air of treatment room and in water supply at spas of various countries.

Location	Radon concentration in water supply (kBq m^{-3})	Radon concentration in air (kBq m^{-3})	Reference
Bad Gastein (Austria)	1480	3.3	Uzunov et al., 1981
Kovary (Poland)	110–1110	0.19–496	Chruscielewski, 1983
Misasa (Japan)	437	0.81	Morinaga et al., 1984

In Table 5 we report the results of a national survey of radon and radon-daughter air concentrations (Sciocchetti et al., 1980, 1988; this paper) at spas on the Isle of Ischia and in other areas of the country. Measurements were carried out during simulation of a typical therapeutical procedure in a bath. The very low equilibrium values measured at the spas on Ischia are typical for short-duration therapy. Under those conditions, concentrations of radon daughters varied from 0.003 to 1.97 WL (0.011 to 7.29 kBq m^{-3} equilibrium equivalent radon [EER]).

EXPOSURE ESTIMATES

Patient exposure depends on the type of therapy. Typical examples of treatments are thermal baths, inhalation of radon-laden air, drinking hot spring water, applying mud packs, swimming in hot-spring water, and physical therapy. Table 6 shows air concentrations of radon, radon decay products, and equilibrium factors in treatment facilities of certain Italian spas. The highest radon levels were measured inside baths at each spa surveyed. For comparison, Table 7 shows values relevant to spas of countries other than Italy.

Table 5. Survey of radon and radon-daughter air concentrations in treatment rooms at selected Italian spas.

Location	Radon concentration (kBq m^{-3})	PAEC[a] (WL)	EER[b] (kBq m^{-3})
Ischia (17 spas)	0.44–514	0.003–1.97	0.011–7.29
Abano Terme (2 spas)	4.07	0.038	0.141
	6.66	0.078	0.289
Lurisia	315	0.160	0.592
Merano Centro Termale	3.96	0.45	1.66
Lucca Bagni di Lucca	1.1–2.9	—	—

[a] Potential aplha energy concentration
[b] Equilibrium equivalent radon factor (EER/Rn concentration)

Table 6. Air concentration of radon, radon decay products, and equilibrium factors in different treatment facilities at selected Italian spas.

Location	Radon Concentration (kBq m^{-3})	EER[a] (kBq m^{-3})	Equilibrium Factor (EER/Rn Conc.)
Bath area			
Merano (Centro Termale)	3.96	1.66	0.42
Ischia (17 spas)	0.44–514	0.01–25.9	0.02–0.05
Ischia (Antiche Terme Comunali)	0.41	0.18	0.45
Ischia (Terme Militari)	6.70	0.34	0.05
Mud Facility			
Merano (Centro Termale)	1.11	0.44	0.40
Ischia (Antiche Terme Comunali)	0.003	—	—
Ischia (Terme Militari)	0.30	0.02	0.07
Pool			
Merano	0.85	0.22	0.26
Ischia (Antiche Terme Comunali)	0.04	—	—
Inhalation Facility			
Merano	0.89	0.11	0.12
Ischia (Terme Militari)	0.11	0.09	0.80
Room Adjacent to Bathroom			
Merano	0.81	0.48	0.59
Ischia (Terme Militari)	2.07	0.02	0.01

[a] Equilibrium equivalent radon

Table 7. Air concentrations of radon, radon decay products, and equilibrium factors in baths at spas of various countries.

Location	Radon concentration (kBq m^{-3})	EER[a] (kBq m^{-3})	Equilibrium Factor (EER/Rn conc.)	Reference
Kowary (Poland)	1.9–496	1.18–120	0.60–0.24	Chruscielewski et. al., 1983
Kamena Vourla (Greece)	—	20	—	
Bad Gastein (Austria)	3.33	1.48	0.45	Steinhäusler, 1988
Misasa (Japan)	0.81	—	—	Morinaga et. al., 1984

[a] Equilibrium equivalent radon

DOSE TO PATIENT

Typical exposure-dose factors recommended by the OECD/NEA Report have been used (OECD/NEA, 1983). The effective dose equivalent was estimated using a conversion coefficient of 5.5 mSv WLM^{-1}. For exposure during inhalation therapy we used a coefficient of 5.4 x 10^{-3} WL kBq^{-1} m^{-3} to correlate the radon-daughter air concentration to the radon concentration in the mineral water of the inhaled aerosol (Uzunov et al., 1981).

Patient doses were evaluated taking into account radon and radon-daughter levels and duration of treatment. Table 8 compares radon-daughter exposures and estimated doses received during treatments in spas of different countries. At spas in Ischia, the effective dose equivalent per treatment varies from 0.001 to 0.39 mSv. Table 9 compares water radon concentrations and doses received during inhalation therapy procedures in different countries.

OCCUPATIONAL DOSE

Spa attendants are also exposed to elevated radon-daughter concentrations. Table 10 estimates occupational exposures and doses of bath attendants in Italy and in other countries. Data show that, in some cases, doses to personnel could exceed the International Commission on Radiation Protection (ICRP)-recommended upper limit of 50 mSv y^{-1} (ICRP, 1982).

Table 8. Comparison of radon-daughter exposures and effective dose equivalents from therapeutic baths at spas of different countries.

Location	PAEC[a] (WL)	Duration (h)	Effective dose equivalent (mSv)	Reference
Ischia (Italy)	0.05–2	6	0.001–0.39	Sciocchetti, 1988
Merano (Italy)	0.45	12	0.17	Sciocchetti, 1988
Ikara (Greece)	—	—	0.5	Kritidis and Probomas, 1988
Bad Gastein (Austria)	0.5–0.9	5	0.02–0.15	Steinhäsler, 1988
Momin Prohod (Bulgaria)	0.4	10	0.13	Uzunov et. al., 1981

[a] Potential aplha energy concentration.

Table 9. Comparison of water radon concentrations and effective dose equivalents at spas of different countries.

Location	Radon concentration in water (kBq m^{-3})	Duration (h)	Effective dose equivalent (mSv)	Reference
Bockstein (Austria)	—	24	7	Steinhäusler, 1988
Momin Prohod) (Bulgaria)	1850	3.75	1.28	Uzunov etl al., 1981
Merano (Italy)	2035	6	2.14	Sciocchetti, 1988

Table 10. Estimated annual occupational exposures and doses of bath attendants in radon spas in Italy and other countries.

Location	PAEC[a] (WL)	Annual exposure time (h)	Effective dose equivalent (mSv)	Reference
Ischia (17 spas) 1988 (Italy)	0.003–2	900	0.01–60.5	This paper
Merano 1988 (Italy)	0.45	1800	26.2	This paper
Bad Gastein (Austria)	0.5–0.9	2000	33–60.5	Steinhäusler, 1988
Momin Prohod 1981 (Bulgaria)	0.4	400	5.2	Uzunov et al., 1981
Kamena Vourla and Ikaria (Greece)	—	—	5–50	Kritidis and Probonas, 1988

[a] Potential aplha energy concentration.
[b] Attendants worked 4 to 5 h daily in thermal bath areas.

Risk Assessment

The assessment of risk from radon-daughter exposure is based both on data from epidemiological studies on underground uranium miners and dosimetric modeling of biological effects. The role of other carcinogens as cofactors in the induction of lung cancer (e.g., ore dust, diesel fumes, or cigarette smoke, inhaled in the mine atmosphere) is uncertain. We must also take into account differences in the physiological characteristics of the miners. The respiratory minute volume for heavy physical work results in a higher dose per unit exposure, by a factor of 2, than that for patients in spa environments.

Many studies have shown that radon-induced lung-cancer risk increases at cumulative exposures resulting in doses greater than 600 mSv. In some spas employees could reach this level in 10 y. An epidemiological study of lung cancer among spa employees would be of interest. However, in most cases, an increase in cancer incidence will not be significant if the number of subjects is relatively small. Investigations in a spa at Badgastein (Austria) revealed a higher incidence of health effects, such as a higher frequency of chromosome aberrations and atypical cells in sputum samples, in employees with increased radiation burdens (Steinhäusler, 1988).

Other investigations revealed a significant relationship between radon exposures and levels of biological markers such as $^{210}Pb/^{210}Po$ skeletal burden and ^{210}Po concentration in urine (Clemente et al., 1979, 1984).

The basis for assessing radiation risk is a detailed knowledge of the mean exposure for various groups of persons. A large variation in exposure situations requires a series of measurements and collection of individual data. Tables 11 and 12 show, respectively, distribution of WL values in all spas surveyed and distribution of annual radon-daughter exposure of spa workers, from the preliminary survey.

CONCLUSIONS

Results confirm that occupational exposure to radon can occur in thermal spas. Therefore, the same radioprotection procedures should be followed there as in other radioactive working areas, e.g., uranium mines. Patient doses for each treatment are generally comparable with those resulting from annual exposure to natural radiation in the environment.

Table 11. Distribution of radon WL values at spas in Italy, estimated from survey results.

WL Levels	No. Spas Surveyed
<0.1	18
0.1–0.3	1
>0.3	1

Table 12. Distribution of radon-daughter exposure of spa workers, estimated from survey results

Annual exposure (WLM y^{-1})	Spa workers	
	No.	%
<1	450	90
1–4	25	5
>4	25	5

The large amount of radon decay products in the air of the workrooms and living areas in thermal spas suggested that radon-daughter inhalation may be responsible for lung-cancer induction even in a nonmining environment.

ACKNOWLEDGMENT

The authors gratefully acknowledge the help of Dr. P. Nascimben in the paragraph concerning the geological characteristics of thermal areas.

REFERENCES

Bogolyubov, VM, II Gusarov, and SV Andreev. **1983**. The "risk-benefit" ratio in radon therapy, pp. 109-114. In: *Proceedings of the International Congress on Natural Radioactivity and Thermal Waters*, Merano (Italy), November 17-19, 1983. (U Solimene, H Fruhauf, G Draetta, eds.).

Chruscielewski, W, T Domanski, and W Orzechowski. **1983**. Concentrations of radon and its progeny in the rooms of Polish spas. Health Phys 45:421-424.

Clemente, GF, A Renzetti, and G Santori. **1979**. Assessment and significance of the skeletal doses due to Po-210 in radon spa workers. Environ Res 18:120-226.

Clemente, GF, A Renzetti, G Santori, F Steinhäusler, and J Pohl-Rüling. **1984** Relationship between the ^{210}Pb content of teeth and exposure to Rn and Rn daughters. Health Phys 47:253-262

ICRP. **1982**. *Limits for Inhalation of Radon Daughters by Workers*, ICRP-32. Pergamon Press, Oxford, UK.

Kritidis, P and M Probonas. **1988**. The Greek radon spas: Hot spots of natural radioactivity in the Mediterranean area, pp. 53-57. In: *International Conference on Environmental Radioactivity in the Mediterranean Area*, Barcelona, May 10-13, 1988.

Morinaga, H, M Mifuna, and K Furuno. **1984**. Radioactivity of water and air in Misasa Spa. Radiat Protect Dosim 7:295-297.

Nascimben, P. **1989**. Cenni informativi sulla genesi delle acque termominerali radioattive di Merano e dell'Isola d'Ischia, pp. 83-112. In: *Aspetti di Radioprotezione nelle Stazioni Termali/Strahlenschutzaspekte der Radontherapie*. (A Gentili, E Moroder, and G Sciocchetti, eds.). Italienisch-Osterreichischer Workshop, Merano (Italy), March 6-7, 1986.

OECD/NEA. **1983**. *Dosimetry Aspects of Exposure to Radon and Thoron Daughter Products*, Nuclear Energy Agency Experts Report. OECD, Paris.

Sciocchetti, G. **1988**. Sorgenti Radioattive ed Esposizione alle Radiazioni in Ambiente Termale, pp. 27-46. In: *Proceedings of Workshop "Aspects of Radiation Protection in Thermal Spas*, Merano (Italy), March 6-7, 1986. (A. Gentili, E. Moroder and G. Sciocchetti, eds.).

Sciocchetti, G, F Scacco, and CF Clemente. **1980**. Evaluation of radon and radon daughter exposure of the Italian population, pp. 167-179. In: *Proceedings of the Specialist Meeting on the Assessment of Radon and Daughter Exposure and Related Biological Effects*, Rome (Italy), March 3-8, 1980.

Steinhäusler, F. **1988.** Radon spas: Source term, doses and risk assessment, pp. 257-259. In: *Proceedings of the Fourth International Symposium on the Natural Radiation Environment*, Lisbon, Portugal, December 7-11, 1987.

UNSCEAR. **1982**. United Nations Scientific Committee on the Effects of Atomic Radiation, Report to the General Assembly. United Nations, New York.

Uzunov, I, F Steinhäusler, and E Polh. **1981**. Carcinogenic risk of exposure to radon daughters associated with radon spas. Health Phys 41:807-813.

QUESTIONS AND ANSWERS

Q: Hopke, Clarkston University. Is there a big enough population of spa workers to be a usable cohort for looking at the risks of radon? I have heard Professor Steinhäusler suggest that the population was too small. You listed a number of others. Are there enough so that you could look for something in these highly exposed people?

A: Sciocchetti. Yes, this is the most important question concerning epidemiological studies. Actually, the number of subjects is not large. In my opinion, in Italy we could identify perhaps several hundred people exposed to abnormal radon levels in radioactive spas.

Q: Richardson, Bristol, UK. The doses you show seem very low. What was your conversion factor for mSv y^{-1} to the concentration level of radon in Bq m^{-3}?

A: Sciocchetti. The conversion factor that we use is that suggested by CRE and the OECD/ENEA. The value is 5.5 mSv WLM^{-1}. You can convert the working level to kBq m^{-3} of equilibrium equivalent radon concentration using the coefficient 3.7. 1 WLM corresponds to 73.9 Bq m^{-3} y.

Q: Harley, New York University. I was going to ask about the dose also. The factor you use, 5.5 mSv WLM^{-1}, is a conventional factor. I think that's a good average, but the atmosphere in a spa might be a little different from, say, a normal home atmosphere, and the dose is very sensitive to particle size. So I wonder if you have plans to carry out particle-size measurements in spas.

A: Sciocchetti. I agree, but this is also a conventional presentation of the first data, and the task of our research is also to characterize the spa atmosphere, measuring all parameters which influence the dose to the lungs.

Q: Knutson, Environmental Measurements Laboratory. I could add that it looks as though the equilibrium ratios were quite small; about 1 in 50 in one case I saw.

A: Sciocchetti. This is the same question we discussed some time ago with Dr. Steinhäusler. The values really cast doubt because the air was sampled at the start, when water was introduced in the bath; it was only radon-laden ("young") air, with very low radon progeny concentration. We reached high equilibrium, fortunately, after the end of the therapeutic cycle, in some cases; this was not true in all cases, but only for the cases that we analyzed. We have also measured, for example, at Merano spa, an equilibrium value of about 0.4.

THE EFFECT OF HOME WEATHERIZATION ON INDOOR RADON CONCENTRATION

G. W. Egert,[1] R. L. Kathren,[2] F. T. Cross,[3] and M. A. Robkin [4]

ABSTRACT

In an effort to reduce energy consumption costs, increasing numbers of homeowners are weatherizing their homes to minimize the loss of heated and cooled air to the outside. These house-tightening measures decrease the natural infiltration rate of fresh air into the house, potentially increasing the concentration of indoor pollutants, including the radioactive gas, radon.

We measured radon concentrations with track-etch detectors in 17 wooden frame homes for a period of 3 months before and after weatherization. An additional 42 homes that were not weatherized were also sampled; they constituted the control group. The measured concentrations in the weatherized homes, the control group, and in both groups combined were described by log-normal distributions. The differences between final and initial concentrations approximate both a log-normal and a normal distribution. Student's *t*-test and the Wilcoxon Rank Sum Test of both log-normal and normal-data distributions at the 0.05 significance level show an increase in the indoor radon concentration following home weatherization of 40% and 60%, respectively. This suggests that standard weatherization techniques may increase indoor radon levels by approximately 50%.

QUESTIONS AND ANSWERS

Q: David Gooden, St. Francis Hospital, Tulsa, Oklahoma. That is the part of the country where we produce natural gases. In that industry we know that there are naturally occurring radioactive materials in those gases: radium, radon. Could that be a possible explanation for what you saw?

A: Robkin (presenter for Egert paper). As I recall, BPA's requirement for weatherization is that the homes must be all-electric. So, there should be no radon from natural gas. I'm not sure how far we are from the gas-supply wells in Richland, but I suspect that we are

[1]Westinghouse Hanford Company, Richland, Washington
[2]University of Washington, Seattle, Washington and Hanford Environmental Health Foundation, Richland, Washington
[3]Pacific Northwest Laboratory, Richland, Washington
[4]University of Washington, Seattle, Washington

quite a long distance. Radon, therefore, would have decayed long before the gas reached here. I don't think it was a factor.

Q: Sterling, Simon Fraser University. Measurements have been made with a 6-in. fan in the crawl space of weatherized houses and we found about a 50% reduction in radon concentration by having this type (crawl space) of ventilation. This is bad news for indoor levels in cold climates, because it is difficult to weatherize a house on one end and then ventilate the basement or crawl space on the other.

A: Robkin. I have no particular argument with that. I wonder if it is possible to seal the flooring above the crawl space beneath the house and then ventilate it without a lot of heat exchange? I don't know how effective this sort of installation might be in a place where the winters are very severe.

Q: Sterling. Well, in Vancouver we don't have severe winters, but the question is really one of either reducing the radon by ventilation or weatherizing the house, and of not doing either.

Q: Morley, Radiation Protection, British Columbia. BPA, I gather, has done a large number of homes in the Pacific Northwest, something like 40,000. How come your sample size is so small in this particular setting? Are the rest of the data from BPA being looked at?

A: Robkin. Not by this student, who was resident here, and was given access by some homeowners who were willing to participate. It was, for want of a better expression, a finite study that he was able to do in the amount of time he had.

Q: Morley. It was my understanding that, in looking at the BPA data in general, they have seen a remarkable increase with weatherization.

A: Robkin. I don't think I want to comment on those data, since I am not entirely familiar with the results.

Q: Harley, NYU. I would like to make a comment. The last I heard of that Bonneville study, the increase they found with weatherization was 25%. I was sitting here thinking about asking whether you felt that 40% was the same number as the 25%. Or, since these data are recent, perhaps there are some actual changes in weatherization techniques that might give a higher value?

A: Robkin. Well, 40% is the result for this particular set of data, and I think that it would be a little presumptuous to make any definite comparison between the two.

Q: Harley, New York University. They are probably the same number.

Q: Schoenberg, New Jersey State Department of Health. I'm sorry that I didn't hear you indicate whether your measurements were made in basements or living areas?

A: Robkin. They were mostly living-area measurements with the detector placed on a shelf or on a wall.

Q: (Speaker unknown.) I'm wondering, what is your rationale for placing a passive detector on a wall?

A: Robkin. I guess I don't know how to answer that. What would be the problem?

Q: (Unknown speaker). Well, I'm the author of the document on which the EPA compendium on indoor air, Appendix C, was based, which includes the placement and use of passive monitors. A wall is specifically not prescribed as the place for a passive device. I am curious as to what your rationale was for placing it on a wall? I believe even Teradex, who supplies these devices states that they should not be placed on walls.

A: Robkin. Since the device is based on radon diffusion, I guess I don't know how to respond to that. I would assume that the gas would diffuse freely.

Harbottle, Brookhaven National Laboratory. I have a commentary. Some years ago, I was reading Dr. Castrén's work along the lines being discussed this morning and decided to undertake some similar studies on Long Island, because we have a moraine there. In Finland, the moraine and eskers have been shown to be related to high radon levels in houses. We have completed the measurements on this. I think the answer, at least as far as Long Island's moraine goes (it is a Wisconsin moraine), is very different, and I think it is because of the local geological state of the moraine. Our moraine has about 3 to 4 parts per million of uranium and somewhat more, 11 to 15, of thorium. But the difference seems to be that the minerals that contain the radioactivity in the soil on Long Island are very resistant to weathering. There has simply not been enough time to weather them into a condition where they can emanate. The result is that the thoron emanation is only about 2% or 3%, and the radon is about 15%, so they are not good sources of radon for houses. The radioactivity is there, but the radon simply does not get out of the soil grains. This is not a transport phenomenon, this is an emanation phenomenon.

PREDICTION OF INDOOR RADON CONCENTRATION BASED ON RESIDENCE LOCATION AND CONSTRUCTION

I. Mäkeläinen, A. Voutilainen, and O. Castrén

Finnish Center for Radiation and Nuclear Safety, Helsinki, Finland

Key words: *Indoor radon, dwellings, model, geology, construction of building*

ABSTRACT

We have constructed a model for assessing indoor radon concentrations in houses where measurements cannot be performed. It has been used in an epidemiological study and to determine the radon potential of new building sites. The model is based on data from about 10,000 buildings. Integrated radon measurements were made during the cold season in all the houses; their geographic coordinates were also known. The 2-mo measurement results were corrected to annual average concentrations. Construction data were collected from questionnaires completed by residents; geological data were determined from geological maps.

Data were classified according to geographical, geological, and construction factors. In order to describe different radon production levels, the country was divided into four zones. We assumed that the factors were multiplicative, and a linear concentration-prediction model was used.

The most significant factor in determining radon concentration was the geographical region, followed by soil type, year of construction, and type of foundation. The predicted indoor radon concentrations given by the model varied from 50 to 440 Bq m^{-3}. The lower figure represents a house with a basement, built in the 1950s on clay soil, in the region with the lowest radon concentration levels. The higher value represents a house with a concrete slab in contact with the ground, built in the 1980s, on gravel, in the region with the highest average radon concentration.

INTRODUCTION

As a rule, measuring of long-term average radon concentrations is so easy and inexpensive to do that no substitute for it is needed. There are, however, situations where indoor radon measurements cannot be performed. In epidemiological studies it may be necessary to estimate the exposure which a person had received while living in a house which has been destroyed. Sometimes, but fortunately not very often, measurement is not permitted by owners. Last but not least, it may be desirable to predict the indoor radon concentrations in houses to be constructed in both old and new building areas. Different methods for assessing the radon risk (radon availability) at the building site have been developed. They often entail onsite measurements of radon in materials such as soil gas, soil samples, or external radiation.

We use indoor radon concentration in existing houses as an indicator of radon availability in soil, but measurements of indoor radon concentrations have to be adjusted for construction factors. This adjustment is made using statistical analysis. We assume the impact of residence location factors and construction factors to be multiplicative. However, because of the nearly lognormal distribution of radon concentrations, a linear model can be used after logarithmic transformation.

Another objective of this study is to obtain information about the effect of the type of house construction on radon concentrations.

MATERIALS AND METHODS

We have a radon data base of about 20,000 detached houses with known geographical coordinates. Alpha-track measurements were made during the cold season over a period of 2 mo. Results were seasonally adjusted using a model for indoor radon variations (Arvela and Winqvist, 1989). Most of the dosimeters were distributed by the local health authorities, who marked the exact sites of the houses on maps. The coordinates were determined from basic maps (scale 1:20,000), and the soil and rock types were ascertained from geological maps, using a mapping system. Construction data were collected from questionnaires completed by the residents. Because the data were incomplete, only 14,000 houses could be included in this analysis.

Indoor radon concentration may be regarded as proportional to the radon concentration of the soil gas, the leakage parameter of the house (which accounts for the resistance made by the foundation and the soil to the leakage air flow), the inverse of the volume of the house, and the inverse of the air exchange rate, among others. These parameters are difficult to measure, and in this study we substituted more accessible data for them. We divided these data roughly into location factors and construction factors.

To classify the data according to location factors, we divided Finland into four zones, high, low intermediate, and Tampere, depending on the uranium concentrations of till samples (Geological Survey of Finland, 1985). The zones (Figure 1) are shown for the 12 administrative districts of the country. Tampere is a separate zone because it has exceptional eskers (Voutilainen and Mäkeläinen, 1991). Inside each zone, the data were classified by soil type (sand and gravel in eskers, sand and gravel in other deposits, till, rock, and clay).

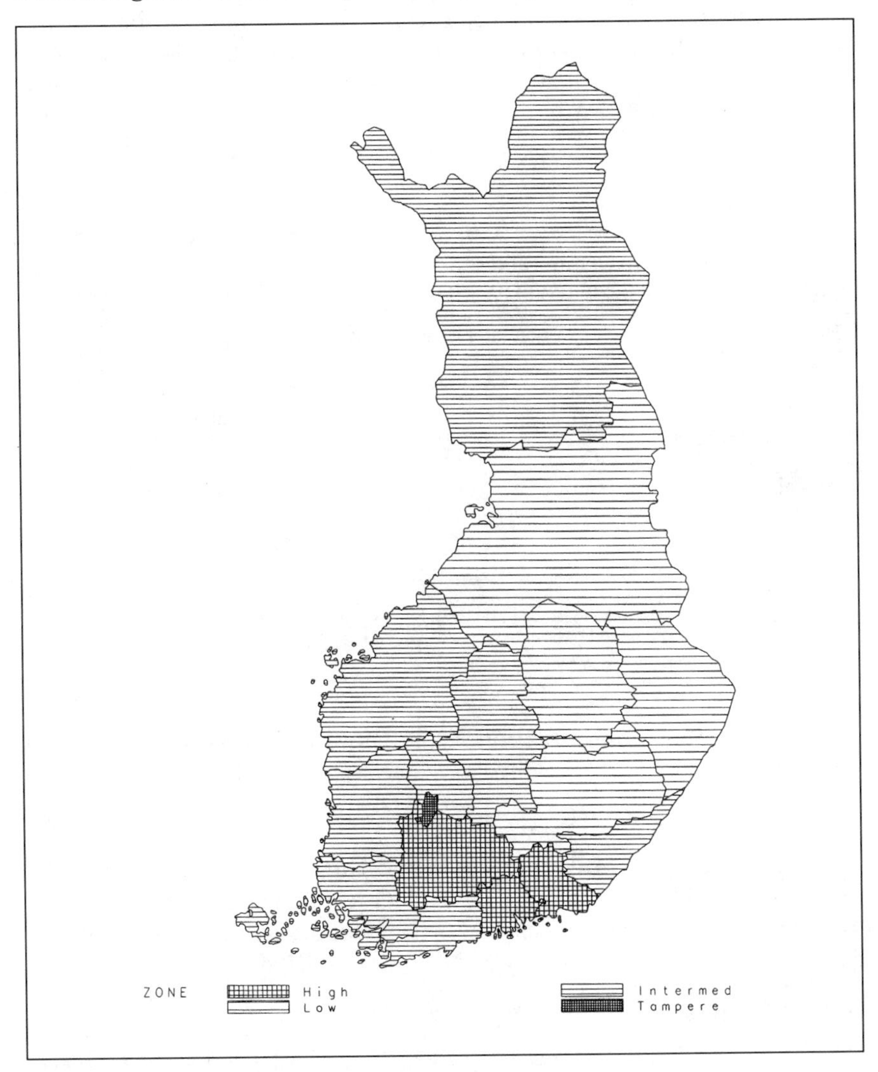

Figure 1. The four zones of Finland, based on the uranium concentrations of till samples. Zones are shown for the country's 12 administrative districts.

Construction factors used were taken from the questionnaire given to owners. Originally, we used seven factors, which were classified into two to four classes. The factors were: foundation type, construction year, ventilation type, heating system, number of stories, house type, and, in houses with a basement, the type of staircase. The foundation type and construction year classes are shown in Table 1. The most common foundation type in Finland is the slab-on-grade laid inside foundation walls. Radon enters the house through the walls and floor joists in the footing. Slabs with thickened edges do not have such joists. Basements in Finland are often used for garages, saunas, storage, and so on.

Table 1. Parameter estimates and their 95% confidence limits for construction factors in the multiplicative model for predicting indoor radon concentration.

Effect	Parameter Estimate	95% Confidence Limits
Foundation type		
Slab-on-grade[a]	1.26	(1.21, 1.31)
Crawl space, etc.[b]	0.71	(0.67, 0.75)
House with basement	1	—
Building year[c]		
Before 1946	0.78	(0.74, 0.83)
1946–1960	0.60	(0.58, 0.62)
1961–1975	0.74	(0.71, 0.77)
1976 and after	1	—

[a] Slab-on-grade laid inside foundation walls.
[b] House supported on a slab with thickened edges, or house having crawl space.
[c] Involves only slab-on-grade and basement houses.

The indoor radon concentration, C, in the dwelling is described as a product:

$$C_{ij...k} = m.a_{1i}a_{2j}...a_{nk}\epsilon_{ij...k}, \qquad (1)$$

where m is the constant (intercept) and the indices 1...n refer to the parameters of the n factors (location or construction) of the model. The indices i, j,..,k refer to classes within factors. The model may also contain interaction terms between a:s. ϵ is relative error.

In logarithmic form:

$$\log(C_{ij...k}) = \log(m) + \log(a_{1i}) + \log(a_{2j}) +...+ \log(a_{nk}) + \log(\epsilon_{ij...k}). \qquad (2)$$

The estimates of the parameters log(m), log(a) and log(ϵ) can be calculated using the linear model. We tried different programs devised by SAS (SAS Institute Inc., 1985) and used their results as a basis for rejecting less significant factors. The final analysis presented here was made with the GLM procedure.

The estimates of the parameters of the multiplicative model (equation) are antilogarithms of the estimates of the additive, i.e., the logarithmic linear model (equation). The estimates of the parameters of the multiplicative model, except for the intercept, are ratios of the class parameter values, the last class having unity value. The GLM procedure also calculated standard errors for the estimates (STEE) for the additive model. The 95% confidence limits for the parameter estimates of the multiplicative model were obtained by dividing and multiplying the parameter estimates with exp(1.96 STEE).

RESULTS

The number of observations was 13,795. The selected model has four statistically significant factors and an interaction factor. These factors were the zone, the soil type, their interaction, the foundation type, and the construction year class. The construction year involves only slab-on-grade and houses with basements. Log(σ) is the estimate of log(ϵ), the root of the residual mean square. In most models studied the remaining factors were also significant.

The square of the multiple correlation coefficient R^2 is 0.30. About 23% of the variation is due to location and 7% to construction. The estimate of σ is 2.29, and the constant term (intercept) is 81 Bq m^{-3}. The parameter estimates and their relative errors for the multiplicative model are shown in Table 1 for construction and in Table 2 for location parameters. Table 3 gives the interaction parameters between zone and soil.

Applying the model with the parameters of Tables 1, 2, and 3 to a house with a crawl space built on clay in the low-radon zone results in a geometric mean concentration of 0.71 x 1.0 x 1.0 x 81 Bq m^{-3} = 58 Bq m^{-3}. Similarly, a house with a basement built in the 1950s in the same zone has a radon concentration of about 0.60 x 1.0 x 1.0 x 1.0 x 81 Bq m^{-3} = 49 Bq m^{-3}. A new house with a slab-on-grade foundation has a concentration of 1.26 x 1.0 x 1.0 x 1.0 x 81 = 102 Bq m^{-3}. Three similar houses on eskers in the high-radon zone have concentrations of 2.45 x 1.71 x 1.02 = 4.27 times higher, that is, 247, 209, and 436 Bq m^{-3}, respectively.

Table 2. Parameter estimates and their 95% confidence limits for the location factors in the multiplicative model for predicting indoor radon concentration.

Effect	Parameter Estimates	95% Confidence Limits
Zone		
Tampere	1.69	(1.40, 2.03)
High	2.45	(2.11, 2.85)
Intermediate	1.14	(1.00, 1.31)
Low	1	—
Soil type		
Sand and gravel in eskers	1.71	(1.44, 2.02)
Sand and gravel in other deposits	1.20	(1.01, 1.42)
Rock	1.19	(1.01, 1.41)
Till	1.02	(0.88, 1.19)
Clay	1	—

Table 3. Parmeter estimates and their 95% confidence limits for the interaction terms of zone and soil type in the multiplicative model for predicting indoor radon concentration. All terms involving the low zone are equal to unity.

Zone	Soil type	Parameter Estimates	95% Confidence Limits
Tampere			
	Esker[a]	1.99	(1.57, 2.53)
	Other sand[b]	1.01	(0.74, 1.38)
	Rock	0.87	(0.66, 1.14)
	Till	1.31	(0.98, 1.75)
	Clay	1	—
High			
	Esker	1.02	(0.85, 1.23)
	Other sand	0.96	(0.80, 1.16)
	Rock	0.89	(0.74, 1.07)
	Till	1.17	(0.99, 1.39)
	Clay	1	—
Intermediate			
	Esker	0.91	(0.75, 1.10)
	Other sand	0.92	(0.76, 1.11)
	Rock	0.97	(0.80, 1.17)
	Till	1.10	(0.93, 1.30)
	Clay	1	—

[a] Sand and gravel in eskers.
[b] Sand and gravel in other deposits.

The most prominent of the statistically significant but rejected factors was hot-water heating, which was associated with a radon concentration about 20% lower. Factors that were combined with a higher radon concentration were an open staircase from the basement to the upper floor, forced ventilation, a one-story house and a one-family house.

DISCUSSION

The value of σ, 2.29, shows quite a large variation that cannot be addressed by the model. Further, the multiple correlation coefficient, R^2, cannot reach values near unity because of the large variation within classes. The radon availability varies considerably within zones and with the permeability of the soil, as do the ventilation and leakage parameters within construction classes. In an analysis of this kind the real effects may also be "smoothed" because of possible errors in the data.

The most prominent feature of the results is the effect of location. The ratios of zone parameters describe quite well the average uranium concentrations of the zones, which range from about 2-3 to 6-7 ppm.

In Finland, eskers are the most critical building sites in terms of radon concentration. Eskers are long, narrow, steep-sided ridges formed by glacial rivers. Their composition of stratified sand and gravel makes them permeable to water and air. As the interaction terms show, the Tampere zone is more critical than other esker areas in Finland. It is among the most radon-prone in the country because of the height of the esker and its high permeability. We have reported on it in more detail elsewhere (Voutilainen and Mäkeläinen, 1991).

The most critical foundation is the slab-on-grade type. The least critical is that in houses with crawl space and in houses supported on a slab with thickened edges. These two lowest groups were combined because of the very similar parameter estimates.

The effect of construction year might be due partly to the air exchange rate and partly to the building materials and technique used. The lowest construction "class" comprises houses built from 1945 to 1960. These houses are probably less airtight and have a higher air exchange rate than houses built after the "energy crisis" of 1974, which may have very low air exchange rates. New houses with basements have permeable ground contact walls, which radon can easily penetrate. In houses with crawl space, radon concentration does not depend on the construction year.

With such a large number of observations, factors may be statistically significant but may increase the square of the multiple correlation coefficient, R^2, very little. On the other hand, many of the construction factors are correlated, so it is possible to select the best model in many ways. We selected a fairly simple model for which, in most cases, it is possible to obtain the data needed to predict radon values. Another reason for rejecting less significant factors was that there was no good physical explanation for them.

This work is based on a large volume of not very accurate data and is not a substitute for studies that are on a smaller scale but yield more accurate data. A building is a complex entity, and interpretation of parameters is not straightforward. We are planning a new questionnaire to obtain more relevant and reliable data.

REFERENCES

Arvela, H and K Winqvist. **1989.** A model for indoor radon variations. Environ Int 15:239-246.

Geological Survey of Finland. 1985. *Geochemical Atlas of Finland.* Geological Survey of Finland, Helsinki.

SAS Institute Inc. 1985. *SAS[R] User's Guide: Statistics, Version 5.* SAS Institute Inc., Cary, NC.

Voutilainen, A and I Mäkeläinen. The use of indoor radon measurements and geological data in assessing the radon risk of soil and rock on construction sites in Tampere. In: *The 1991 International Symposium on Radon and Radon Reduction Technology* vol. 2 (in press).

QUESTIONS AND ANSWERS

Q: Schlesinger, Israel. I was wondering about making so many measurements. How many do you do in one home? In an earlier presentation we saw big differences between different rooms in the home, so which one represents a house? The variations in your parameters are even less than variations between rooms.

A: Mäkeläinen. We have made only one measurement per house, because we wanted to measure many houses. I think there are differences in different rooms, but we wanted to get a good geographical distribution and as many houses as possible.

Q: Harley. These modeling exercises are always very interesting. Even though the factors are fairly well known about bringing radon into a home, the models usually have a rather poor correlation. You mentioned your overall r^2 was about 0.3, or 30%. This was for a 2-month measurement, is that correct? [Yes.] Was there any particular type of home where the r^2 seemed to be much better?

A: Mäkeläinen. I didn't stratify this.

Q: Harley. But it is interesting that even though the factors are there, when you try to build a model, it doesn't seem to account for everything.

Q: (Speaker unknown.) Using your model, you predict what the concentration will be in a building just by calculating. Have you some good figures where you have predicted numbers and then tested a house? How close do they actually come to what the predictions were?

A: Mäkeläinen. No, this is a method we use for towns and large areas, sort of a first step to advise local authorities where they have to be concerned about radon in new building areas. We haven't tested our model like that.

PLATE-OUT RATES OF RADON PROGENY AND PARTICLES IN A SPHERICAL CHAMBER

Y. S. Cheng[1], Y. F. Su[2], and B. T. Chen[1]

[1]Inhalation Toxicology Research Institute, Lovelace Biomedical and Environmental Research Institute, Albuquerque, New Mexico

[2]Pacific Northwest Laboratory, Richland, Washington

Key words: *Plate out, wall deposition, ^{212}Pb, turbulent intensity*

ABSTRACT

Theoretical deposition models show that turbulence and natural convection in a room are the major factors that influence plate-out rates. Here we describe plate-out measurements for radon progeny and aerosol particles in a spherical chamber under controlled laboratory conditions. The temperature and velocity profiles in still and turbulent air were monitored. A 161-L spherical aluminum chamber was used to study mixing. A laboratory mixer with variable speeds and speed control was mounted on the chamber. Temperature profiles inside the chamber for both still air and during mixing were essentially constant within the accuracy of the temperature probe (±0.5°C). No movement inside the chamber was detected in the still air. During mixing, air velocity was detected when rotational speeds were higher than 500 rpm. Monodisperse silver aerosols and polystyrene latex particles in the range of 5 nm to 2 μm diameter were used in the deposition study. Radon-220 (thoron) progeny were generated by passing ^{220}Rn gas into the chamber and letting the gas decay into ^{212}Pb. The deposition rates of the particles and thoron progeny (^{212}Pb) in the chamber were determined by monitoring the concentration decay of the aerosol as a function of time. Higher deposition rates were observed during increased air mixing. The higher rates were more significant for particles smaller than 1.0 μm, indicating that the turbulence produced by mixing caused an increased deposition by turbulent diffusion. Because radon progeny in the indoor environment are predominantly ultrafine particles smaller than 0.2 μm, their plate-out rates may be greatly affected by air motion.

INTRODUCTION

In indoor and mining environments, deposition or "plate out" of radon progeny on walls occurs simultaneously with their attachment to airborne particles. Attachment and plate-out processes affect the atmosphere in which radon exposure takes place by reducing concentrations and shifting activity size distributions. Both processes have important consequences in determining the deposition pattern and initial dose of inhaled radon progeny.

The deposition of particles in a photochemical reaction chamber may be used as a surrogate for studying aerosol dynamics under indoor conditions. The reported plate-out rates for radon progeny indoors, or in experimental chambers, have a wide range of values that may reflect room or chamber conditions (George et al., 1983; Rudnick et al., 1983; Bigu, 1985; Rudnick and Maher, 1986; Holub, 1984; Holub et al., 1988; Van Dingenen et al., 1989; Vanmarcke et al., 1991). For example, Rudnick et al. (1983) measured very high plate-out rates for ^{218}Po in a chamber when a fan was in operation. Theoretical deposition models show that turbulence and natural convection in a room are the major factors that influence the transport of aerosols; sedimentation, diffusion, and electrostatic attraction are the main mechanisms affecting deposition (Crump and Seinfeld, 1981; McMurry and Rader, 1985; Nazaroff and Cass, 1987, 1989). These models describe plate-out rates for particles ranging from 0.001 to 20 μm in diameter, which includes the sizes of unattached and attached radon progeny.

Laboratory studies in smaller chambers (36 to 400 L) under controlled conditions are used to test theoretical models and to study factors that may influence plate-out rates. Crump et al. (1983) and Okuyama et al. (1986) have shown that the deposition of aerosol particles between 0.006 and 2 μm follows the homogeneous turbulence model in still and turbulent air (mixing by fan or ventilation). Holub et al. (1988), Van Dingenen et al. (1989), and Vanmarcke et al. (1991) reported that the homogeneous deposition model also applies to the deposition of unattached radon progeny and submicrometer-size particles. McMurry and Rader (1985) showed enhanced deposition for charged particles in an electrical field, especially for submicrometer-size aerosols. In general, results from laboratory studies support the homogeneous turbulence model.

This model assumes that at the center of the room the air is well mixed by turbulence and that the aerosol concentration is uniformly distributed. The model further assumes that near the chamber wall is a boundary layer with a concentration gradient, where aerosols transfer from the chamber to the wall. In this model, the diffusional deposition is enhanced by the air turbulence. The magnitude of the turbulence diffusion is characterized by the eddy diffusivity, defined as $D_e = k_e x^n$, where k_e is the coefficient of eddy diffusivity, and x is the distance from the wall. The coefficient of eddy diffusivity and n are functions of flow patterns inside the chamber and are obtained by curve-fitting procedures from the plate-out data. Values of n range from 2 to 3 and those of k_e from 0.0064 to 2.7 x 10^2 (sec^{-1}) when the value of n is set at 2 (Crump et al., 1983; Holländer et al., 1984; McMurry and Rader, 1985; Okuyama et al., 1986; Holub et al., 1988, Van Dingenen et al., 1989). These wide ranges of parameters indicate that the flow conditions in these chambers or rooms may vary considerably. However, in these studies, no temperature or velocity profiles were reported; therefore, we are not certain how these parameters are related to room conditions. For example, in still air, one study reported values of n = 2.0 and k_e = 0.0064 (McMurry and Rader, 1985), while another study reported n = 2.6 and k_e = 0.035 (Van Dingenen et al., 1989). Clearly, there is a need to characterize and control the flow conditions in a chamber in order to fully understand the plate-out processes and to estimate the parameters involved in the prediction model.

In this paper we describe the plate-out measurements in a spherical chamber for thoron progeny and particles (0.001 to 2.0 μm). The temperature and velocity profiles in the chamber with still air and with mixed air were measured; preliminary results of plate out are presented. Chamber temperatures were controlled, and both temperature and velocity profiles were monitored. A cylindrical chamber with top and bottom plates under separate temperature control was used to study the effect of the temperature gradient on plate out, and a spherical chamber with the same volume was used to study the effect of mixing on plate out. The results of the temperature effect are described elsewhere (Chen et al., submitted to *Aerosol Sci. Technol.*).

EXPERIMENTAL METHODS

Chamber

A 161-L spherical aluminum chamber was used. A laboratory mixer (Labmaster Model TS 2010, Thomas Scientific, Swedesboro, NJ) with variable speeds and speed control was mounted on the chamber. A turbine propeller with six flat blades was used to produce turbulence inside the chamber, which was electrically grounded to minimize the electrical field. The chamber was located in an exposure room with temperature and humidity controls. The temperature profiles inside the chamber for both still and turbulent air were monitored with a thermocouple. The velocity profile was measured using a velocity transducer (Model 8470, TSI, Inc., St Paul, MN). The temperature and velocity probes were located at distances between 3 and 21 cm from the wall in the radial direction. Figure 1 is a schematic diagram of the experimental setup.

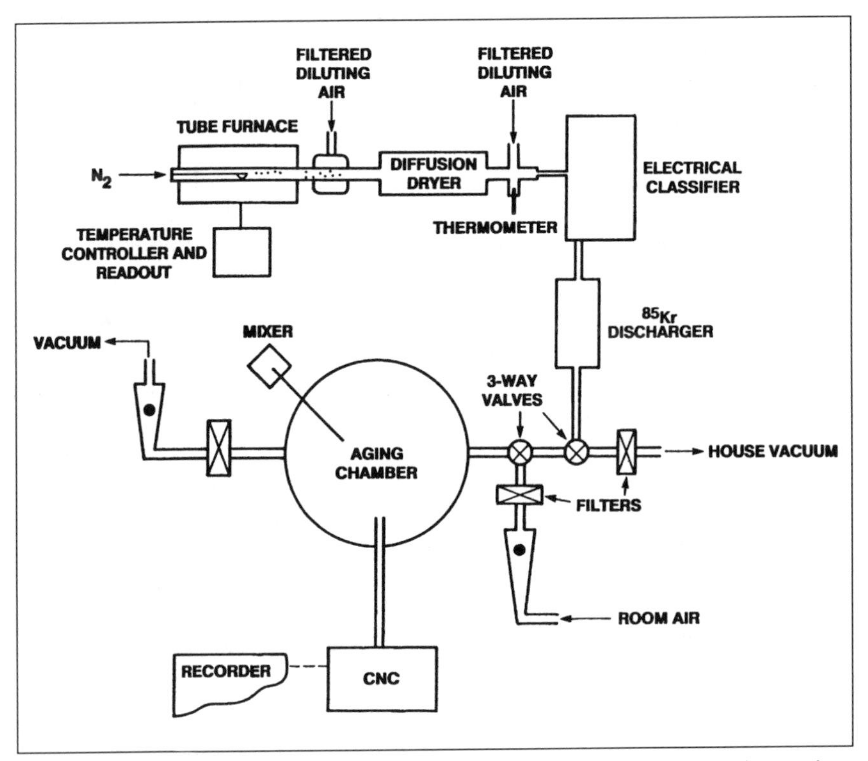

Figure 1. Schematic diagram of experimental apparatus; CNC = condensation nucleus counter.

Aerosol Generation and Monitoring

Monodisperse silver aerosols (density = 10.5 g cm^{-3}) and polystyrene latex (PSL) particles (density = 1.05 g cm^{-3}) were used in the deposition study. Ultrafine silver aerosols (diameter = 5 to 140 nm) were produced by evaporation/concentration, and then classified into monodisperse fractions by an electrostatic classifier (Cheng et al., 1990).

Aerosol concentrations in the chamber were monitored by a condensation nucleus counter (CNC, Model 3020, TSI, Inc., St. Paul, MN). The flow rate was 300 ml min^{-1}, and the sampling duration was 2 min. The aerosol concentration was kept below 800 particles cm^{-3} so that the counting mode of the CNC could be used. The aerosol concentration was also corrected for coincidence errors. Aerosol size was ascertained by determining penetration through a screen diffusion battery (Cheng and Yeh, 1983). Mean size was calculated using screen diffusion battery theory (Cheng et al., 1990).

Monodisperse PSL particles between 0.1 and 2 μm were generated from a Retec nebulizer. For particles smaller than 1 μm, the aerosol was passed through an electrical classifier to remove aggregates and other satellite particles. Aerosol concentrations were then determined, using a CNC as described in the previous paragraph. For micrometer-size particles, aerosol concentrations were monitored by an aerodynamic particle sizer (APS; TSI, Inc., St. Paul, MN). The sampling flow rate for the APS was 5 L min^{-1}. One L min^{-1} of aerosol was withdrawn from the chamber and diluted with 4 L min^{-1} of filtered air before entering the APS. Sampling time was set at 1 min.

In all cases, the aerosol flow passed through a ^{85}Kr discharger to achieve Boltzmann equilibrium. The aerosol was drawn into the chamber continuously for about 1 h; in another hour we ascertained that the aerosol was distributed uniformly before taking the first sample. Usually, five or six samples were taken at intervals. For rapid plate-out rates, the interval was 20 to 30 min; for slow plate-out rates, intervals were longer, often overnight.

Radon-220 (thoron) progeny were generated by passing ^{220}Rn gas into the chamber and letting the gas decay into ^{212}Pb. After the chamber was filled with thoron gas (20 min), we waited 4 min before taking the first sample. This allowed more than 95% of the radon gas to decay into radon progeny. Concentration and size of radon progeny were determined by taking samples through a 50-mesh stainless steel screen followed by a 47-mm filter. The sampling flow rate was 10 L min^{-1}, and the sampling duration was 2 min. Gamma activity on the screen and filter were counted by a NaI detector. The diffusion coefficient of the radon progeny was then estimated from the screen penetration, using the diffusion battery theory (Cheng and Yeh, 1980). Because of the rapid decay of thoron progeny concentration in the chamber, only three samples were taken, at 5-min intervals.

Plate-out Rate and Model

The deposition rate of the particles and radon progeny (^{212}Pb) in the chamber was determined by monitoring the concentration decay of the airborne material as a function of time. The concentration decay can be fitted into an exponential function, $C/C_o = 1\text{-}\exp(-\beta t)$, where C_o is the initial concentration, and C is the concentration at time t. The concentration at each time point was corrected for the removal of aerosol in the sample drawn to the monitor. A nonlinear, least-squares curve-fitting procedure in the RS/E software (BBN Research Systems, Cambridge, MA) was used to obtain the estimated value of the plate-out constant, β, for each experiment.

The homogeneous turbulence deposition model for a spherical chamber has been solved by Crump and Seinfeld (1981) for n = 2 and by Van Dingenen et al. (1989) for arbitrary values of n:

$$\beta = \frac{3n \sin(\frac{\pi}{n}) \sqrt[n]{k_e D^{n-1}}}{2\pi R} \left[2D_1(y) + \frac{1}{2} y\right]; \quad (1)$$

where

$$y = \frac{\pi V_s}{n \sin(\frac{\pi}{n}) \sqrt[n]{k_e D^{n-1}}}; \quad (2)$$

D_1 is the Debeye function (Abramowitz and Stegun, 1970)

$$D_1(y) = \frac{1}{y} \int_o^y \frac{t\, dt}{e^t - 1}; \quad (3)$$

V_s is the particle sedimentation velocity defined as:

$$V_s = \frac{\rho_p\, d_p^2\, g\, C(d_p)}{18\eta}; \quad (4)$$

D is the particle diffusion coefficient defined as:

$$D = \frac{k\, T\, C(d_p)}{3\, \pi\, \eta\, d_p}; \quad (5)$$

g is the gravitational acceleration, k is the Boltzmann constant; C is the slip correction factor (Ramamurthi and Hopke, 1989); and d_p is the particle density. In this equation, no electrostatic deposition was included, because the metal chamber was grounded, and aerosols with a Boltzmann-charged equilibrium were used. Equation (1) was used to fit the experimental data, which included data for PSL (ρ_p = 1.05), silver (ρ_p = 10.5), and ^{212}Pb (ρ_p = 11.3).

The SimuSolv simulation software (Steiner et al., 1990) was used to fit the homogeneous turbulence model with the experimental data and to estimate the best-fit values for parameters k_e and n.

RESULTS

Temperature and Velocity Profiles

Temperature profiles inside the chamber for still and turbulent air conditions were essentially constant within the accuracy of the temperature probe (between 22 ± 0.5°C and 24 ± 0.5°C). No movement was detected in still air or when the mixer was operated at 300 rpm. Air movement was detected when the mixer was operated at more than 500 rpm. Figure 2 shows the measured velocity profile as a function of time at a fixed location. The fluctuation of the velocity indicated that the turbulence was caused by mixing. In Figure 3, the mean velocity (in cm sec^{-1}) is plotted as a function of distance from the wall. As expected, velocity was higher for the higher mixing speed. Also, there were velocity gradients in the radial direction, with the highest velocity near the core of the chamber.

Plate-Out Rates

The concentration decay as a function of time for submicrometer-size particles in still air is shown in Figure 4. A straight line in the semilog scale indicates that the concentration in the chamber followed a single-exponential decay curve. The slope of the line is, then, the decay constant, β. In Figure 5 we plot the plate-out rate constant (β, 1 sec^{-1}) as a function of particle size from 1.3 nm to 2.02 μm (2020 nm) in still air. The deposition rate was dominated by diffusion deposition for particles smaller than 100 nm and by sedimentation for particles larger than 500 nm. Minimal deposition occurred in the size range 100 to 500 nm, where the deposition rates for both mechanisms were within the same order of magnitude. Also shown in the figure are the fitted curves for the turbulence deposition model [equation (2)] for particle densities of 1.05 and 10.5. The density effect was significant only for particles larger than 70 nm, where sedimentation deposition was dominant. The estimated values (mean and standard deviation) are n = 2.58 ± 8.16 x 10^{-8} and k_e = 9.83 x 10^{-3} ± 7.4 x 10^{-4}. The value of n was close to the values of 2.7, estimated by Okuyama et al. (1986); 2.6, by Holub (1984); and 2.6, by Van Dingenen et al. (1989).

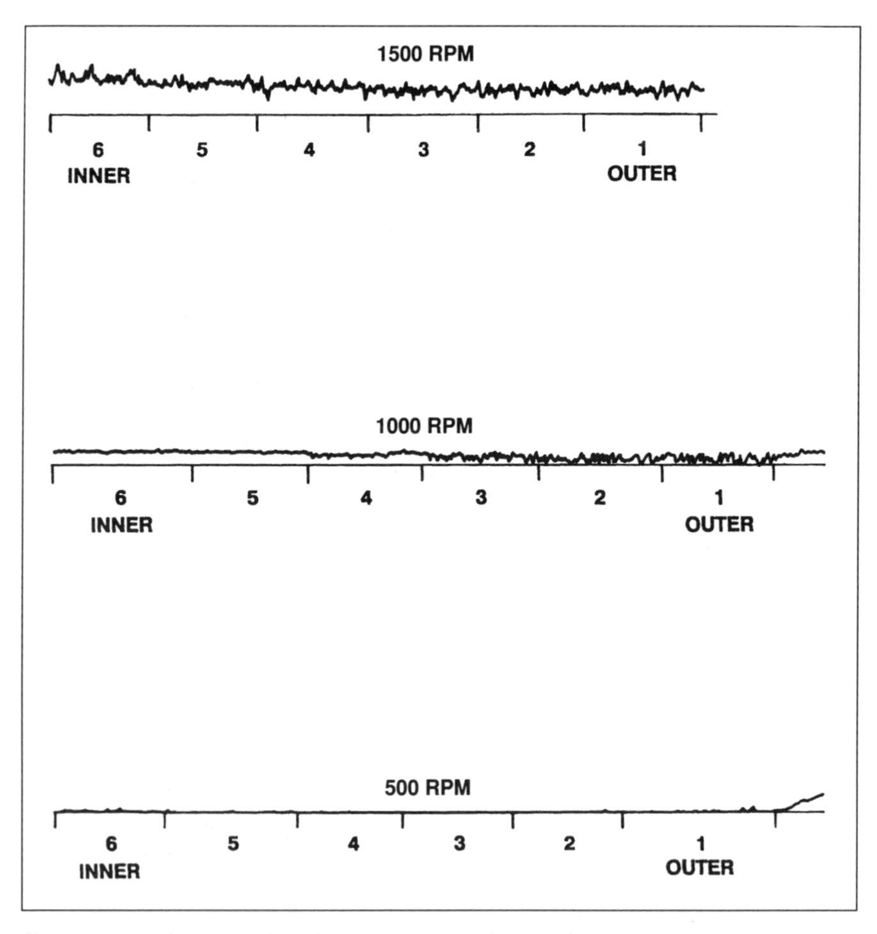

Figure 2. Velocity as function of time obtained in chamber for mixing speeds of 500, 1000, and 1500 rpm, respectively.

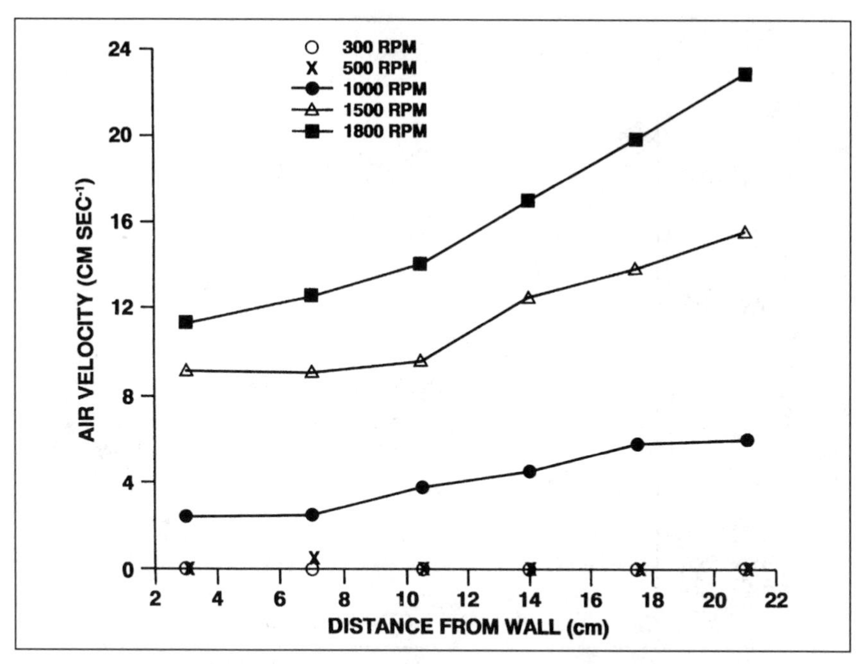

Figure 3. Velocity profile in chamber as a function of distance from wall and mixing speed.

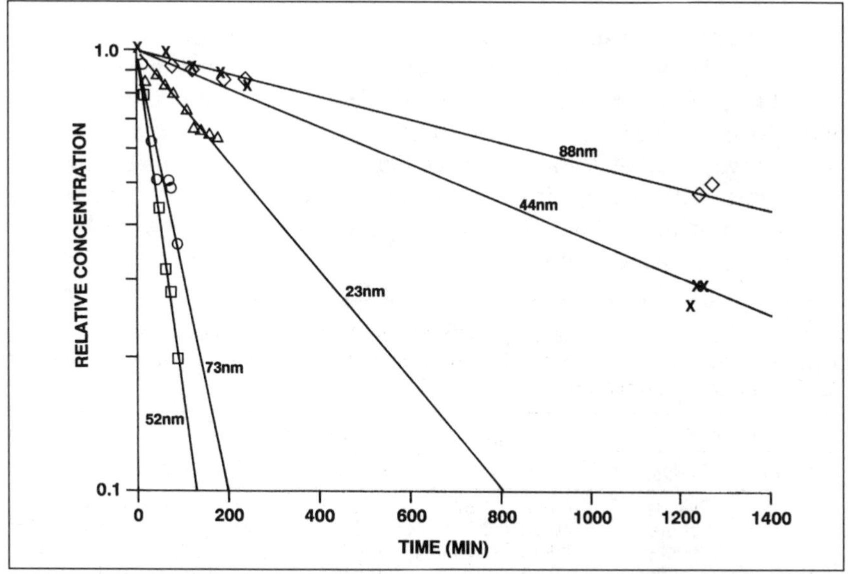

Figure 4. Decay of aerosol concentration as function of time for still air condition.

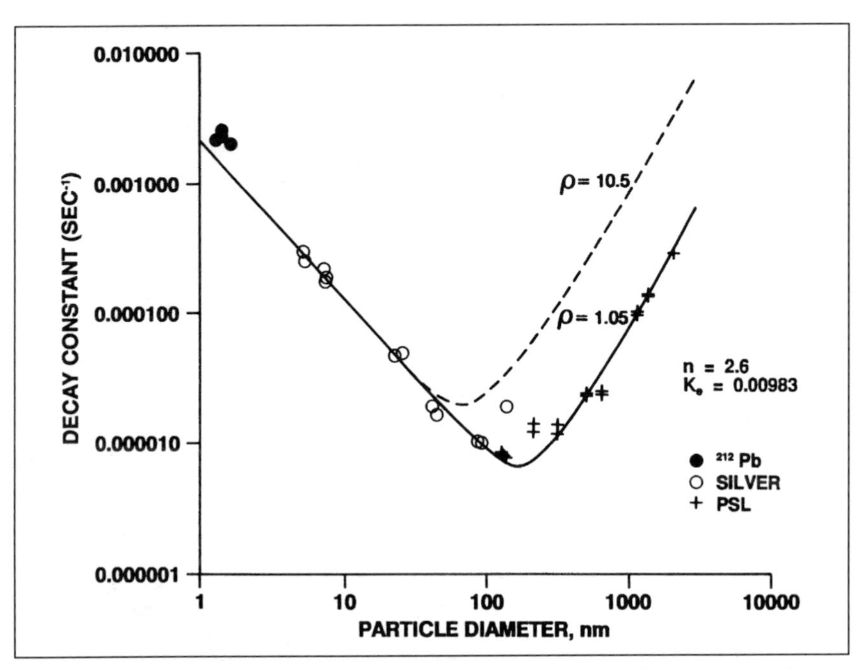

Figure 5. Plate-out rate constant as a function of particle size for still air condition. Curves are best fits using equation (2). PSL = polystyrene latex.

To compare the results obtained in still air chambers from available experiments, we list the estimated values of k_e under the condition of n = 2.6 in Table 1. Some data were recalculated by using the same simulation procedure. Three studies, including this one, gave values of k_e between 0.0093 and 0.012 sec^{-1} $cm^{-0.6}$, which we considered to be the same within experimental error. The other values reported by Okuyama et al. (1986) and Van Dingenen et al. (1989) were 5 and 3 times larger, respectively.

Table 1. Comparison of plate-out rate constants in chambers with still air. Turbulent intensity, k_e, was best-fitted value calculated from the Crump and Seinfeld (1981) theory, assuming n = 2.6.

Chamber		Size range	k_e	
Vol (L)	Shape	(μm)	($sec^{-1}cm^{-0.6}$)	Reference
2.65	Cylinder	0.006 to 1.87	0.0551	Okuyama *et al.*, 1986
250	Sphere	0.02 to 0.3	0.012	McMurry and Rader, 1985
230	Sphere	0.00085 to 0.2	0.035	Van Dingenen *et al.*, 1989
165	Cylinder	0.04 to 3.0	0.0119	Chen *et al.*, 1991
161	Sphere	0.0013 to 2.0	0.00983	Our work

We are in the process of measuring deposition rates at 300, 900, and 1800 rpm mixing rates. Preliminary data show that for 624-nm latex particles the plate-out rates increased with increased mixing speed (Figure 6). Figure 7 shows the experimental data at rotational speeds of 1000 and 1800 rpm using PSL particles, indicating that for 1-μm and larger particles, the plate-out rates were not greatly affected by mixing in the chamber. However, higher plate-out rates were observed for particles smaller than 1 μm in diameter. Additional data are being obtained before substitution in the simulation procedure.

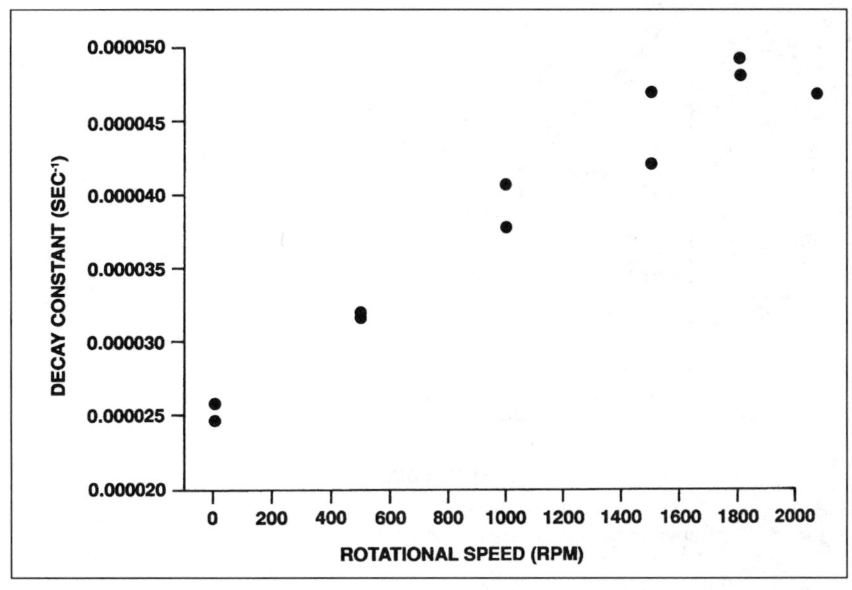

Figure 6. Plate-out rate constant as a function of mixing speed for 0.624-μm-dia polystyrene latex particles.

DISCUSSION AND CONCLUSIONS

This experiment is, perhaps, one of the few chamber deposition studies in which the room temperature was well regulated and both temperature and velocity profiles were measured. No temperature gradients were detected in the spherical chambers that were kept at ambient temperature. We did not detect any air movement in the chamber under still air conditions or when the mixer operated at 300 rpm. However, this does not mean that there was absolutely no air movement in the chamber. It means that we could not detect the slight air movement with the velocity probe, given the limit of detection, estimated at about 1 cm sec^{-1}. The other limitation of our method was that the response of the velocity probe was not measured in real time; therefore, the turbulence intensity from the velocity measurement could not be detected.

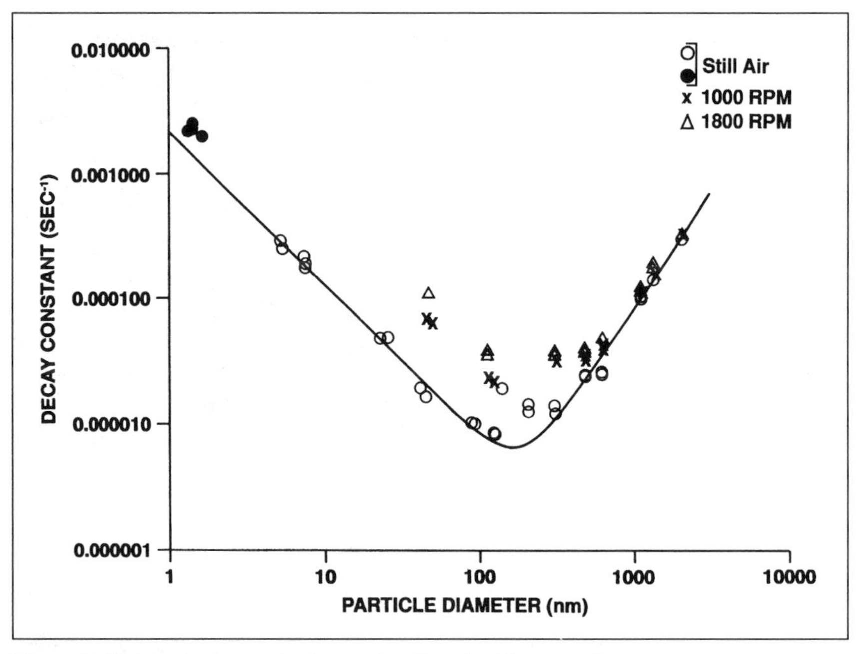

Figure 7. Plate-out rate constants as a function of mixing speed.

Under still air conditions, the plate-out rates for particles ranging between 1 nm and 2.02 μm were best described by the homogenous turbulence deposition model, despite the fact that no air movement or turbulence was detected in the chamber. Similar conclusions were obtained from other studies (McMurry and Rader, 1985; Van Dingenen et al., 1989). Holländer et al. (1984) suggested that a 0.05°C temperature gradient in a room could produce natural convection and, therefore, some mixing of air. With an accuracy of 0.5°C in our temperature measurement, we cannot rule out the possibility that a slight temperature gradient resulted in some turbulence in the chamber.

Our results showed that particle density can enhance the sedimentation deposition of particles greater than 70 nm (0.07 μm). On the other hand, the increased turbulence (by mixing or other mechanisms) in a room can enhance plate-out rates for submicrometer-size particles because of the turbulent diffusion mechanism. Therefore, one way to remove attached and unattached radon progeny from the air is to increase air circulation by using a fan or other means of moving the air.

Eddy diffusivity in turbulent flow is determined by two parameters, n and k_e. Conventionally, the eddy diffusivity is defined as $D_e = k_e x^n$ with n = 2, as supported by Crump et al. (1983), Holländer et al. (1984), McMurry and Rader (1985), and Chen et al. (submitted). However, our results and the results of several other studies showed that the best-fit estimate of n was equal to 2.6 from plate-out measurements (Okuyama et al., 1986; Holub et al., 1988; Van Dingeneny et al., 1989). The reasons for the difference are not clear.

The value of k_e in still air was about 0.01 (assuming n = 2.6), based on several studies, including ours (McMurry and Rader, 1985; Chen et al., submitted). The other two studies had values that were several times higher. The temperature profiles and controls in those studies were not reported, and there could well be some temperature gradient in their chamber that would result in a higher intensity of turbulence (Chen et al., submitted).

Our planned future work includes completion of experiments on different rotational speeds and particle sizes, and evaluation of various deposition models to devise a useful predictive model for the plate out of radon progeny and particles in indoor environments.

ACKNOWLEDGMENTS

The authors are indebted to the technical help of T. Holmes, G. Feather, and S. Brodrick; R. F. Holub for discussion; P. Bradley for editing, and several of our colleagues for reviewing the paper. This research was sponsored by the Office of Health and Environmental Research, U.S. Department of Energy, under Contract No. DE-ACO4-76EV01013.

REFERENCES

Abramowitz, M and A Stegun. **1970**. *Handbook of Mathematical Functions*. Dover, New York.

Bigu, J. **1985**. Radon daughter and thoron daughter deposition velocity and unattached fraction under laboratory conditions in underground uranium mines. J Aerosol Sci 16:157-1651.

Chen, BT, HC Yeh, and YS Cheng. Evaluation of a photochemical reaction chamber. Aerosol Sci Technol (submitted).

Cheng, YS and HC Yeh. **1980**. Theory of a screen-type diffusion battery. J Aerosol Sci 11:313-320.

Cheng, YS and HC Yeh. **1983**. Performance of a screen-type diffusion battery, pp. 1077-1094. In: *Aerosols in the Mining and Industrial Work Environments*, VA Marple and BYH Liu (eds.). Ann Arbor Science Publishers, Ann Arbor, MI.

Cheng, YS, Y Yamada, HC Yeh, and YF Su. **1990**. Size measurement of ultrafine particles (3 to 50 nm) generated from electrostatic classifiers. J Aerosol Res 5:44-51.

Crump, JG and JH Seinfeld. **1981** Turbulent deposition and gravitational sedimentation of an aerosol in a vessel of arbitrary shape. J Aerosol Sci 12:405-415.

Crump, JG, RC Flagan, and JH Seinfeld. **1983**. Particle wall loss rates in vessels. Aerosol Sci Technol 2:303-309.

George, AC, EO Knutson, and KW Tu. **1983**. Radon daughter plateout—I. measurements. Health Phys 45:439-444.

Holländer, WB, W Koch, and G. Pohlmann. **1984**. Design and performance of an aerosol reactor for photochemical studies, pp. 309-319. In: *Proceedings of the 3rd European Symposium, Physico-chemical Behavior of Atmospheric Pollutants*. Reidel, Dordrecht, FRG.

Holub, RF. **1984**. Turbulent plateout of radon daughters. Radiat Protect Dosim 7:155-158.

Holub, RF, F Raes, R Van Dingenen, and H Vanmarcke. **1988**. Deposition of aerosols and unattached radon daughters in different chambers; theory and experiments. Radiat Protect Dosim 24:217-2209.

McMurry, PH and DJ Rader. **1985**. Aerosol wall losses in electrically charged chambers. Aerosol Sci Technol 4:249-268

Nazaroff, WW and GR Cass. **1987**. Particle deposition from a natural convection flow onto a vertical isothermal flat plate. J Aerosol Sci 18:445-455.

Nazaroff, WW and GR Cass. **1989**. Mathematical modeling of indoor aerosol dynamics. Environ Sci Technol 23:157-166.

Okuyama, K, Y Kousaka, and S Yamamoto. **1986**. Particle loss of aerosols with particle diameters between 6 and 2000 nm in stirred tank. J Colloid Int Sci 110:214-223.

Ramamurthi, M and PK Hopke. **1989**. On improving the validity of wire screen "unattached" fraction Rn daughter measurements. Health Phys 56:189-194.

Rudnick, SN and EF Maher. **1986**. Surface deposition of 222-Rn decay products with and without enhanced air motion. Health Phys 51:283-293.

Rudnick, SN, WC Hinds, EF Maher, and MW First. **1983**. Effect of plate out, air motion and dust removal on radon decay product concentration in a simulated residence. Health Phys 45:463-470.

Steiner, EC, TD Rey, and PS McCroskey. **1990**. *SimuSolv-Modelling and Simulation Software*. Dow Chemical Co., Midland, MI.

Van Dingenen, R, F Raes, and H Vanmarcke. **1989**. Molecule and aerosol particle wall losses in smog chambers made of glass. J Aerosol Sci 20:113-122.

Vanmarcke, H, C Landsheere, R Van Dingenen, and A Poffijn. **1991**. Influence of turbulence on the deposition rate constant of the unattached radon decay products. Aerosol Sci Technol 14:257-265.

QUESTIONS AND ANSWERS

Q: Hopke, Clarkson University. Did you assume the 2.6, or did you fit it?

A: Cheng. This is a best fit from the data.

Q: Hopke. Okay, because there is a theoretical argument that it ought to be 2 or 3, and all the experiments find 2.6. So you are empirically finding 2.6. [Yes.]

A: Cheng. There is evidence in the literature that the value of n is between 2 and 3. A value of 2.6 is in agreement with other values reported in the literature.

AN EXPERIMENTAL FACILITY TO SIMULATE RADON-PROGENY BEHAVIOR IN DWELLINGS

P. Eklund and M. Bohgard

Lund Institute of Technology, Box 118, S-221 00 Lund, Sweden

Key words: *Radon exposure room, unattached fraction*

ABSTRACT

We have built a room (volume, 20 m^3) to simulate domestic exposure conditions for studying the interaction between radon daughters and ambient aerosols. The room has its own ventilation system, in which the air exchange rate can be adjusted to between 0 and 5 air changes h^{-1} (ach). The air supply can accommodate low particle concentrations (less than 1 particle cm^{-3} for ventilation rates less than 1 ach). Aerosols can be introduced at any specified particle concentration from 1 to 10^5 cm^{-3}.

Experiments so far have focused on (1) measuring the size distribution of radioactive aerosols by using a differential mobility analyzer combined with track-etch film; (2) studies of the unattached fraction of radon daughters and the impact of the aerosol and induced motion on this fraction, as measured with the single-screen/filter technique; (3) plate out under various conditions, such as different air flow rates, various aerosol size-distributions and concentrations, and ranges of temperature and humidity, etc. Plate out is measured using a large-area, flow-through pulse ionization chamber. Using this facility, plate out can be measured even at high particle concentrations.

This large-scale experimental facility is useful for studying the aerosol parameters relevant to domestic radon-daughter exposure.

INTRODUCTION

The presence of ^{222}Rn and its short-lived decay products in the air of dwellings yields a significant radiation dose to the public (ICRP, 1987; NAS, 1988). The decay products (radon daughters, RnD) are either attached to particles of the room aerosol, or are "unattached" or "free." After inhalation, the probability and site of deposition of a radon daughter in the human airways depends on whether or not the daughter is attached to an aerosol particle. Unattached RnD are assumed to be several times more hazardous than aerosol-attached RnD (James, 1988).

Detailed knowledge about the interaction between RnD and aerosol particles in dwellings is necessary to estimate the risk of exposure to RnD and to decide which remediation strategies should be applied. One way to obtain such knowledge is to perform experimental studies in well-controlled aerosol/RnD atmospheres.

The fact that the unattached RnD have high diffusivities (0.01-0.10 $cm^2 s^{-1}$) places certain requirements on an experimental facility. If the chamber is small, with a high area-to-volume ratio, the unattached fraction of RnD will be small compared to a more realistic environment, because of the high deposition rates of the unattached RnD on the surfaces of the chamber. The goal of obtaining realistic air movements is another reason for choosing a "full-scale" experimental room. This paper describes an experimental setup for RnD-aerosol studies and some preliminary results.

EXPERIMENTAL FACILITY

A schematic view of the experimental facility is given in Figure 1.

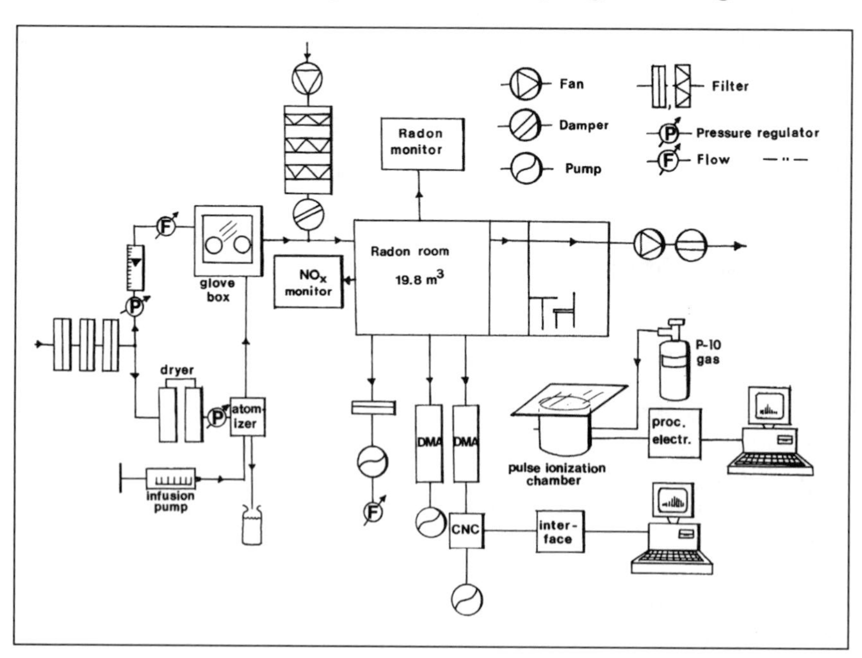

Figure 1. Schematic view of the experimental radon exposure facility. DMA = differential mobility analyzer; CNC = condensation nucleus counter.

RADON EXPOSURE ROOM

The radon exposure room is in a trailer (dimensions, 7.5 x 2.3 x 2.2 m^3). The walls are made of urethan, covered with glass fiber, plastic, and plastic paint. The volume of the radon room is 19.8 m^3. The trailer also contains a control room (volume, 11 m^3), and an air lock (volume, 5 m^3).

Filtered air is forced into the room through a 1-m^3 glove box from a pressurized air supply. A fan extracts air from the room so that a balanced mechanical ventilation is maintained. The specific ventilation rate can be varied up to 1 h^{-1}. Ventilation rates up to 5 h^{-1} can be achieved by pushing ambient air through a three-stage filtration unit. The room has 12 supply and 12 extract terminals along its long sides; by regulating these terminals, different flow conditions can be achieved.

RADON AND AEROSOL GENERATION

A ^{226}Ra source is placed inside the room, attached to an ion-exchange resin (Bjurman et al., 1987), and placed inside a stainless steel cylinder. The ends of the cylinder are made of porous glass, which facilitates the emanation of radon gas. Air is continuously pushed through a stainless steel container that holds this cylinder. At a ventilation rate of 0.5 h^{-1}, a ^{222}Rn concentration of about 3500 $Bq \cdot m^{-3}$ is achieved; it is monitored with a continuous radon monitor (Gammadata, Uppsala, Sweden ATMOS 10).

NaCl aerosols are generated with a nebulizer (TSI, St. Paul, MN, collision atomizer 3075). The glove box facilitates simulation of domestic aerosol production from pollutants such as cigarette smoke, dust, etc.

The background particle concentration in the ventilated room is less than 5 cm^{-3}, measured with a condensation nucleus counter (TSI 3760). By generating an NaCl aerosol, particle concentration can be varied from 10^2 to 10^6 cm^{-3}. During some experiments, a small fan is operated at low speed inside the room to create a homogeneous aerosol.

RADON AND AEROSOL MONITORING

The following parameters are considered important when studying the behavior of the RnD in the exposure room:

- concentration of ^{222}Rn,
- concentration of attached and unattached RnD,
- size-distribution of the background aerosol,
- size-distribution of the NaCl aerosol,
- temperature, relative humidity, and
- concentration of nitrogen oxides, NO_x.

UNATTACHED FRACTION

The unattached fraction is measured with the single-screen technique (George, 1972). The screen has 24 mesh counts cm^{-1} and is used in combination with a backup filter (Millipore AA [St. Quentin Yvelines, France], 0.8 μm thick). Less than 1% of the aerosol-attached RnD and about 80% of the unattached RnD are collected by the screen (Reineking and Porstendorfer, 1990).

We have measured the unattached fraction present at different aerosol concentrations (0-20,000 cm^{-3}). Figure 2 presents the fraction of unattached RnD at equilibrium-equivalent radon (EER) concentration, which is calculated from the following formula:

$$f_p = \frac{0.105 \text{ x } C_A^f \cdot 0.516 \cdot C_B^f + 0.379 \cdot C_C^f}{0.105 \cdot C_A + 0.516 \cdot C_B + 0.379 \text{ x } C_C} \quad (1)$$

in terms of the individual concentrations of RaA (C_A), RaB (C_B) and RaC (C_C). The superscript f refers to the unattached activity concentrations.

Figure 2 also shows the ratio of unattached activity concentration to total activity concentration of RaA (f_A). The results clearly indicate that the unattached fraction decreases as the particle concentration increases. The unattached fraction is from 7-9% at low and moderate aerosol concentrations.

The equilibrium factor, F, is defined as the ratio of *EER* to the radon concentration, C_{Rn}:

$$F = EER/C_{Rn}. \quad (2)$$

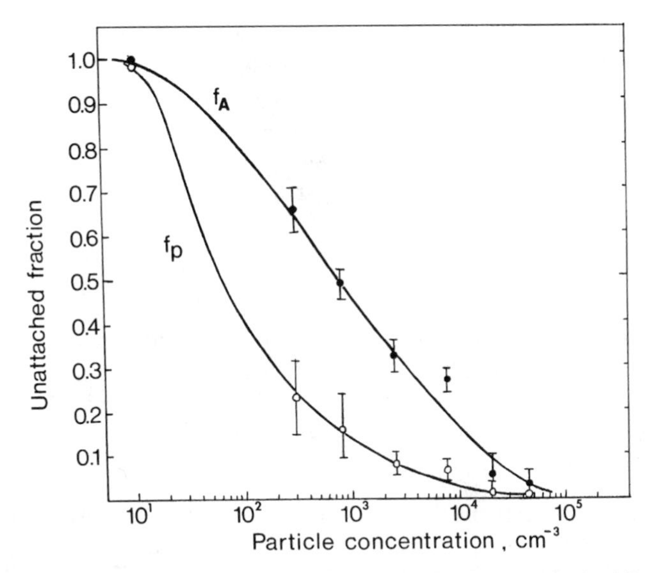

Figure 2. Unattached fraction of radon daughters, f_p, and ratio of unattached RaA to total activity of RaA, f_A, at various aerosol concentrations. Each point is a mean of 4-6 measurements. Uncertainties equal ±1 SD of distribution.

Figure 3 shows the equilibrium factor, F, vs. particle concentration. At low and moderate concentrations, F is between 0.3 and 0.4; at higher concentrations it reaches a value of about 0.5.

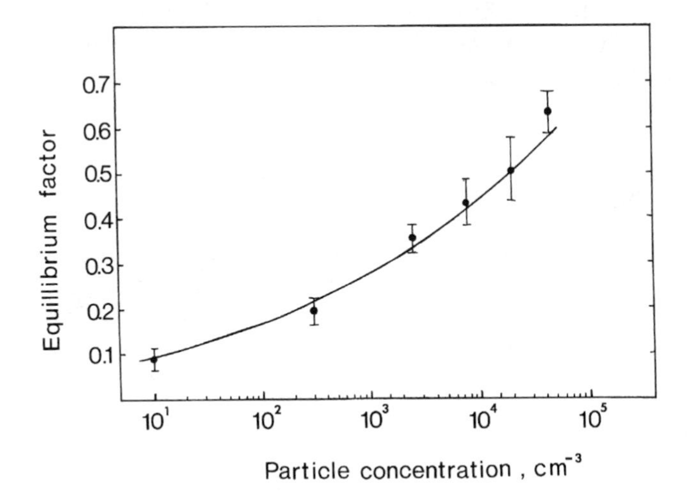

Figure 3. Equilibrium factor F at various aerosol particle concentrations. Each point is a mean of 4 to 6 measurements. Uncertainties equal ±1 SD of distribution.

An NaCl aerosol (geometric mean diameter, 100 nm; activity median diameter, 250 nm; σ_G, 2.0) was used during all experiments. The room ventilation rate was 0.5 h^{-1}. The size-distribution of aerosol-attached activity was measured with a differential mobility analyzer (TSI 3071) combined with track-etch film for measuring the activity-weighted size distribution (Johansson et al., 1988).

PLATE OUT

We have studied the plate-out rate of the radon progeny on glass plates (30 x 25 cm^2) placed inside the radon chamber. Using a flow-through pulse ionization chamber with a 280-cm^2 window, it is possible to measure the deposition of RnD even at high particle concentrations (>20,000 cm^{-3}).

So as not to disturb the air of the room by entering and exiting during low particle-concentration measurements, we use metal disks of 47 mm dia. that can be read from the outside. The room has four ports for plate-out measurements, four outlets for aerosol/radon samples, and one for screen/filter samples.

Plate-out velocity for unattached RaA was measured during exposure to an unattached fraction of 100% (i.e., 10 particles per cc). Aerosol was then added and, during medium and high aerosol concentrations, plate out from both unattached and attached fractions was measured (since we cannot separate them). The contribution from the unattached fraction was subtracted, and the plate-out velocity for attached RaA was calculated.

DISCUSSION

It is our experience that this large-scale facility is very useful for studying the behavior of radon progeny and its interaction with the aerosol. Existing and future room models of radon progeny behavior can be tested with this kind of exposure room.

REFERENCES

BEIR. 1988. *Health Risks of Radon and Other Internally Deposited Alpha-emitters*, Committee on the Biological Effects of Ionizing Radiation, BEIR-IV. National Academy Press, Washington, DC.

Bjurman, B, B Erlandsson, and S Mattsson. **1987**. Efficiency calibration of Ge spectrometers for measurements on environmental samples. Nucl Instrum Meth Phys Res, A26 2:548-550.

George, AC. **1972**. Measurements of the uncombined fraction of radon daughters with wire screens. Health Phys 23:390-392.

ICRP. 1987. Lung cancer risk from environmental exposures to radon daughters, ICRP 50. Ann ICRP 17:2.

James, AC. **1988**. Radon and its decay products in indoor air, pp. 259-309. Lung *Dosimetry*, WW Nazaroff and AV Nero (eds.). Wiley & Sons, New York.

Johansson, GI, KR Akselsson, P Eklund, HC Hansson, and G Jonsson. **1988**. A new technique for the determination of activity distributions of radon daughters. J Aerosol Sci 19:1027-1029.

Reineking, A and J Porstendorfer. **1990**. Unattached fraction of short-lived Rn decay products in indoor and outdoor environments: An improved single-screen method and results. Health Phys 58:715-727.

QUESTIONS AND ANSWERS

Q: McLaughlin, University College, Dublin. On your very last line you showed polonium-218 and polonium-214 with both peaks exactly the same height. How do you account for this? Surely there was recoil taking place off the glass after the decay of the polonium-218. How can the polonium-214 effectively be in equilibrium on the glass?

A: Eklund. Well, I am not sure I quite understand the question. These were taken with a very short exposure to radon daughters, and I don't see the problem of getting equal heights of those two peaks.

Q: McLaughlin. If it was a very short exposure you wouldn't have much polonium-214 building up anyway. During the decay of the polonium-218, some of the daughter products formed would presumably recoil off and then polonium-214 could not grow into full equilibrium.

RECOIL-DEPOSITED Po-210 IN RADON-EXPOSED DWELLINGS

C. Samuelsson

University Hospital, Lund, Sweden

Key words: *Radon, ^{210}Po, retrospective, dwellings, glass*

ABSTRACT

Short-lived decay products of ^{222}Rn plate out on all surfaces in a house containing radon gas. Following the subsequent alpha decays of the mother nuclei, the daughter products ^{214}Pb and ^{210}Pb are superficially and permanently absorbed. Because of its long half-life (22 y), the activity of absorbed ^{210}Pb is cumulative. The activity of ^{210}Pb, or its decay product, ^{210}Po, on the surface can thus reflect the past radon-daughter and plate-out history of a house over several decades.

Our results and experience from measuring ^{210}Po and ^{222}Rn in 22 dwellings are presented. In these studies, the ^{210}Po subsurface activity of one plain glass sheet per dwelling (window panes were not used) was compared with the period of exposure multiplied by the mean radon concentration measured over a 2-mo period. Considering the large uncertainty in the estimated radon exposure, the surface ^{210}Po correlates well ($r = 0.73$) with accumulated radon exposure. The ^{210}Po activity of the glass samples was measured nondestructively, using an open-flow pulse ionization chamber, and this detector has also been successfully used in field exercises.

Vitreous glass, a nearly ideal substrate for deposition of radioactivity, can be found in all dwellings. Migration on the surface activity caused by diffusive transport is thought to be negligible. Results indicate that exploitation of the ^{210}Po activity in glass surfaces as an estimator of lung-cancer risk is feasible, but the method must be utilized with great care to minimize the influence of non-dose-related factors. The accuracy of the glass-polonium method and the conversion factors between surface ^{210}Po and lung-cancer risk have not yet been assessed.

PROLOGUE

George Formby (1906-1961) was a popular filmstar, singer, and ukulele player. One of his greatest hits, "When I'm Cleaning Windows," was first heard in the movie "Keep Your Seats, Please," from 1936. The song was such a great success that the tune was re-recorded in 1937 under the title "Window Cleaner No. 2." It started off with these lyrics:

"You've heard about my capers when windows I've to clean,
Now I'd like to tell you of a few more things I've seen.
I've seen Miss Thompson in her flat
Take off her shoes, her coat and hat,
I've seen her take off more than that
When I'm cleaning windows."

Partly inspired by Mr. Formby, we have also joined the window-cleaning business. Unluckily, and despite our use of 96% ethanol, we haven't been able to spot Miss Thompson. Instead, we have observed the young granddaughter of Miss (!) Lea D. Longlived, Miss Polonia, dressed up in a very thin (50 nm) and transparent glassy fabric. Our assistant, Mr. Pulse Ionchamber, is especially sensitive to Miss Polonia, and he claims to see her radiant beauty in most dwellings.

INTRODUCTION

In searching for the answer to the question in the title of this conference (Indoor Radon: Reality or Myth?), we are hampered by not being able to accurately assess the radon exposure history of a person. Studying lung-cancer incidence today, it is evident that radon-daughter levels of several years ago, rather than current levels, are relevant. Even if we do not know the mechanism or time parameters for the carcinogenicity of radon, it is interesting to investigate whether retrospective assessment of radon exposure (RARE) is feasible. ("Radon" in the acronym is used in a broad sense to identify the source term, not the active lung-cancer agent. Our primary interest is, of course, to assess the accumulated lung-cancer risk, and in doing this, airborne short-lived progenies and their size distribution are much more important than the radon gas itself.)

The basic principles behind RARE, the method proposed by Samuelsson (1988), are presented schematically in Figure 1. All extant radon atoms will, within a few weeks, become ^{210}Pb atoms. All indoor surfaces accumulate long-lived radon daughter atoms and, in hard materials, a more or less permanent record of past radon-daughter concentration levels is created. Unfortunately, because ^{210}Pb decays, "radon memory" is limited in time.

Embedding long-lived nuclides in the surfaces of a room is a two-step process, consisting of surface plate out of short-lived progeny, and alpha decay, which causes daughter nuclei to recoil into the surface layer. The range of distribution of the recoiling nuclei in hard materials is such that most of them will be safe from usual surface-cleaning procedures.

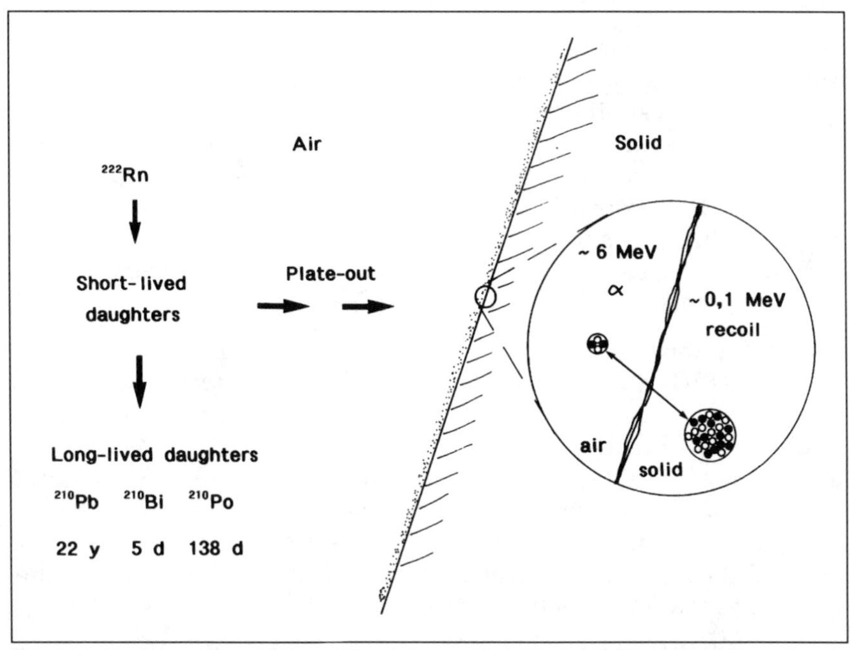

Figure 1. Surface plate out followed by alpha recoils embeds long-lived daughter activity in a surface. Range of recoiling nuclei in glass is about 50 nm (calculations from Landsheere, 1989).

Typically, the embedded surface activity of ^{210}Pb will be very low. A surface exposed to 100 Bq m^{-3} (2.7 pCi L^{-1}) of radon for 10 y exhibits roughly only 1 Bq of ^{210}Pb m^{-2}. In view of this low surface activity, a discussion concerning detection methods seems appropriate.

DETECTION METHODS

The superficially embedded, long-lived radon daughters provide a very thin alpha/beta source that we will take advantage of. Direct determination of beta-emitting ^{210}Pb or its daughter ^{210}Bi is not feasible. Alpha-spectrometry is feasible because ^{210}Po emits a 5.3-MeV alpha particle while decaying. However, the usual solid-state devices of today, with their excellent background and energy resolution characteristics, are intended for small-area samples only. The ideal detector is the alpha spectrometer of the fifties: the pulse ionization chamber, a detector which can easily accommodate samples with surface areas of several square decimeters (1 dm^2 = 0.1 ft^2).

Because of their complexity and cost, "electronic" detectors are less suitable for large-scale in-situ measurements, for instance, extensive epidemiological studies. Autoradiographic techniques using track-etch alpha detectors may also be feasible, but they would require some sort of energy resolution or discrimination against the alpha background of the substrate material. Depending on origin and manufacturing technique, sheets of glass, for instance, may exhibit quite different alpha background count rates, making it difficult to apply any type of total-alpha detector.

Figure 2 compares the energy resolution of a 30-L pulse ionization chamber (PIC) and a 300-mm^2 surface barrier detector (SBD). The sample is a sheet of glass taken from a window exposed to roughly 6 kBq y^{-1} m^{-3} during a period of 25 y. The ^{210}Po peak is easily detectable, even with this small SBD, but in order to achieve reliable counting statistics, measurement must be made over a rather long period, 5 d in this case. In our closed type of PIC spectrometer, sample size is about 200 cm^2, and net count rate in the ^{210}Po peak is about a hundred times that of the 300-mm^2 SBD.

Our closed PIC accommodates sheets of glass with diameters less than 180 mm. Large samples have to be cut down in order to be measured. In field measurements in private dwellings, a destructive procedure is a definite drawback. A special PIC with the ability to analyze glass sheets nondestructively was therefore constructed. It is an open type, and needs a continuous supply of counting gas. Its light weight makes it appropriate for field use.

MATERIALS AND METHODS

The Akarp Study

With the objective of investigating the relationship between radon concentration levels and long-lived daughter activity on surfaces, a group of detached houses in the village of Akarp, near Lund, was investigated. All dwellings in the area have high indoor radon levels because of the material used to build them: alum shale lightweight concrete ("blue concrete"). Of about 60 households asked by mail to help in a RARE research project, 25 answered favorably. Participating home owners loaned us a glass-covered photograph or similar article for a few days; in return we provided them with a radon measurement in their houses, free of charge. The National Institute of Radiation Protection (SSI) in Stockholm supplied and evaluated the track-etch films (personal communication, Hans Mellander). The radon cup holding the CR-39 film strip is the same type as used by the National Radiological Protection Board (NRPB) (Bartlett and Bird, 1987).

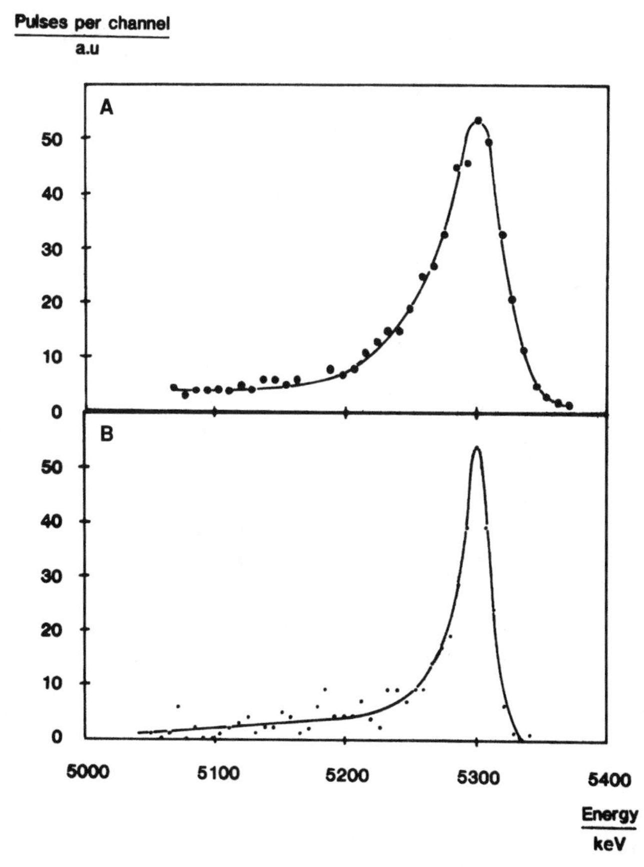

Figure 2. Polonium-210 5.3-MeV peak measured by A, closed pulse ionization chamber and B, a 300-mm^2 surface barrier detector. Sample is sheet of glass exposed from 1963 to 1988 to roughly 6 kBq y^{-1} m^{-3}. Net pulse count rate in A is more than 100 times that in B. Full width at half maximum and sample size are 56 keV and 212 cm^2, respectively, in A, and 26 keV and collimated to 2.5 cm^2 in B.

All dwellings in the study are about 23 y old, and only a few have undergone major reconstruction. Most of the houses, which have no basements, are ventilated by natural convection and have individual oil-fired heating systems, which distribute hot water to radiators beneath the windows.

In each dwelling a radon cup was placed in the room from which the glass was taken, and another cup was exposed in the bedroom, from March to April, 1990. Weather conditions were warmer than normal for southern Sweden, with very few nights below freezing. Heating systems were in use throughout most of the exposure period.

When returning the radon cups, homeowners answered questions about the age and exposure position of the glass sample and other, more general questions about the construction, heating, and ventilation of their houses. Most glass samples supplied were analyzed nondestructively by means of the open PIC. Two or three glass items were usually analyzed in our laboratory each day.

In-House Variations of ^{210}Po

In a separate study we investigated the homogeneity of superficially trapped long-lived daughters in houses. To check the variability of ^{210}Po activity in window surfaces and, at the same time, test the open PIC under field conditions, the necessary equipment was installed in a van and parked outside the dwelling to be investigated. This was a "blue concrete" house, with a basement, situated in a town 300 km north of Lund. The moderately enhanced radon levels indicate that radon from the ground is of little importance, and that the source was predominantly the "blue concrete." From this house (designated No. 10 below), mirrors, some cupboard glass, and most of the windows were brought to the van one at a time and analyzed for ^{210}Po.

RESULTS AND DISCUSSION

The Akarp Study

Of the participating 25 dwellings, 22 have furnished useful glass samples. For each house, the accumulated radon exposure was calculated by multiplying the age of the glass (i.e., exposure time in the house) by the radon concentration value obtained from the single track-etch film exposed in the same room as the glass sample. The measured ^{210}Po activity of the glass samples was corrected for decay (old samples) or nonequilibrium with ^{210}Pb ("young" samples), and the corrected values were plotted against the estimated radon exposure (Figure 3).

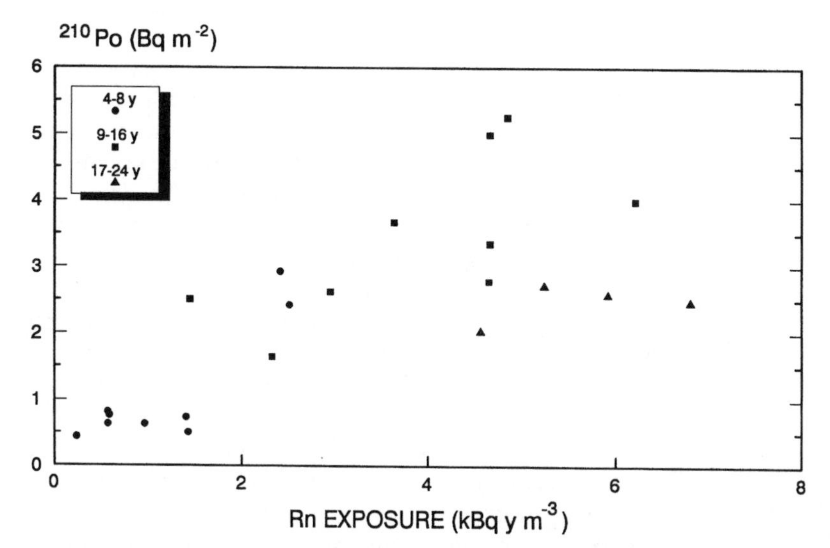

Figure 3. Surface activity of all glass items in Akarp project, corrected for decay and lack of equilibrium with ^{210}Pb. Radon concentration was measured with CR-39 radon film exposed during March and April, 1990, in room from which glass sample was taken. Radon exposure was calculated by multiplying measured radon concentration by age of glass.

The uncertainty in the individual CR-39 measurements of radon concentration depends on the number of tracks counted. This statistical error is insignificant for the data shown in Figure 3, corresponding to a relative standard deviation of less than 1% in most cases. The relative standard deviation from counting statistics obtained when analyzing ^{210}Po is, in most cases, about 5%. The assumption that the measured radon level has continued for the whole exposure period (up to 24 y), can involve a substantial error. Also, the estimated exposure of the sampled item depends on how well the homeowner remembers the history of the item. Realistically, large error bars must be attached to the accumulated radon exposure, and because of the possibility of faulty memories, outliers cannot be excluded. Considering the uncertainty in the estimated radon exposure, which is based on a single radon measurement, and the possible variability in the plate-out phenomenon, the spread in Figure 3 is not surprising.

Old glass samples tend show a lower ^{210}Po value than expected from the linear trend. Presumably, radon levels were lower several years ago than today because of current energy-saving measures. The coefficient of correlation (r^2) is equal to 0.54 in Figure 3.

In-House Variations of ^{210}Po

Predictions of different air-mixing models indicate substantial variation in indoor plate-out velocities. Several factors (e.g., air movement patterns, temperature differences, particle size distribution) can have significant influence on the plate-out rate of short-lived radon daughters. A relationship between absorbed surface activity and accumulated radon exposure can easily be overwhelmed in local in-house variations of short-lived daughter plate-out rate. In a real home environment, several parameters that govern plate-out rate are difficult, if not impossible, to quantify. One would expect that the inward-facing side of a windowpane is subject to substantial variations in both plate-out rates and in radon concentration levels. The air-tightness of a window frame, ventilation habits, proximity to fresh-air vents, etc. affect the local radon concentration. The air-movement patterns and turbulence caused by curtains, radiators, air vents, and depth of a window recess are important in determining both the supply of airborne daughters to the window surface and the actual deposition velocity of the decay products. In addition, any temperature gradient between the surface and room air may cause thermophoresis: i.e., a cold surface attracts, and a warm surface repels, the airborne decay products.

With all these factors in mind, the small spread (less than a factor of 2) in values among samples from House No. 10 (Figure 4) is somewhat surprising. It is obvious that several of the short-term, diurnal, and seasonal variations of the plate-out rate leveled out during the long exposure times involved. It should be noted that: (1) radiators are situated beneath all windows except the one in the terrace door. (2) The two low values (marked “vent,” Figure 4) are for windows having special window-catches for ventilation purposes. According to the owners, the left window (“vent”) in the large bedroom, for instance, is kept slightly opened throughout most of the year during sleeping hours. The absence of a radiator may explain the low polonium/radon ratio of the terrace door (Figure 4). Excluding the door and the windows used for ventilation, the spread in the window values (Figure 4) is only ±10%, presumably a lucky coincidence. The corresponding variation for another house now under evaluation is, from preliminary data, ±30%.

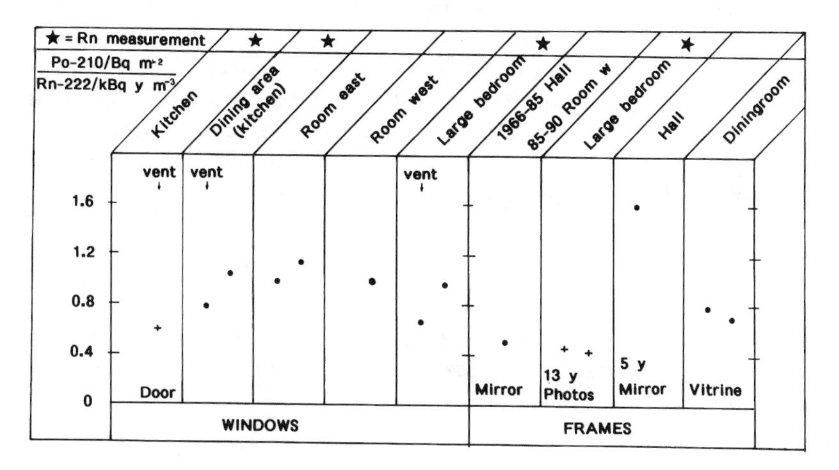

Figure 4. Normalized ^{210}Po surface activity of 14 different glass samples taken from House No. 10. Radon measurements were performed in rooms marked (*). Sheets of glass were exposed for 22 or 23 y prior to measurement, except when noted. Samples marked + were analyzed in January, 1989; all other measurements were made at the house in July, 1990.

Nearly all glass samples analyzed were exposed in House No. 10 since it was built. Exceptions are a mirror (exposed for only 5 y), a photograph frame (exposed for 13 y), and a cupboard door from a vitrine in the dining room that is a few years older than the house. At the moment, we have no explanation for the high ^{210}Po/^{222}Rn value of the mirror in the hall, which received a 5-y exposure. A possible cause, already mentioned in connection with the results from the Akarp study (Figure 3), is the overestimated accumulated radon exposure for old glass samples. When calculating the ^{210}Po/^{222}Rn ratio for the vitrine glass, we did not account for the unknown radon exposure prior to 1966.

The general trend for this house is that windows, at least those that are not usually opened, show a higher ^{210}Po value than other indoor glass samples. This is probably because of the enhanced mixing of room air close to a window, caused by the warm radiator beneath. Since the surface of the windowpanes during most of the year is cooler than room temperature, thermophoretic effects also contribute to the enhanced surface activity of windows.

Another contributing factor for low radon levels is that if dust and grease are allowed to build up on a glass surface, the alpha recoils will increasingly be trapped in the contaminant rather than in the glass surface. Consequently, the nonremovable part of the ^{210}Po activity will be reduced compared to that on a clean piece of glass. Practices in House No. 10 are such that mass load and/or cleaning effects are insignificant and cannot explain the differences in Figure 4. This has been verified by measuring some of the glass items before cleaning, whereas normal procedure is to clean all samples with 95% ethanol before measuring.

The data presented in Figure 4 are from a naturally ventilated house situated in a cold temperate region in which the building material constitutes the main radon source. The trends and values in Figure 4 are not necessarily applicable or typical for differently equipped houses in other climates.

CONCLUSIONS

This study has proven that levels of ^{210}Po in glass surfaces can be measured nondestructively in the field using an open PIC. The measured ^{210}Po activity in the 22 dwellings measured is positively correlated ($r = 0.73$) with the estimated radon exposure. We do not know how well the ^{210}Po activity correlates with the true integrated radon concentration, nor do we know the tumorigenic potential of the indoor atmosphere. Therefore, it is difficult to judge the feasibility and accuracy of the polonium-glass method as a retrospective radon/lung-cancer risk monitor. It is clear that windows situated above radiators normally showed more ^{210}Po activity than other exposed glass surfaces away from radiators. Improved air-mixing during the winter season is thought to be responsible for the enhancement, but thermophoretic forces may also play a part.

The practical exploitation of ^{210}Po deposited in glass as a RARE monitor must proceed with great care, and the influence from modifying and non-dose-related factors must be taken into account. It is certainly advisable to analyze more than one glass sample in a dwelling to increase the precision of measuring accumulated radon levels.

ACKNOWLEDGMENT

This work is supported by the National Institute of Radiation Protection, SSI, Stockholm, grant No. P 505.88, and the U.S. Department of Energy, grant No. DE-FG02-89ER60911.

REFERENCES

Bartlett, DT and TV Bird. **1987.** *Technical Specification of the NRPB Radon Personal Dosemeter,* NRPB-R208. National Radiological Protection Board, Chilton, Didcot, Oxon, UK.

Landsheere, C. 1989. Experimental and Theoretical Investigation of ^{210}Po Activity Absorbed in Glass. Ghent State University, Ghent, Belgium.

Samuelsson, C. 1988. Retrospective determination of radon in houses. Nature 334:338-340.

QUESTIONS AND ANSWERS

Q: (Speaker unknown.) I was going to suggest that perhaps the glass in the mirror has a much higher natural radioactivity, and although that, of course, is distributed through the glass, the alpha energies are higher than that of polonium-210, and so you would get an increased baseline. I wonder if you can correct for this in any way.

A: Samuelsson. We have an uncertainty when analyzing for polonium because of this (baseline), but normally we take a left background and a right background beside the peak, and that should take care of most of the problem with different levels of the background. But, of course, we have to decide the interval for integration. It will give us some uncertainty, but not too much.

Q: Schoenberg, New Jersey. Do you have any estimate of the lowest detectable concentration by your method, and have you also made duplicate measurements so you have some estimate of the error in your reading?

A: Samuelsson. Yes, the lower detection limit, as you say, is below 600 Bq m^{-3} y. It depends somewhat on the behavior of the chamber. Pulse ionization chambers are sometimes a bit tricky. André Poffijn and I were in Copenhagen, outside of Niels Jonassen's lab, and we analyzed the glass from Belgium. In that case we have an increased background and you can, perhaps, increase that low detection limit by a factor of 2. But I should say that around 1 kBq m^{-3} is what you should see, depending on the chamber, of course, if you increase the surface. We now use a surface of around 200 cm^2. Our new chambers are even larger, so maybe we can go down further.

Q: Suess, World Health Organization. You wrote in your abstract, and you may have used it when presenting your paper, about absorption, not adsorption. That means that cleaning the windows with whatever detergent or other cleaning material would not remove anything. Now I assume most of you have verified this experimentally. Can you elaborate on this a little?

A: Samuelsson. All samples that were discussed here were cleaned with alcohol before we made the measurements. But normally you don't seem to take away too much, because a housewife normally cleans windows perhaps once a month, or every half a year, and then the removable activity is very little compared to the nonremovable. We associate the adsorbed activity with the plateout activity, and the absorbed with the imbedded activity. But, of course, you can run into trouble taking a glass that sat for 10 years in a house and was never cleaned. Then, most of the activity is in the grease or on the dust, and you are in trouble. Ideally, the glass should be cleaned once in a while, but not too often.

EXPERIMENTAL AND THEORETICAL STUDY OF THE FRACTION OF ^{210}Po ABSORBED IN GLASS

J. Cornelis,[1] C. Landsheere,[1] A. Poffijn,[1] and H. Vanmarcke[2]

[1]State University of Ghent, Ghent, Belgium

[2]Belgian Nuclear Research Center, Mol, Belgium

Key words: *Radon, glass, plate-out, attachment rate*

ABSTRACT

The ^{210}Po activity embedded in glass may be an appropriate indicator of past radon exposures. Based on the theory of Lindhard, it can be concluded that, after alpha decay of the absorbed ^{214}Po fraction, some 30% of the recoiling ^{210}Pb atoms escape from within the glass. Measurements of the alpha activity of ^{214}Po show that repeated cleaning of vitreous glass removes about 80% of the deposited activity. In this paper we show how the expected ^{210}Po activity is calculated as a function of the attachment rate, being the most important influencing parameter of the room model.

INTRODUCTION

In 1987, Lively proposed using the long-lived radon decay product ^{210}Pb ($T_{1/2}$ = 22.3 y) deposited on glass surfaces as an estimate for long-term indoor radon concentrations (Lively, 1987). In 1988 Samuelsson developed a technique in which the measured alpha decay of absorbed ^{210}Po is used as an indicator of the radon exposure level of previous decades (Samuelsson, 1988). This technique could become an important tool in epidemiology in view of the long latency period for tumor manifestation following exposure to radon progeny.

Figure 1 shows how radon decay products can become embedded in a macroscopic surface (glass). Of the ^{218}Po and ^{214}Po activity deposited on surfaces such as vitreous glass, 50% is absorbed in the material after alpha decay. The maximum penetration depth of the recoiling nuclei is about 0.055 μm (Landsheere, 1988).

As a result of the long half-life of ^{210}Pb, the outermost layer of glass contains information on the integrated radon levels in the house over the last few decades. Alpha spectrometric analyses (Samuelsson, 1988) have shown that diffusion of the long-lived decay products into glass is negligible, so plate out of the short-lived decay products is the main source of activity in the surface layer.

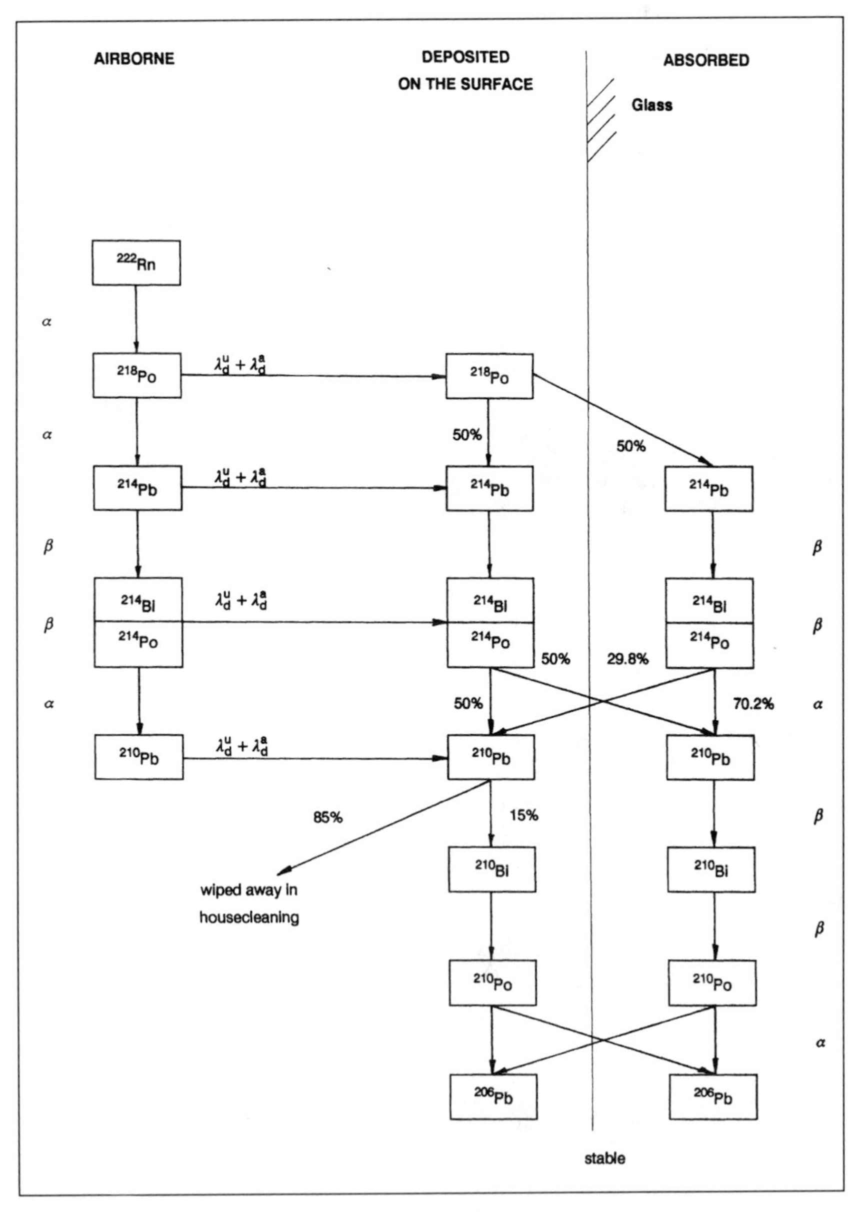

Figure 1. Processes leading to absorption of ^{210}Po by glass.

Cleaning vitreous glass is generally supposed to remove all of the deposited atoms (Landsheere, 1988). However, our experiments show that a fraction of the deposited activity (some 20%) remains attached to the glass. In any case, measurements of the alpha activity of the absorbed *and* remaining deposited ^{210}Po can then be used for estimating the cumulative radon exposure.

INVESTIGATION OF EFFECTS OF GLASS-CLEANING ON ^{214}Po

In order to test the hypothesis that cleaning a glass sheet completely removes all deposited radon-daughter activity, an experiment was arranged in the facility shown in Figure 2. A circular glass sheet was exposed on one side to air (50% humidity) in a radon chamber (1000-L volume) filled with radon in equilibrium with its progeny. The chamber also contained some artificially generated aerosol, supplied by an atomizer. The concentration of this aerosol was measured with a condensation nucleus counter. Turbulence was standardized using a resistor wire, dissipating 43.5 W.

After reaching steady state, the alpha spectrum of the glass sheet was recorded (period $[t_1,t_2]$ in Figure 3A). Thereafter, we rubbed the sheet with a cloth and alcohol, and measured the remaining ^{214}Po alpha-activity (period $[t_3,t_4]$ in Figure 3A).

A first test of the cleaning hypothesis consists of calculating the remaining fraction of ^{214}Po, i.e., the ratio of the measured number of counts $\alpha_g(3)$ (g = index for glass) of ^{214}Po in the observation period $[t_3,t_4]$ to the total expected number (αt) if no cleaning had taken place [dashed curve in Figure 3A]. If the hypothesis holds, this ratio, $\alpha_g(3)/\alpha_t$, equals the fraction of embedded ^{214}Po:

$$\frac{\text{Absorbed } ^{214}\text{Po}}{\text{Absorbed + Deposited } ^{214}\text{Po}},$$

which is found by solving the differential equations according to the compartment model for the sampling and subsequent decay period.

In the compartment model, the attachment rate, X, to aerosols is one of the unknown parameters. The aerosol concentrations measured at the end of the sampling period were used to calculate the value of the attachment rate according to the following formulas (Bricard, 1977):

$$X = \Sigma_{Dp}\, dx\,(Dp)$$

and

$$dX(D_p) = \frac{2\pi.D.D_p}{\frac{8.D}{v.D_p} + \frac{1}{H}} \times dN(D_p), \qquad (1)$$

with $H = 1 + 2.\Delta/D_p$,

where $dX(D_p)$ stands for the attachment rate of aerosol particles in diameter interval $[D_p],[D_p + dD_p]$, D_p for the aerosol particle diameter, $dN(D_p)$ for the number of particles in diameter interval $[D_p, D_p + dD_p]$ and D, v, Δ, respectively, for the mean diffusion coefficient, thermal velocity, and free path of the unattached daughters. (Reference values: 0.068 cm^2 s^{-1}, 2.74.10^4 cm s^{-1}, 6.53.10^{-6} cm.)

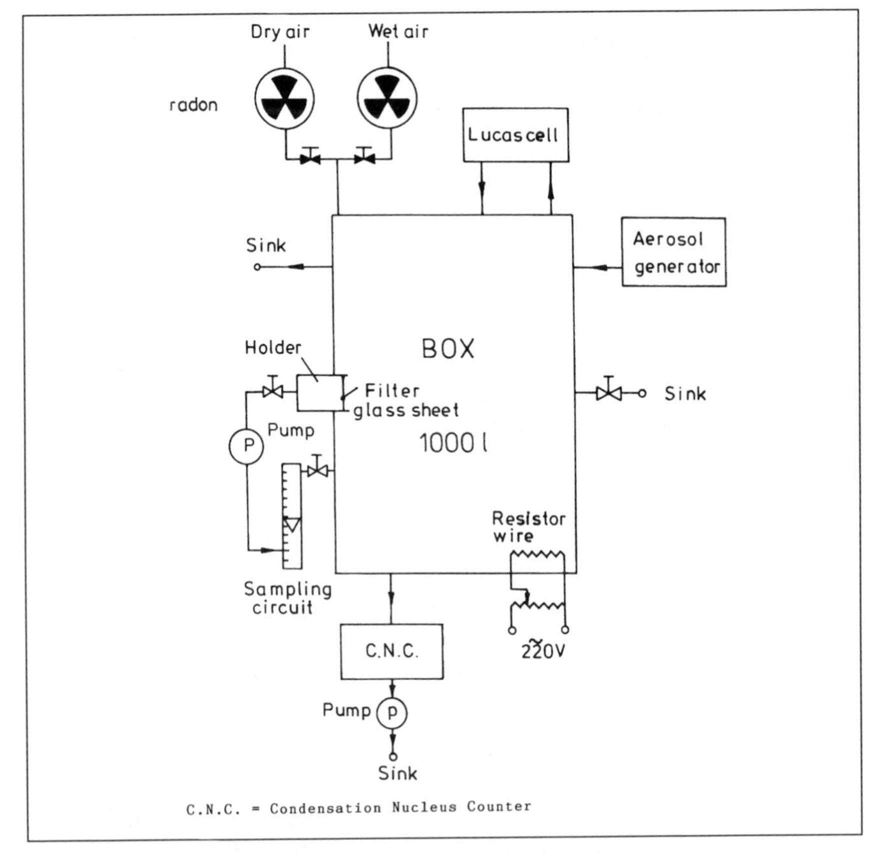

Figure 2. Experimental facility for radon exposure of a glass sheet.

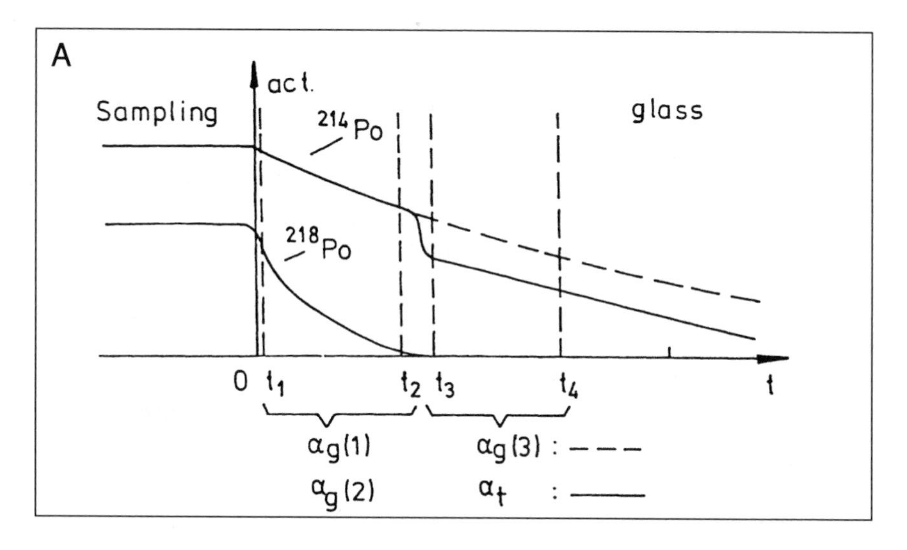

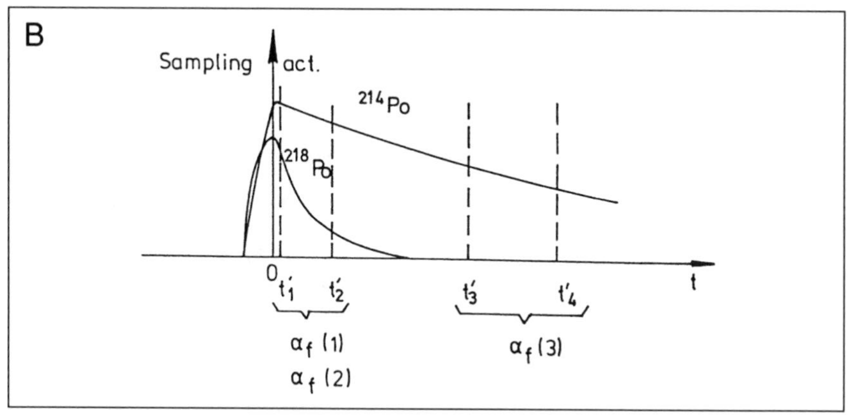

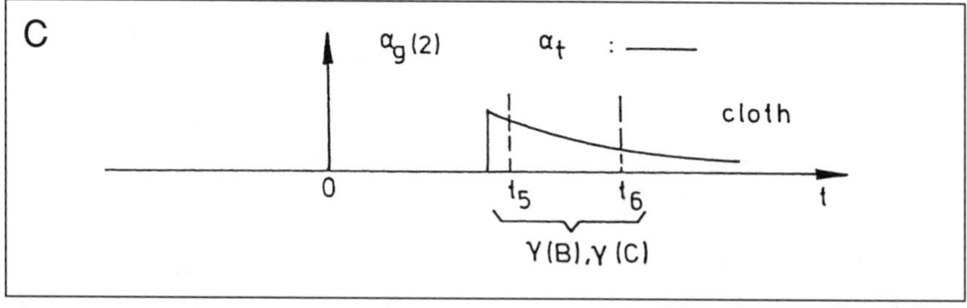

Figure 3. A, alpha-activity time course following experimental exposure of glass to radon progeny; B, alpha-activity time course of filter; C, alpha-activity time course of cleaning cloth.

The distribution of the aerosol was measured with an electrostatic classifier and appeared to be monodisperse, so that one value for the attachment rate holds over the whole distribution diameter range.

Figure 4 shows the results for the experiments carried out at different aerosol concentrations. Embedded fractions were calculated with two sets of deposition constants (in h^{-1}) for the unattached short-lived daughters, namely (11,11,11) (full curve) and (11,5.5,5.5) (dashed curve). From this plot we concluded that the fraction remaining after cleaning was higher than the embedded fraction, indicating that a part (about 20%) of the deposited activity of ^{214}Po is not removed by cleaning a glass sheet after exposure.

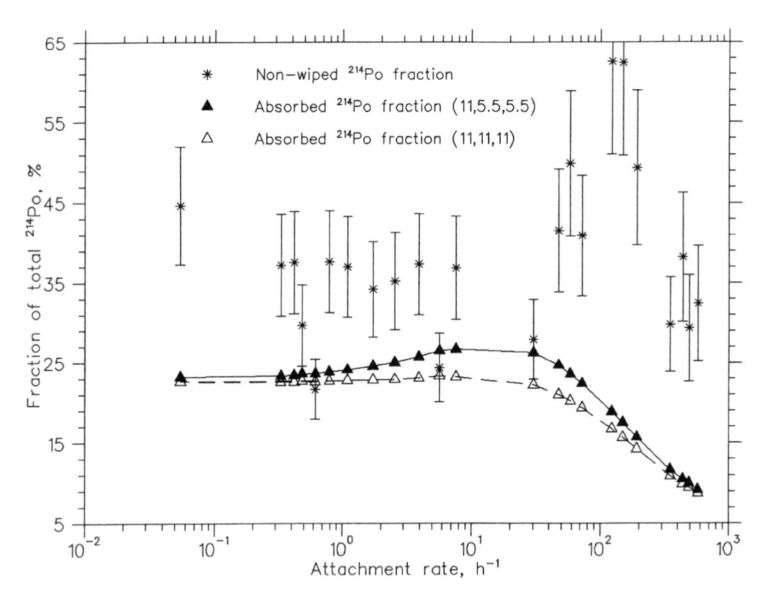

Figure 4. Alpha activity of glass after exposure to radon and cleaning. Theoretical results with equal (△) and different (▲) values of radon-daughter deposition constants; *, experimental results.

The ^{214}Po activity curve without cleaning (dashed line in Figure 3A) was calculated starting from the air activity collected on a filter. Alpha spectra were recorded during two intervals (see Figure 3A). The three sets of counts, α_F, were used for calculating, by means of ordinary Bateman equations, the number of ^{214}Po counts $\alpha_i(2)$ one would measure in the period $[t_1,t_2]$. We actually “imitated” the uncleaned glass sheet by using an imaginary filter, the alpha-activity time course (Figure 3B) of which is a mismatching overlay of Figure 3A. The ratio $\alpha_T/\alpha_i(2)$ for the period $[t_1,t_2]$ yields the conversion factor from filter to glass, correcting for the differences in exposure time and sampling mechanism. Multiplying the expected number of ^{214}Po filter-counts, $\alpha_i(3)$, for the interval $[t_3,t_4]$ with this conversion factor yields the value of α_t for the same interval.

A complementary test of the cleaning hypothesis started from gamma-activity measurements of ^{214}Pb and ^{214}Bi on the cloth used for cleaning the glass sheet. The activity time course is shown in Figure 3C. The number of measured counts is $\gamma(B)$ of ^{214}Pb and $\gamma(C)$ of ^{214}Bi. In the same way as for the alpha measurements, we considered the ratios of the measured number of (gamma) counts to the expected ones if no cleaning has occurred; this ratio, $\gamma(B)/\gamma'(B)$ for ^{214}Pb and $\gamma(C)/\gamma'(C)$ for ^{214}Bi, represents the fraction of activity that has been wiped off by cleaning.

In Figures 5 and 6, the deposited fraction (solid line) is compared to the wiped-off fraction. In contrast to the first test, experimental and theoretical results agree quite well, indicating that the cleaning hypothesis might be valid. However, the large error bars on the experimental points make such a conclusion quite speculative.

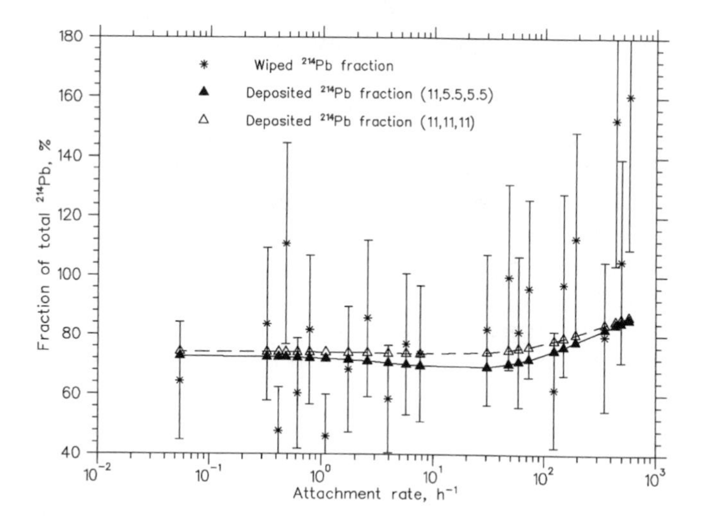

Figure 5. Comparison of theoretical and experimental values (asterisks) of ^{214}Pb fraction (solid line) deposited on glass.

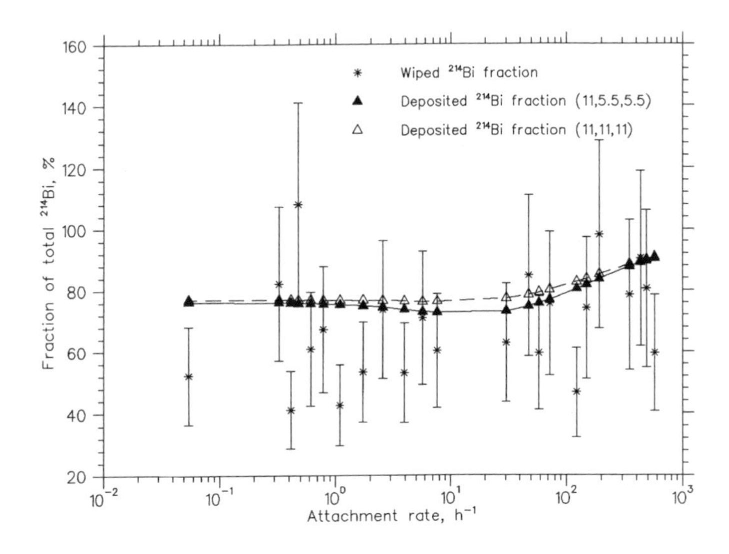

Figure 6. Comparison of theoretical and experimental values (asterisks) of ^{214}Po fraction (solid line) deposited on glass.

CALCULATING THE ^{210}Po ACTIVITY OF GLASS

With the result of the tests on ^{214}Po in mind, we assumed that the cleaning hypothesis was not valid. Next, we assumed that if the test had been executed on long-lived ^{210}Po, we would have found the same result as for ^{214}Po. From the compartment model we calculated the embedded ^{210}Po activity per radon concentration (Bq m^{-3}) as a function of the attachment rate, X, for zero ventilation (solid line, Figure 7). We added to this absorbed activity the deposited fraction of ^{210}Po that would not be wiped off by cleaning. It is obvious that in real conditions, cleaning (for example, at a rate of once a week) affects only the deposited long-lived radon daughters, not the short-lived ones. The final result (* in Figure 7) represents the activity to be expected on a glass sheet exposed to 1 kBq m^{-3} of radon for 20 y with weekly cleaning. The surface-to-volume ratio was estimated to be 3 m^{-1}.

DISCUSSION

Figure 7 shows that over the usual range of values for the attachment rate in dwellings (10 to 100 per hour), the calculated ^{210}Po activity varies by more than a factor of 3. Thus, accurate retrospective determination of the cumulative radon activity from measuring the ^{210}Po glass activity involves estimating the (local) time-averaged attachment rate. Furthermore, the contribution of ^{210}Po fraction that was not wiped off is small compared to the uncertainties of the aerosol concentration and ventilation rate. In real conditions the absorbed ^{210}Po fraction may be a little higher because of electrostatic deposition.

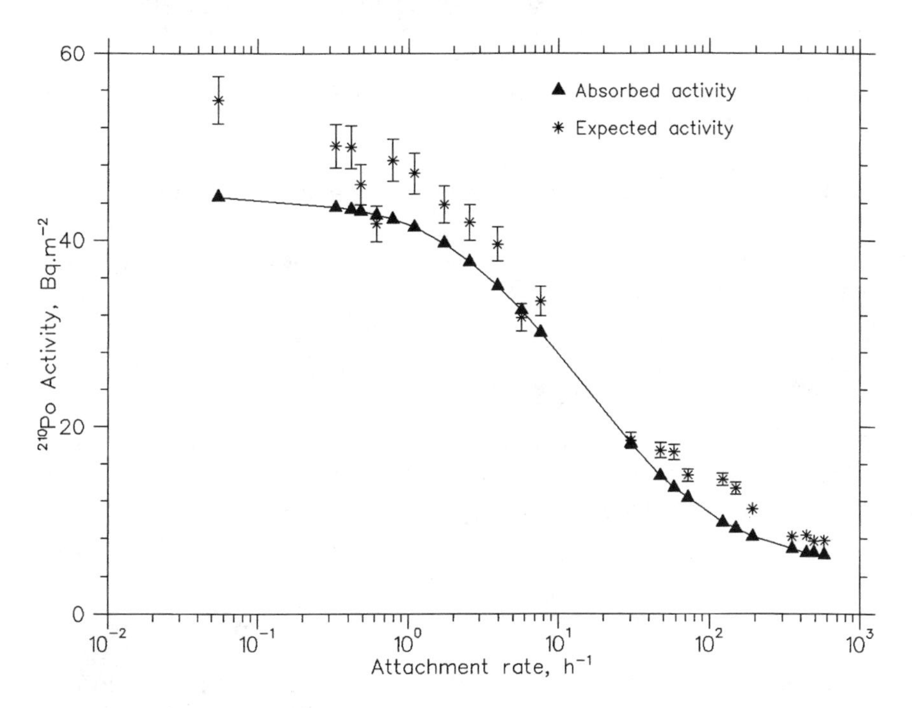

Figure 7. Comparison of ^{210}Po surface activity on glass after 10 y radon exposure and complete cleaning (solid line), or incomplete cleaning (asterisks).

It should be mentioned that similar preliminary experiments made with almost no aerosol resulted in a much lower fraction of ^{214}Po remaining after cleaning twice after sampling (less than 10%), so that theoretical and experimental values overlapped within the overall uncertainty. The results of the experiments described are therefore somewhat surprising. They have led to another series of tests, carried out as accurately as possible, with two cleaning sessions after exposure or, if possible, by testing on ^{218}Po.

ACKNOWLEDGMENT

This work is partly funded by the Commission of the European Communities (proposal No. 0211).

REFERENCES

Bricard, J. **1977**. *Physique des Aerosols. II. Nucléation, Condensation, Ions, Électrisation, Propriétés Optiques*, R-4831(a). Commissariat à l'Energie Atomique, Fontenay-Aux-Roses, France.

Landsheere, C. **1988**. *Experimentele en theoretische studie van de fractie van ^{210}Po-activiteit geabsorbeerd in glas*, Thesis. State University of Ghent, Belgium.

Lively, RS. **1987**. Surface radioactivity resulting from the deposition of ^{222}Rn daughter products. Health Phys 52:411-415.

Samuelsson, C. **1988**. Retrospective determination of radon in houses. Nature 334:338-340.

QUESTIONS AND ANSWERS

Q: Harbottle, Brookhaven. Concerning the absorbed atom, when it recoils, about half of the lead-214 atoms will go into the glass, and will end up penetrating the glass. Is the discrepancy you found between your measurements and model due to the escape of recoiling lead-210 atoms from the glass surface?

A: Poffijn (presenter for Cornelis paper). In the modeling, recoil was taken into account, and so finally some 30% of the ^{210}Pb could escape from the glass. Our calculations were based on these data.

Q: Harbottle. There is a phenomenon which was studied in the 1920s, I guess, involving the recoil of alpha projected atoms in vacuoles into glass surfaces. And there was a number calculated, called the "sticking coefficient" for want of a better term. Some of the atoms bounce off without penetrating; even under ideal conditions they will bounce off. I did some work some years ago along these same lines. We found that the efficiency is substantially below the 50% that you would calculate. I wonder if that is what is explaining this kind of effect? There is another phenomenon that is akin to sputtering with heavy atoms. A heavy atom moves in, heats up a small volume of the surface, and then distills itself back out into the gas phase. So there are several mechanisms: scattering, sputtering, etc. by which you can lose these incoming atoms, even though they are extremely high in energy.

Q: Knutson, Environmental Measurements Laboratory. I think, on your last viewgraph, you indicated that the surface-to-volume ratio was 1.5 m^{-1} Well, if you said that it was 4.5 m^{-1}, wouldn't that explain the discrepancy between measurement and model predictions?

A: Poffijn. The outcome is quite dependent on the surface-to-volume ratio, but we did not measure that in this room because, in our original calculation, we took 3, that is, S over V. But these were measured. We considered the range of values 1.5 to 2.0 realistic for this particular room. Even for 2 one never gets down to 12 Bq m^{-2} surface activity.

Q: Harley, New York University. It wasn't very clear to me whether these electrostatic effects are significant. Could you discuss that?

A: Poffijn. To avoid all discussion we said maybe it could be higher. It could be low, I think, if electrostatic effects play a role. It should be high in this application area, but I'm not convinced about that either.

Q: Harley. Have you measured the charge on any of these glasses?

A: Poffijn. Yes, we just measured this and there didn't seem to be any static electricity.

MEASUREMENT OF THE SIZE DISTRIBUTIONS OF RADON PROGENY IN INDOOR AIR

P. K. Hopke, M. Ramamurthi,[a] and C.-S. Li [b]

Clarkson University, Potsdam, New York

Key words: *Activity-weighted size distributions, radon decay products, unattached fraction, equilibrium factors*

ABSTRACT

A major problem in evaluating the health risk posed by airborne radon progeny in indoor atmospheres is the lack of available information on the activity-weighted size distributions that occur in the domestic environment. With an automated, semicontinuous, graded screen array system, we made a series of measurements of activity-weighted size distributions in several houses in the northeastern United States. Measurements were made in an unoccupied house, in which human aerosol-generating activities were simulated. The time evolution of the aerosol size distribution was measured in each situation. Results of these measurements are presented.

INTRODUCTION

Substantial interest in the properties and occurrence of radon and its progeny has arisen in recent years because it has been recognized that the infiltration of radon into the indoor environment may constitute a significant human health hazard. The inhalation and subsequent lung deposition of the short-lived radon progeny ^{218}Po, ^{214}Pb, and $^{214}Bi/^{214}Po$ have thus warranted the study of their diffusivity and association with molecular cluster aerosols in the ultrafine cluster size range (0.5-5 nm) and larger mode aerosols. Traditionally, the ultrafine cluster and accumulation modes of activity size distribution have been termed "unattached" and "attached" fractions, respectively, in view of the significant difference in their diffusivities. Unattached ^{218}Po has been assumed to have a single, constant diffusion coefficient, typically 0.054 cm^2 s^{-1} (Chamberlain and Dyson, 1956), and samplers

(a) Current Address: Battelle, Columbus Laboratories, 505 King Avenue, Columbus, OH 43201-2693.

(b) Current Address: Institute of Public Health, College of Medicine, National Taiwan University, Taipei, Taiwan, ROC.

have been developed that provide operationally defined estimates of these two modes of activity (Ramamurthi and Hopke, 1989). Experimental measurements of activity-weighted size distributions in recent years have shown that the unattached fraction is, in reality, an ultrafine cluster fraction in the 0.5- to 5-nm range, whose diffusivity and characteristics depend on the nature of the indoor atmospheric environment (Reineking and Porstendörfer, 1986).

In view of the important contribution of the ultrafine cluster mode to the estimated lung alpha dose (see, for example, James, 1988), efforts have been made to develop size distribution measurement techniques that overcome the lack of sensitivity of conventional methods for particle size $d_p < 5$ nm. The alternative techniques developed, adaptations of the wire-screen diffusion battery concept, have been called graded screen arrays (GSA) (Holub and Knutson, 1987; Ramamurthi and Hopke, 1991). The GSA systems consist of varying mesh number, single/multiple wire-screen stages operated either in series or in parallel, with a choice of a wide range of wire-screen parameters and sampling flow rates, coupled with a technique for determining the radioactivity associated with the particle size distribution.

The automated, semicontinuous GSA system used in this study is described in detail in Ramamurthi and Hopke (1991). It provides semicontinuous estimates of ^{218}Po, ^{214}Pb, and ^{214}Bi activity size distributions and concentrations. The system was then used in a series of field studies whose results are presented in this paper.

MEASUREMENT SYSTEM

The measurement system involves the use of six compact sampler-detector units (Figure 1) operated in parallel. Each sampler-detector unit couples wire-screen penetration, filter collection, and activity detection in such a way as to minimize depositional losses. At the same time, it is sufficiently rugged for field operations.

The system samples air simultaneously in all units through the sampler slit between the detector and filter sections. The filter section consists of a filter holder assembly that is inserted from the base of the lower aluminum block, allowing easy filter replacement. Air is drawn through a 25-mm Millipore (0.8 μm, Type AA) filter supported by a stainless steel support screen. One of the sampler-detector units is operated with an uncovered sampler slit to measure the total ambient radon progeny concentration. The sampler slits on the remaining units are covered with single or multiple wire screens of differing wire-mesh number. The upper section of each sampler-detector unit holds a surface barrier alpha-detector. The detector is positioned concentrically with the filter and detects the alpha particles emitted by the ^{218}Po and ^{214}Po atoms collected or formed on the filter. Signals from the alpha-detectors are connected through amplifiers into a multiplexer and routed to a computer-based multichannel analyzer. A dedicated microcomputer controls acquisition of the alpha spectra, sampling pump operation, sample time, and data analysis. A sequence of sampling, counting, and analysis permits automated, semicontinuous operation of the system with a frequency of from 1.25 to 3 h.

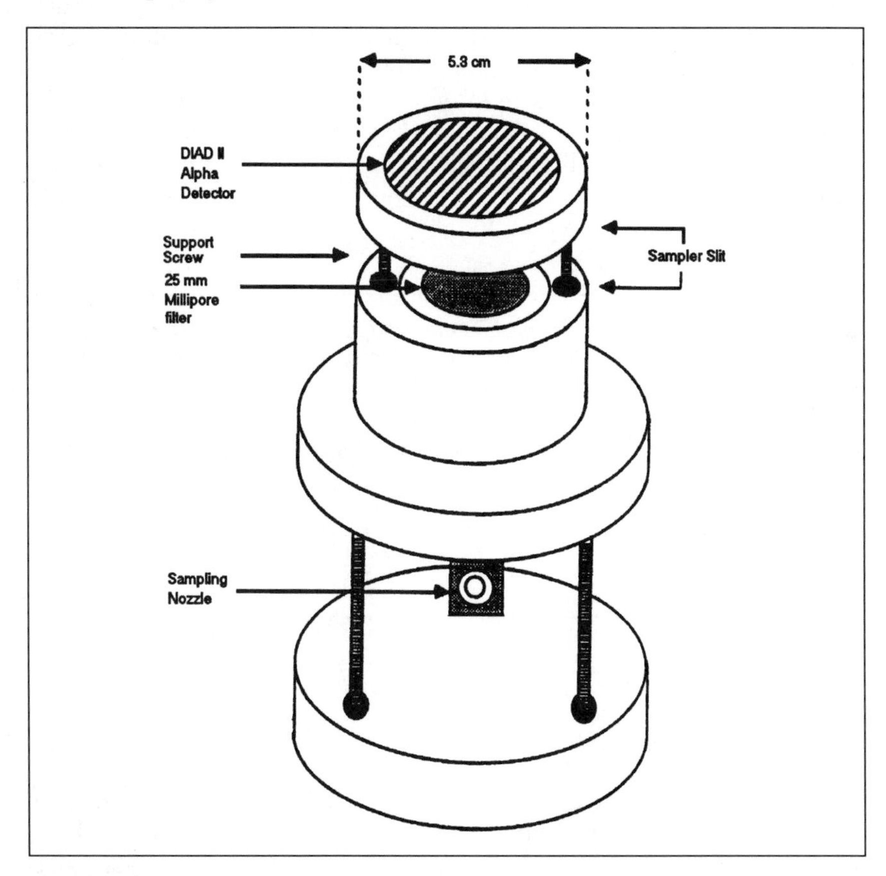

Figure 1. Sampler-detector unit in the system for measuring size distributions of radon progeny in indoor air.

The alpha counts from ^{218}Po and ^{214}Po detected by each detector in the two counting intervals are used to calculate the radon decay-product concentrations penetrating each unit (Tremblay et al., 1979). The observed concentrations of ^{218}Po, ^{214}Pb and ^{214}Bi are used to reconstruct the corresponding activity-weighted size distributions, using the expectation-maximization (Maher and Laird, 1985) or Twomey (1975) algorithms. The penetration characteristics of the five stages with screens are calculated using the Cheng-Yeh penetration theory (Cheng and Yeh, 1980; Cheng et al., 1980).

Table 1 presents details of six optimized sampler-detector units. The number and width of the size intervals used in the reconstruction process was dictated by considerations of size distribution accuracy and stability. We selected an optimum number of six inferred size intervals in geometric progression within the 0.5- to 500-nm size interval. This progression of size intervals maximizes the differences in penetrability through the various stages, insuring solution accuracy and stability while yielding sufficient size resolution in the inferred activity distribution.

Table 1. Design and operating parameters for optimized sampler-detector unit in system for measuring size distributions of radon progeny in indoor air. Sampling flow rate = 15 L m^{-1} for each unit; detector-filter separation ≈ 0.8 cm for all units.

Unit	Sampler Slit Width, cm	Sampler Diameter, cm	Wire Screen Mesh × Turns	$d_p(50\%)$ (0.5–500 nm range), nm
1	0.5	5.3	—	—
2	0.5	5.3	145	1.0
3	0.5	5.3	145 × 3	3.5
4	0.5	5.3	400 × 12	13.5
5	1.0	12.5	635 × 7	40.0
6	1.0	12.5	635 × 20	98.0

FIELD MEASUREMENTS

Measurements were made in the Princeton, New Jersey, area, in a one-story residence with living room, dining room, kitchen, two bedrooms, a study room, two bathrooms, and basement. Activity size distributions were measured in the living room and one bedroom over a 2-wk period (1/16-1/31/90). A total of about 10 measurements were made in the living room and more than 100 measurements in the bedroom with different types of particle generation. Aerosols were generated from candle-burning, cigarette-smoking, vacuuming (electric motor), cooking, and opening the door during normal activities in domestic environments. Particle concentrations were measured using a Gardner manual condensation nucleus counter (Gardner Associated, Inc., Schenectady, NY). The concentration and size distribution of radon progeny were determined using the semicontinuous graded-screen array system described above.

Figure 2 presents four activity-weighted size distributions (A,B,C,D) determined in the living room, with no active aerosol sources operating in the house except for the pilot lights of the gas range in the adjoining kitchen. It can be seen that three of the size distributions are very similar. However, in that obtained at 2230 h, activity is observed in the size interval from 1.5 to 5 nm. This peak was observed at erratic intervals over the 2-wk period in which these measurements were made. A possible source of these particles is leakage from the oil-fired furnace used to heat the house. The majority of the measurements were made in the master bedroom. Figures 3 and 4 present examples of the activity-weighted size distributions (A,B,C,D) obtained in the bedroom with the bedroom door open and closed, respectively. When the door was open, there was a higher particle concentration, and the median diameter of the attached mode shifted downward in size from the distributions observed in the closed room. These changes in activity size that occurred with the door open suggest that there is a source of particles elsewhere in the house and that these particles are transported into the bedroom by the higher ventilation rate produced by the open door. However, the distribution is smaller in size than that observed in the living room, which is nearer to the kitchen. Thus, the source of the particles is not obvious.

To observe the nature of the activity size distributions produced by active particle sources as well as to test the effectiveness of an air filtration system (Hopke et al., this volume), size distributions were obtained during periods when particles were being actively generated. The activity size distributions observed during cooking, vacuuming the rug, burning a candle, and while a cigarette was smoldering in an ashtray are shown in Figure 5 (A,B,C,D).

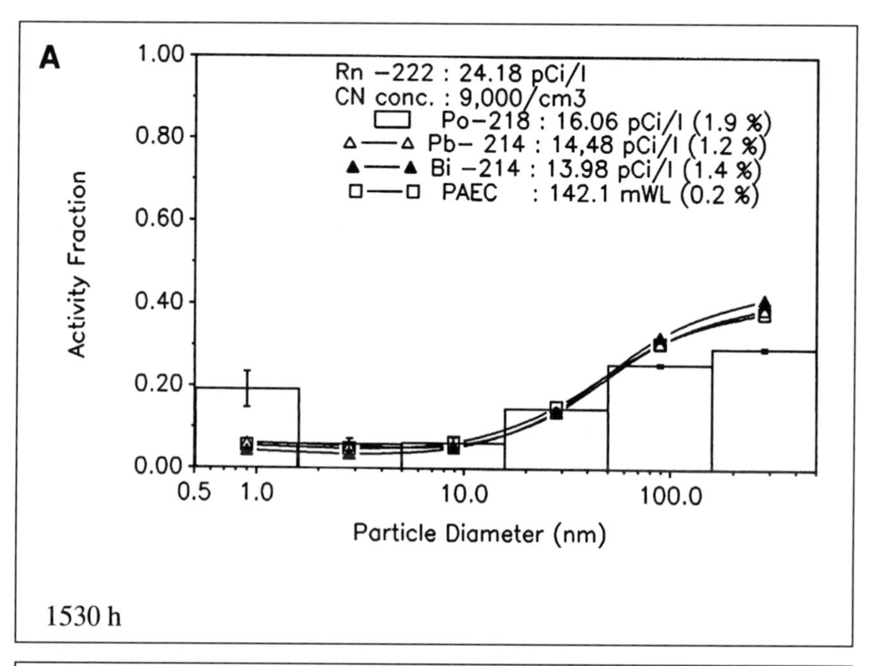

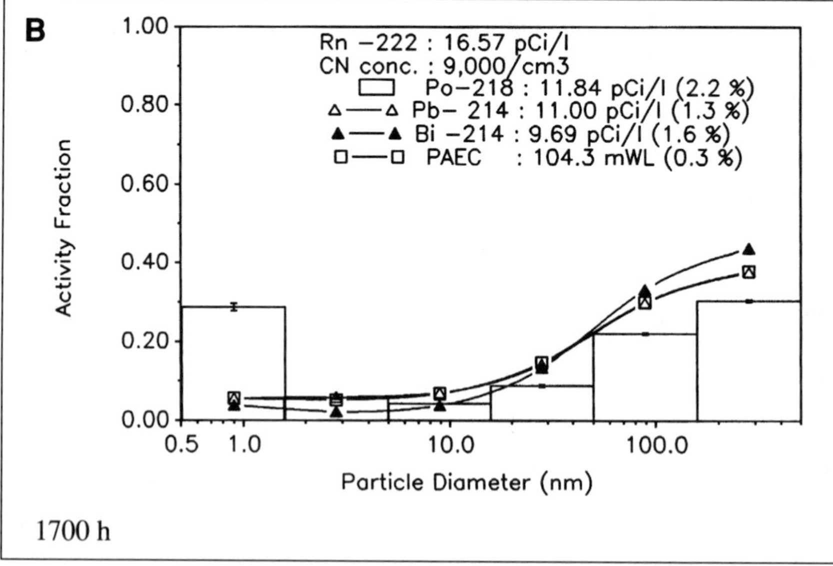

Figure 2. Activity-weighted size distributions of radon progeny measured in the living room of Princeton House 21 at various times on January 17, 1990: A, 1530 h; B, 1700 h; (continued on the following page)

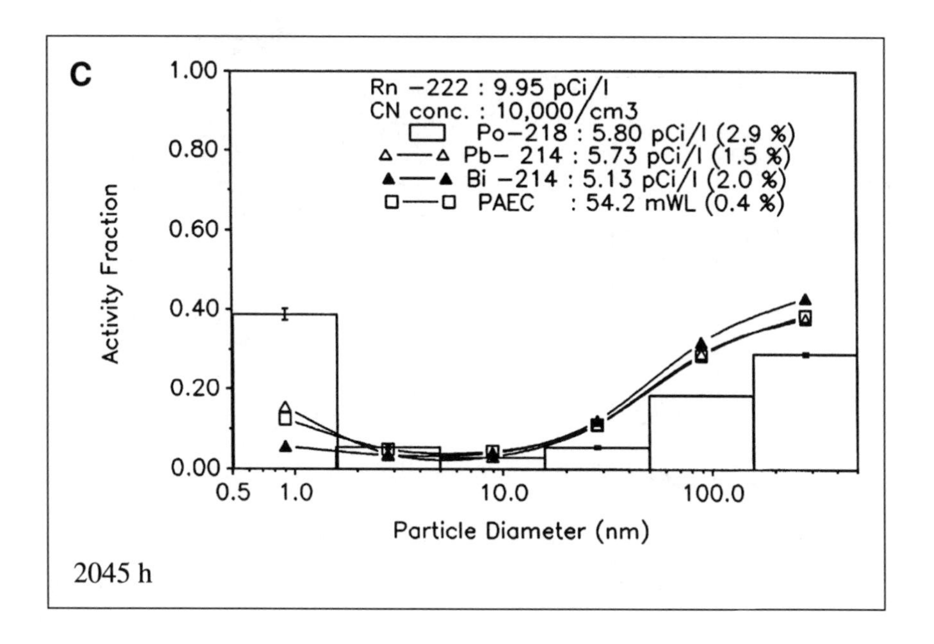

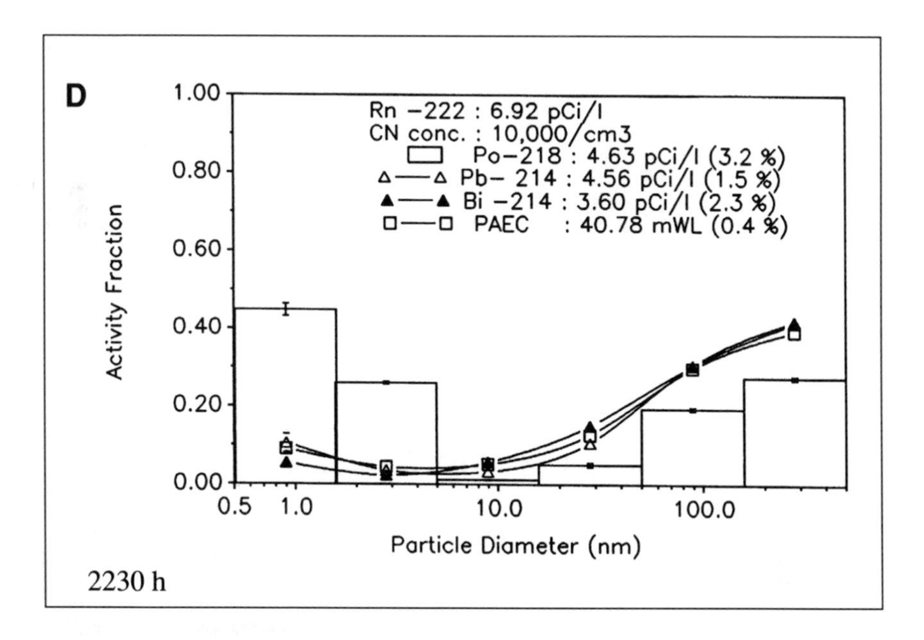

Figure 2 continued. C, 2045 h; D, 2230 h.

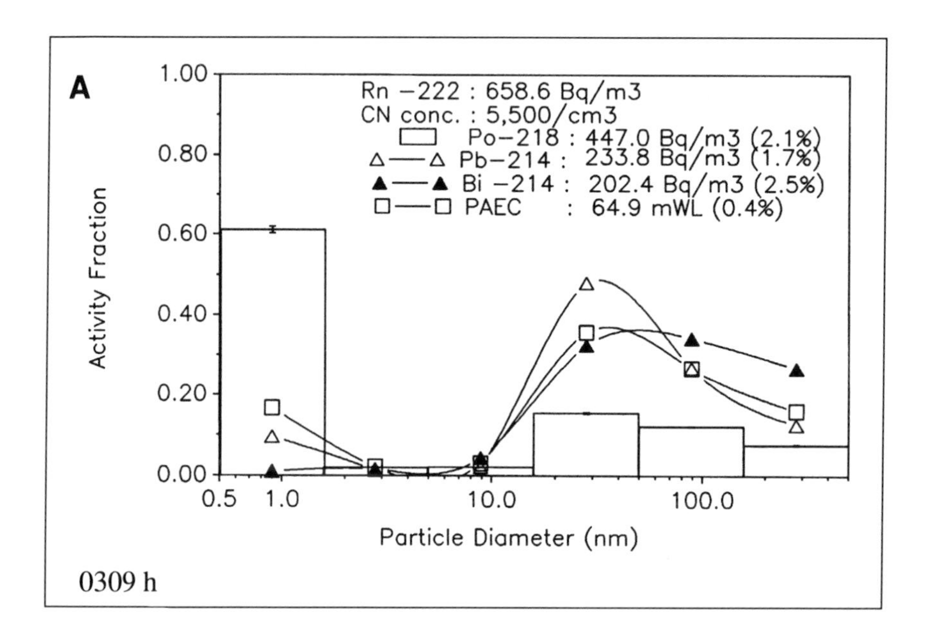

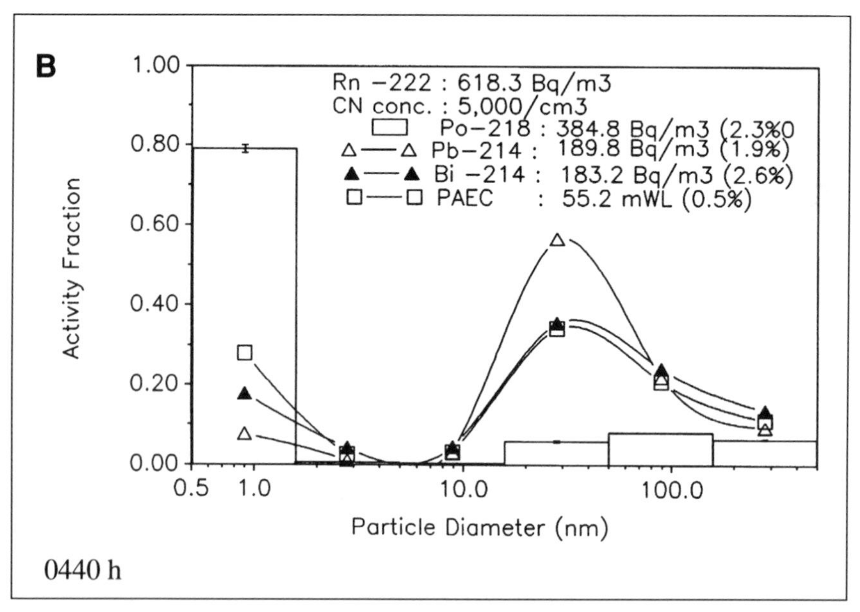

Figure 3. ^{218}Po, ^{214}Pb, ^{214}Bi/^{214}Po, and potential alpha energy concentration (PAEC) activity size distributions measured under typical conditions in the bedroom of Princeton House 21 with door open on January 21, 1990: A, 0309 h; B, 0440 h; (continued on next page)

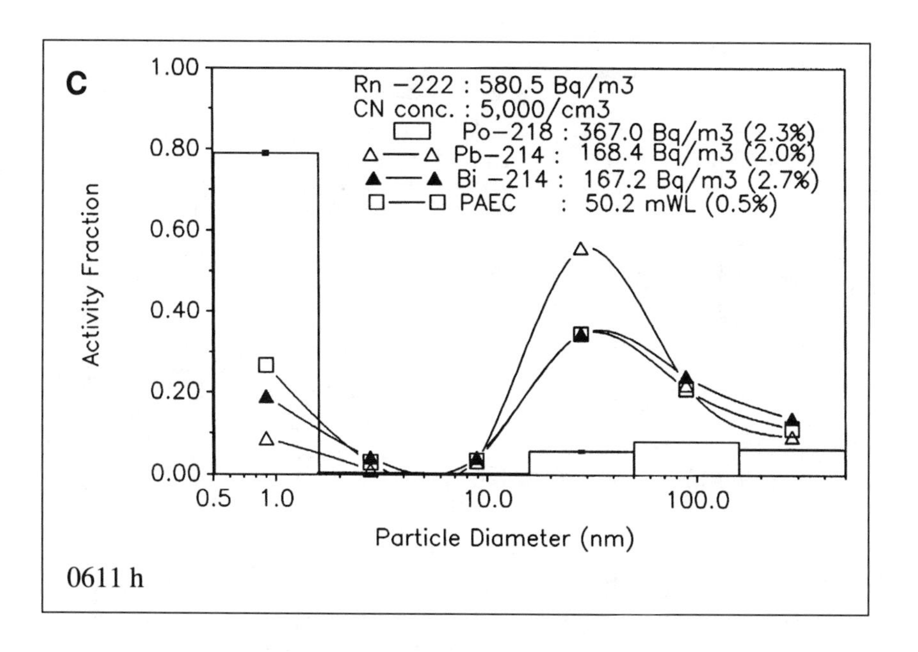

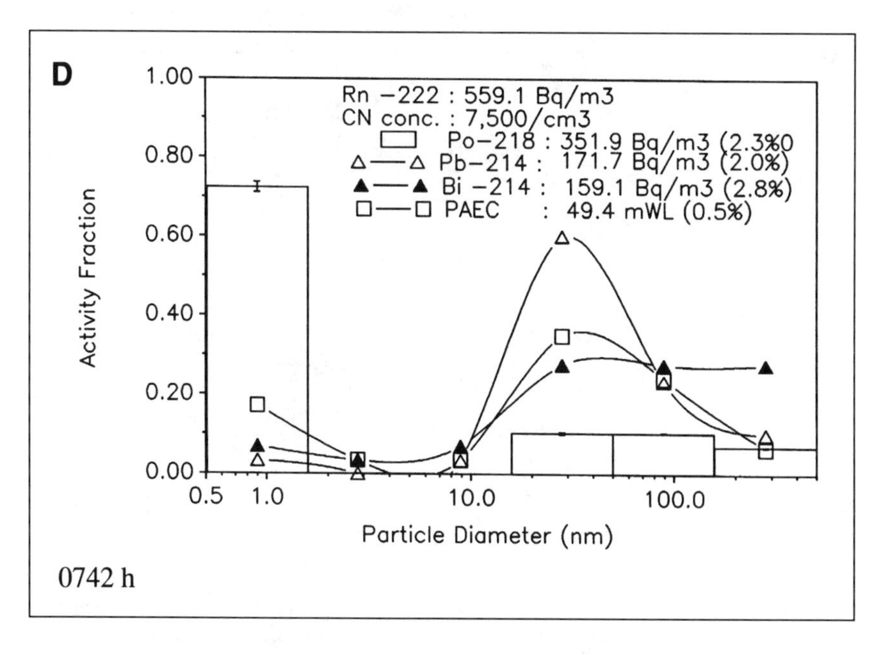

Figure 3 continued. C, 0611 h; D, 0742 h.

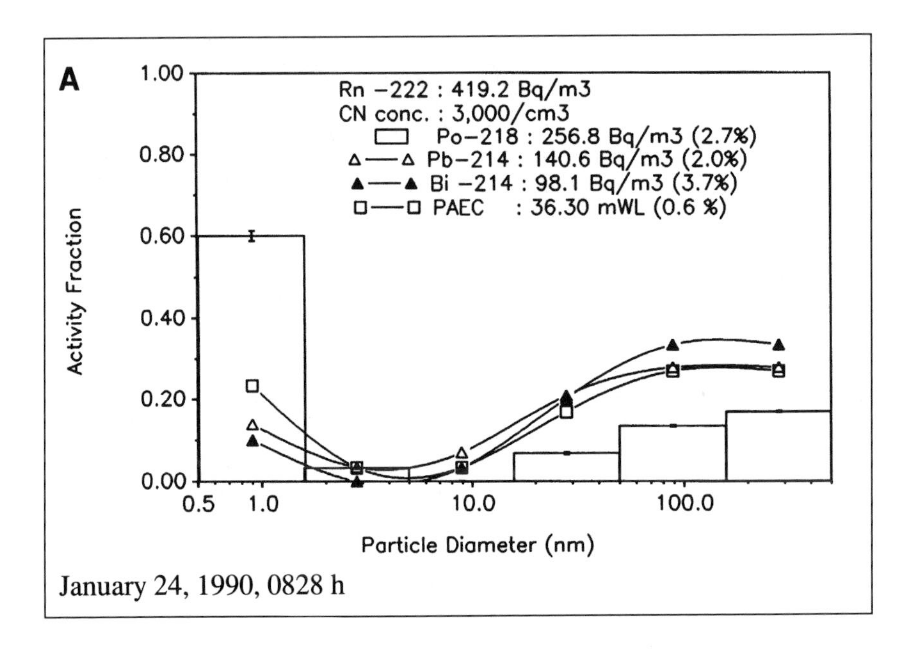

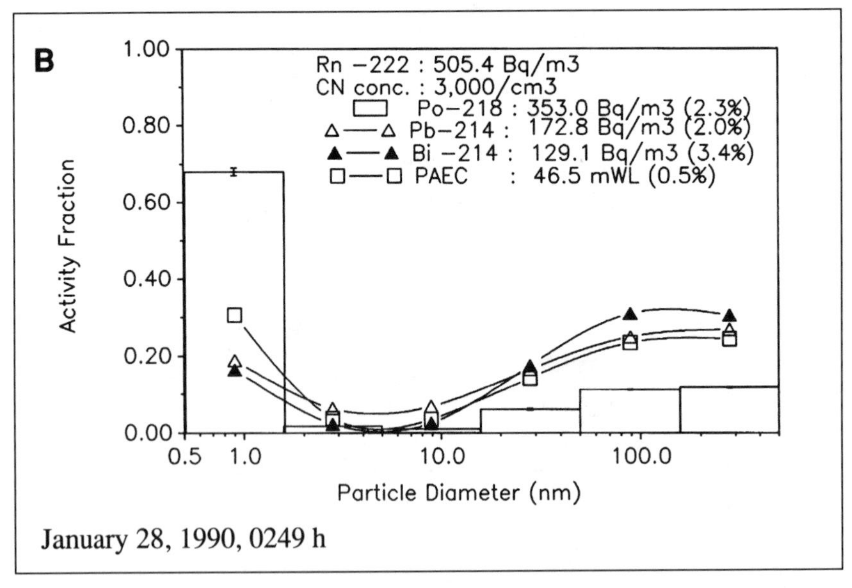

Figure 4. ^{218}Po, ^{214}Pb, ^{214}Bi/^{214}Po, and PAEC activity size distributions measured under typical conditions in the bedroom of Princeton House 21 with door closed: A, on January 24, 1990, 0828 h; B, January 28, 0249 h; (continued on next page)

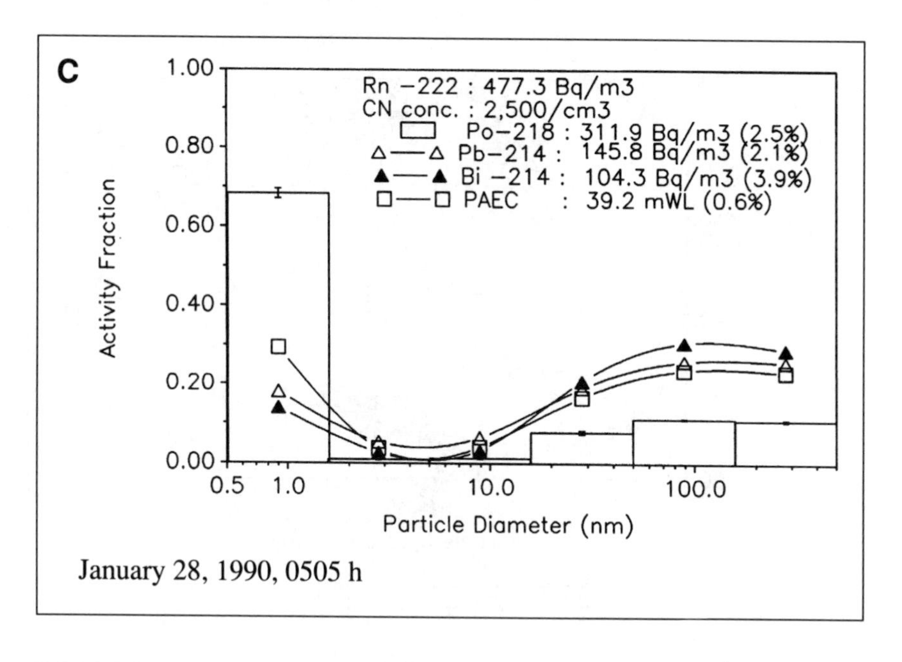

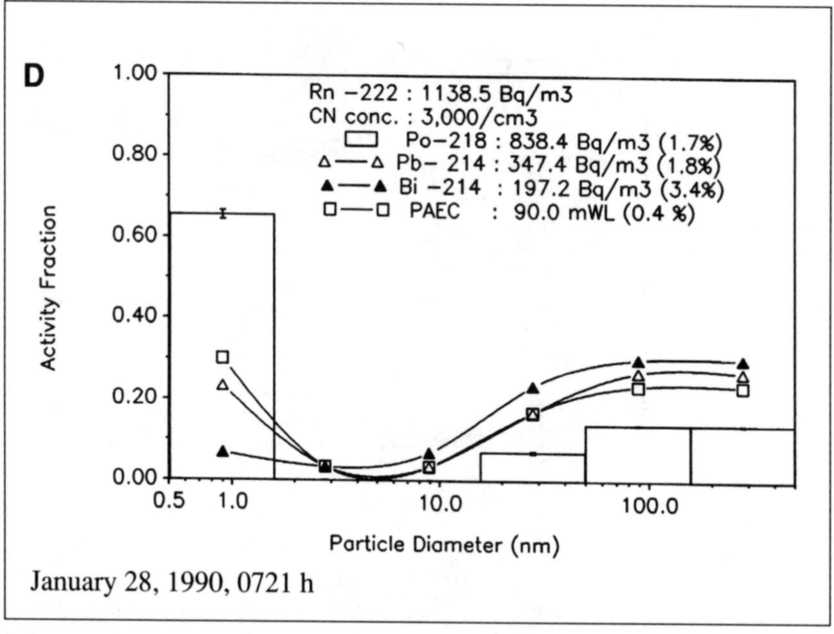

Figure 4 continued. C, January 28, 1990 at 0505 h; D, January 28, 1990 at 0721 h.

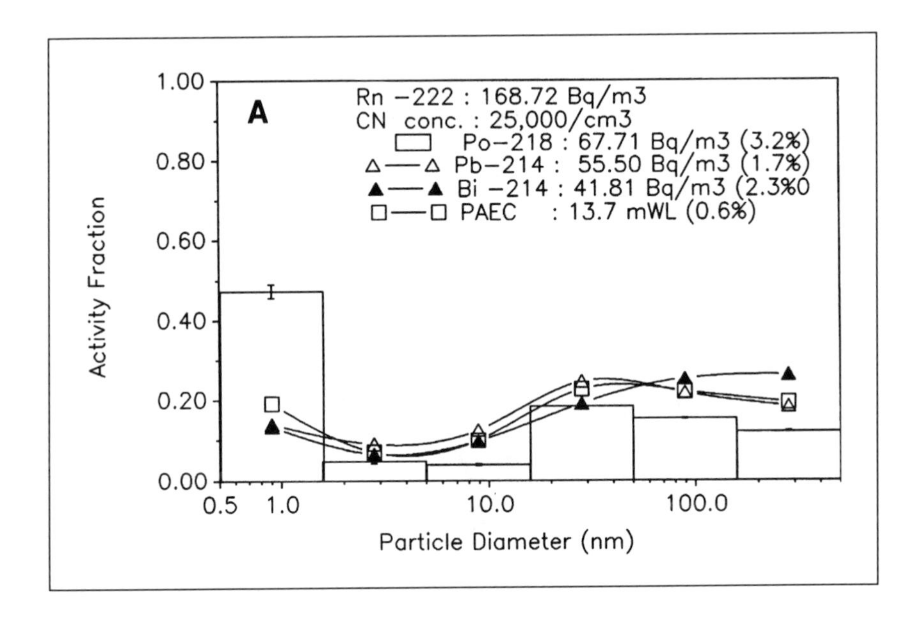

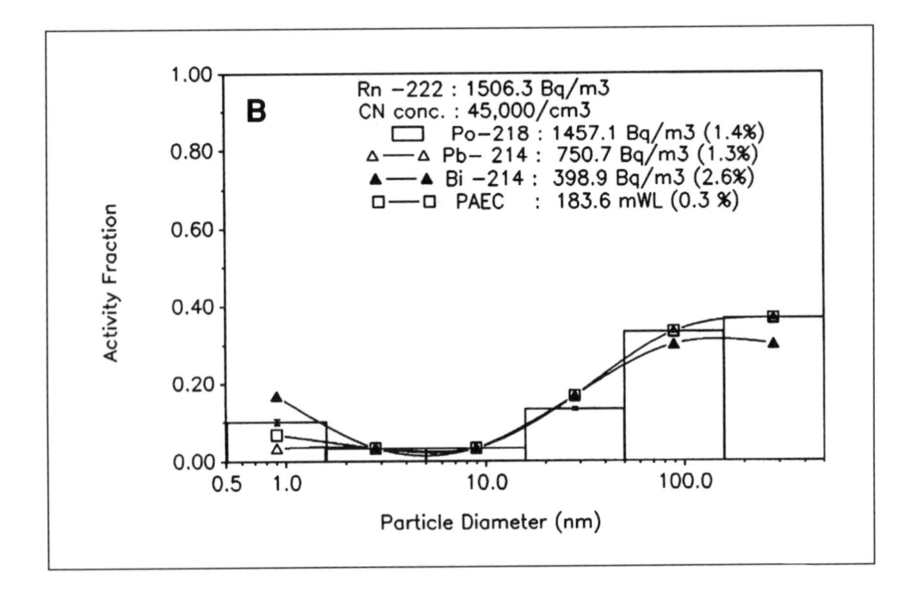

Figure 5. ^{218}Po, ^{214}Pb, ^{214}Bi/^{214}Po, and PAEC activity size distributions measured in the bedroom of Princeton House 21 during the period when aerosol sources were actively producing particles: A, vacuuming; B, cigarette smoldering; (continued on next page)

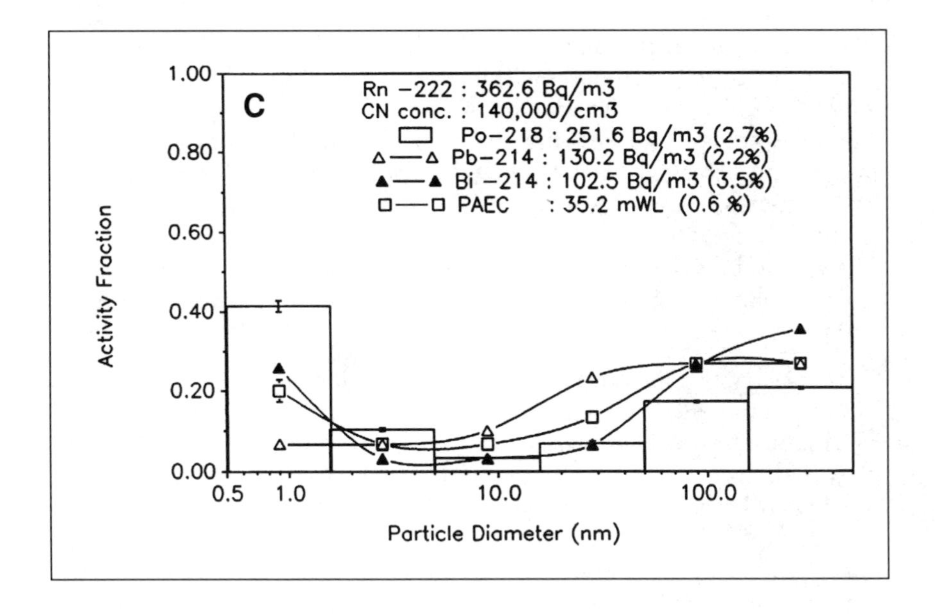

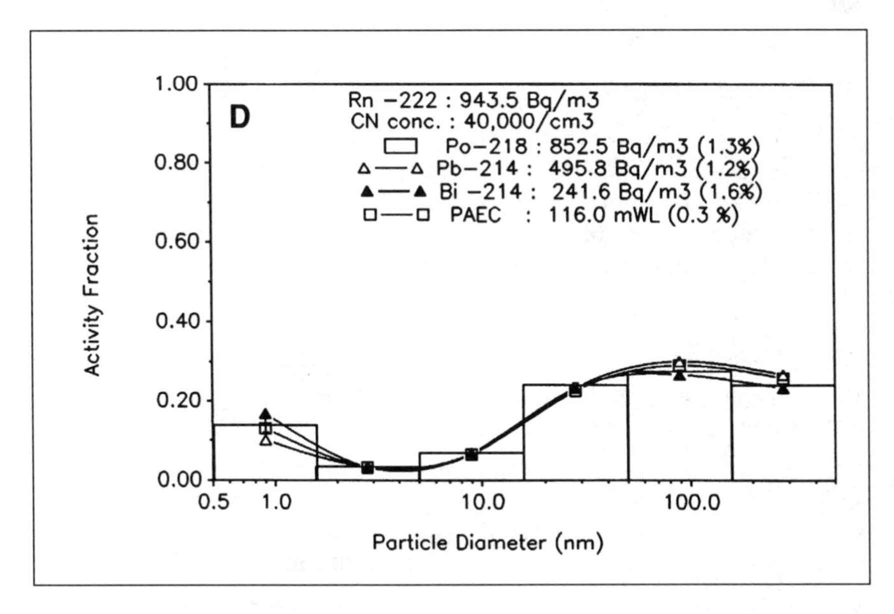

Figure 5 continued. C, candle burning; D, cooking.

Because of the large number of particles generated by such activities in the domestic environment, the working level increased for a time, while the unattached fraction decreased. The particles generated from cigarette smoke and cooking dramatically shifted almost all radon progeny to attached fractions, where they remained for a long time. The particles produced from candle-burning and vacuuming were much smaller; the average attachment diameter was approximately 15 nm. The candle-burning and vacuuming particles decreased the unattached fraction but returned to the original background distributions about 150 min later.

In summary, a measurement system was developed to characterize indoor radon decay-product radioactivity on an automated, semi-continuous basis. The system was designed and calibrated for sampling both the ultrafine cluster and attached modes of radioactivity present in indoor air, with the optimized design based on the results of both experimental and numerical simulation studies. The measurement system is capable of monitoring changes in activity concentrations and size distributions as a function of time and indoor events. Further measurements with the system should permit an improved estimation of the health hazards from radon decay products in indoor air.

ACKNOWLEDGMENTS

This work was supported by the U.S. Department of Energy under contract DE-FG02-89ER61029, and the New Jersey Department of Environmental Protection under contract P32108. We also gratefully acknowledge the collaboration of K. Gadsby, A. Cavallo, and R. Socolow of Princeton University in the field experiments.

REFERENCES

Chamberlain, AC and ED Dyson. **1956**. The dose to the trachea and bronchi from the decay products of radon and thoron. Br J Radiol 29:317-325.

Cheng, YS and HC Yeh. **1980**. Theory of screen type diffusion battery. J Aerosol Sci 11:313-319.

Cheng, YS, JA Keating, and GM Kanapilly. **1980**. Theory and calibration of a screen-type diffusion battery. J Aerosol Sci 11:549-546.

Holub, RF and EO Knutson. **1987**. Measuring polonium-218 diffusion-coefficient spectra using multiple wire screens, pp. 341-356. In: *Radon and its*

Decay Products: Occurrence, Properties and Health Effects, PK Hopke (ed.). American Chemical Society, Washington DC.

James, AC. **1988**. Lung Dosimetry, pp. 259-309. In: *Radon and its Decay Products in Indoor Air*, WW Nazaroff and AV Nero (eds.). Wiley-Interscience, New York.

Maher, EF and NM Laird. **1985**. EM algorithm reconstruction of particle size distribution from diffusion battery data. J Aerosol Sci 16:557-570.

Ramamurthi, M and PK Hopke. **1989**. On improving the validity of wirescreen unattached fraction radon daughter measurements. Health Phys 15:189-194.

Ramamurthi, M and PK Hopke. **1991** An automated, semicontinuous system for measuring indoor radon progeny activity-weighted size distributions, d_p: 0.5-500 nm. Aerosol Sci Technol 14:82-92.

Reineking, A and J Porstendörfer. **1986**. High-volume screen diffusion batteries and α-spectroscopy for measurement of the radon daughter activity size distributions in the environment. J Aerosol Sci 17:873-879.

Tremblay, RJ, A Leclerc, C Matthieu, R Pepin, and MG Townsend. **1979**. Measurement of radon progeny concentration in air by α-particle spectrometric counting during and after air sampling. Health Phys 36:401-411.

Twomey, S. **1975**. Comparison of constrained linear inversion and an iterative nonlinear algorithm applied to the indirect estimation of the particle size distribution. J Comput Phys 18:188-200.

QUESTIONS AND ANSWERS

Q: (Speaker unknown.) I think that this is a very important question, because you say when you use a vacuum cleaner, your activity distribution moved to about 10 to 20 or 10 to 30 nanometers. It seems to me that when we use a vacuum cleaner, it also produces some large particles by mechanical processes. So, I wonder, if you only use a screen diffusion battery, with our impactors and some other things, the large particles could also be trapped on the screen by other mechanisms, and your conversion programs and most other conversion programs could not tell the difference.

A: Hopke, Clarkson University (presenter for Li paper). Yes, we could not properly identify large particles although, if anything, that would then give us an artifact in the smallest bin size. The thing is, that although we will suspend some large particles, the total number relative to the 90 or 100 thousand cm^{-3} that we are seeing directly off of the source doesn't really seem to change the activity size distribution.

Q: (Speaker unknown.) Well, if you have another instrument to measure the indoor aerosol size distribution, that would help to clear up the problem.

A: Hopke. Yes.

Q: N. Harley, New York University. It wasn't clear to me why the particle concentration was so low in that bedroom. Were you air-cleaning, or is that the marvelous clean air in Princeton?

A: Hopke. That is the marvelous clean air in Princeton. We couldn't believe how clean this house was.

Q: Harley. It is like a "clean" room. What is the cutoff size on the condensation nuclei count?

A: Hopke. It's a Gardner, so it is probably 16 or 20.

UNATTACHED FRACTION AND SIZE DISTRIBUTION OF AEROSOL-ATTACHED RADON AND THORON DAUGHTERS IN REALISTIC LIVING ATMOSPHERES AND THEIR INFLUENCE ON RADIATION DOSE

A. Reineking, G. Butterweck, J. Kesten, and J. Porstendörfer

Universität Göttingen, Federal Republic of Germany

Key words: *Unattached fraction, radon and thoron daughters, activity size distribution, radiation dose*

ABSTRACT

In all dosimetric models, the aerosol particle size and the unattached fractions of radon and thoron daughters are important parameters for estimating the radiation dose to humans. Using various measuring techniques (high-volume impactors, low-pressure cascade impactors, and screen diffusion batteries) to cover particle sizes ranging from 0.5 nm to 10,000 nm, we measured activity size distributions of radon and thoron progeny in air and the influence of different aerosol sources. We made these measurements under realistic living conditions in houses and in the open air. The unattached fraction, f_p, was measured with a single-screen technique in conjunction with alpha-spectroscopy.

Indoor measurements showed that the mean average value of activity median aerodynamic diameter (AMAD) for the main activity mode (accumulation mode) for aged aerosols was significantly smaller (AMAD, ~200 nm) compared with outdoor measurements (AMAD, ~400 nm). In closed rooms, aerosol sources such as a gas heater or a burning candle added an additional condensation nucleus mode with diameters between 10 and 100 nm. Outdoors, resuspension, combustion, and nucleation processes resulted in activity in the nucleus size mode and also in the coarser size ranges (up to a few micrometers).

Our results show that most published calculations of the natural radiation dose received by the public are based on incorrect values. Under normal conditions (low ventilation, without aerosol sources), we found that the mean f_p value, 0.095, was three times higher than that proposed in the literature.

Finally, the implications of these experimental data on particle sizes and f_p for radiation dose are presented.

INTRODUCTION

In dosimetric models for estimating tissue doses from inhalation of short-lived radon and thoron decay products, the size characteristics of the unattached and aerosol-attached activities in the ambient air are important, since deposition in different parts of the respiratory tract depends strongly on particle size.

We recently measured the concentrations of radon and its decay products, activity size distributions, and the fractions of unattached and aerosol-attached activity under realistic living conditions in various dwellings and in the open atmosphere (Reineking and Porstendörfer, 1990; Reineking et al., 1988). In ongoing studies we are also measuring thoron gas and thoron decay products.

These experimental data affect radiation dose calculations. Dose coefficients were derived from a modified model of James-Birchall (James, 1988) and a model of Jacobi-Eisfeld (Jacobi and Eisfeld, 1980). Other effects include the influence of different aerosol characteristics, mouth and nasal breathing, the influence of particle growth in the humid air of the respiratory tract during inhalation, and the radiation dose to epithelial cells and to basal cells.

Mean annual effective dose equivalents of radon and thoron are estimated, for the public in the Federal Republic of Germany, from these dose models, using representative activity concentrations.

EXPERIMENTAL METHODS

Activity size distribution of radon and thoron decay products were measured with a low-pressure cascade impactor and high-volume screen diffusion batteries. The impactor, which consists of eight stages and a backup filter, operates at a flow rate of 1.8 $m^3\ h^{-1}$. Calibration measurements yield 50% cut-off diameters in the range from 80 nm to 6000 nm. For aerosol-attached activities, interstage losses are less than 2% of the total activity. However, 75% of the unattached activity is lost in the entrance and on the walls of the first three stages. To determine activity size distributions, the activities of ^{214}Pb, ^{214}Bi and ^{212}Pb were measured after air sampling, using low-level pure germanium detectors, or NaI detectors.

The particle diameters of the unattached radon daughters (0.5 to 5 nm) and of aerosol-attached daughter activity up to 400 nm were measured with calibrated screen diffusion batteries, with different screen and mesh numbers, at a volumetric flow rate of 2 $m^3\ h^{-1}$. To do so, we took four parallel samples: three with screen diffusion batteries, and one without, as an absolute sample. This procedure was repeated twice with different screen numbers. Radon daughters were collected on membrane filters, and alpha emissions were detected and counted with surface barrier detectors during and after air sampling. From the measured alpha counts, the concentrations of the radon daughters (^{218}Pb, ^{214}Pb, $^{214}Bi/^{214}Po$) and the corresponding proportions of airborne daughter activities penetrating each screen were calculated.

Activity size distributions were obtained from the size-fractionated activities by comparing measured with simulated activities. True size distributions were approximated by the sum constituent lognormal distributions. In addition, impactor data were evaluated by a graphical (cumulative) method. A detailed description of these methods is given in Reineking and Porstendörfer (1986) and Reineking et al. (1988).

For separating unattached from aerosol-associated radon and thoron daughters, a single screen with a 50% penetration characteristic for 4.0-nm-diameter particles was used. Radon-daughter activities were measured by alpha-spectroscopy during and after air sampling (see size distribution measurements). A newly developed data evaluation method (Reineking and Porstendörfer, 1990) was used to correct the measurement of unattached radon progeny for aerosol-attached activity deposited on the screen by impaction and interception.

To determine unattached ^{212}Pb activity, the amount deposited on the screen was measured by gamma-spectroscopy after air sampling. This amount was corrected for deposition of aerosol-attached ^{212}Pb by comparison with simultaneous measurements of the activity size distribution of attached ^{212}Pb.

Radon and thoron gas in air were measured continuously by electroprecipitation of positively charged ^{218}Po and ^{216}Po ions in an electric field (10 to 18 kV) onto a surface barrier detector (Porstendörfer et al., 1980).

The aerosol particle concentration was registered by means of a calibrated condensation nuclei counter (General Electric). With this instrument, the counting efficiency is only a few percent for particles of 3 to 4 nm diameter, but the efficiency is 100% for particles larger than about 10 nm diameter (Scheibel and Porstendörfer, 1986).

RESULTS

Average values of measured indoor radon and thoron daughter activity size distributions are summarized in Table 1. Results of measurements in low or moderately ventilated rooms ($v < 0.5\ h^{-1}$) without aerosol sources are compared with measurements in the presence of typical aerosol sources.

For an aged aerosol, all activity size distributions of ^{218}Po can be approximated by bimodal lognormal distributions, whereas size distributions of $^{214}Bi/^{214}Po$ are approximately unimodal. In our measurements (depending on the particle concentrations), about one-third of the ^{218}Po activity was unattached, and the diffusion equivalent diameters ranged from 0.5 to 2.0 nm, with a mean diameter of 1.2 nm.

The activities of radon daughters attached to aerosol particles (accumulation mode) are the dominant part of all size distributions, and this mode of ^{218}Po distributions agrees with those of the ^{214}Pb and $^{214}Bi/^{214}Po$ distributions measured with the diffusion batteries and the low-pressure cascade impactor. In the diameter size range larger than 100 nm, aerodynamic equivalent diameters can be calculated from diffusion or thermodynamic equivalent diameters by multiplying with the square root of the particle density. Typical environmental atmospheres with mean aerosol particle densities of 1.4 g cm^{-3} yield deviations of less than 20% between the measured average AMD and AMAD of the accumulation mode of an aged aerosol (steady-state conditions, without additional aerosol sources). This difference is small compared to the uncertainties concerning the data evaluation procedure, the calibration of size-fractionating devices, the approximation of the true size distribution by lognormal distributions, and the normal variation of the size distribution under realistic living conditions. For these reasons, the AMD/AMAD of 181 to 199 nm and the mean geometric standard deviation of 2.0 to 2.6 show that the size distributions of the aerosol-attached activity are identical for all short-lived radon decay products. The impactor measurements show that, in addition to the activity of the accumulation mode, about 3% of the activity is associated with aerosol particles smaller than 80 nm (nucleus mode). There is no evidence for another coarser aerosol mode in the size range of some micrometers. The extreme values of the parameters $AMAD_i$, σ_i, and f_i derived from lognormal distributions fitted to single measurements (values are shown in parentheses in Table 1) vary by a factor of two compared to the average of all measurements.

Table 1. Mean values of lognormal activity size distributions of radon and thoron decay products in low-ventilation or moderately ventilated rooms ($v < 0.5\ h^{-1}$) with an aged aerosol and with typical aerosol sources. Extremes of single measurements in parentheses. DB = screen diffusion batteries; LPI = low-pressure cascade impactor.

			Unattached Mode			Nucleus Mode			Accumulation Mode		
Aerosol condition	Method, number of measurements[a]	Nuclide	AMD_1 (nm)	σ_1	Activity fraction, f_1	AMD_2 (from DB) or $AMAD_2$ (from LPI) (nm)	Growth, σ_2	Activity fraction, f_2	AMD_3 or $AMAD_3$ (nm)	σ_3	Activity fraction f_3
Aged aerosol	DB 19	^{218}Po	1.2 (0.5–2.0)	2.0 (1.3–2.6)	0.34 (0.16–0.54)	—	—	—	181 (97–265)	2.0 (1.5–2.7)	0.66 (0.46–0.84)
	DB 19	^{214}Bi/ ^{214}Po	2.8 (0.7–8.4)	2.3 (1.6–3.6)	0.03 (0–0.11)	—	—	—	198 (115–313)	2.1 (1.4–3.1)	0.97 (0.89–1.00)
	LPI 29	^{214}Pb/ ^{214}Bi	—	—	—	<80	?	0.03 (0–0.17)	199 (132–285)	2.6 (1.9–4.2)	0.97 (0.83–1.00)
	LPI 10	^{212}Pb	—	—	—	<80	?	0.14 (0.06–0.20)	217 (175–273)	1.8 (1.5–2.1)	0.86 (0.80–0.94)
Burning candle	LPI 3	^{214}Pb/ ^{214}Bi	?	?	?	53 (49–61)	4.4 (4.0–4.9)	1.00	—	—	—
During/ after smoking	LPI 6	^{214}Pb/ ^{214}Bi	?	?	?	<80	?	0.04 (0–0.09)	126 (92–160)	3.0 (2.4–3.7)	0.96 (0.91–1.00)
Gas heating	LPI 3	^{214}Pb/ ^{214}Bi	?	?	?	<80	?	0.12 (0.09–0.16)	227 (204–246)	2.6 (2.4–2.8)	0.88 (0.84–0.91)

[a] DB = Screen diffusion batteries; LPI = Low-pressure cascade impactor.

Preliminary results for the thoron decay product ^{212}Pb yielded size distributions similar to those shown in Table 1 for short-lived radon decay products, with mean median diameters of 217 nm and a smaller mean geometric standard deviation of 1.8. However, for the ^{212}Pb aerosol, on average, about 14% of airborne activity was associated with aerosol particles in the nucleus mode (AMAD < 80 nm).

The influence of particles from typical aerosol sources on the activity size distribution was studied with the low-pressure impactor. Besides the unattached mode and aerosol-attached activity of the accumulation mode, a nucleus-mode size range (<80 nm) was sometimes found. Very small primary particles produced by a burning candle resulted in activity size distributions with an average AMAD of 53 nm and an average σ_2 of 4.4. During and after cigarette-smoking, an average AMAD of 130 nm was measured. Gas heating produced an additional nucleus mode (12%) with an AMAD smaller than 80 nm.

In the open air (in the vicinity of Göttingen), activity size distributions of ^{214}Pb and ^{212}Pb were measured with the impactor (Table 2). The activity size distributions of both decay products were similar. About 90% of the activity was associated with aerosol particles of the accumulation mode with mean median diameters of 300 to 400 nm. Because of outdoor aerosol sources (resuspension and combustion processes) and meteorological conditions (turbulent exchange, precipitation), activities in the nucleus-mode size range (<80 nm: 0-15%) and coarse-mode size range (>1500 nm: 0-25%) were sometimes measured.

In closed rooms without additional aerosol sources, a mean unattached fraction, (f_p, 0.096) of potential alpha energy was measured at a mean aerosol particle concentration of 6100 cm^{-3} and a mean equilibrium factor, F, of 0.30 (see Table 3). The f_p value is about three times higher than the values used in the literature for calculating the radiation exposure of the public (NEA, 1983; ICRP, 1987; UNSCEAR, 1988). In closed rooms with additional aerosol sources (cigarette smoke, heating systems, aerosol from a burning candle), particle concentrations ranged up to 10^6 cm^{-3}, f_p values decreased (sometimes below the detection limit of 0.005), and F values increased to as high as 0.77. In outdoor air, a mean unattached fraction of 0.02 and a mean aerosol particle concentration of 3.4 x 10^4 cm^{-3} were measured at 1 m above ground. The mean equilibrium factor, F, was 0.7. From preliminary indoor measurements of thoron gas and thoron decay products, a mean f_p value of 0.06 and a mean F of 0.09 were estimated.

Table 2. Mean values of lognormal activity size distributions of radon and thoron decay products outdoors (near Göttingen). Extremes of single measurements in parentheses. DB = screen diffusion batteries; LPI = low-pressure cascade impactor.

		Unattached Mode			Nucleus Mode			Accumulation Mode			Coarse Mode			
Method[a]	Nuclide	AMD_1 (nm)	σ_1	f_1	$AMAD_2$ (nm)	σ_2	f_2	$AMAD_3$ (nm)	σ_3	f_3	$AMAD_4$ (nm)	σ_4	f_4	Number of measurements
DB	^{218}Po	2.7 (2.5–3.0)	2.0 (1.8–2.2)	0.14 (0.04–0.22)	?	?	?	?	?	?	?	?	?	3
LPI	^{214}Pb ^{214}Bi	?	?	?	<80	?	0.08 (0–0.15)	386 (173–645)	2.3 (1.6–4.4)	0.90 (0.69–1.00)	2730 (1660–4430)	1.7 (1.2–1.9)	0.02 (0–0.25)	21
LPI	^{212}Pb	?	?	?	<80	?	0.11 (0-0.35)	330 (149–519)	2.0 (1.5–3.6)	0.87 (0.65–1.00)	4240 (1580–6460)	1.6 (1.1–2.5)	0.02 (0–0.23)	44

[a] DB = Screen diffusion batteries; LPI = Low-pressure cascade impactor.

Table 3. Mean values of aerosol particle concentrations (Z), unattached fractions (f_p), and equilibrium factor (F), measured indoors and outdoors. Extremes of single measurements in parentheses.

Conditions	Nuclide	Number of measurements; comments	Z ($\times 10^3$ cm^{-3})	f_p	F
INDOORS					
Aged aerosols	Radon daughters	79 10 rooms	6.1 (1.1–18)	0.096 (0.016–0.246)	0.30 (0.15–0.49)
Closed room ($V < 0.5$ h^{-1})	Thoron daughters	10 3 rooms	9.5 (3.79–19)	0.057 (0.012–0.137)	0.085 (0.012–0.193)
OUTDOORS					
1 m above ground	Radon daughters	36 daytime	34 (7.5–93)	0.02 (0–0.22)	0.67 (0.23–1.19)
	Thoron daughters	continuously during 9 weeks	?	?	≈0.01
INDOORS					
Burning candles Closed room	Radon daughters	13 3 rooms	385 (16–1000)	0.045 (0–0.139)	0.33 (0.26–0.44)
INDOORS					
During/after smoking Closed room	Radon daughters	16 4 rooms	170 (15–255)	0.006 (0–0.60)	0.56 (0.37–0.77)
INDOORS					
Gas heating Closed room	Radon daughters	3 1 room	165 (46–231	0.084 (0.086–0.092)	0.23 (0.21–0.24)

DOSE CALCULATIONS

The following are based on a model described by James (1988) and on a model published by Jacobi and Eisfeld (1980). An average breathing rate of 0.75 m^3 h^{-1} is assumed, as are the so-called "Weibel A" airway dimensions (Weibel, 1963). In both models, clearance processes were considered.

Calculated regional doses per WLM as a function of particle size, expressed by AMD, are adapted from James (1988) (Figure 1). In this figure, doses averaged over all epithelial cells in the tracheobronchial (T-B) and alveolar-interstitial (A-I) regions in the lung are compared with doses for basal cells only. For the following dose calculations for epithelial cells (nose breathing), the lower-bound curve was used. The size dependence of the dose conversion factors for the (thoron) decay products is approximated by comparing values given by James (1988) for unattached ^{212}Pb and thoron decay products with median diameters in the range 200 to 300 nm with corresponding values given for radon decay products.

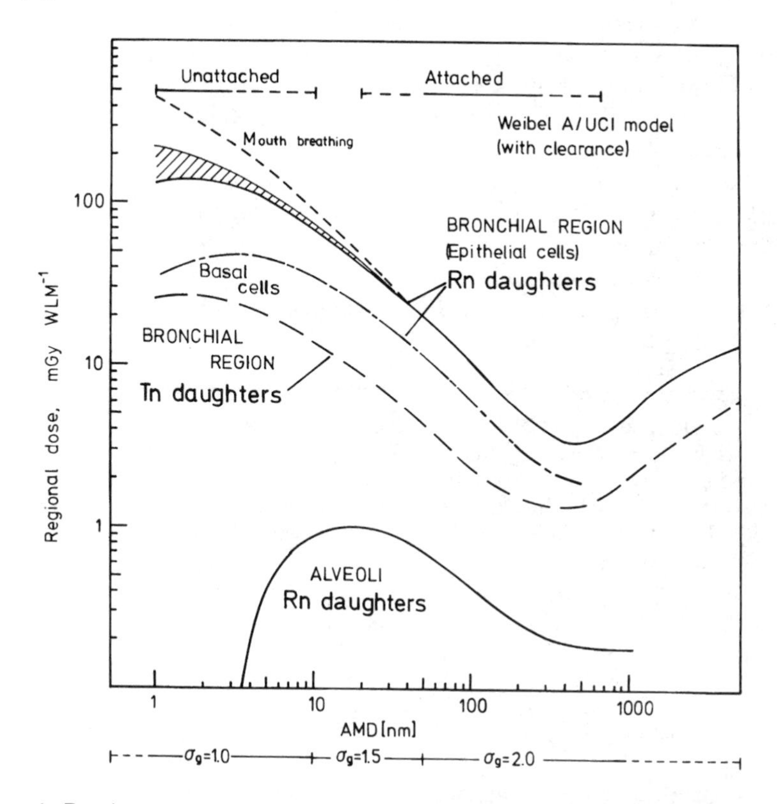

Figure 1. Respiratory tract regional doses per WLM, averaged over all epithelial cells and for basal cells only in the bronchial region, and averaged over all epithelial cells in the alveolar region, for radon daughters and thoron daughters as a function of AMD. Values adapted from James (1988). In diameter size range between 1 and 5 μm, regional dose was extrapolated from deposition values published by Heyder et al. (1986). Size dependence of tracheobronchial dose for thoron daughters extrapolated from doses for unattached activity and for aerosol-attached activities with AMD between 200 and 300 nm (James, 1988). Breathing rate = 0.75 $m^3 h^{-1}$.

Dose calculations in the Jacobi-Eisfeld model (Jacobi and Eisfeld, 1980) were based on the assumption that the radiation dose to basal cells alone is relevant to assessment of lung-cancer risk. The dose calculations in the original work of Jacobi-Eisfeld were based on bronchial deposition values of 4% for a polydisperse aerosol with an AMAD of 250 nm. In this paper, these dose calculations were modified, using actual deposition values based on experimental results of Heyder and coworkers (Stahlhofen, 1984; Schiller, 1985; Heyder et al., 1986) and considering our measured activity in distributions. For representative indoor and outdoor conditions (see Tables 1 and 2) the deposition probabilities obtained for the James model and the modified Jacobi-Eisfeld model are summarized in Table 4. The fractional probabilities of radon progeny deposition in the human respiratory tract of these two models are different, because James (1988) applied theoretical values based on the work of Taulbee and Yu (1975). Both models assume that about half of the inhaled unattached progeny is deposited in the nose. For aerosol-attached activities with AMAD of 200 to 400 nm, the James model calculated that about 4% is deposited in the T-B region; for the modified Jacobi-Eisfeld model, a value of 2% was used.

Table 4. Deposition probabilities of aerosol particles in respiratory tract regions during inhalation (breathing rate = 0.75 $m^3 h^{-1}$) for typical indoor and outdoor radon-daughter activity size distributions. Values for modified Jacobi-Eisfeld model based on results published by Stahlhofen (1984), Schiller (1985), Heyder et al. (1986). For James (1988) model, theoretical values based on Taulbee and Yu (1975) were adapted.

Condition	Aerosol size	Naso-pharyngeal region	Tracheo-bronchial region	Alveolar Interstitial region	Total
		James, 1988			
Unattached activity	Ambient	0.5	0.5	—	1.00
INDOORS (Aerosol-attached activity)	Ambient	0	0.04	0.11	0.15
Closed room Aged aerosol	Doubled	0	0.03	0.08	0.11
OUTDOORS	Ambient	0	≈0.04	≈0.11	≈0.15
Aerosol-attached activity	Doubled	?	?	?	?
		Modified Jacobi and Eisfeld, 1980			
Unattached activity	Ambient	0.5	0.5	—	1.00
INDOORS Aerosol-attached activity	Ambient	0.02	0.02	0.17	0.21
Closed room Aged aerosol	Doubled	0.04	0.02	0.18	0.24
OUTDOORS	Ambient	0.10	0.02	0.16	0.28
Aerosol-attached activity	Doubled	0.15	0.03	0.22	0.40

The regional doses for exposure to 1 WLM are summarized in Tables 5A, B, and C for radon-daughter size distributions and unattached fractions measured. Also included in these tables are the influence of particle growth during inhalation and the dose calculations for mouth-breathing. For indoor exposure (aged aerosol, for instance, in a bedroom) more than half of the dose to the bronchial region results from unattached radon decay products. Decreasing f_p values in the outdoor atmosphere and in rooms with higher aerosol particle concentration (where there is cigarette-smoking, for instance) yielded a decrease of T-B (and total) dose by a factor of two. Decreasing AMAD of the accumulation or nucleus mode (burning candle) yielded a slight increase in T-B dose.

Table 5A. Respiratory tract regional dose calculated for measured indoor radon daughter (aged aerosol) size distributions. Calculations based on models from James (1988) and Jacobi and Eisfeld (1980). Breathing rate = 0.75 m^3 h^{-1}.

		Regional dose (mGy WLM^{-1})			
		James, 1988			
		Tracheobronchial region			Alveolar Interstital region
	Aerosol size	Epithelial cells Nasal breathing	Epithelial cells Mouth breathing	Basal cells Nasal breathing	
AMD_1 = 1.2 nm (0.5–2.0)	Ambient	13.0 (2.1–33.2)	40.3 (4.8–108.2)	3.6 (0.6–11.3)	—
f_p = 0.096 (0.016–0.246)	Doubled	13.0 (1.8–33.2)	26.9 (3.0–108.2)	4.6 (0.6–11.8)	—
$AMAD_2 \approx$ 50 nm	Ambient	0.60	0.60	0.35	0.02
f_2 = 0.03 (0-0.15)	Doubled	0.32	0.32	0.20	0.01
$AMAD_3$ = 199 nm (132–285)	Ambient	4.49 (3.15–7.28)	4.49 (3.15–7.28)	2.82 (1.95–4.19)	0.25 (0.17–0.32)
f_3 = 0.88 (0.75–0.91)	Doubled	2.99 (2.78–3.37)	2.99 (2.78–3.37)	1.76 (?)	0.17 (0.13–0.29)
Complete size distribution	Ambient	18.1 (5.9–41.1)	45.4 (8.6–116.1)	6.8 (2.9–15.8)	0.27 (0.19–0.34)
	Doubled	16.4 (4.9–36.9)	30.2 (6.1–111.9)	6.6 (2.6–13.8)	0.18 (0.14–0.30)
		Modified Jacobi and Eisfeld, 1980*			
Complete size distribution	Ambient	—	—	3.5 (1.6–7.0)	0.42 (0.35–0.45)
	Doubled	—	—	3.5 (1.6–7.0)	0.44 (0.37–0.47)

Table 5B. Respiratory tract regional dose calculated for measured indoor (closed room) radon-daughter size distributions with additional aerosol sources. Calculations based on models from James (1988). Breathing rate = 0.75 $m^3 h^{-1}$.

		Dose (mGy WLM^{-1})			
		James, 1988			
		Tracheobronchial region			Alveolar Interstital region
	Aerosol size	Epithelial cells Nasal breathing	Epithelial cells Mouth breathing	Basal cells Nasal breathing	
Burning candle	Ambient	25.3	38.1	11.5	0.77
	Doubled	16.3	22.8	6.9	0.48
During/after smoking	Ambient	9.7	11.4	5.4	0.38
	Doubled	5.5	6.4	3.2	0.26
Gas heating	Ambient	17.5	41.7	7.0	0.30
	Doubled	15.3	27.5	6.3	0.20

Aerosol growth during inhalation changed the dose by less than a factor of two. Calculations from the James model show that the regional dose to the A-I region was always less than 10% of that to the bronchi. However, the Jacobi-Eisfeld model predicts, for outdoor conditions, a dose to the A-I region up to 25% of that to the bronchi. Particles from typical aerosol sources may change the dose to the A-I region by a factor of three compared to the dose under indoor conditions with an aged aerosol. However, these influences are minor compared with the effects of the sensitivity of different cells (epithelial, basal) to radiation exposure, or the effects of mouth- or nose-breathing.

In Table 6 the effective dose equivalents, H_{eff}, for each radon and thoron gas concentration are summarized for both indoor and outdoor conditions. These calculations include measurements of the equilibrium factor F (see Table 3) and assume that an adult male spent 80% of his time indoors and the remaining time outdoors. In indoor air, the dose conversion factors for thoron daughters are similar to those of radon progeny.

Representative radon gas concentrations in survey studies in the Federal Republic of Germany (BMI, 1985) are used to estimate the mean annual dose equivalent (Table 7). The thoron gas and decay product concentrations were recently measured outdoors and in moderately ventilated rooms. The modified Jacobi-Eisfeld model predicts a mean annual dose, H_{eff}, of 1 mSv, with a fraction of about 12% from thoron decay products. The James model predicts a doubled dose to the basal cells. The difference can be explained by the different deposition probabilities in the T-B region involved in the two models (see Table 4). For the epithelial cells, the annual dose increases to 4 mSv. For mouth-breathing, unrealistic values up to 10 mSv y^{-1} are predicted.

Table 5C. Respiratory tract regional dose calculated for measured outdoor radon-daughter size distributions. Calculations based on models from James (1988) and Jacobi and Eisfeld (1980). Breathing rate = 0.75 m^3 h^{-1}.

		Regional dose (mGy WLM $^{-1}$)			
		James, 1988			
		Tracheobronchial region			Alveolar Interstital region
	Aerosol size	Epithelial cells	Epithelial cells Mouth breathing	Basal cells	
AMD_1 = 2.7 nm (2.5–3.0)	Ambient	2.6 (0–29.7)	5.0 (0–57.2)	0.96 (0–10.6)	—
f_p = 0.02 (0.0–0.22)	Doubled	2.0 (0–23.1)	3.0 (0–37.4)	0.94 (0–10.6)	—
$AMAD_2$ ≈ 50 nm	Ambient	1.6	1.6	0.92	0.06
f_2 = 0.08 (0–0.15)	Doubled	0.84	0.84	0.52	0.04
$AMAD_3$ = 386 nm (173–645)	Ambient	3.2 (2.6–5.9)	3.2 (2.6–5.9)	1.94 (1.29–3.53)	0.18 (0.12–0.17)
f_3 = 0.88 (0.68–0.98)	Doubled	3.7 (2.7–6.4)	3.7 (2.7–6.4)	≈1.76	0.13
$AMAD_4$ = 2730 nm (1660–4430)	Ambient	≈0.2 (0–3.5)	≈0.2 (0-3.5)	?	≈0
f_4 = 0.02 (0–0.25)	Doubled	≈0.3?	≈0.3	?	≈0
Complete size distribution	Ambient	7.6 (2.6–43.3)	10. (2.6–68.2)	≈3.8	≈0.24
	Doubled	6.8 (2.7–30.6)	7.8 (2.7–44.9)	≈3.2	≈0.17
		Modified Jacobi and Eisfeld, 1980			
Complete size distribution	Ambient	—	—	1.7 (1.2–6.4)	0.43 (0.30–0.44)
	Doubled	—	—	2.2 (1.8–7.0)	0.59 (0.47–0.60)

Table 6. Annual effective dose equivalents H_{eff}, for each radon and thoron gas concentration, C_0 (aged aerosol), indoors and outdoors. Calculations based on models of James (1988) and Jacobi and Eisfeld (1980). Breathing rate = 0.75 m^3 h^{-1}; n = residence factor).

				$\frac{H_{eff}}{C_0}\left[\frac{mSv\ y^{-1}}{Bq\ m^{-3}}\right]$			
				James, 1988			Jacobi and Eisfeld, 1980 (modified)
Measurement conditions	Nuclides	F,n	Aerosol size	Epithelial cells Nasal breathing	Epithelial cells Mouth breathing	Basal cells Nasal breathing	Basal cells Nasal breathing
INDOORS Aged Aerosol Closed room	Radon daughters	F = 0.30 n = 0.8	Ambient	0.074 (0.024–0.166)	0.183 (0.035–0.467)	0.028 (0.012–0.065)	0.016 (0.008–0.030)
($v < 0.5\ h^{-1}$)			Doubled	0.067 (0.020–0.149)	0.122 (0.025–0.450)	0.027 (0.011–0.057)	0.016 (0.008–0.030)
	Thoron daughters	F = 0.09 n = 0.8	Ambient	0.050 (0.021-0.090) tissues outside lung 0.001	?	?	0.015 tissues outside lung 0.009
			Doubled	0.046 (0.019–0.086) tissues outside lung 0.001	?	?	0.015 tissues outside lung 0.009
OUTDOORS 1 m above ground	Radon daughters	F = 0.67 n = 0.2	Ambient	0.018 (0.006–0.098)	0.023 (0.006–0.153)	≈ 0.009	0.005 (0.003–0.015)
			Doubled	0.016 (0.006–0.069)	0.018 (0.006–0.101)	≈ 0.008	0.006 (0.005–0.017)
	Thoron daughters	F = 0.01 n = 0.2	Ambient	8.3×10^{-4} (5.1–12.7)*10^{-4} tissues outside lung <1* $\times 10^{-4}$	?	?	2.7×10^{-4} tissues outside lung 3.2×10^{-4}
			Doubled	7.8×10^{-4} (5.1–12.7)*10^{-4} tissues outside lung < 1 $\times 10^{-4}$	?	?	2.7×10^{-4} tissues outside lung 3.2×10^{-4}

Table 7. Estimated annual effective dose equivalent, H_{eff}, to the general public in the Federal Republic of Germany resulting from inhalation of radon and thoron decay products. Doses calculated from models of James (1988) and Jacobi and Eisfeld (1980) using typical air activity concentrations. Breathing rate = 0.75 $m^3 h^{-1}$.

				H_{eff} [mSv y^{-1}]			
				James, 1988			Jacobi and Eisfeld 1980 (modified)
Condition	Nuclides	C_0 (Bq m^{-3})	Aerosol size	Epithelial cells Nasal breathing	Epithelial cells Mouth breathing	Basal cells Nasal breathing	Basal cells
INDOORS Closed room Aged aerosol	Radon daughters	50	Ambient	3.70 (1.2–8.3)	9.15 (1.8–23.4)	1.40 (0.60–3.3)	0.80 (0.40–1.50)
			Doubled	3.35 (1.0–7.5)	6.10 (1.3–22.5)	1.35 (0.55–2.9)	0.80 (0.40–1.50)
OUTDOORS	Radon daughters	10	Ambient	0.18 (0.021–0.090)	0.23 (0.06–1.0)	≈ 0.09	0.05 (0.03–0.15)
			Doubled	0.16 (0.06–0.98)	0.18 (0.06–1.0)	≈ 0.08	0.06 (0.05–0.17)
INDOORS Closed room Aged aerosol	Thoron daughters	5	Ambient	0.25 (0.11–0.45)	?	see epith. cells	≈ 0.12
		C(Pb–212) = 0.45	Doubled	0.23 (0.006–0.069)	?	see epith. cells	≈ 0.12
OUTDOORS	Thoron daughters	20	Ambient	0.02 (0.01–0.03)	?	see epith. cells	≈ 0.01
		C(Pb–212) = 0.20	Doubled	0.02 (0.01–0.03)	?	see epith. cells	≈ 0.01
TOTAL	Radon and thoron daughters		Ambient	4.15 (1.4–9.8)	≈9.38 (1.9–24.9)	1.8 (0.8–3.9)	0.98 (0.56–1.78)
			Doubled	3.76 (1.2–8.7)	≈ 6.53 (1.5–24.0)	1.7 (0.7–3.4)	0.99 (0.58–1.80)

CONCLUSION

Measurements of aerosol size characteristics, unattached fractions, and equilibrium factors of short-lived radon and thoron decay products under realistic living conditions are summarized. Dose calculations based on these results show that aerosol characteristics are of minor influence on the calculated dose compared to model assumptions concerning the location and type of cells at risk (epithelial, basal), uncertainties in extrathoracic filtration (nasal and oral filtration efficiencies), and model estimates of regional deposition probabilities.

ACKNOWLEDGMENT

This work was partly supported by Commission of the European Communities contract BI-F-130-D.

REFERENCES

BMI. 1985. Radon in Wohnungen und im Freien, Erhebungsmessungen in der Bundesrepublik Deutschland. Bundesministerium des Innern Bericht, Bonn.

Heyder, J, J Gebhardt, G Rudolf, CF Schiller, and W Stahlhofen. **1986.** Deposition of particles in the human respiratory tract in the size range 0.005-15 μm. J Aerosol Sci 17:811-825.

ICRP. 1987. Lung cancer risk from indoor exposure to radon daughters, ICRP Publication 50. Ann ICRP 17:1-60.

Jacobi, W and K Eisfeld. **1980.** Dose to tissues and effective dose equivalent by inhalation of radon-222, radon-220 and their short-lived daughters, GSF Report S-626. Gesellschaft für Strahlen- und Umweltforschung, München-Neuherberg.

James, AC. **1988.** Lung dosimetry, pp. 259-309. In: *Radon and Its Decay Products in Indoor Air.* John Wiley and Sons, New York.

NEA. 1983. *Dosimetry Aspects of Exposure to Radon and Thoron Daughter Products*, NEA Report. Nuclear Energy Agency, Paris.

Porstendörfer, J, A Wicke, and A Schraub. **1980.** Methods for a continuous registration of radon, thoron and their decay products indoors and outdoors, pp. 1293-1307. In: *Natural Radiation in the Environment, III.* Technical Information Center, U.S. Department of Energy, New York.

Reineking, A and J Porstendörfer. **1986.** High-volume screen diffusion batteries and alpha spectroscopy for measurement of the radon daughter activity size distributions in the environment. J Aerosol Sci 17:873-879.

Reineking, A, KH Becker, and J Porstendörfer. **1988.** Measurements of activity size distributions of the short-lived radon daughters in the indoor and outdoor environment. Radiat Protect Dosim 24:245-250.

Reineking, A and J Porstendörfer. **1990.** "Unattached" fraction of short-lived Rn decay products in indoor and outdoor environments, an improved single-screen method and results. Health Phys 58:715-727.

Scheibel, HG and J Porstendörfer. **1986.** Counting efficiency and detection limit of condensation nuclei counters for submicrometer aerosols. II. Measurements with monodisperse hydrophobic Ag and hygroscopic NaCl aerosols with particle diameters between 2 and 100 nm. J Colloid Interface Sci 109:275-291.

Schiller, OF. **1985.** Diffusionsabscheidung von Aerosolteilchen im Atemtrakt des Menschen, Dissertation. University of Frankfurt, Frankfurt/Main.

Stahlhofen, W. **1984.** Human data on deposition, pp. 39-62. In: *Lung Modelling for Inhalation of Radioactive Materials,* EUR-9384. Commission of the European Communities, Brussels.

Taulbee, DB and CP Yu. **1975.** A theory of aerosol deposition in the human respiratory tract. J Appl Physiol 38:77-85.

UNSCEAR. 1988. *Sources, Effects and Risks of Ionizing Radiation,* Annex A, pp. 413-134. United Nations Scientific Committee on the Effects of Atomic Radiation, New York.

Weibel, ER. **1963.** *Morphometry of the Human Lung*. Springer-Verlag, Berlin.

QUESTIONS AND ANSWERS

Q: James, Pacific Northwest Laboratory. It is not a question—more a comment. The main difference between the model you call James 88 and Jacobi-Eisfeld is, in fact, in the treatment of the dissolution characteristics of unattached daughters. Jacobi assumes that these are highly soluble, and so the component of dose from unattached daughters is very much lower. There isn't any evidence in support of that assumption, and all other calculations treat clearance of deposited unattached daughters the same as attached daughters.

Q: Cheng, Inhalation Toxicology Research Institute. In your calculations you show a large difference between nose- and mouth-breathing in a dose calculation. So, what percentage of nose deposition do you assume?

A: Reineking. I think that in both models it is assumed that about 50% is deposited in the nose region.

Q: Cheng. If there is mouth breathing, do you assume that there is a lower percentage of deposition in the mouth?

A: Reineking. Yes, I think that is right.

C: Cheng. I think, in some of our most recent measurements with some oral casts we saw substantial deposition in the mouth, even though this is much lower than nasal deposition in nose-breathing; perhaps 50 to 60% of that in nose breathing.

C: James. I would just like to confirm that those calculations did assume no deposition in the mouth during mouth-breathing. I think the data we now have show that this is wrong.

Q: Harley. When you apportion your time between outdoor and indoor, how do you deal with this? Do you, for instance, account for what happens in the office, and particle differences there, or particle differences in the home at night, say, while sleeping. What sort of weighting do you do for the 24-h daily cycle?

A: Reineking. Could you repeat that?

Q: Harley. I am curious, since you weighted, for example, outdoor exposure versus indoor exposure, what kind of indoor exposure and outdoor exposure did you assume? Did you assume, say, that one spends 4 hours . . .

A: Reineking. I assumed that 80% is spent in a room, and 20% is spent in outdoor air.

Q: Harley. I see. And then, for the indoor aerosol, did you have any differences between day and night?

A: Reineking. For my dose calculations I used the values of an aged aerosol; differences between day and night can be ignored.

Q: Rundo, Argonne National Laboratory. In following up on your answers to the last question, did you use the same breathing rate for the whole day? It seemed to me that this was very high. It is more than 12 L min^{-1}.

A: Reineking. I used a breathing rate of about 0.75 m^3 h^{-1} for all these calculations. I do not distinguish that, perhaps, in the outdoor atmosphere we have to use another breathing rate.

Q: Rundo. I think my point is that it is very much lower when you are resting.

A: Reineking. Yes, okay, you are correct.

Q: Rundo. It is lower still when you are asleep.

Q: Schlesinger, Israel. If, according to your numbers, the effective dose equivalent to the people of Germany is 4 mSv, you will have to find about 200 lung cancers per million people per year just from radon.

A: Reineking. If the value of 4 mSv is correct, then we will have to find a lot of lung cancers in Germany, but you . . .

Q: Schlesinger. In reality, what is the death rate for lung cancer in Germany?

A: Reineking. I don't know; I can't say.

Q: Schlesinger. I think it is 400 per million per year, so half of it will be from radon in this case.

C: Harley. Well, it depends on which dose model you use. There are lots of dose models coming on line. Thank you, Dr. Reineking. We will hear more about dose models later from Dr. James in his invited talk.

REANALYSIS OF DATA ON PARTICLE SIZE DISTRIBUTION OF RADON PROGENY IN URANIUM MINES

E. O. Knutson and A. C. George

Environmental Measurements Laboratory, U.S. Department of Energy, New York, New York

Key words: *Uranium mines, particle size, dose conversion factor, attached fraction, unattached fraction*

ABSTRACT

We reanalyzed 26 samples from radon progeny particle-size measurements made in 1971 by George et al. in four New Mexico uranium mines. These data were obtained with parallel disk diffusion batteries (still in use at our laboratory), together with a Mercer-type diffusion sampler. Seventeen additional samples taken with a cascade impactor were not reanalyzed.

The original data analysis, reported in two separate publications, was based on the assumption that the progeny consist of distinct attached and unattached species that can be sampled and analyzed separately. In the new analysis, we treated the progeny as a continuous spectrum of particle sizes, ranging from 1 to 1000 nm, covering both attached and unattached progeny. This was achieved by combining the diffusion battery and diffusion sampler data, and making one calculation. In the new analysis, the assumption of a unimodal lognormal size distribution was dropped.

The new calculations showed that 9 of the 26 distributions were unimodal, agreeing closely with the original analyses. Eleven of the new distributions had a bimodal structure, with widely separated modes evoking the classical idea of attached and unattached radon progeny. The remaining six cases had bimodal structures that were not distinct and were therefore not consistent with the classical picture.

A dose conversion factor has been computed for each particle size spectrum, yielding values from 1.6 to 7.8 Gy m^3 J^{-1} h^{-1}, generally higher than the previously cited values. The highest values correspond to mine locations with low equilibrium factors (<0.1) and severe disequilibrium among progeny nuclides.

INTRODUCTION

Data on the incidence of lung cancer in uranium miners are an important source of information concerning the carcinogenic potential of breathing the short-lived progeny of radon. To fully use this information, one must be able to extrapolate data from the mining environment to others. For this purpose, any data that help to quantify the in-mine relationship between exposure and dose are valuable. One key factor in this relationship is the size of the aerosol particles which carry the radon progeny.

In June and July of 1971, personnel from the U.S. Atomic Energy Commission's Health and Safety Laboratory (HASL; now the Environmental Measurements Laboratory, EML) measured particle sizes in four New Mexico uranium mines. The main tool for this work was a newly developed diffusion battery. In addition, a Mercer-type diffusion sampler was used to measure the unattached progeny, especially ^{218}Po. A Mercer-type, low-flow-rate impactor was used to measure radon progeny size for comparison with the diffusion battery results.

Although two reports (George et al., 1975, 1977) give the results of the 1971 measurements, we decided to further study and analyze these data using refined calculational techniques because aerosol size is a key factor in the evaluation of dose per unit exposure, and thus, the extrapolation of risk from mines to homes, where millions of people are affected. Furthermore, the concepts of attached and unattached progeny have changed in recent years, and it seemed appropriate to reinterpret the old data in light of this.

Sampling and alpha-counting equipment used in the 1971 measurements are described in George et al. (1975, 1977) and in the references cited in those reports. We will add a few details on diffusion battery dimensions that are needed for calculations.

Much of the data from Tables 1 to 4 of George et al. (1977) have been reproduced here as Table 1. Units have been changed to SI, and a column has been added for equilibrium ratio. Also, Table 1 contains a few values of f_a and f^* that were omitted from the original tables. Impactor data, indicated by single- and double-prime marks, are included here, although they were not reanalyzed.

Table 1. Sampling conditions during measurement of radon progeny particle-size distributions in New Mexico uranium mines (adapted from George et al., 1977).

Test No.[a]	Mine Code	Position in Mine	Temp., °C	Relative Humidity, %	Aerosol Conc., cm^{-3}	Radon Conc., $kBq\ m^{-3}$	PAEC, $\mu J\ m^{-3}$[b]	Eq. Factor	^{218}Po*	^{214}Pb*	^{214}Bi*	AMD, nm[c]	GSD[d]	f_a[e]	f*[f]
1	A	Slushers Posit. 1	17	85	120000	13.3	14.4	0.19	0.45	0.18	0.13	240	2.2	0.010	0.004
2	A	Main Drift	16	95	66000	12.2	14.6	0.23	0.66	0.24	0.10	230	2.1	0.055	0.036
3	A	Machine Shop	18	70	140000	4.1	5.0	0.22	0.57	0.23	0.12	170	2.2	0.035	0.020
3'	A	Machine Shop	18	70	17000	5.2	4.6	0.16	—	—	—	150	3.1	—	—
3"	A	Machine Shop	18	70	90000	4.1	6.6	0.29	0.90	0.28	0.09	170	2.8	0.047	0.042
4	A	Slushers Posit. 2			110000	10.0	6.9	0.12	0.34	0.13	0.05	240	2.9	0.043	0.015
5	A	Main Drift	17	88	93000	6.3	6.4	0.18	0.63	0.18	0.06	160	2.3	0.039	0.024
5'	A	Main Drift	17	88	56000	6.7	6.0	0.16	0.29	0.16	0.13	200	2.5	0.065	0.019
6	B	Stope 1	15	97	200000	26.6	20.8	0.14	0.40	0.15	0.07	110	2.0	0.020	0.008
6'	B	Stope 1	15	97	36000	44.4	52.0	0.21	0.65	0.21	0.10	120	2.7	0.070	0.045
7	B	Secondary Drift 1	14	96	25000	5.9	2.1	0.06	0.40	0.06	0.02	310	4.0	—	—
7'	B	Secondary Drift 1	14	96	330000	9.6	15.2	0.28	0.90	0.22	0.17	320	1.9	0.04	0.036
8	B	Secondary Drift 2	14	97	210000	26.3	8.3	0.06	0.27	0.05	0.01	110	3.1	0.027	0.007
9	C	Machine Shop	15	79	17000	3.7	4.2	0.20	0.33	0.22	0.16	110	3.0	0.208	0.069
9'	C	Machine Shop	17	79	65000	3.0	6.0	0.36	0.92	0.39	0.16	170	2.3	0.012	0.010
10	C	Slushing Posit. 1	14	88	7000	4.8	4.0	0.15	0.38	0.12	0.12	270	3.1	0.190	0.075
10'	C	Slushing Posit. 1	14	88	6000	7.7	5.2	0.12	0.21	0.12	0.09	220	2.2	0.260	0.053
11	C	Main Drift 1	14	88	260000	6.3	6.2	0.18	0.54	0.18	0.06	280	3.2	0.009	0.004
12	C	Stope 1	18	90	3000	7.0	6.7	0.17	0.33	0.15	0.14	210	1.5	0.390	0.125

* Relative to Rn
**Averages are based on data from the 26 diffusion battery tests only; data from impactor tests not included.
[a] ', " = measurements made with the impactor
[b] PAEC = potential alpha energy concentration
[c] AMD = activity median diameter
[d] GSD = geometric standard deviation
[e] f_a = ratio of unattached ^{218}Po to total ^{218}Po
[f] f* = ratio of unattached ^{218}Po to radon

Continued on next page

Table 1. *Continued.*

Test No.[a]	Mine Code	Position in Mine	Temp., °C	Relative Humidity, %	Aerosol Conc., cm^{-3}	Radon Conc., $kBq\ m^{-3}$	PAEC, $\mu J\ m^{-3}$[b]	Eq. Factor	^{218}Po*	^{214}Pb*	^{214}Bi*	AMD, nm[c]	GSD[d]	f_a[e]	f*[f]
13	C	Main Drift 2	16	96	400000	14.1	17.9	0.23	0.49	0.21	0.18	200	1.3	0.013	0.005
14	C	Stope 2	15	100	320000	17.8	22.9	0.23	0.66	0.25	0.11	170	1.5	0.018	0.012
15	C	Slush Posit. 2	21	78	1000000	7.8	6.0	0.14	0.41	0.14	0.06	200	3.7	0.065	0.025
16	C	Stope 3	17	98	100000	20.7	13.9	0.12	0.32	0.11	0.08	160	2.5	0.040	0.013
17	D	Stope 1	18	76	40000	13.0	19.1	0.27	0.48	0.25	0.22	190	2.4	0.120	0.055
17'	D	Stope 1	18	76	26000	18.8	20.8	0.20	0.36	0.21	0.14	130	2.7	0.130	0.046
18	D	Drift 1	17	73	12000	17.4	18.7	0.19	0.32	0.18	0.17	170	3.0	0.340	0.100
18'	D	Drift 1	17	76	3000	15.5	20.8	0.24	0.41	0.25	0.20	110	3.2	0.370	0.150
19	D	Drift 2	19	69	5000	18.5	20.8	0.20	0.52	0.19	0.13	170	3.1	0.310	0.160
19'	D	Drift 2	19	68	6000	18.8	19.1	0.18	0.31	0.16	0.16	140	3.7	0.280	0.086
20	D	Drift 3	17	55	15000	19.6	16.6	0.15	0.20	0.14	0.14	170	2.7	0.320	0.065
20'	D	Drift 3	17	55	110000	16.3	25.0	0.27	0.68	0.27	0.16	120	2.8	0.040	0.027
20"	D	Main Haulage 1	18	55	18000	20.0	20.8	0.19	0.46	0.18	0.14	110	2.1	0.100	0.046
21	D	Stope 2	17	57	80000	19.6	16.0	0.15	0.37	0.14	0.09	120	3.5	0.095	0.035
21'	D	Stope 2	17	57	130000	20.3	15.8	0.14	0.37	0.14	0.10	100	2.4	0.030	0.010
22	D	Stope 3	16	70	70000	16.7	25.0	0.27	0.67	0.27	0.17	140	1.9	0.047	0.030
22'	D	Stope 3	16	70	15000	17.4	27.0	0.28	0.67	0.29	0.18	110	2.5	0.220	0.150
23	D	Stope 4	18	70	40000	19.2	20.8	0.19	0.43	0.18	0.14	110	2.4	0.049	0.020
23'	D	Stope 4	18	70	120000	18.5	25.0	0.24	0.48	0.23	0.18	110	2.4	0.027	0.013
24	D	Main Haulage 2	19	60	26000	9.3	4.0	0.08	0.25	0.07	0.04	100	3.9	0.185	0.047
24'	D	Main Haulage 2	19	59	230000	10.7	4.4	0.07	0.23	0.06	0.06	110	2.7	0.067	0.015
25	D	Exhaust Draft	17	66	70000	22.2	21.6	0.18	0.50	0.17	0.10	160	3.2	0.069	0.035
25'	D	Exhaust Draft	17	66	110000	22.2	21.8	0.18	0.34	0.17	0.14	130	2.7	0.072	0.024
26	D	Near Shaft	19	53	10000	5.6	1.7	0.05	0.21	0.05	0.01	85	3.3	0.326	0.068
		Averages**	17	80	132000	13.4	12.5	0.17	0.43	0.16	0.10	176	2.7	0.121	0.042

* Relative to Rn
** Averages are based on data from the 26 diffusion battery tests only; data from impactor tests not included.
[a] ', " = measurements made with the impactor
[b] PAEC = potential alpha energy concentration
[c] AMD = activity median diameter
[d] GSD = geometric standard deviation
[e] f_a = ratio of unattached ^{218}Po to total ^{218}Po
[f] f* = ratio of unattached ^{218}Po to radon

CALCULATIONS

Calculation of Radon Progeny Concentrations

The protocol used in the 1971 diffusion battery measurements involved collecting progeny onto five filters and one metal disk. The filters were from the diffusion battery, and the metal disk was from a Mercer-type diffusion sampler. The sampling flow rate was 1.1 L min^{-1} for each substrate, and the sampling time was 10 min. After sampling, these six substrates were removed and alpha-counted simultaneously according to the Thomas (1972) protocol to determine the individual radon progeny concentrations and the potential alpha energy concentration (PAEC).

The file folder from the 1971 measurements contains the original data sheets, as well as a computer printout showing the results of the first stages of data reduction for the alpha counts from the six substrates. For each substrate, the printout gives the corresponding concentrations of the individual progeny nuclides and the PAEC, together with one-sigma uncertainty estimates.

We spot-checked the calculations of progeny concentrations and uncertainty terms and found agreement within 2 to 3%. We used these numbers as input to the particle-size calculations.

In this paper, we used only PAEC as the measure of radon progeny concentration. The 1971 data file gives also the activity concentrations of the old, individual nuclides, but the PAEC is preferred for calculation due to its relatively smaller uncertainty.

Treatment of Diffusion Sampler Data

In the previous reports, the data from the Mercer-type diffusion sampler were used only as a measure of the unattached fraction of ^{218}Po. Experiments had shown that for freshly formed ^{218}Po, the collection efficiency of the sampler's lower disk was 55%, and this value was used in calculating the unattached fraction from data obtained in the field. In recent years, evidence has shown that the term unattached progeny is often a misnomer. In some situations, unattached progeny may consist of one progeny atom surrounded by thousands of molecules of some condensable species. Compared to the bare atoms, these clusters would have considerably smaller diffusion coefficients, and the rate of bronchial deposition would consequently be reduced. Therefore, it is prudent to adopt a more general point of view regarding the particle-size distribution of radon progeny. In the reanalysis, we combined the diffusion sampler data with the diffusion battery data, permitting calculation of a single size distribution to fit the full range from atomic size to 1000 nm.

To incorporate the Mercer-type diffusion sampler data in the diffusion battery calculation, it was necessary to develop an equation giving the collection efficiency of the sampler as a function of particle size. The tests mentioned above showed that for particles with diffusion coefficient of 0.05 $cm^2\ s^{-1}$, the collection efficiency (lower disk only) is 55%. For many collection geometries and flow conditions, the theory (Hidy and Brock, 1970, p. 174, eqn 7.131) is that the collection rate is proportional to the diffusion coefficient to the two-thirds power. We assumed that this applies to the diffusion sampler; thus the efficiency was computed from the equation

$$\text{efficiency} = 0.55(D/0.05)^{2/3} \quad , \tag{1}$$

where D is the diffusion coefficient ($cm^2\ s^{-1}$) for the particle size in question.

Parallel Disk Diffusion Battery

Table 2 gives the diffusion battery dimensions needed to make particle-size calculations. It enlarges on Table 1 of George et al. (1975), which gave only the total effective length of each of the four sections.

Table 2. Key dimensions of parallel disk diffusion battery, cm.[a]

Battery Section 1	Section 2	Section 3	Section 4
0.34	0.64	1.31	2.55
	0.31	1.03	2.61
		1.88	2.56
		1.29	2.58
		2.57	2.56
			2.58
			2.60
			2.55
			2.56
			2.61

a Each disk has 14,500 nearly cylindrical holes with diameters of 0.0229 cm.

We used the traditional Gormley-Kennedy equations (Fuchs, 1964, p. 205) to calculate the theoretical penetration of different particle sizes through each diffusion battery section. They replace the Thomas equation that was referenced in the previous paper; however, the two equations give virtually identical results.

As indicated in Table 2, each diffusion battery section consists of a number of disks; each has 14,500 nearly cylindrical holes. Penetration through each section is calculated by applying the penetration equation to each disk separately, then multiplying these values for the separate penetrations. Computer listings from the mid-1970s verify that this procedure was used in the previous reports.

Data Unfolding

The data from each sample taken in the mines consist of six values with units $\mu J\ m^{-3}$, together with the uncertainty estimate for each. Also available are the theoretical penetrations (or efficiency, in the case of the Mercer-type diffusion sampler). To extract particle-size distributions, we used an unfolding method called the expectation-maximization algorithm (Maher and Laird, 1985). A Pascal-language microcomputer program called Spectra3, a descendant of an earlier program called ExMaxDB, was used (Knutson, 1989). In the previous reports, the diffusion battery data were processed using a prototype version of the nonlinear, weighted, least-squares procedure developed by Raabe (1978). In applying the algorithm, it was assumed that the particle size distribution can be fitted to a unimodal lognormal distribution.

RESULTS AND DISCUSSION

The PAEC-weighted particle-size distributions were calculated for the 26 diffusion battery samples that were described by George et al. (1977). In the three cases (identified as 7, 12, and 15 in Tables 1 and 3) where one of the six input data items was missing, the calculation was based on the five remaining data points.

Types of Activity-Size Spectra

Of the 26 spectra, 9 were unimodal (or had a second mode of insignificant amplitude) and are indicated by "U" in Table 3. The spectrum from test 22, shown in Figure 1, is typical. Also shown in Figure 1 is a unimodal, lognormal distribution constructed from the parameters activity median diameter (AMD) = 140 nm and geometric standard deviation (GSD) = 1.9, taken from Table 1 of this report.

Table 3. Particle-size distribution characteristics and inferred dose conversion factors from reanalysis of 1971 Environmental Measurements Laboratory data from four New Mexico uranium mines.

Test No.	Sample Date & Time[a]	Type[b]	AMD, nm[c]	AGMD, nm[d]	GSD[e]	DCF, Gy per J h m^{-3}[f]
1	7106221250	U	206	200	3.1	2.4
2	7106221345	U	212	190	3.2	2.6
3	7106230845	U	118	161	2.7	2.3
4	7106231030	O	181	162	3.4	2.8
5	7106231340	C	138	143	3.1	2.8
6	7106240930	U	78	111	2.5	2.8
7	7106241310	C	211	82	8.0	7.1
8	7106241410	U	100	91	3.7	4.1
9	7106250850	O	143	81	4.5	5.1
10	7106251130	C	191	97	7.2	6.8
11	7106251300	O	176	125	5.0	4.2
12	7106280905	C	193	134	5.3	5.2
13	7106281015	U	183	195	2.0	1.6
14	7106281125	U	164	158	1.9	1.8
15	7106281305	O	124	100	4.9	4.6
16	7106281415	U	161	145	3.0	2.7
17	7106290930	C	172	149	3.4	3.1
18	7106291040	C	150	107	5.3	5.4
19	7106291150	C	113	105	5.8	5.8
20	7106291300	C	136	114	4.8	4.8
21	7106301015	C	106	84	4.9	5.3
22	7106301120	U	136	142	2.3	2.2
23	7106301245	O	118	96	3.1	3.5
24	7107010920	C	109	61	5.8	6.9
25	7107011030	O	144	107	3.7	3.7
26	7107011135	C	106	57	5.8	7.8
Averages			149	123	4.2	4.1

[a] Date & Time is coded as YYMMDD HHMM; i.e., 7106221250 means 1971, June 22, at 12:50 hours.
[b] U = unimodal spectrum, C = classical bimodal spectrum, O = other bimodal spectrum.
[c] AMD = activity median diameter.
[d] AGMD = activity geometric mean diameter.
[e] GSD = geometric standard deviation.
[f] DCF = expected dose to tracheobronchial region of adult male at work, per hour of exposure to potential alpha energy concentrations = 1.0 J m^{-3}.

The AMD for the new curve in Figure 1 is 136 nm. We also calculated the activity geometric mean diameter (AGMD) and the GSD: AGMD = 142, and GSD = 2.3. Figure 1 shows that when the size distribution is indeed unimodal, and particularly when the distribution is narrow, the old and new calculations give nearly the same results.

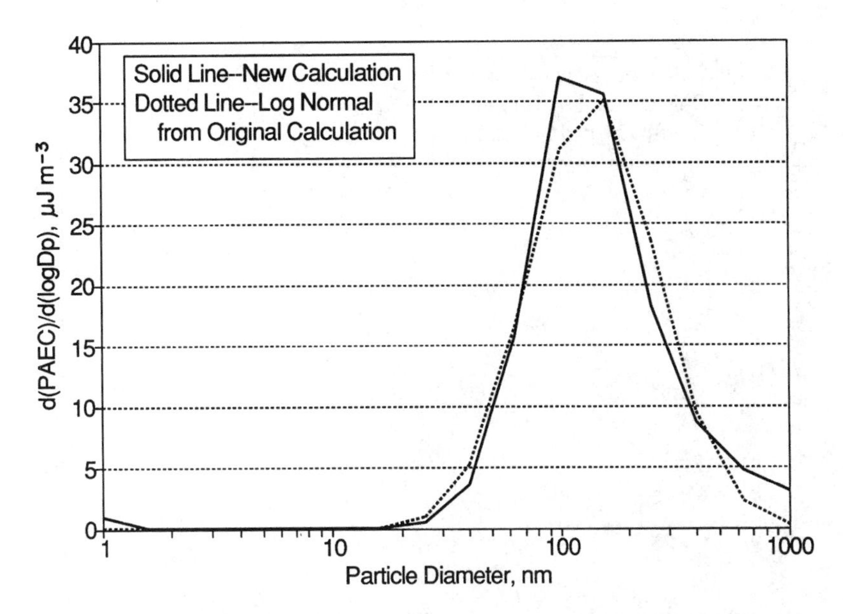

Figure 1. Particle-size distribution, weighted by potential alpha energy concentrations; typical of 9 distributions showing a single mode. PAEC = potential alpha energy concentration.

Eleven of the 26 spectra had the classical bimodal structure in which one mode is at 1 nm, and the more prominent mode is near 100 nm. Figure 2, corresponding to test 20, is typical of these 11 cases. In Figure 2, the original values for AMD and GSD were 170 nm and 2.7, respectively; for the new values, the AMD, AGMD, and GSD were 136 nm, 114 nm, and 4.8, respectively. (Note that the AGMD and GSD are well defined and can be calculated even though the distribution is not lognormal.) The increased complexity of this spectrum, compared to that from test 22, is reflected in the larger differences between AMD and GSD from the new and old calculations.

The six remaining spectra did not fit into either of the two categories discussed above. One example is given in Figure 3, corresponding to test number 11. The spectrum is bimodal, but the less prominent mode is located at 6 nm rather than at 1 nm. For this distribution, the AMD, AGMD, and GSD are 176 nm, 125 nm, and 5.0, respectively. The parameters for the original distribution are AMD = 280 nm and GSD = 3.2. The difference between the new and old results, seen in Figure 3, is the most prominent difference found in this reanalysis.

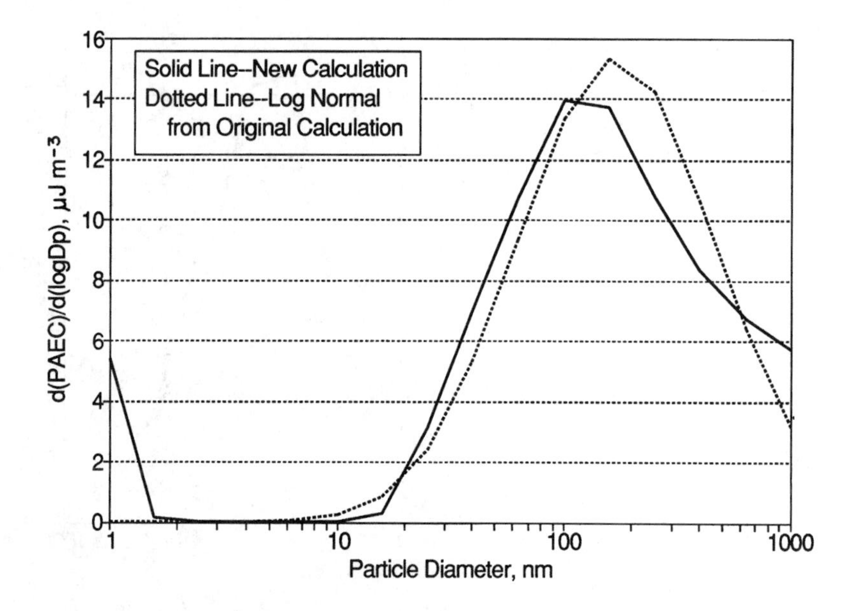

Figure 2. Particle-size distribution, weighted by potential alpha energy concentrations; typical of 11 distributions showing two well-separated modes.

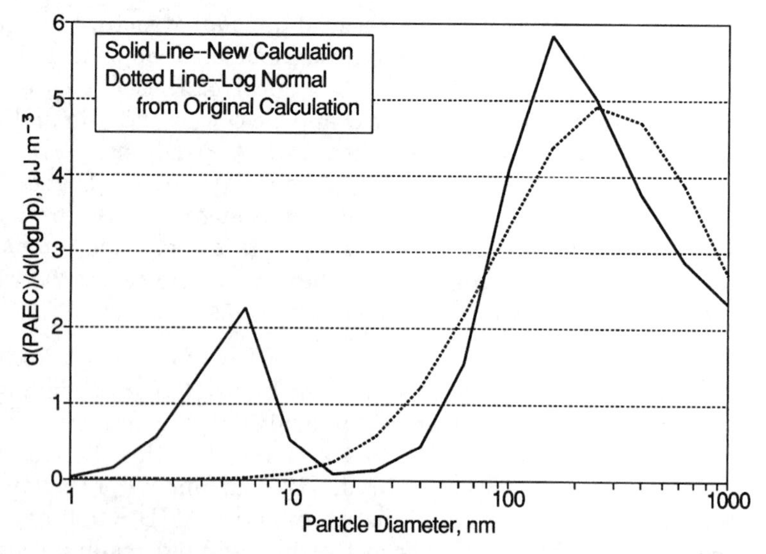

Figure 3. Particle-size distribution, weighted by potential alpha energy concentrations; typical of 6 distributions showing growth of unattached particles.

The evidence from the recalculated size distributions suggests that clustering is uncommon in uranium mines. Only 6 of the 26 distributions show some clustering, indicating that the classical bimodal picture is appropriate for uranium mine aerosols. A unimodal distribution can be considered a special case of the classical bimodal distributions; i.e., one which has negligible unattached progeny.

Summary Statistics

Table 3 gives summary statistics from the reanalysis of the 1971 data. The columns AMD, AGMD, and GSD are standard statistics for describing size distributions. Also given are the date and time each sample was taken and a code to indicate the type of spectrum found. The final column will be discussed below.

As seen in Table 3, the average value of the AMD from the new calculation is 149 nm, which is somewhat smaller than the value of 176 nm from the old calculation (Table 1). As already mentioned, the new and old calculations agree quite well for unimodal spectra. The major difference between the new and old calculations is in the GSD. However, this comparison is somewhat moot because single values of the AGMD and the GSD are not adequate for describing multimodal spectra. None of these parameters were used in the calculation described next.

Dose Conversion Factors

The final column in Table 3 shows the dose conversion factor (DCF), calculated from each particle-size spectrum using the following assumptions. These calculations were made by Professor Naomi Harley of New York University Medical Center. The basic assumptions were: breathing rate = 18 L min^{-1}; nuclide ratios (Po/Pb/Bi) = 61/29/21 for particles >25 nm, and (Po/Pb) = 10/1 for particles <25 nm; nasal deposition according to Cheng et al. (1988), (but capped at 60%) for particles <200 nm, 30% for larger particles; Cohen and Asgharian (1990) equations for diffusional deposition; target cell depth = 22 μm; particle density = 1.5 g cm^{-3}; lung branching angle = 25°; dose averaged over seven airways. Dose factors were calculated for each of the 16 particle sizes from the size distribution calculations. These factors were folded into each of the 26 size distributions to yield the final column in Table 3. The DCF is a direct indication of the health implications of the various measured size distributions. As presented here, the DCF takes into account the full size distribution of the progeny, from atoms to 1-μm particles. As seen in Table 3, the DCF range from 1.6 to 7.8 Gy m^3 J^{-1} h^{-1} (5.7 to 28 mGy WLM^{-1}), with a grand average of 4.1 Gy m^3 J^{-1} h^{-1} (14 mGy WLM^{-1}). Several different averages of the dose conversion factor for these various uranium mine conditions are shown in Table 4 by type of activity, size distribution, mine, degree of work activity, and level of exposure.

Table 4. Summary of dose conversion factors from reanalysis of New Mexico uranium mine particle-size data.

Type of Average	Value, Gy per J h m^{-3}
Grand average	4.1
Average for 9 unimodal spectra	2.5
Average for 11 classical-bimodal spectra	4.8
Average for 6 other bimodal spectra	4.0
Average for Mine A	2.6
Average for Mine B	4.7
Average for Mine C	4.0
Average for Mine D	4.9
Average based on 15 most active areas[a]	3.9
PAEC-weighted average[b] overall	3.6
As above, for 15 most active areas	3.2

[a] Includes data from stopes, slushing positions, machine shops, and entrance shaft.
[b] Computed using the potential alpha energy concentration as a weight factor for each dose conversion factor equivalent to assuming that miners rotate among all positions, spending equal time in each.

The newly calculated DCF values in Table 3 are from 1 to 5 times larger than the value of 0.5 rad per working level month (1.4 Gy m^3 J^{-1} h^{-1}, 5 mGy WLM^{-1}) given in the National Council on Radiation Protection and Measurements Report 78 (NCRP, 1984) as typical for mining environments. No single reason for this difference is apparent from the data. Table 4 shows that, for those distributions we have designated unimodal, the average is 2.5 Gy m^3 J^{-1} h^{-1} (8.8 mGy WLM^{-1}), which is nearly twice the NCRP value. Clearly, the ultrafine or unattached progeny also contribute greatly to the higher DCF, as seen from the average value of 4.8 Gy m^3 J^{-1} h^{-1} (Table 4) for the classical/bimodal group of spectra.

We examined the circumstances leading to the largest values (7.8, 7.1, and 6.9, from tests 26, 7, and 24, respectively). Test 26 was of a sample taken near the fresh-air intake of Mine D. As seen in Table 1, the equilibrium factor for this test, 0.05, was very low, indicating a fresh source of radon gas. Plate out of the radon progeny can be ruled out as a reason for this low equilibrium factor because the aerosol concentration was moderately high: 10,000 cm^{-3}.

The newly calculated size spectrum for test 26 has 12% of the PAEC in the smallest (1-nm) size class. The older analysis (see Table 1) shows a correspondingly large amount of unattached ^{218}Po for test 26 (f^* = 6.8%, implying 13% unattached PAEC). Clearly, this high value of unattached progeny contributed to the high DCF in test 26. Tests 7 and 24 were carried out under circumstances similar to those of test 26, and the same discussion applies. However, neither equilibrium factor nor f^* is a perfect indicator of DCF. For example, in test 8, the equilibrium factor was 0.06, but f^* and DCF were only moderately high. Also, test 19 had by far the largest value of f^* (16%), but the DCF was not correspondingly large.

CONCLUSIONS

From the reanalysis of the 1971 uranium mine data, we found that the classical dichotomous concept in which radon progeny consist of distinct, well-separated, attached and unattached species is appropriate for the uranium mines. Only 6 of the 26 particle-size distributions were inconsistent with this concept.

For the 26 samples, the AMD, based on PAEC, ranged from 78 to 212 nm, and the grand average was 149 nm. These values are slightly smaller than those found in the previous analysis of these data.

Dose conversion factors, computed from the spectra, ranged from 1.6 to 7.8 Gy m^3 J^{-1} h^{-1}, with an average of 4.1 Gy m^3 J^{-1} h^{-1}. This range is approximately 1 to 5 times the "benchmark" value given in NCRP Report 78 (1985).

REFERENCES

Cheng, YS, Y Yamada, HC Yeh, and DL Swift. **1988.** Deposition of ultrafine aerosols in a human nasal cast. J Aerosol Sci 19:741-751.

Cohen, BS and B Asgharian. **1990.** Deposition of ultrafine particles in the upper airways: an empirical analysis. J Aerosol Sci 21:789-798.

Fuchs, NA. **1964.** *The Mechanics of Aerosols.* Macmillan Company, New York.

George, AC, L Hinchliffe, and R Sladowski. **1975.** Size distribution of radon daughter particles in uranium mine atmospheres. Am Ind Hyg Assoc J 36:484-490.

George, AC, L Hinchliffe, and R Sladowski. **1977.** *Size Distribution of Radon Daughter Particles in Uranium Mine Atmospheres,* Technical Report No. HASL-326. Health and Safety Laboratory, U.S. Energy Research and Development Administration, New York.

Hidy, GM and JR Brock. **1970.** *The Dynamics of Aerocolloidal Systems.* Pergamon Press, New York.

Knutson, EO. **1989.** *Personal Computer Programs for Use in Radon/Thoron Progeny Measurements,* Technical Report No. EML-517. Environmental Measurements Laboratory, US Department of Energy, New York.

Maher, EF and NM Laird. **1985.** EM algorithm reconstruction of particle size distributions from diffusion battery data. J Aerosol Sci 16:557-570.

NCRP. 1984. *Evaluation of Occupational and Environmental Exposures to Radon and Radon Daughters in the United States,* Technical Report No. NCRP 78. National Council on Radiation Protection and Measurements, Bethesda, MD.

Raabe, OG. **1978.** A general method for fitting size distributions to multicomponent aerosol data using weighted least-squares. Environ Sci Technol 12:1152-1167.

Thomas, JW. **1972.** *Measurement of Radon Daughters in Air by Alpha Counting of Air Filters,* Technical Report No. HASL-256. Health and Safety Laboratory, US Energy Research and Development Administration, New York.

QUESTIONS AND ANSWERS

Q: Johnson, Pacific Northwest Laboratory. This question is probably directed more to the chair than to the speaker. Could you give us a little more background as to why there is such a big difference between the benchmark NCRP number and the numbers that you calculated?

A: Harley, New York University. In a word, it is the particle size distribution.

Q: Johnson. I wasn't quite familiar with the benchmark. That was the whole impact— what is the particle size?

A: Harley. The benchmark was based on the small mode and the large mode average values. Then this calculation resolves the spectra into a much finer particle size bin. The dose factor is much bigger for the smaller-size particles, so, when you put it all together, there were 16 particle-size bins, and the conversion factors come out much higher. I must add that it is very important to look at the way a particle size is convoluted from these calculations. This is not necessarily gospel, but when you look at the way the particle size was deduced here, it is probably done better than the original calculations.

C: Knutson. I am sure that deconvolution is an ongoing discussion in aerosol sciences; it has been going on for 20 years and probably will continue for another 20. I do agree that the new results are more near the truth than the old ones.

C: Harley. Science marches on relentlessly. There is nothing like facts.

Q: Caswell, National Institute of Standards and Technology. Can you comment about how these results relate to what you might expect to see in the home or in a building?

A: Knutson. Yes, although my memory kind of fails me. In the house, sizes were a little bit bigger, I guess, leading to a slightly smaller...

C: Harley. That is a good question. In summary, the size bins, again, tend to be shifted to lower groups, so the conversion factors come out quite similar, in other words, higher than you would assume for the so-called benchmark atmosphere. You must remember that the benchmark atmosphere was based on very scant data, so this is a very important point, that there is a great need for this type of input data.

Q: Kirkland, Scotland. You said you averaged your data over seven generations. I wonder which seven generations these were, and why you did that, rather than 15 or 16 generations as Tony James does. Also, could you comment on whether you think there is a dose variation within the seven generations you chose.

A: Harley. At one time there was a significant difference among generations, but now, with the data on deposition in the hollow casts of human lungs, these variations have disappeared. So now, dose-averaging over many generations is the way people do it. The cancers are more proximal, around generations 2 and 3 to 6, so I think that this is the pertinent region for looking at dose.

Q: James. You mentioned that you had taken into account Beverly Cohen's enhanced deposition factor. How much impact did that have on the dose conversion?

A: Harley. Very little.

Q: James. Yes. I find the same.

Q: Wheeler, Department of Energy-Nevada. Were these working mines that you were measuring?

A: Knutson. Yes, they were active mines. Of course, you should speak to Andy George about it. He is here, and he will tell you exactly the conditions. From reading the paper, it seems they couldn't be sure that conditions hadn't already changed from the 60s. I guess there was already increased ventilation in mines at that time, but they were active mines.

Q: Wheeler. Is there any effect of the ventilation rates on your particle size distributions, or was that studied?

A: Knutson. I don't know that there is anything in our data that could tell us the answer. There were some samples taken near the fresh-air input, at least one that I can remember. It really had a dramatic effect, because in that area the equilibrium factor was ~0.05. It was like raw radon and the polonium-218 in that air sample. It was one of those dose conversion factors at the upper end.

DOSIMETRY MODELING

DOSIMETRY OF RADON AND THORON EXPOSURES: IMPLICATIONS FOR RISKS FROM INDOOR EXPOSURE[a]

A. C. James

Pacific Northwest Laboratory, Richland, Washington

Key words: *Radon dosimetry, lung, cancer risk, other organs*

ABSTRACT

Current estimates of lung-cancer risks due to the inhalation of radon and its progeny in homes are based on extrapolations of excess mortality observed in populations of underground miners. To project lung-cancer risk to the general public it is necessary to account for any effects that different exposure conditions may have on doses received by critical target cells in the respiratory tract. This paper summarizes the results of a review of aerosol parameters, physiological and biological factors, and cells at risk that are involved in comparing doses between mine and indoor environments.[b] The dose received by sensitive cells in the bronchi from exposure to a given amount of potential alpha-energy (commonly given the unit Working Level Month, WLM) is found to be approximately 20% lower indoors than for healthy underground miners. However, this estimate of the ratio of dose per unit exposure in homes compared to that in mines (termed the "*K*-factor") is sensitive to the assumed hygroscopic properties of radon-progeny aerosols. The estimate of *K* varies from about 0.9, if radon-progeny aerosol particles are assumed not to grow hygroscopically in the respiratory tract, to about 0.6 if the ambient particles are assumed to double in size by hygroscopic growth.

The estimated *K*-factor is up to 40% higher for young children than for adults. However, according to the BEIR-IV Committee's risk projection model, such higher dose rates in childhood make a negligible contribution to the overall lifetime risk of lung cancer. We also conclude that the uranium miners who were studied to

(a) Work performed for the U.S. Department of Energy under Contract No. DE-AC06-76-RLO 1830.

(b) The review of "Comparative Dosimetry of Radon in Mines and Homes" was carried out for the National Research Council (NRC, 1991) by a Scientific Panel. Members were Jonathan M. Samet (Chairman), University of New Mexico; Roy E. Albert, University of Cincinnati; Joseph D. Brain, Harvard School of Public Health; Raymond A. Guilmette, Inhalation Toxicology Research Institute; Anthony C. James, Battelle, Pacific Northwest Laboratories; and David G. Kaufman, University of North Carolina. The Panel's report was completed several months after the author's presentation at the Hanford Life Sciences Symposium. This paper has been amended to reflect the Panel's final, published conclusions.

develop the underlying relationship between exposure to radon progeny and lung-cancer risk did not receive significant additional dose from their exposure to uranium ore dust. Thus, the BEIR-IV Committee's risk-projection model can be used with an estimated *K*-factor of 0.8 (based solely on the comparative doses from radon progeny) to yield estimates of the lifetime excess lung-cancer risk among the U.S. population attributable to indoor radon exposure: approximately 0.7% for males and 0.3% for females. At the current U.S. Environmental Protection Agency (EPA) Action Level of 4 pCi L^{-1} (150 Bq m^{-3}), the estimated excess lung-cancer risks from lifetime exposure are approximately 2% for males and 0.8% for females.

In addition to these developments in converting radon-progeny exposures into doses absorbed by target cells, the International Commission on Radiological Protection (ICRP) Task Group on Human Respiratory Tract Models for Radiological Protection will propose that the bulk of the tissue weighting factor used to represent the risk of lung cancer in the ICRP system of dosimetry should be apportioned to the dose received by bronchial tissue. Furthermore, based on improved dosimetry for the atomic-bomb survivors, the ICRP has increased approximately threefold their recommended estimate of the lifetime risk associated with irradiating the lung as a whole. Taken together, these changes in both the calculated effective doses per unit exposure to radon progeny and the lung-cancer risk attributed to unit effective dose lead to unrealistically high estimates of radon-induced lung cancer. This is true for both the general population and for underground miners. Therefore, the effective dose calculated for exposure to radon (or thoron) progeny would need to be corrected by a factor substantially less than unity, in order to bring dosimetric estimates of lung-cancer risks into line with projections from epidemiological studies.

The paper also provides updated estimates of the bronchial and effective doses per unit of exposure to thoron progeny in mines and homes, and of the doses received by tissues other than the lung from exposures to radon, thoron, and their short-lived progeny.[a]

INTRODUCTION

Epidemiological studies of underground miners of uranium and other minerals have provided reasonably firm estimates of the risk of lung cancer associated with exposure to radon progeny in underground mines (Lubin, 1988; National Research Council [NRC], 1988). A scientific panel appointed by the NRC has reviewed the basic information

(a) At the request of the editors of these proceedings, the author has added estimates of tissue doses to the material presented at the Hanford Symposium. These new estimates were developed jointly by the author and his PNL colleagues, Drs. Karla Thrall and Edmond Hui, and will be published in detail elsewhere.

and modeling procedure needed to extend these data to radon exposure of the general population in the home environment. The process of risk extrapolation must take into account characteristic differences in exposure conditions between mines and homes. Also considered are physiological differences between working miners and men, women and children exposed in the home that affect the dose of alpha-radiation to target cells in the respiratory epithelium. This paper presents a synopsis of the findings of the NRC panel with respect to the comparative dosimetry of radon and thoron in mines and homes. These findings are taken further to examine, using the BEIR-IV Committee's risk projection model (NRC, 1988), the expected lung-cancer risks from exposure to radon and thoron in the home.

The direct epidemiological observations of lung cancers in underground miners that are related to radon exposure, together with the improved lung dosimetry, provide a means to test the applicability of the International Commission on Radiological Protection's (ICRP) concept of effective dose and its implied lifetime cancer risks. In this regard, ICRP has recently increased threefold the risk factor applied to irradiation of the lung (ICRP, 1991). Also, the ICRP Task Group on Human Respiratory Tract Models for Radiological Protection (Bair, 1991) considers that a larger proportion of the weighting factor used to represent the risk of lung cancer should be assigned to the dose received by bronchial tissue. For radon and thoron progeny this bronchial dose is much higher than the dose to lung tissue as a whole. The paper applies the new dosimetric analyses developed by the NRC panel to draw critical comparisons with lung-cancer risks implied by the calculation of effective dose.

CONVERTING EXPOSURES TO RADON PROGENY INTO DOSES RECEIVED BY SENSITIVE CELLS

The reviews carried out by the ICRP Task Group (Bair, 1991) and by the NRC's scientific panel on Comparative Dosimetry in Mines and Homes (NRC, 1991) both concluded that the target tissue at risk for the induction and development of human lung cancer is comprised principally of the populations of basal and secretory cell nuclei in the bronchial epithelium. For risk assessment purposes, both studies concluded that it is appropriate to use the average dose received by these combined populations of cell nuclei. In order to calculate the relevant doses, the NRC panel adopted the morphological model of bronchial epithelium and the depth-distributions of target cell nuclei proposed by the ICRP Task Group (as described by James et al., 1991a). The detailed methods and assumptions employed to calculate radon-progeny deposition and clearance behavior in the respiratory tract, and the resulting doses absorbed by target cell populations in the bronchial, bronchiolar, and alveolar-interstitial epithelia are described fully in the NRC panel's report (NRC, 1991). They will not be reviewed here, but a summary is provided elsewhere in these proceedings (see paper by Fisher et al.).

Figure 1 shows the conversion coefficient between exposure to potential alpha-energy and dose to bronchial target cells that is calculated using NRC's (1991) dosimetry model. For convenience, this coefficient is referred to as the bronchial dose conversion factor (DCF) and is expressed in terms of the familiar unit of exposure, the working level month (WLM). The bronchial DCF is derived by calculating separately the mean doses received by basal and secretory cell nuclei, then averaging these values. The figure shows the resulting variation of the DCF with the characteristic size of the radon-progeny aerosol (represented by the activity median thermodynamic diameter, AMTD) and with the exposed subject's level of physical exertion. It is seen that, for a particular level of physical exertion, the DCF may vary over a 10- to 20-fold range with the radon-progeny aerosol size. The highest values of the DCF apply to so-called "unattached" radon progeny that may be present in the air as "clusters" with other contaminant molecules and thus have an equivalent diameter of several thousandths of a micrometer (i.e., several nanometers). The aerosol size at which the DCF has a minimum value depends quite strongly on the subject's breathing rate (determined by the level of exertion).

The figure also shows that nasal filtration of unattached radon progeny (of about 0.001 μm, or 1 nm, particle diameter), and of the larger molecular clusters, is a critical factor in determining the bronchial DCF. Earlier dosimetric studies (for example, Harley and Pasternack, 1982; NEA, 1983; NCRP, 1984; James, 1988) generally assumed values of nasal filtration efficiency derived from the experimental results of George and Breslin (1969). However, a recent study of the deposition of unattached thoron progeny in a hollow cast of the human nose indicated a substantially higher filtration efficiency (Cheng et al., 1989; see also the paper by Cheng et al. in these proceedings). This higher filtration efficiency has since been confirmed by several investigators, in several different laboratories (Swift et al., 1992),[(a)] and is assumed here.

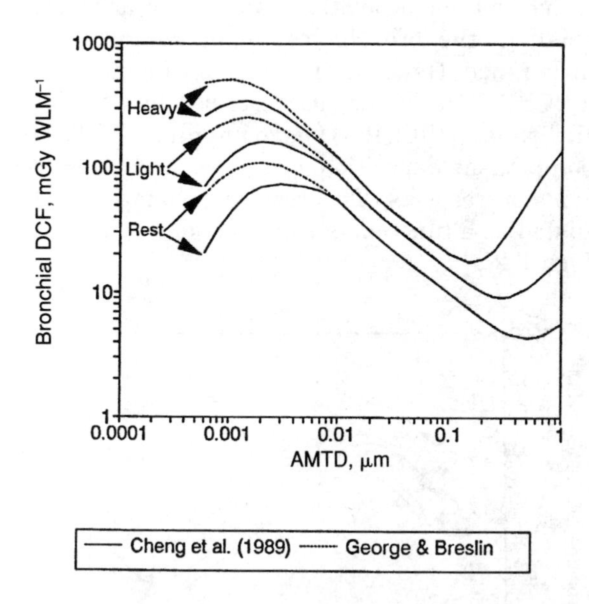

Figure 1. Bronchial dose conversion factor for exposure of adult male to radon progeny, calculated as a function of aerosol activity median thermodynamic diameter (AMTD) using the NRC (1991) dosimetry model. Shown are (1) effect of assumed level of physical exertion and (2) substantial effect for ultrafine aerosol particles of assumed nasal filtration efficiency (according to George and Breslin, 1969, or Cheng et al., 1989).

Bronchial DCF for Underground Miners

The DCF for exposures to radon (and thoron) progeny currently recommended by ICRP (ICRP, 1981) and UNSCEAR (UNSCEAR, 1982) are based on the dosimetric modeling study carried out by NEA (1983). This assumed that sensitive cells are distributed throughout both the

(a) At the time the NRC panel prepared their report, this experimental confirmation of the high filtration efficiency of the human nose for ultrafine radon progeny was not available. The panel therefore based its calculations of the bronchial DCF on the generally accepted values of nasal filtration efficiency derived from the work of George and Breslin. However, the panel did address the current uncertainty in this factor, and illustrated in footnotes to their report the effects on their calculated values and recommendations of a higher than assumed nasal filtration efficiency.

bronchi and the finer, bronchiolar airways of the lungs. The target cell nuclei were assumed to be located basally in all airways, at characteristic depths below the epithelial surface, which are large in the bronchi and small in the bronchioles compared to the radon-progeny alpha-particle range. However, by incorporating more-recent developments, the ICRP Task Group (as described by James et al., 1991b) and the NRC panel (NRC, 1991) have improved NEA's (1983) calculation of radon progeny deposition, clearance, and target cell dosimetry in several major respects. The resulting changes in the bronchial DCF, calculated as a function of radon-progeny aerosol size, are illustrated in Figure 2.

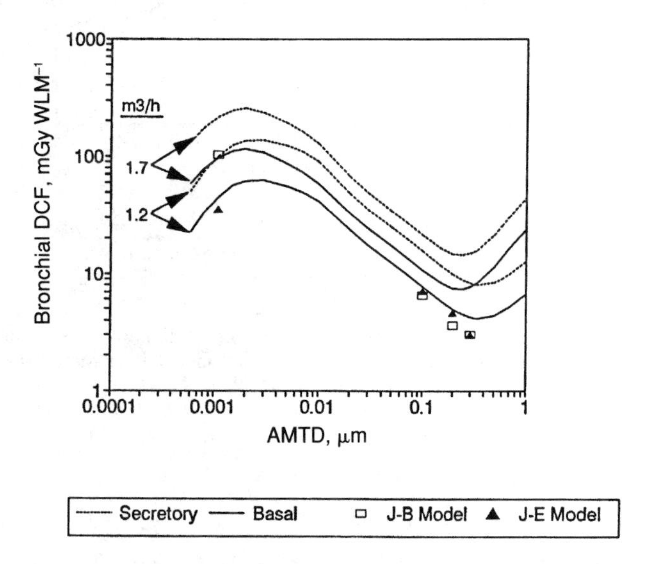

Figure 2. Dose to secretory and basal cell nuclei in bronchi of uranium miner from exposure to radon progeny, calculated using NRC (1991) dosimetry model compared with values derived for basal cells by NEA, 1983 (using "J-B" and "J-E" models).

For an underground miner, NEA (1983) assumed a mean breathing rate of 1.2 m^3 h^{-1}. Figure 2 compares values of the bronchial DCF calculated for this breathing rate using the NRC (1991) model with NEA's (1983) values. It is seen that, for basal cell targets and for so-called "attached" radon-progeny aerosols (with AMTD larger than 0.1 μm), the NRC model gives marginally higher values of the DCF than NEA's model. For "unattached" radon progeny, the DCF calculated by NEA depended very strongly on the assumed model of deposition and clearance [in that study, on the "J-B" model (James et al., 1980) or the "J-E" model (Jacobi and Eisfeld, 1980)].

The figure also shows that the dose received by secretory cell nuclei in bronchial epithelium, which the NRC model also calculates, is approximately twofold higher than that calculated for basal cell nuclei. Thus, the "reference" bronchial dose considered by NRC (1991), which is obtained by taking the average of basal and secretory cell doses, is approximately 50% higher than the dose to just basal cells. A further significant factor that was reconsidered by the ICRP Task Group (Roy and Courtay, 1991) and also by NRC (1991) is the representative breathing rate for an underground miner. The value of 1.2 m^3 h^{-1} used earlier applies strictly to a sedentary worker (ICRP, 1975). By contrast, to represent the physical labor involved in underground mining, an average breathing rate of 1.7 m^3 h^{-1} is appropriate (NRC, 1991). This can be subdivided further into a rate that is higher still (assumed to be 1.9 m^3 h^{-1}) for active mining and 1.5 m^3 h^{-1} for the lighter work involved in ore haulage and other miscellaneous mining tasks. Figure 2 shows that, by taking account of both secretory cells as sensitive targets and the higher mean breathing rates of underground miners than those of workers in general, the "reference" bronchial DCF is increased by a factor of approximately three.

Representative Dose Conversion Factors for Underground Miners

It was seen in Figures 1 and 2 that the characteristic size of the radon-progeny aerosol and the fraction of potential alpha-energy inhaled in the unattached or ultrafine forms are critical in determining the dose conversion factor. However, reliable data on which representative values of these parameters for particular exposure environments can be based are extremely sparse. NEA (1983) assumed that the unattached fraction of potential alpha-energy, f_p, in a uranium mine was typically about 2.5%. According to NRC's (1991) assessment, in the uranium mine atmospheres of the 1950s and '60s era which are relevant to the epidemiological studies of underground miners, the value of f_p during active mining was typically only 0.5%, but that applicable to less dusty operations such as ore "haulage" was typically 3%. For the attached radon-progeny aerosol, NEA (1983) assumed a typical AMTD of 0.2 μm. NRC's (1991) reassessment yielded similar estimates of the attached aerosol AMTD: 0.25 μm for dusty operations and 0.15 μm in cleaner air. However, the NRC panel considered that these attached radon-progeny aerosols are likely to have been hygroscopic, and that individual particles were most likely to have grown, in the respiratory tract, to double their size in ambient air (NRC, 1991). The effects of these various assumptions about the radon-progeny activity-size distribution in mine air, and its stability in the respiratory tract, are shown in Figure 3.

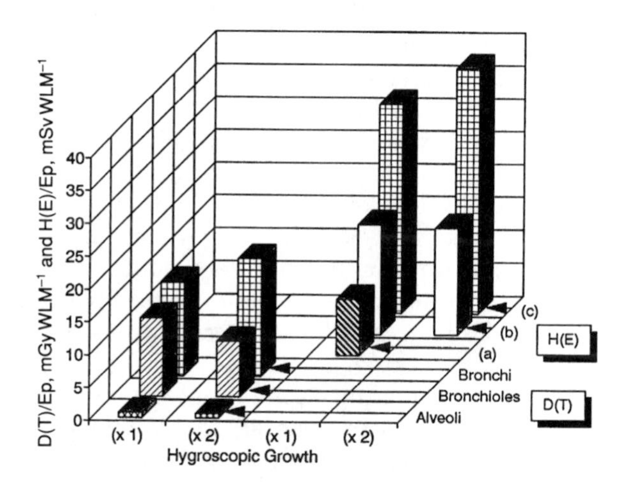

Figure 3. Conversion factors between exposure to radon progeny and mean tissue dose, D(T), to the bronchi, bronchioles, and alveoli, calculated for uranium miner using NRC (1991) dosimetry model compared with corresponding values in terms of effective dose, H(E). The latter are calculated by several methods: (a) that given by NEA (1983) using the tissue weighting factors recommended by UNSCEAR (1982); (b) NRC (1991) dosimetry model with UNSCEAR (1982) weighting factors; (c) NRC (1991) model with weighting factors proposed by ICRP Lung Modeling Task Group (Bair, 1991). Figure shows effects of assumed hygroscopic growth of radon-progeny aerosol particles in respiratory tract.

The figure shows (to the left and in the foreground), in terms of bronchial, bronchiolar, and alveolar tissue doses, the average conversion factors calculated for 1950s and '60s uranium miners according to the NRC (1991) model. If the attached radon-progeny aerosol particles are assumed not to grow in the respiratory tract, the calculated doses to bronchial and bronchiolar target cells are similar. However, hygroscopic growth (by a factor of two) is calculated to increase the bronchial dose significantly while decreasing that to the bronchioles. The estimated bronchial DCF is about 14 mGy WLM^{-1} on the assumption of no particle growth, and 18 mGy WLM^{-1} if particles are assumed to double in size on inhalation. These values are to be compared with the substantially lower reference value of 6.3 mGy WLM^{-1} derived by NEA (1983).

To the right and rear of Figure 3 are shown resulting estimates of the effective dose. In NEA's (1983) calculation, the effective dose was obtained by applying half of ICRP's (1977) tissue weighting factor of 0.12 for the lung as a whole to the doses calculated separately for the "tracheobronchiolar" and "pulmonary" (i.e., alveolar) tissue regions. The resulting recommended value of the DCF, expressed in terms of effective dose, was 8.5 mSv WLM^{-1}. When this same weighting procedure is applied to the tissue doses calculated using the NRC (1991) dosimetry model, however, Figure 3 shows that the calculated effective dose is doubled, at about 17 mSv WLM^{-1}. By averaging the effective dose over both the bronchial and bronchiolar tissues in this way, the assumed hygroscopicity of the attached radon-progeny aerosol has virtually no effect on the calculated DCF.

The ICRP Task Group on Human Respiratory Tract Models for Radiological Protection has proposed a significantly different apportionment of radiation risk between the bronchial, bronchiolar, and alveolar-interstitial epithelial tissues (Bair, 1991). The proposed reapportionment reflects the observation that cancers of the human lung arise predominantly in the bronchi (in both smokers and nonsmokers) and not from alveolar epithelium. In turn, it is assumed that radiation sensitivity reflects the regional distribution of lung cancers observed in "nonexposed" subjects. The Task Group considers that 80% of the overall risk from irradiating the lung uniformly is likely to arise from bronchial epithelium, 15% from bronchiolar epithelium, and only 5% from the alveolar interstitium. Figure 3 shows that the effect of this re-apportionment of risk factors is to increase the calculated effective dose by approximately a factor of two. The resulting effective dose is then more directly related to the physical dose absorbed by the bronchial epithelium. Hence the effective dose is, again, sensitive to the assumed hygroscopicity of the attached radon-progeny aerosol. If we take the average of the two values of effective dose calculated for the assumptions of no hygroscopic growth or particles doubling in size within the respiratory tract (to reflect the fact that the actual degree of hygroscopic growth of uranium mine aerosols is highly uncertain), then we obtain a single reference value of the conversion factor to an effective dose of about 35 mSv WLM^{-1}. This is almost fourfold higher than the reference value of 10 mSv WLM^{-1} recommended for a uranium miner by ICRP (1981).

Representative Dose Conversion Factors for Indoor Exposure

Figures 4 and 5 draw the equivalent dosimetric comparisons for exposure of an adult male in the home, based on the NRC (1991) and NEA (1983) dosimetry models. Figure 4 shows that the bronchial DCF calculated using the NRC (1991) dosimetry model for basal cells are generally higher than the values calculated by NEA (1983). The breathing rate of 1.02 m^3 h^{-1} used by NRC (1991) represents the average value for an adult male exposed in the living room of a home (NRC, 1991; Roy and Courtay, 1991). The breathing rate of 0.45 m^3 h^{-1} represents a sleeping subject (night-time exposure). Figure 5 shows the overall dose conversion factors calculated for the exposure conditions assumed by NRC (1991) to "typify" home environments. It is assumed that f_p is normally 8%, and that the AMTD of the attached radon-progeny aerosol is normally 0.15 μm in both living rooms and bedrooms. NEA (1983) assumed the same ambient size distribution for the attached aerosol but a significantly lower typical value for f_p of only 2.5%. It is seen from Figure 5 that NRC's (1991) assumption that the attached aerosol doubles in size within the respiratory tract decreases both the average bronchial and bronchiolar doses estimated for typical indoor exposure. The reference value of the bronchial DCF is approximately 12 mGy WLM^{-1}. This is to be compared with the threefold lower value of 4 mGy WLM^{-1} derived by NEA (1983).

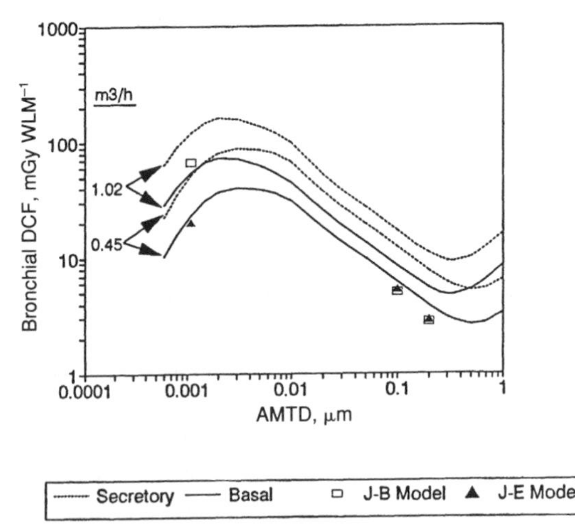

Figure 4. Dose to secretory and basal cell nuclei in bronchi of adult male exposed to radon progeny indoors, calculated using NRC (1991) dosimetry model compared with values derived for basal cells by NEA, 1983 (using the "J-B" and "J-E" models) for a mean breathing rate of 0.75 m^3 h^{-1}. DCF = dose conversion factor.

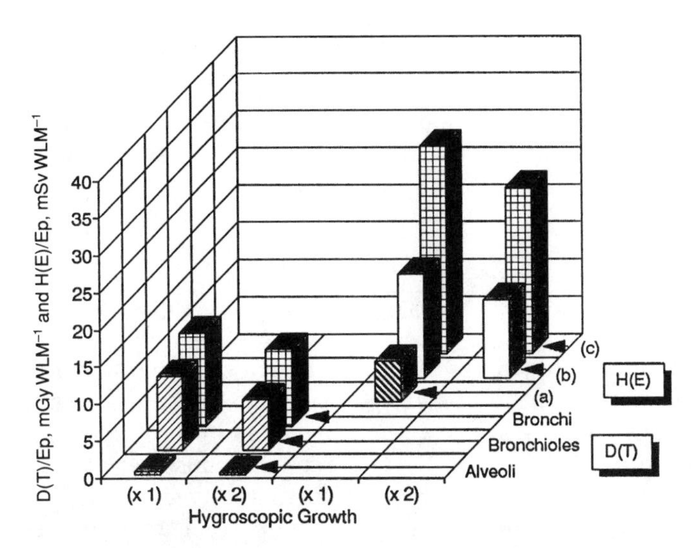

Figure 5. Conversion factors between indoor exposure to radon progeny, mean tissue doses, D(T), and effective dose, H(E), calculated for adult male, as described for Figure 3.

Figure 5 also shows (to the right and rear of the figure) the resulting estimates of effective dose from exposure in the home. NEA (1983) derived a reference DCF of 5.5 mSv WLM^{-1}. The equivalent value derived from the NRC (1991) model when risk is apportioned equally to the doses received by tracheobronchiolar and alveolar-interstitial tissues is approximately 12 mSv WLM^{-1}, i.e., approximately twofold higher. Re-apportionment of the regional risk factors in the manner proposed by the ICRP Lung Modeling Task Group yields a reference DCF in terms of effective dose of approximately 25 mSv WLM^{-1}. This is approximately fourfold higher than the value of 6.3 mSv WLM^{-1} recommended by ICRP (1987). We will consider later the implications of this marked increase in the estimated dose conversion factor for the implied lifetime risks from indoor exposure.

COMPARATIVE DOSIMETRY FOR RADON EXPOSURES IN MINES AND HOMES

Following the approach described in the BEIR-IV report (NRC, 1988), the NRC dosimetry panel proposed that the risk per unit exposure in the home, $(Risk)_h/(WLM)_h$, can be related to that in mines, $(Risk)_m/(WLM)_m$, by a dimensionless factor, K:

$$K = \frac{(\text{Risk})_h/(\text{WLM})_h}{(\text{Risk})_m/(\text{WLM})_m} \quad . \tag{1}$$

On the premise that risk is primarily related to the doses received by the relevant cellular targets, the value of the risk extrapolation factor K is given by the ratio of doses that result from unit exposure in each environment:

$$K = \frac{(\text{Dose})_h/(\text{WLM})_h}{(\text{Dose})_m/(\text{WLM})_m} \quad . \tag{2}$$

Figure 6 shows the values of the factor K calculated using the NRC (1991) dosimetry model for various subjects (a man; a woman; children of ages 10 y, 5 y, and 1 y; and a 1-mo-old infant) exposed to radon progeny in a typical home. Several reference doses are used for this comparison: (a) the mean physical doses received by target cells in the bronchial epithelium of a uranium miner (per unit exposure) compared to the bronchial doses received by the various subjects in the home; (b) the relative values of effective dose per unit exposure in each environment, calculated by apportioning most of the risk to bronchial dose; and (c) the relative values of effective dose calculated according to ICRP's (1981) and UNSCEAR's (1982) equal apportionment of risk between tracheobronchiolar and pulmonary doses. It is seen that the estimated value of K is not sensitive to the precise definition of the reference dose (i.e., to whether or not this is defined by the physical dose absorbed by target cells in bronchial epithelium or by an effective dose which also takes account of the doses received by target cells in the bronchioles and alveoli).

The calculated K-factor is highest for 1-y-old children (Figure 6). These values are approximately 40% higher than the K-factors calculated for adults. However, by the age of 5 y, the K-factor approaches the adult values. Therefore, provided that young children are not intrinsically more sensitive than adults to the induction of lung cancer, the effect of a higher K-factor in the first few years of life on the lung-cancer risk from exposure over the whole lifetime is trivial. According to the BEIR-IV risk projection model, in which the lung-cancer risk from exposure to radon progeny diminishes with time since exposure, the effect of a higher dose per unit exposure in childhood is further reduced.

It is seen from Figure 6 that the K-factor depends rather critically on the assumptions made about the hygroscopicity of the attached radon-progeny aerosols in mines and homes. On the assumption that attached radon-progeny aerosol particles in both environments do not grow in size within the respiratory tract, the *K*-factor applicable to the bulk of a lifetime's exposure is approximately unity for both men and women. However, if it is assumed that the attached radon-progeny aerosol in both mines and homes is hygroscopic (and doubles in size within the respiratory tract), then the *K*-factor becomes 0.6. In the absence of more specific information on the hygroscopic behavior of radon-progeny aerosols, it can be argued that the best current estimate of the *K*-factor for exposure in the home would be an intermediate, rounded value of 0.8.

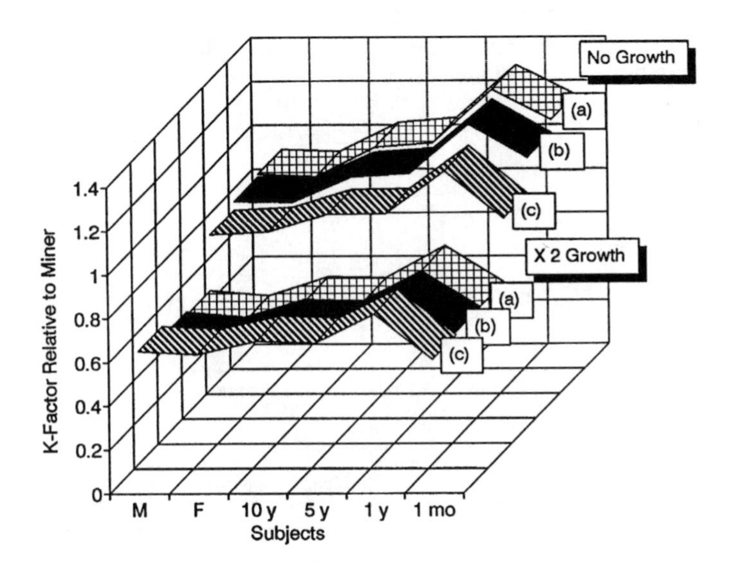

Figure 6. Values of dosimetric risk extrapolation factor, *K*, relative to healthy uranium miner, calculated for subjects exposed to radon progeny at home. Figure shows effect on calculated *K*-factor of assumed hygroscopic growth of radon-progeny aerosol particles in respiratory tract. *K*-factor is derived in several ways: (a) in terms of mean dose received by bronchial target cells; (b) effective dose, according to ICRP Lung Modeling Task Group's proposed tissue weighting factors; (c) effective dose according to UNSCEAR (1982) tissue weighting factors.

Effect on the *K*-Factor of Concomitant Exposure to Uranium Ore Dust

Since uranium-ore particles emit alpha particles, the possibility must be considered that the uranium miners included in the epidemiological studies may have received a significant bronchial dose from this source in addition to that from radon progeny. Harley and Fisenne (1985) assessed the average airborne concentration of ^{238}U in a "dusty" uranium mine atmosphere to be typically 6.6 dpm m^{-3} (0.7 α-Bq m^{-3}), with a likely maximum value of 22 dpm m^{-3} (3.6 α-Bq m^{-3}). Table 1 compares Harley and Fisenne's estimates of the annual dose to target cells in the bronchial epithelium of a uranium miner from exposure to ore dust at these concentrations with values derived from the ICRP Task Group's proposed lung dosimetry model (James et al., 1991a; James and Birchall, 1991).

Table 1. Estimates of annual dose to bronchial epithelium from exposure to uranium ore and equivalent annual exposure to ^{222}Rn progeny.

	Annual Dose ($mGy\ y^{-1}$)		Equivalent Annual Exposure[a] ($WLM\ y^{-1}$)	
Model	Average	Maximum	Average	Maximum
Harley and Fisenne (1985)	1	5 100[b]	0.06	0.4 6[b]
ICRP Task Group[c]				
Fast bronchial clearance[d]	0.3	1.4	0.02	0.09
Partially slow bronchial clearance[e]	40	190	2.5	13

[a] Based on 16 mGy WLM^{-1} derived from NRC (1991) dosimetry model.
[b] Assuming 0.1 mg of 1%-U ore were to reside for 1 y on 1 cm^2 of epithelial surface.
[c] Reference, James et al. (1991a); James and Birchall (1991).
[d] Assuming that ore particles deposited in bronchial airways (or cleared into them from the bronchioles) are all cleared rapidly (with a half-time of 2 hours).
[e] Assuming that 20% of ore particles deposited in bronchial airways (or cleared into them from the bronchioles) is retained in bronchial mucus (to be cleared with a half-time of 20 days).

Harley and Fisenne (1985) examined the possibility that a small fraction of deposited ore particles may be retained for a long time on the surface of bronchial epithelium; they considered the example of 0.1 mg of 1% uranium ore retained for 1 y on a 1-cm^2 area of epithelial surface. In this case, the annual dose to target basal cells was estimated to be on the order of 100 mGy y^{-1}, corresponding to 6 WLM y^{-1} exposure to radon progeny. The ICRP Task Group also considers that not all particles deposited in the bronchial airways are removed rapidly by bronchial mucus (as described by Bailey et al.,

1991). For dosimetry purposes, the Task Group proposes the cautious assumption that 20% of material deposited directly in the bronchi, or entering the bronchi from distal airways, is retained with a half-time of 20 days. In this case, the estimated average exposure in a dusty, high-grade uranium mine (i.e., 0.7 α-Bq m^{-3}, corresponding to an annual intake of about 2400 α-Bq y^{-1} at a mean breathing rate of 1.7 m^3 h^{-1}) would be equivalent to an annual radon-progeny exposure of approximately 2.5 WLM y^{-1}. However, this value is substantially less than the actual radon-progeny exposure rate for the lowest-exposed uranium miners found to have excess lung cancers, i.e., approximately 10 WLM y^{-1} for the medium exposure group of Eldorado Beaverlodge uranium miners (NRC, 1988). It can be concluded that exposure of uranium miners to ore dust has made a negligible contribution to the observed excess incidence of lung cancers among the epidemiological study groups. Therefore, it is appropriate to base the projection of lung-cancer risks from underground miners to home dwellers on the comparative doses from radon-progeny exposure alone.

Effect on the *K*-Factor of Cigarette Smoke in the Home

Figure 7 compares the *K*-factors calculated for subjects exposed in homes in which the atmosphere is not contaminated with cigarette smoke, with the values estimated for homes occupied by cigarette-smokers. The AMTD of the attached radon-progeny aerosol in the "smoking" home is assumed to be 0.25 μm, and the unattached fraction of potential alpha-energy, f_p, is assumed to average 3% (NRC, 1991). For night-time exposure (assumed to be 50% of the total exposure time indoors), when each subject is assumed to be asleep, the radon-progeny aerosol is taken to be the same as that in a "non-smoking" living room, i.e., that f_p is 8% and the attached aerosol AMTD is 0.15 μm. Figure 7 shows that the estimated *K*-factor is lower for subjects exposed in a smoking home compared to the values calculated for nonsmoking homes, by approximately 20%. However, in reality, the implied lower relative risk from radon exposure in a cigarette-smoke atmosphere may not reduce the overall lifetime risks of lung cancer for subjects who do not themselves smoke. The lower relative risk coefficient may well apply to a higher baseline lung-cancer risk from passive exposure to cigarette smoke.

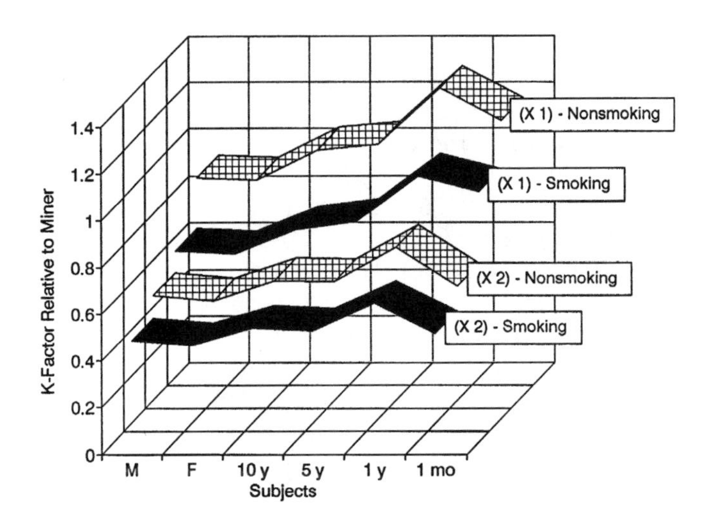

Figure 7. Comparison of *K*-factors calculated for subjects exposed in smoking and nonsmoking households, showing effects of assumed hygroscopic growth of radon-progeny aerosol particles in respiratory tract.

LIFETIME LUNG-CANCER RISKS INDOORS, DERIVED FROM THE BEIR-IV MODEL

According to the BEIR-IV Committee's risk projection model, the lifetime excess risks of lung cancer in males and females attributable to lifetime exposure at the EPA Action Level of 4 pCi L^{-1} (150 Bq m^{-3}) are reduced approximately in proportion to the *K*-factors calculated for adults. For the baseline lung cancer risks applicable to the U.S. public (with $K = 0.8$ for adults), the projected values range from 0.2% for female nonsmokers to 3.5% for male smokers. The overall risks for the combined populations of smokers and nonsmokers exposed at the EPA Action Level are estimated to be about 0.8% for females and 2% for males.

COMPARISON OF DOSIMETRIC AND EPIDEMIOLOGY-BASED ESTIMATES OF LUNG-CANCER RISKS

Underground Miners

The risk projection model developed by the BEIR-IV Committee from their combined analysis of four major study groups of radon-exposed underground miners gives the following estimates of lifetime excess risk of lung cancer from exposure to 100 WLM of radon-progeny potential alpha-energy over a working life (NRC, 1988):

Smokers	8.8%
Nonsmokers	0.9%
Population average	5.2%.

The population average value of 5.2% is to be compared with the lifetime risks implied by the calculated effective dose, together with ICRP's recommended risk factor of 4% per sievert (ICRP, 1991; UNSCEAR, 1988). Based on ICRP's (1981) and UNSCEAR's (1982) recommendation to assign half the overall weighting factor for the lungs to bronchial-bronchiolar dose, the total effective dose received from 100 WLM exposure is 1.7 Sv (i.e., 17 mSv WLM^{-1}). The implied lifetime excess risk is therefore 4 x 1.7 = 6.8%. This value is similar to the BEIR-IV Committee's epidemiology-based estimate. However, when proper account is taken of the higher relative sensitivity of bronchial and bronchiolar epithelia compared with alveolar tissue (Bair, 1991), the dose conversion factor is increased to about 35 mSv WLM^{-1}, and the implied lifetime excess risk to about 14% from exposure over a working life to 100 WLM. The value of 14% is almost threefold higher than the estimate based directly on the epidemiological data.

Indoor Exposure

Jacobi (1989) has compared estimates of the relative lifetime risk of lung cancer from chronic indoor exposure to radon progeny that are projected from lung dosimetry and ICRP's risk factors with those projected directly from the underground miners. Table 2 compares values of the relative lifetime risk of lung cancer that are projected for the U.S. population using both the dosimetric and epidemiological approaches (i.e., the lifetime probability ratios for bronchial cancer induced by radon-progeny exposure compared to lung cancer from all causes). According to ICRP's (1987) proportional hazard model, adjusted for the estimated K-factor of 0.8, the average exposure to radon progeny in the U.S. (approximately 0.27 WLM y^{-1}) is expected to increase the lung-cancer risks for both nonsmokers and smokers by 22%. The comparable value given by the BEIR-IV model is approximately 11% (Lubin and Boice, 1989) after correction for the estimated K-factor of 0.8. This is less than half the estimate based on the ICRP model, primarily because of the time since exposure factors incorporated in the BEIR-IV model.

Table 2. Projected[a] excess lifetime risk of lung cancer from chronic indoor exposure to 25 Bq m^{-3} equilibrium equivalent concentration of radon progeny[b].

Risk Projection Model	Relative Lifetime Risk of Risk of Lung Cancer
From Rn-exposed Miners	
ICRP 50 (1987)[c]	0.22
BEIR IV (1988)[d]	0.11
From Atomic-Bomb Survivors/Lung Dosimetry	
$W_{T-B} = W_P = 0.06$[e]	0.25
$W_{bronchi} = 0.096$	
$W_{bronchioles} = 0.018$	0.53
$W_{alveolar\text{-}interstitial} = 0.006$[f]	

[a] Average for U.S. males and females; adapted from Jacobi (1989) and Lubin and Boice (1989). Based on U.S. population average lung-cancer rates of 6.7% for males and 2.5% for females.
[b] 25 Bq m^{-3} EER_n is equivalent to 6.7 mWL and 0.27 WLM y^{-1} (assuming 7000 h y^{-1} occupancy). These average values apply to single-family dwellings (approximately 70% of the U.S. population). Concentrations and exposures for 30% of the population living in multi-unit buildings are likely to be lower (Nero et al., 1990).
[c] Multiplicative model; ICRP (1987) assumption that risk per unit exposure in home is 64% of that in mine, modified so that $K = 0.8$.
[d] Multiplicative model; NRC (1988) assumption of equal risk per unit exposure in home and mine, modified so that $K = 0.8$.
[e] Effective dose per unit exposure taken to be 12 mSv WLM^{-1} based on NRC (1991) dosimetry model and UNSCEAR (1982) apportionment of lung tissue risks.
[f] Effective dose per unit exposure taken to be 25 mSv WLM^{-1} based on NRC (1991) dosimetry model and ICRP Task Group's proposed apportionment of lung tissue risks (Bair, 1991).

According to NRC's (1991) dosimetry, if lung-tissue risk is apportioned equally between the tracheobronchial and pulmonary regions, the probability ratio of radon-induced bronchial cancer is more than twofold higher than the BEIR-IV Committee's radon-epidemiology-based projection. Taking the more realistic apportionment of risk primarily to bronchial epithelium that is proposed by the ICRP Task Group, the dosimetry-based probability ratio is almost fivefold higher than that supported by the BEIR-IV Committee's analysis of the epidemiological data. Thus, a correction factor of approximately 0.2 is needed to make the dosimetric risk estimate congruent with the epidemiological data.

Annual Number of Lung-Cancer Deaths Caused by Radon

Let us assume that (1) the 11% probability ratio for radon-induced bronchial cancer compared to lung cancers from all causes, which is estimated for the 70% of the U.S. population who live in single-family dwellings (Table 2), also applies to (2) families living in apartments. We then predict that approximately 10,000 lung cancers in men and 4500 lung cancers in women are caused by indoor exposure to radon in the United States. Lubin and Boice (1989) estimated a 95% confidence interval in the probability ratio for radon-induced lung cancer from an analysis of the BEIR-IV Committee's radon risk coefficient and projection model, and the uncertainty in the exposure distribution for the U.S. population. Their estimated uncertainty bounds correspond to a 95% confidence interval in the probability ratio of 6% to 20%. The corresponding bounds in the estimated total number of radon-induced lung-cancer deaths are 8000 to 27,000 in the United States each year.

COMPARATIVE DOSIMETRY FOR THORON PROGENY

The aerosol size information available to characterize exposures to thoron progeny in mines and in homes is even sparser than that for radon progeny. However, for underground mine atmospheres, NRC (1991) estimated that the unattached fraction of thoron-progeny potential alpha-energy, f_p, is typically on the order of 0.1% during active mining, and 1% during less dusty operations. For thoron-progeny exposure in homes, the NRC panel assumed that f_p is typically 2%. In all environments, the AMTD of the attached thoron-progeny aerosol is likely to be at least as large as that of radon progeny, and NRC (1991) assumed a typical value of 0.25 μm.

Figure 8 shows the DCF calculated using NRC's (1991) dosimetry model for thoron-progeny exposures in mines and homes. The conversion factor is expressed here as the effective dose per unit exposure, assuming (a) that attached thoron-progeny aerosol particles are not hygroscopic, or (b) that these particles double their ambient size on entering the respiratory tract. The figures show separately the contributions to effective dose made by irradiation of the respiratory tract and other body organs. To calculate doses to body organs, we have assumed the metabolic and dosimetric models for ^{212}Pb and ^{212}Bi described by ICRP (1980), in conjunction with an absorption half-time from the lung for thoron progeny of 10 h (Jacobi et al., 1957; Booker et al., 1969; Hursh and Mercer, 1970; Bianco et al., 1974). We have also assumed the tissue weighting factors recommended by ICRP (1991) in conjunction with the lung-tissue risk apportionment proposed by the ICRP Task Group.

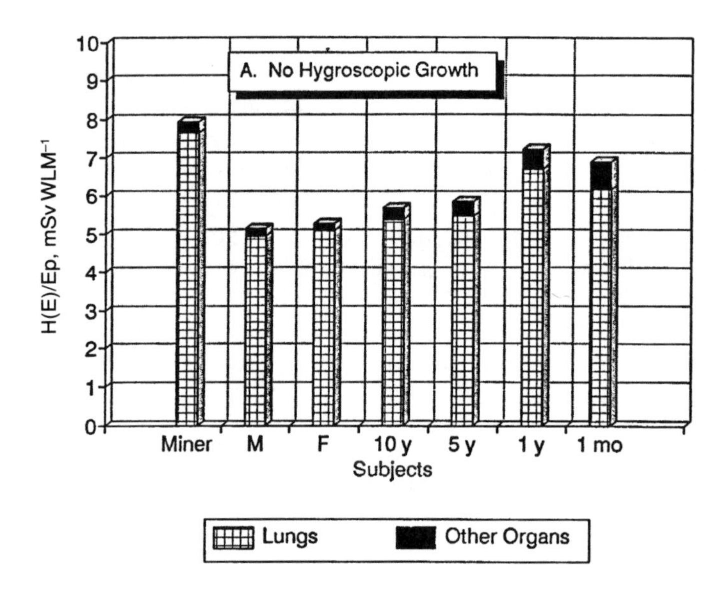

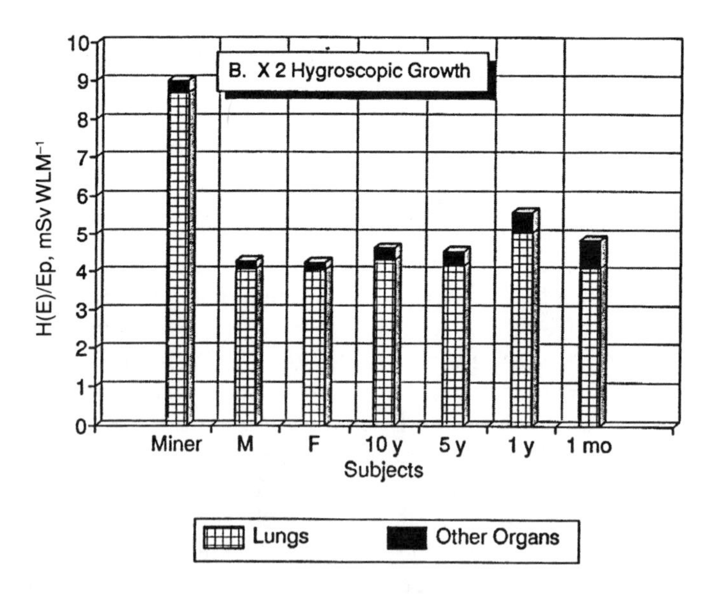

Figures 8 A and B. Effective dose per unit exposure to thoron progeny calculated for underground miner and for subjects exposed in home. Fractional contributions to effective dose made by exposure of lungs and other organs are shown separately, as are effects of assumed hygroscopic growth of thoron-progeny aerosol particles in the respiratory tract.

It is seen from Figure 8 that only a small fraction of the effective dose is calculated to arise from body organs other than the lungs (i.e., approximately 4% in adults, rising to about 10% in very young children). The assumption of particle growth by a factor of two in the respiratory tract has the effect of increasing the DCF calculated for a miner (by about 13%) but decreasing those calculated for subjects exposed in the home (by approximately 20% for adults and children, and 30% for a 1-mo-old infant). Taking the average values, the calculated DCF are, respectively, 8.5 mSv WLM^{-1} for a miner; approximately 5 mSv WLM^{-1} for adults and children through age 5 y; 6.4 mSv WLM^{-1} for a 1-y-old child; and approximately 6 mSv WLM^{-1} for a 1-mo-old infant. These values are to be compared with estimates of 35 mSv WLM^{-1} for a miner exposed to radon progeny, 25 mSv WLM^{-1} for adults exposed in the home, 28 mSv WLM^{-1} for 10- and 5-y old children, 34 mSv WLM^{-1} for a 1-y-old, and 29 mSv WLM^{-1} for a 1-mo-old infant. The resulting *K*-factors for exposures to thoron progeny relative to the radon-progeny dose per unit exposure for a uranium miner are shown in Figure 9. For underground miners, the thoron-progeny *K*-factor is found to be approximately 0.25, irrespective of the assumptions made about hygroscopicity of attached radon- and thoron-progeny particles in the respiratory tract. For adults and children through 5 y exposed to thoron progeny in the home, the *K*-factor is estimated to be approximately 0.15, with somewhat higher values in younger children. The assumption that attached thoron-progeny particles are hygroscopic reduces the *K*-factors calculated for exposure in the home by 30% to 40%.

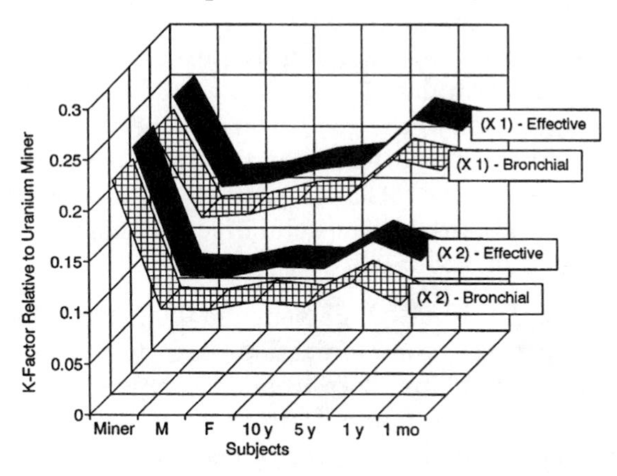

Figure 9. *K*-factors calculated for underground miner and for subjects exposed in home to thoron progeny relative to dose conversion factors calculated for miners exposed to radon progeny.

LIFETIME LUNG-CANCER RISKS FROM INDOOR EXPOSURE TO THORON PROGENY

The calculated *K*-factors for indoor exposure to thoron progeny can be applied to estimate the lifetime lung-cancer risks from the values projected earlier for exposure to radon progeny. If we assume that the average indoor concentration of thoron progeny in U.S. homes is 3.4 mWL (Nero et al., 1990), then for 80% occupancy the average annual exposure is 0.14 WLM y^{-1}. The corresponding estimate of the lifetime excess risk of lung cancer in males attributable to thoron progeny is approximately 0.15%, and that for females approximately 0.06%. Thus, on the average, exposure to thoron progeny is estimated to cause on the order of one-tenth the number of lung cancers in the U.S. population that are caused by radon progeny, i.e., the estimated number of bronchial cancers caused in the United States each year by exposure to thoron progeny is, very approximately, 1000.

DOSES TO OTHER ORGANS OF THE BODY

Peterman and Perkins (1988) developed a kinetic model of the tissue uptake and washout of inert gases from the human body, which they showed to be consistent with the measured washout of ^{133}Xe. Tissue concentrations are determined by a combination of the blood flow through the volume of a tissue and the partition coefficient for the inert gas between blood and the tissue. To model the uptake and retention of radon and thoron in human tissues, Peterman and Perkins used the partition coefficients derived by Bernard and Snyder (1975) from Nussbaum and Hursh's (1957) classic study of radon solubility in tissues of rats exposed to the gas for various periods. The solubility coefficient for radon in blood was found to be 0.43. Values for body tissues ranged from 4.8 for fat, to about 0.15 for the skeleton (minus marrow), the skin and muscle. The solubility coefficient for most other tissues was found to be about 0.3.

Thrall et al. (K. D. Thrall, A. C. James, and T. E. Hui, Pacific Northwest Laboratory, personal communication) have tested Peterman and Perkins' kinetic model against data from exposures of human subjects by inhalation. On applying appropriate physiological factors to allow for the subjects' body weights (degree of obesity) and ventilation and perfusion rates (minute volume and blood flow through tissues appropriate for the level of physical exertion), the physiologically based model is found to predict closely the human data and will also predict a larger body of experimental animal data. We have used Peterman and Perkins' model here, in conjunction with new biokinetic models for lead and bismuth in the lung and body organs. The latter were developed from a review of the experimental animal and human data (K. D. Thrall, Pacific Northwest Laboratory, personal communication), in order to calculate the doses to body organs from exposure to airborne radon, thoron, and their short-lived progeny, which are discussed later.

Red Bone Marrow

Richardson et al. (1991) proposed that uptake of radon gas by fat cells in the red bone marrow is a key factor in irradiating the hematopoietic stem cells. This finding hinges on the high solubility of radon in fat that is strongly indicated by the work of Nussbaum and Hursh (1957; 1958); i.e., on the value of the fat/blood partition coefficient, which is on the order of 10. Table 3 compares Richardson et al.'s calculated doses to stem cells in bone marrow with the values calculated here by applying improved biokinetic models for the lead and bismuth progeny.

Table 3. Estimates of annual average equivalent dose to active bone marrow from exposure of U.S. population to radon (^{222}Rn), thoron (^{220}Rn), and their short-lived progeny.

Annual Equivalent Dose for Indicated Subjects ($\mu Sv\ y^{-1}$)						
	Subject					
Model	Man	Woman	10 y	5 y	1 y	3 mo
Richardson et al. (1991)						
^{222}Rn	154	136	128	92	—	31
Progeny	11	20	33	55	—	26
Total	165	156	160	147	—	57
Richardson et al. (1991)						
^{220}Rn progeny	75	115	193	301	—	125
Thrall et al. (personal communication)						
^{222}Rn	145	121	126	111	89	64
Progeny	0.5	0.5	0.8	1.0	1.5	1.1
Total	145	121	127	112	90	65
(Excluding activity in fat)	(26)	(32)	(35)	(44)	(62)	(58)
Thrall et al. (personal communication)						
^{220}Rn	3	3	4	5	6	6
Progeny	11	13	21	30	53	40
Total	14	16	24	34	59	46
(Excluding activity in fat)	(5)	(6)	(8)	(11)	(18)	(15)

In Table 3 we consider adults and children exposed in the home to radon and thoron concentrations of 55 and 50 Bq m^{-3}, respectively, the estimated average values for U.S. homes. For radon, we generally confirm the doses derived by Richardson et al. (1991). However, these calculated values depend critically on the assumptions made about the fat content of active marrow, the size distribution of fat cells, and the degree of retention in the fat of ^{214}Pb, produced by the decay of ^{222}Rn and ^{218}Po (^{214}Bi is fat-soluble). The values shown in Table 3 must therefore be considered preliminary. In the absence of more specific experimental data, Richardson et al. appear to have made reasonable assumptions about the fat content and fat-cell size in active bone marrow as a function of age.

However, we find that two of the assumptions made by Richardson et al., those concerning the biokinetic behavior of radon progeny, are not well-supported. First, the available data from experimental animals do not support the assumption that all ^{214}Pb produced by decay of radon gas in body fat is retained there. The data indicate that ^{214}Pb ions are washed into the blood circulation (and redistributed to other tissues) at a rate that depends on blood flow through the fat (K. D. Thrall, personal communication). Secondly, we find that Richardson et al.'s assumption that one-third of the radon-progeny atoms that are deposited in the lung are absorbed into the blood and distributed to other tissues [based on the work of Pohl and Pohl-Rüling (1967)] is not supported by other studies. The half-time of absorption of radon progeny from the lungs is on the order of 10 h (Albert and Arnett, 1955; Jacobi et al., 1957; Hursh et al., 1969; Booker et al., 1969; Hursh and Mercer, 1970; Bianco et al., 1974). Thus, Table 3 shows that our estimates of dose to marrow stem cells (and their dependence on subject age) differ in detail from those of Richardson et al., particularly in respect to the doses contributed by exposure to radon and thoron progeny.

Bone Surfaces, Liver, and Kidney

Table 4 summarizes the annual dose rates to the bone surfaces, liver, and kidney, calculated for an adult male from exposure to the average U.S. indoor concentrations of radon and thoron and their short-lived progeny. Expressed in terms of equivalent dose, the bone surfaces are calculated to receive approximately 0.4 mSv y^{-1}, the liver 0.25 mSv y^{-1}, and the kidneys, 1.4 mSv y^{-1}. Approximately 60% of the dose to each tissue is calculated to arise from the exposure to radon gas (^{222}Rn) at the estimated average concentration of 55 Bq m^{-3}. Most of the remainder is calculated from intake of thoron progeny (principally ^{212}Pb), at the estimated average indoor concentration of 3.4 mWL.

Table 4. Estimated annual equivalent doses to tissues from the average U.S. exposures to radon, thoron, and their short-lived progeny.

	Annual Equivalent Dose to Following Tissues (μSv y^{-1})		
Source	Bone Surfaces	Liver	Kidneys
^{222}Rn	230	130	730
Progeny	7	9	57
Total	240	140	790
^{220}Rn	9	10	42
Progeny	170	91	570
Total	180	100	620
Grand Total	420	240	1400

Summary of Tissue Doses and Risks

Table 5 shows the effective doses (as defined by ICRP, 1991) calculated for the lungs and other body organs from the average combined exposure to radon and thoron and their short-lived progeny in U.S. homes. The table also shows the implied contributions to lifetime cancer risk. Lung cancer is expected to dominate the additional risk of cancer from indoor radon and thoron, contributing approximately 98% of the lifetime excess risk. The bulk of these lung cancers (approximately 90%) are expected to be caused by the radon exposure. On the basis of the preliminary dosimetry of active bone marrow applied here (and the recommended ICRP risk estimate), approximately 1 of 80 of the additional cancers caused by indoor radon and thoron is expected to be a leukemia.

Table 5. Estimated contributions of body organs to effective dose and lifetime risks from the average U.S. exposures to radon, thoron, and their short-lived progeny.

Tissue	Annual Effective Dose[a] ($\mu Sv\ y^{-1}$)	Lifetime Cancer Risk	
		Total (%)	% Due to ^{222}Rn
Lungs	1490[b]	0.52	91
Red Bone Marrow	19	0.007	91
Bone Surfaces	4	0.001	57
Liver	12	0.004	58
Kidneys	(35)	(0.012)	(56)
All Tissues	1525	0.53	89
Lungs Only (%)	98	98	—

[a] Calculated using the tissue weighting factors recommended by ICRP (1991).
[b] Calculated values of the annual effective dose from exposure to radon and thoron progeny are multiplied by a correction factor of 0.2 to make the implied lifetime risks consistent with the epidemiological data from underground miners.
[c] Kidneys are not specified by ICRP as an organ at risk. They are here assigned a nominal tissue weighting factor of 0.025 for the most highly irradiated organ other than the lung. According to ICRP's (1991) recommendations, the kidney dose would not be included in the assessment of overall risk.

CONCLUSIONS

Summarized below are: (1) the principal implications of current dosimetric extrapolations of radon-induced lung-cancer risk from mines to homes; (2) the comparative status of epidemiologically based and dosimetric estimates of radon lung-cancer risks; and (3) current estimates of cancer risks for other organs of the body.

- Based on the modeling study carried out by NRC (1991), a "best estimate" of the bronchial DCF for uranium miners of the 1950s and '60s is approximately 16 mGy WLM^{-1}. The comparable value of the DCF for exposure in the home is approximately 12 mGy WLM^{-1}. The ratio of these DCF, which is termed the *K*-factor, can be used with the BEIR-IV Committee's risk model to project the lifetime risks of lung cancer from exposure in the home. A rounded value for this dosimetric risk extrapolation factor of 0.8 is derived for this purpose.
- The calculated value of *K* is sensitive to the assumptions made about hygroscopic growth of radon-progeny aerosols within the respiratory tract. Uncertainty in this factor alone results in a possible range of *K* from approximately 0.6 to unity. Further uncertainty is introduced by a lack of comprehensive data needed to define typical

radon-progeny unattached fractions and activity-size distributions in mine and home atmospheres, and their variability.

- Based on an estimated average indoor radon concentration of 55 Bq m^{-3} (which corresponds to an annual exposure to radon progeny of about 0.27 WLM y^{-1}), and an estimated *K*-factor of 0.8, the projected lifetime excess lung-cancer risks are about 0.7% for the U.S. male population and 0.3% for the female population. The radon-related excess risk may be up to twofold higher for smokers but only one-fifth of the total population values for nonsmokers.
- The effective dose to the lungs as a whole is calculated by applying the 80% apportionment of the tissue weighting factor to bronchial epithelium, 15% to bronchiolar epithelium, and 5% to alveolar epithelium proposed by the ICRP Task Group on Human Respiratory Tract Models for Radiological Protection. The DCF is then estimated for exposure to radon progeny in underground mines as 35 mSv WLM^{-1} and that for exposure in homes is 25 mSv WLM^{-1}.
- According to ICRP's (1991) risk factors, the implied lifetime risk of lung cancer in miners attributable to radon is 1.4×10^{-3} per WLM of exposure. However, according to the BEIR-IV Committee's model, the equivalent risk projected for a normal mixed population of smoking and nonsmoking miners is approximately 5×10^{-4} per WLM of exposure. This is approximately threefold lower than the estimate of radon risk based on effective dose.
- For the U.S. population as a whole, and according to the calculated effective dose, the probability of bronchial cancer from radon progeny is a factor of 0.53 (53%) of that for lung cancer from all causes. However, according to the BEIR-IV Committee's risk projection model, corrected by the estimated *K*-factor of 0.8, the probability ratio is 0.11 (11%). This is approximately fivefold lower than the probability ratio implied by the effective dose.
- To bring the lung-cancer risk for the general public that is implied by the calculated effective dose from indoor exposures to radon and thoron progeny into congruence with the risk projected from the epidemiological studies of underground miners, it is necessary to apply a provisional "risk-normalization" factor of approximately 0.2.
- Dosimetry of the respiratory tract and other body organs implies that, for exposure to the same amount of potential alpha-energy, the lifetime cancer risk from airborne thoron progeny in a mine is

approximately one-quarter of that from radon progeny. For indoor exposure, the projected lifetime cancer risk per unit exposure to thoron progeny is approximately one-sixth of that from radon progeny.

- According to current dosimetric models and risk factors recommended by the ICRP, the projected lifetime risks for cancers in tissues other than the lungs from indoor exposure to radon and thoron amount to only a few percent of the bronchial-cancer risk. Of these other cancers, the highest risks are estimated to be those of kidney cancer and leukemia. However, the leukemia risk is currently estimated to be only about one-eightieth of that for bronchial cancer.

REFERENCES

Albert, RE and LC Arnett. **1955**. Clearance of radioactive dust from the human lung. AMA Arch Ind Health 12:99-106.

Bailey, MR, A Birchall, RG Cuddihy, AC James, and M Roy. **1991**. Respiratory tract clearance model for dosimetry and bioassay of inhaled radionuclides. Radiat Protect Dosim 38:179-188.

Bair, WJ. **1991**. Overview of ICRP respiratory tract model. Radiat Protect Dosim 38:147-152.

Bernard, SR and WS Snyder. **1975**. *Metabolic Models for Estimation of Internal Radiation Exposure Received by Human Subjects from the Inhalation of Noble Gases*, ORNL Report 5046 C.1, pp. 197-204. Oak Ridge National Laboratory, Oak Ridge, TN.

Bianco, A, FR Gibb, and PE Morrow. **1974**. Inhalation study of a submicron size lead-212 aerosol, pp. 1214-1219. In: *Proceedings of the 3rd International Congress of the International Radiological Protection Association,* CONF-730907, U.S. Atomic Energy Commission, Washington, DC.

Booker, DV, AC Chamberlain, D Newton, and ANB Stott. **1969**. Uptake of radioactive lead following inhalation and ingestion. Br J Radiol 42:457-466.

Cheng, YS, DL Swift, YF Su, and HC Yeh. **1989**. Deposition of radon progeny in human head airways. In: *Proceedings of the DOE Technical Exchange Meeting on Assessing Indoor Radon Health Risks,* September 18-19, 1989, Grand Junction, CO. CONF-8909190, NTIS, Springfield, VA.

George, AC and AJ Breslin. **1969**. Deposition of radon daughters in humans exposed to uranium mine atmospheres. Health Phys 17:115-124.

Harley, NH and BS Pasternack. **1982**. Environmental radon daughter alpha dose factors in a five-lobed human lung. Health Phys 42:789-799.

Harley, NH and IM Fisenne. **1985**. Alpha dose from long lived emitters in underground uranium mines, pp. 518-522. In *Occupational Radiation Safety* *in Mining*, Volume 2, H. Stocker (ed.). Canadian Nuclear Association, Toronto, Canada.

Hursh, JB, A Schraub, EL Sattler, and HP Hofmann. **1969**. Fate of ^{212}Pb inhaled by human subjects. Health Phys 16:257-267.

Hursh, JB and TT Mercer. **1970**. Measurement of ^{212}Pb loss rate from human lungs. J Appl Physiol 28:268-274.

ICRP. 1975. *Report of the Task Group on Reference Man*, Publication 23. Pergamon, Oxford.

ICRP. 1977. *Recommendations of the International Commission on Radiological Protection*, ICRP Publication 26. Ann ICRP 1(3).

ICRP. 1980. *Limits for Intakes of Radionuclides by Workers*, ICRP Publication 30, Part 2. Ann ICRP 4(3/4).

ICRP. 1981 *Limits for Inhalation of Radon Daughters by Workers*, ICRP Publication 32. Ann ICRP 6(1).

ICRP. 1987. *Lung Cancer Risk from Indoor Exposures to Radon Daughters*, ICRP Publication 50. Ann ICRP 17(1).

ICRP. 1991 *1990 Recommendations of the International Commission on Radiological Protection*, ICRP Publication 60. Ann ICRP 21(1-3).

Jacobi, W. **1989**. Risk estimates for lung cancer from inhaled radon daughters, pp. 283-288. In: *Radiation Protection—Theory and Practice*, E. P. Goldfinch (ed.), proceedings of the Fourth International Symposium of the Society of Radiological Protection, Malvern, 4-9 June, 1989. IOP Publishing Ltd., Bristol and New York.

Jacobi, W, A Aurand, and A Schraub. **1957**. The radiation exposure of the organism by inhalation of naturally occurring radioactive aerosols, pp. 310-318. In *Advances in Radiobiology.* Oliver and Boyd, Edinburgh.

Jacobi, W and K Eisfeld. **1980**. *Dose to Tissues and Effective Dose Equivalent by Inhalation of Radon-222, Radon-220 and Their Short-lived Daughters*, GSF Report S-626. Gesellschaft für Strahlen und Umweltforschung, Munich-Neuherberg, Federal Republic of Germany.

James, AC. **1988**. Lung dosimetry, pp. 259-309. In *Radon and Its Decay Products in Indoor Air*, W. W. Nazaroff and A. V. Nero (eds.). Wiley Interscience, New York.

James, AC and A Birchall. **1991** Implications of the ICRP Task Group's Proposed Lung Model for Internal Dose Assessments in the Mineral Sands Industry, pp. 201-222. In: *Minesafe International, 1990*. Proceedings of an International Conference on Occupational Health and Safety in the Minerals Industry, Perth, Western Australia, September 10-14, 1990. Chamber of Mines and Energy of Western Australia, Perth, Australia.

James, AC, JR Greenhalgh, and A Birchall. **1980**. A dosimetric model for tissues of the human respiratory tract at risk from inhaled radon and thoron daughters, pp. 1045-1048. In: *Radiation Protection. A Systematic Approach to Safety*, Vol. 2. Proceedings of the 5th Congress of IRPA, Jerusalem, March, 1980. Pergamon Press, Oxford.

James, AC, P Gehr, R Masse, RG Cuddihy, FT Cross, A Birchall, JS Durham, and JK Briant. **1991a** Dosimetry model for bronchial and extrathoracic tissues of the respiratory tract. Radiat Protect Dosim 37:221-230.

James, AC, W Stahlhofen, G Rudolf, MJ Egan, W Nixon, P Gehr, and JK Briant. **1991b** The respiratory tract deposition model proposed by the ICRP Task Group. Radiat Protect Dosim 38:159-165.

Lubin, JH. **1988**. Models for the analysis of radon-exposed populations. Yale J Biol Med 61:195-214.

Lubin, JH and JD Boice. **1989**. Estimating Rn-induced lung cancer in the United States. Health Phys 57:417-427.

NCRP. **1984**. *Evaluation of Occupational and Environmental Exposures to Radon and Radon Daughters in the United States*. National Council on Radiation Protection and Measurements, Bethesda, MD.

NEA. **1983**. *Dosimetry Aspects of Exposure to Radon and Thoron Daughter Products*. Organization for Economic Cooperation and Development, Paris. France.

Nero, AV, AJ Gadgil, WW Nazaroff, and KL Revzan. **1990**. *Indoor Radon and Decay Products: Concentrations, Causes, and Control Strategies*, Report prepared for the U.S. Department of Energy, Office of Health and Environmental Research, Washington, D.C., November 1990. DOE/ER-0480P, NTIS, Springfield, VA.

NRC. **1988**. *Report of the BEIR IV Committee: Health Risks of Radon and Other Internally Deposited Alpha-Emitters*. National Academy Press, Washington, DC.

NRC. **1991** *Comparative Dosimetry of Radon in Mines and Homes*. National Academy Press, Washington, DC.

Nussbaum, E and JB Hursh. **1957**. Radon solubility in rat tissues. Science 125:552-553.

Nussbaum, E and JB Hursh. **1958**. Radon solubility in fatty acids and triglycerides. J Phys Chem 62:81-84.

Peterman, BF and CJ Perkins. **1988**. Dynamics of radioactive chemically inert gases in the human body. Radiat Protect Dosim 22:5-12.

Pohl, E and J Pohl-Rüling. **1967**. The radiation doses received by inhalation of air containing ^{222}Rn, ^{220}Rn, ^{212}Pb (ThB) and their decay products. Ann Acad Brasil Ciên 39:393-404.

Richardson, RB, JP Eatough, and DL Henshaw. **1991**. Dose to red bone marrow from natural radon and thoron exposure. Br J Radiol 64:608-624.

Roy, M and C Courtay. **1991**. Daily activities and breathing parameters for use in respiratory tract dosimetry. Radiat Protect Dosim 35:179-186.

Swift, DL, N Montassier, PK Hopke, K Karpen-Hayes, YS Cheng, YF Su, HC Yeh, and JC Strong. **1992**. The deposition of ultrafine particles in human nasal replicate models. J Aerosol Sci 23:65-72.

UNSCEAR. **1982**. *Ionizing Radiations: Sources and Biological Effects*, Publication No. E.82.IX.8., (with annex D: Exposures to radon and thoron and their decay products). United Nations, New York.

UNSCEAR. **1988**. *Sources, Effects and Risks of Ionizing Radiation. Annex F. Radiation Carcinogenesis in Man*, Publication No. E.88.IX.7. United Nations, New York.

QUESTIONS AND ANSWERS

Q: (Speaker unidentified). I didn't quite get the answer to the earlier question. Which generation, then, was it? The seventh? Was it not to seven you averaged it?

A: Harley, New York University. It was the trachea to 6.

Q: Sterling, Simon Fraser University. I don't understand why there should be a higher lifetime risk for smokers due to radon exposure when in most if not all studies of uranium miners, the relative risk for smokers is actually less than the relative risk for nonsmokers.

A: James. The BEIR IV Committee concluded that the epidemiological data from underground miners are consistent with a multiplicative interaction between radon exposure and the baseline risk of lung cancer, although they did find that submultiplicative interaction gave the best fit to the data. The submultiplicative relationship does imply a somewhat lower relative risk for smokers than for nonsmokers, but because of their order-of-magnitude higher baseline lung-cancer rate, the excess risk from radon exposure remains much higher for smokers.

Q: Sterling. Yes, but you are talking about risk for radon exposure. We all know that cigarette-smoking is bad for you and causes lung cancer, and it is also true, according to Lundin, for the Colorado Plateau miners. I know of no study, no late study of uranium miners, in which the relative risk for radon-exposed smokers compared to control smokers is not smaller than that of exposed nonsmokers compared to control nonsmokers. I find that this is a statement that simply does not correspond to reality.

A: James. The number of lung cancers observed among nonsmoking radon-exposed miners is small, and the statistical power of the estimated relative risk from radon exposure is correspondingly low.

Q: Sterling. A model must yield to reality.

Q: Brown, Saskatchewan. Right at the end you mentioned possible variations over quality factor. What are current thoughts on how quality factor may vary, depending on what type of cell is being irradiated?

A: James. I can't give you a specific answer on that. Current experimental work in which mammalian cells in suspension culture are irradiated in vitro with alpha particles from radon and its progeny seems to be pointing to a "quality factor," Q of substantially less than 20. However, the value of Q may depend on the effect being measured as well as on the type of low-LET radiation used for reference. It will be interesting to see if comparable results are obtained with lung epithelial cells in vitro and in vivo.

Studies with Airway Casts

DEPOSITION OF UNATTACHED RaA ATOMS IN THE TRACHEOBRONCHIAL REGION

M. Shimo[1] and A. Ohashi[2]

[1]Nagoya University, Nagoya, Japan

[2]Ministry of Transport, Tokyo, Japan

Key words: *Deposition, unattached RaA, tracheobronchial region, cast model*

ABSTRACT

To accurately evaluate a human lung dose, it is important to know the fraction of inhaled unattached radon daughters deposited on the surfaces of the trachea and bronchi as well as their distribution. Our experimental results are described for determining the deposition of unattached RaA on a model of the tracheobronchial tree, based on Weibel's model of airway dimensions. The airway cast, which simulated generations 0 to 4 of the Weibel model, was placed in a glass cell connected to a chamber filled with radon gas.

After radon gas (concentration, ~150 kBq m^{-3}) passed through the model at a flow rate of 15 L min^{-1} for 20 min, the model was removed and sectioned in small pieces. Alpha activity on the inner surface of each piece was counted by a small detector with a ZnS(Tl) scintillator. "Hot spots," which had activity several times higher than the surrounding area, were especially observed on the inner side of each generation. The deposition efficiency measured at generation 0 was 2.1 to 2.7 times higher than the value calculated from the Gormley and Kennedy formula, and at generation 3 the efficiency was lower by one-half. Values at other generations agreed with those predicted theoretically.

INTRODUCTION

In accurately evaluating a human lung dose, it is very important to know the fraction of unattached and attached radon daughters deposited on the surfaces of the trachea and bronchi as well as the degree of uniformity of those depositions. Important calculations were made using a numerical model of a human lung to determine the deposition fraction of unattached radon daughters. Based on a correction factor from Martin and Jacobi (1972), Jacobi and Eisfeld (1980) estimated the deposition fraction using the Gormley and Kennedy (1949) formula. Shimo et al. (1981) also calculated and illustrated the deposition fraction in each generation for unattached radon daughters using the Gormley and Kennedy formula, but without correction. The Nuclear Energy Agency (NEA) Expert Report (1983) adopted a result calculated by James and Birchall from the Gormley and Kennedy formula, also without correction. In related work, experimental results for deposition of unattached radon daughters were obtained by Chamberlain and Dyson (1956), using a model which duplicated a human lung, and by James (1977), with a pig lung model. Because the models used in these experiments were different from those used in the numerical calculations, it is difficult to compare directly the results obtained.

In our study, experiments were performed to determine the deposition of unattached RaA on a model of the tracheobronchial tree, based on Weibel's model (Weibel, 1963) of airway dimensions. The experimental results obtained were compared with values calculated using the Gormley and Kennedy formula for laminar air flow in the same airway model.

MATERIALS AND METHODS

Human Lung Model

The two main anatomical models used for estimating the human lung dose are the Weibel and the Yeh and Schum model (Yeh and Schum, 1980). The former is represented by a cylindrical tube with symmetrical branches in two directions, and the latter is represented by a nonsymmetrical tube, which is similar to the human lung. Although the former was not similar to a human lung, it was used because a cast model of the latter was unavailable for these experiments. Branching-angle data obtained by Horsfield and Cumming (1967) were used.

First, a form was made from acrylite (acrylic plastic). Next, a model was molded from room-temperature vulcanite silicone rubber, using the acrylite form. Replicate hollow airways (Figure 1) for the experiment were duplicated in the molded model (Ohashi and Shimo, 1990).

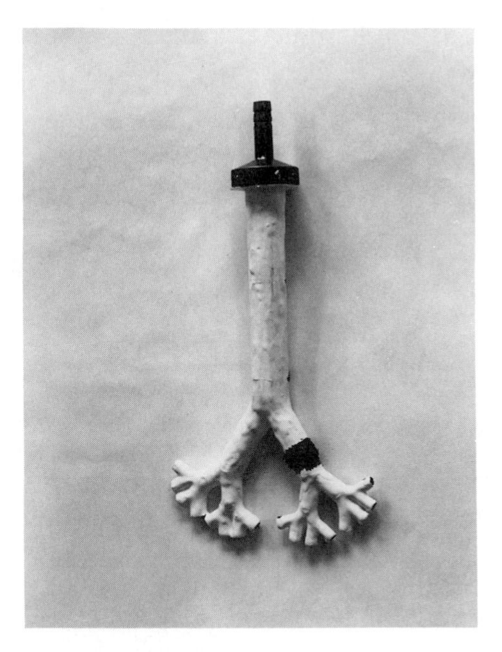

Figure 1. Human lung model molded from acrylite.

Deposition Experiment

The measuring system for the deposition experiment is shown in Figure 2. The airway model, placed in a glass cell, was connected to a chamber filled with radon gas. The radon gas passed through both the airway model and the filter, A, which determined the concentration of unattached RaA to which the lung model was exposed. The experiments were carried out in two runs under conditions indicated in Table 1. After radon gas passed through the model at a flow rate of 15 L min^{-1} for 20 min, the model was removed and sectioned, each piece having an area of about 0.41 cm^2. Alpha activity on the inner surface of each piece was counted with a ZnS(Tl) scintillator position detector (Iida, 1988). This detector, an ARGUS model made by the Hamamatsu Photonics Co., is designed to detect alpha contamination on filter paper.

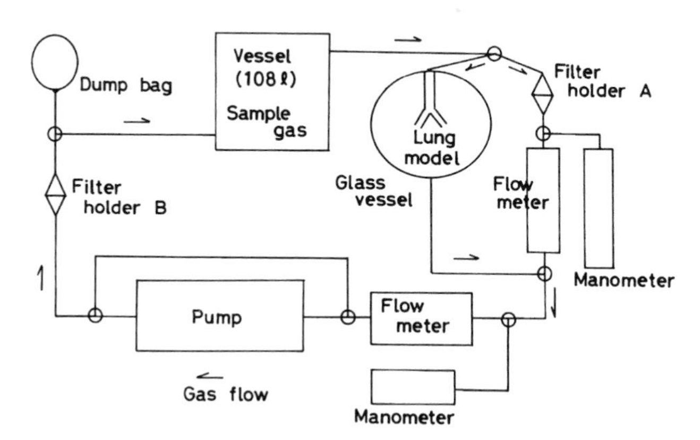

Figure 2. System for measuring deposition of inhaled radon gas.

Table 1. Experimental conditions for measuring deposition of unattached RaA fraction in human airway model.

Particle and concentration	
Rn	1.5×10^5 Bq m^{-3}
Free RaA	3000 Bq m^{-3} (2%)
Room temperature	17–22°C
Relative humidity	58–59%
Radon flow rate	15.0 L min^{-1}
Flow time	20 min
Counting area	~0.41 cm^2
Hot spot	0.104 cm^2 (3.6 mm diameter circle)
Total counting time	~30 min
Number alpha particles counted	10 to 1000

RESULTS AND DISCUSSION

Figure 3A shows the deposition fractions, in percent, of unattached RaA in each generation. About 20% of the inspired RaA atoms were deposited in generation 0, and a total of 36% were deposited in generations 0 through 4.

The surface deposition rates and the deposition within areas in each generation are shown in Figure 3B. Results suggested that the positions most susceptible to deposition are those at the inlet of generation 0 and at the inner corner areas of generations 2, 3, and 4, the so-called "hot spots" (Figure 3C). The hot spots, which had an area of about 0.104 cm^2, were sectioned for radioactive measurement

(Figure 3C). Hot spots are of special interest because they indicate where unattached particles are preferentially deposited. Hot-spot locations varied between experiments, but were generally consistent with areas where tracheal and bronchial cancers are frequently found during fiberoptic bronchoscopy (personal communication, N. D. Ito, Nagoya University, 1987).

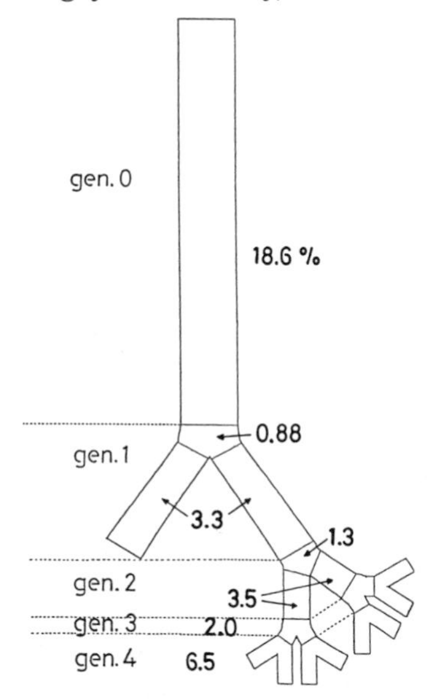

Figure 3A. Deposition fractions (percent) of inhaled atoms deposited.

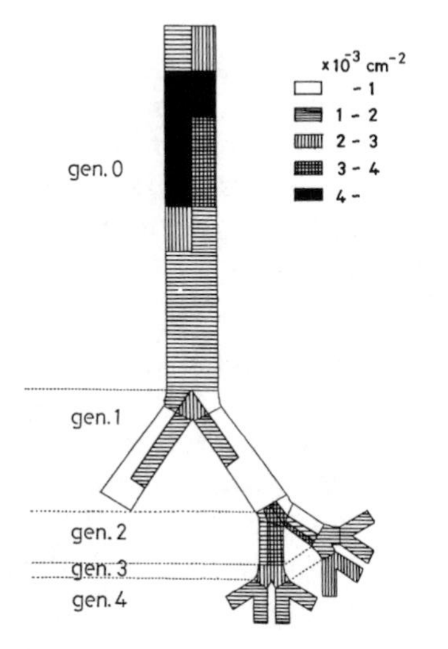

Figure 3B. Surface deposition rates and deposition of inhaled radon gas in generations 0 through 4 of human lung model.

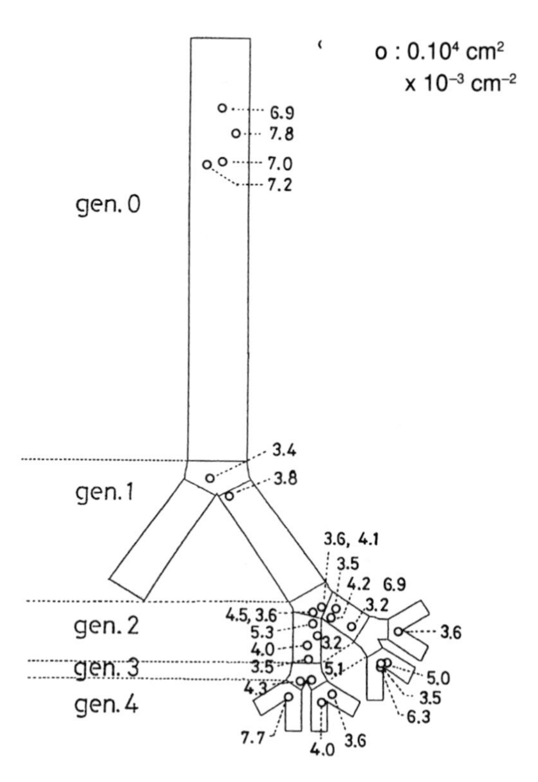

Figure 3C. Locations of hot spots in model described and surface deposition rates for inhaled radon gas.

It is difficult to forecast the location of hot spots in the trachea and bronchi by applying laminar flow theory, using the Gormley and Kennedy formula. Takahashi (1979), in a study of aerosol deposition in respiratory tracts, indicated that no ideal flow exists at the inner-corner areas of generations 2, 3 and 4. The authors, therefore, concluded that no ideal laminar flow occurred in the airway model at a flow rate of 15 L min^{-1}.

Table 2 shows the number of RaA atoms that are expected, according to these experimental results, to deposit each day in the trachea and bronchi of an adult male exposed at home to a radon concentration of 30 Bq m^{-3}, assuming 24-h occupancy, and that the airborne concentration of unattached RaA atoms is 10% of the radon concentration. This table lists the atoms deposited on the surface of the generations, the mean surface deposition, and the surface deposition at hot spots. The results show that the surface density of atoms deposited in hot spots is generally three to four times the average density.

Table 2. Estimated daily deposition of unattached RaA atoms in the trachea and bronchi of an adult male exposed indoors to a radon concentration of 30 Bq m^{-3}.

		Deposited atoms		
			Surface density, cm^{-2}	
Generation	Deposition, %	Total	Mean	Hot Spot
0	18.6	1600	24	75–85
1	4.2	360	10	38–42
2	4.8	410	20	35–60
3	2.0	170	17	48–55
4	6.5	550	17	39–85
Total	36.1	3090		

Minute volume: 7.5 L min^{-1}, Rn concentration: 30 Bq m^{-3}, activity ratio of unattached RaA to Rn: 0.1.

The calculated deposition of unattached RaA atoms in the trachea and bronchi under laminar-flow conditions, using the Gormley and Kennedy formula, is:

$$\eta = 2.56\,\mu^{2/3} - 1.2\,\mu - 0.177\,\mu^{4/3} \qquad \mu \le 0.0312 \tag{1}$$

$$\eta = 1 - 0.8191\exp(-3.657\,\eta) - 0.0975\exp(-22.3\,\eta) - 0.0325\exp(-57.0\,\eta) \qquad \eta \ge 0.0312 \tag{2}$$

$$\mu = \pi DL/q,$$

where η = deposition fraction,

D = diffusion coefficient of unattached atoms, $cm^2\ s^{-1}$,

L = length of tube, cm,

q = flow rate, $cm^3\ s^{-1}$.

The calculations were performed for two cases: with diffusion coefficients of 0.05 $cm^2\ s^{-1}$ (Busigin et al., 1981) and 0.035 $cm^2\ s^{-1}$ (Raabe, 1968). Figure 4 shows that the measured deposition in the trachea (generation 0) is 2.1 times that calculated for an assumed diffusion coefficient of 0.05 $cm^2\ s^{-1}$, and 2.7 times that for a diffusion coefficient of 0.035 $cm^2\ s^{-1}$.

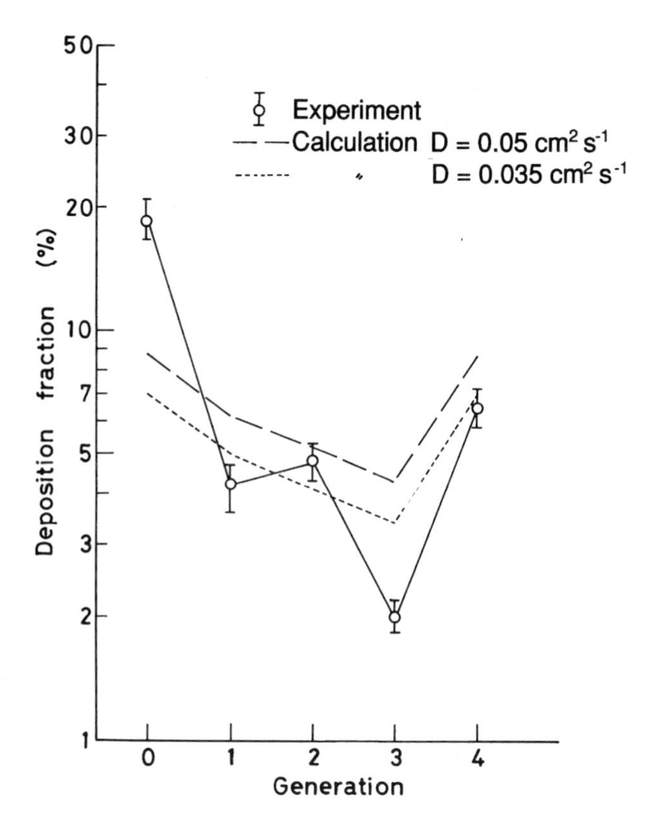

Figure 4. Comparison of experimental results with calculations based on Gormley and Kennedy (1949) formula.

The results also show that the experimental data correlated better with the values calculated using the 0.035 $cm^2\ s^{-1}$ diffusion coefficient than with those calculated using the 0.05 $cm^2\ s^{-1}$ coefficient. However, the experimental value did not match either assumed diffusion coefficient at generations 0 or 3.

Table 3 compares correction factors derived from this study with the assumptions made by James and Birchall (NEA, 1983) and Jacobi and Eisfeld (1980). Our results were closer to the assumptions made by James and Birchall, except at generations 0 and 3. Our result at generation 3 corresponded to one-half the deposition efficiency assumed by James and Birchall, and at generation 0 was 2.5 times greater.

Table 3. Correction factors for values calculated from Gormley and Kennedy (1949) formula.

Generation	Location	Ratio[a]		Correction factor		
		A[b]	B[c]	1[d]	2[e]	3[f]
0	Trachea	2.1	2.7	2.5	1	7.5
1	Main bronchi	0.7	0.8	1	1	5.5
2	Secondary bronchi	0.9	1.2	1	1	5.0
3	Secondary bronchi	0.5	0.6	0.5	1	4.0
4	Tertiary bronchi	0.8	1.0	1	1	2.0
5	Tertiary bronchi				1	1.5
All succeeding generations					1	1.0

[a] Ratio of experimental value to that calculated from Gormley and Kennedy (1949) formula
[b] A, Diffusion coefficient, 0.05 $cm^2\ s^{-1}$
[c] B, Diffusion coefficient, 0.035 $cm^2\ s^{-1}$
[d] Our work
[e] NEA, 1983
[f] Jacobi and Eisfeld, 1980

REFERENCES

Busigin, A, AW van der Vooren, and CR Phillips. **1981**. Measurement of the total and radioactive aerosol size distributions in a Canadian uranium mine. Am Ind Hyg Assoc J 42:310-314.

Chamberlain, AC and ED Dyson. **1956**. The dose to the trachea and bronchi from the decay products of radon and thoron. Br J Radiol 24:317-325.

Gormley, PG and M Kennedy. **1949**. Diffusion from a stream flowing through a cylindrical tube. Proc R Irish Acad 52A:163-169.

Horsfield, K and G Cumming. **1967**. Angles of branching and diameters of branches in the human bronchial tree. Bull Math Biophysics 29:245-259.

Iida, T. **1988**. Charged-particle imaging video monitor system. Rev Sci Instrum 59:2206-2210.

Jacobi, W and K Eisfeld. 1980. *Dose to Tissues and Effective Dose Equivalent by Inhalation of Rn-222, Rn-220 and Their Short-lived Daughters*, GSF-Report S-626. Gesellschaft für Strahlen-und-Unweltforschung mbH, München.

James, AC. **1977**. Bronchial deposition of free ions and submicron particles studied in excised lung, pp. 203-218. In: *Inhaled Particle and Vapours IV*, WH Walton (ed.). Pergamon Press, New York.

Martin, D and W Jacobi. **1972**. Diffusion deposition of small-sized particles in the bronchial tree. Health Phys 23:23-29.

NEA. **1983**. *Dosimetry Aspects of Exposure to Radon and Thoron Daughter Products*, OECD/Nuclear Energy Agency Expert Report. OECD, Paris.

Ohashi, A and M. Shimo. **1990**. Deposition of free radon daughters to respiratory tract—Experiment with tracheobronchial tree model. Hoken-Butsuri 25:23-28.

Raabe, OG. **1968**. Measurement of the diffusion coefficient of radium A. Nature 217:1143-1145.

Shimo, M, T Torn, and Y Ikebe. **1981**. Measurements of unattached and attached RaA atoms in atmosphere and their deposition in human respiratory tract. J Atomic Energy Soc (Japan) 23:851-861.

Takahashi, K. **1979**. A Study on the Deposition of Aerosol Particles in the Human Respiratory Tract, Report of Scientific Research, Ministry of Education, Culture and Science, No.249014. Kyoto University, Kyoto.

Weibel, ER. **1963**. *Morphometry of the Human Lung*. Springer-Verlag, Berlin.

Yeh, HC and GM Schum. **1980**. Models of human lung airways and their application to inhaled particle deposition. Bull Math Biol 42:461-480.

QUESTIONS AND ANSWERS

C: James. I think the results that Dr. Shimo has are consistent with the pattern that Jacobi-Eisfeld found. He didn't have a larynx, and he had a very much higher deposition in the trachea than in the first generation. The right-angle bend probably made things worse. In the experiments that we did with pig lungs at NRPB, the deposition of efficiencies in the trachea are actually decreased, when you put a larynx in, for unattached daughters.

Q: Curtis, Lawrence Berkeley Laboratory. There have been questions about hot spots from time to time over the years, and the last I heard was that people did not think that hot spots were important in this problem. I was very interested to hear your results, and I was wondering if there are other people, perhaps from the floor, who might comment on these very interesting hot spots showing up in your results, although in many of the other results they were not apparent. Is there any comment? Is this inconsistent with other data in the same field?

A: Harley. Yes, I think I could comment a bit on that. Even if there are hot spots, say, a factor of 3 or 4, that's deposition. But then, clearance would tend to dilute this effect, so for the overall dose it probably wouldn't make much difference. In Dr. Beverly Cohen's work, using casts of the hollow human lung, she never got down to this small a particle size for deposition. But she has gone down to about 20 nanometers size, and she does not find this type of hot spot. This lung has a larynx and is breathed cyclically, but there is a slight enhancement of about 20% deposition at bifurcations.

C: James. It is also interesting that in Beverly's results, there is a trend, for different particle sizes, for the enhancement factor that she measures to become less as the particle size decreases. The importance in this work is that you have not taken it all the way down to the size that we are really interested in, which is unattached daughters, and you find that the enhancement factor disappears, on average.

Q: Knutson, Department of Energy. Can you be quite sure that those spots aren't in the electrostatic effect from some part of the apparatus that is touching the cast of the lung? You can get a very well focused deposition of radium A if you have sort of an electrode or even a . . . Are you quite sure that it is not explainable by some electrostatic effect that has to do with the equipment surrounding the cast?

DESIGN, CHARACTERIZATION AND USE OF REPLICATE HUMAN UPPER AIRWAYS FOR RADON DOSIMETRY STUDIES

D. L. Swift,[1] Y.-S. Cheng,[2] Y.-F. Su,[2] and H.-C. Yeh[2]

[1]The Johns Hopkins University, Baltimore, Maryland

[2]Lovelace Inhalation Toxicology Research Institute, Albuquerque, New Mexico

Key words: *Aerosol, deposition, dosimetry, nose, radon*

ABSTRACT

The size distribution of inhaled radon progeny aerosols is a significant factor in dosimetry. The role of the airways above the trachea is an important determinant of the respiratory distribution of both attached and unattached progeny aerosols. In order to provide information on the effect of particle size and breathing conditions on the overall and local deposition, we have developed a method to produce a replicate airway model from an in vivo magnetic resonance imaging coronal scan. The model consists of a sandwich of methacrylate elements, each element having the thickness of the scan interval. The transition between successive scan outlines traced on the front and back surfaces of each element is hand-sculpted in the plastic. The hollow model of the nasal passages thus produced has been characterized both morphologically and fluid-mechanically and has a flow resistance typical of a normal adult.

The model has several distinct advantages for studies of radon progeny aerosol deposition. After exposure to a radioaerosol (or to an aerosol of an otherwise measurable substance) the individual elements can be separated to determine local deposition. The dimensions of specific upper-airway regions can be changed by replacing a small number of elements. The model has been incorporated in an exposure system for determining overall and regional deposition of aerosols whose median diameter is approximately 1.7 nm. Measurements at several flow rates are presented to demonstrate use of the model in radon dosimetry. The model should also be useful for determining the airway deposition of other environmental aerosols.

INTRODUCTION

The distribution of deposited radon progeny aerosols in the human respiratory tract is an important factor in the development of radon dosimetry and risk models. Much attention has been paid to the bronchial distribution of aerosol deposition because the bronchi are the most common site of tumors. However, the airways above the trachea may severely limit the accessibility of radon progeny aerosols to the bronchi. Also, deposition in the nasal, oral, pharyngeal, and laryngeal (NOPL) compartment may result in significant doses to these tissues.

Three approaches to understanding the deposition of aerosols in the human NOPL region have been employed in past studies: in vivo studies with human subjects, studies in experimental animals, and studies employing replicate airway physical models. While in vivo studies are the most direct route, there are practical and ethical limitations to such studies; for example, it is not possible to carry out studies in very small children; also, the use of radioactive tracers must not result in unacceptable radiation dose. Furthermore, rapid clearance in most areas of the NOPL compartment make detailed initial distribution studies difficult. Most commonly used experimental animals have NOPL geometry so different from that of humans that extrapolation from such species to man is very difficult. Replicate model studies in the past have employed postmortem casts whose dimensional characteristics may differ from those of living individuals (Itoh et al., 1985).

With the development of magnetic resonance (MR) and computerized tomography (CT) technology, the ability to visualize interior regions of a living human body has been greatly enhanced. Both methods make it possible to obtain high-resolution, planar "slice" images over a narrow width in any orientation. The MR technology is more attractive for normal individuals because no ionizing radiation exposure is necessary. We will describe the use of MR in constructing replicate NOPL airway models for radon deposition studies.

MR IMAGES AND MORPHOMETRIC CHARACTERIZATION

Our first upper-airway replicate model was made using a series of coronal nasal passage images of the first author (DLS), taken at 3-mm intervals extending from the nose tip to the posterior wall of the nasopharynx. In all, 34 images were obtained. The mucosa/air boundary could be readily identified in these images, and a series of spatially indexed, full-sized airway outlines were obtained; several are displayed in Figure 1. The perimeter and cross-sectional area of each airway image (two values where two passages exist) were obtained by a manually operated trace planimeter. From these data a plot of cross-sectional area and perimeter as a function of distance from the nasal tip was obtained (Figure 2), as well as a cumulative plot of passage volume and lateral surface area (Figure 3), obtained by multiplying each value by the slice thickness.

Figure 1. Three of a series of nasal-passage outlines from magnetic resonance coronal scan.

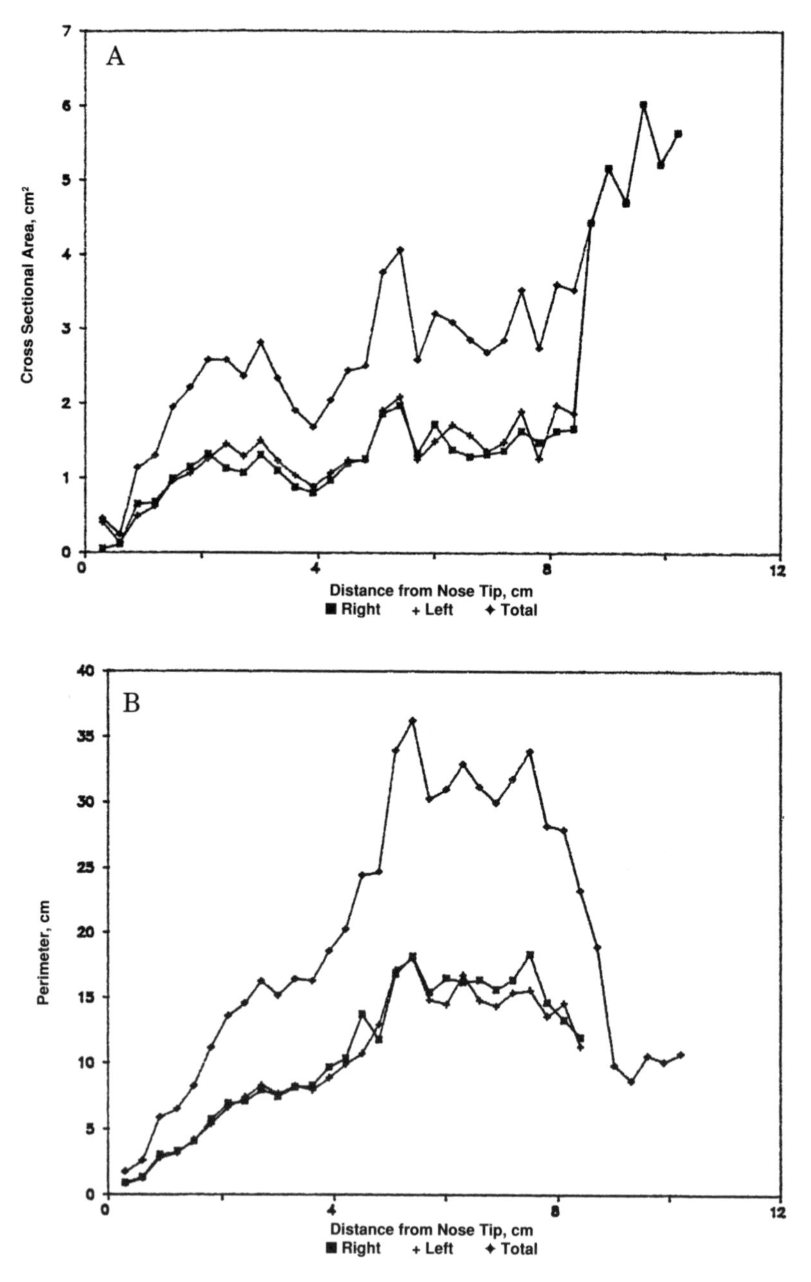

Figure 2. Plots of A, nasal passage area and B, perimeter for each magnetic resonance coronal scan.

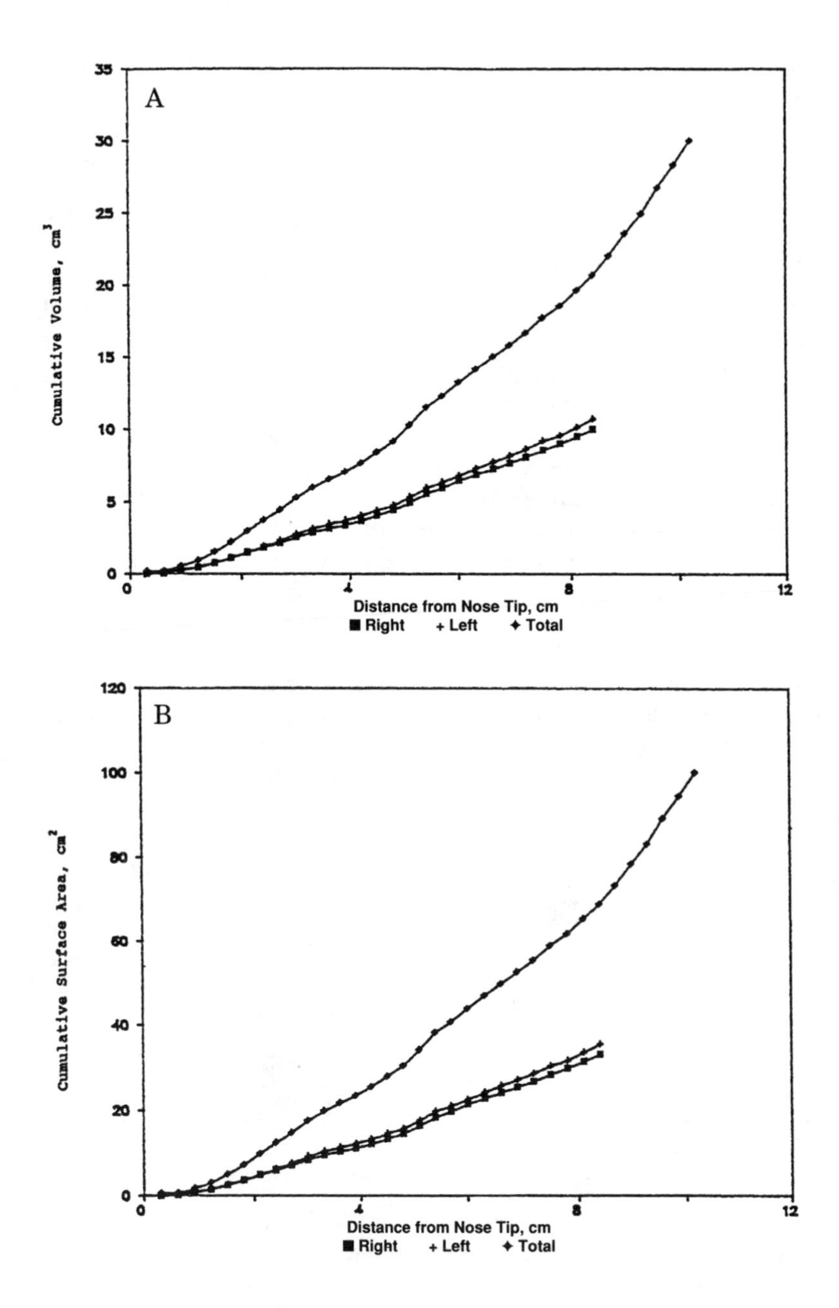

Figure 3. Plots of cumulative volume and lateral area of nasal passage, calculated from magnetic resonance scans.

This morphometric analysis is basically similar to that described by Guilmette et al. (1989), who also used MR coronal images of a human subject. Their subject had a mild degree of bilateral asymmetry, whereas our morphometric analysis (Figures 2 and 3) shows a higher degree of symmetry between the right and left passages. This was particularly marked in the anterior region, where cross-sectional areas described by Guilmette et al. differed by as much as a factor of two.

The cross-sectional areas can be used to calculate the average velocity through the nasal passage only where the flow direction is perpendicular to the coronal plane. In the anterior and far posterior regions, as discussed by Swift and Proctor (1977), the flow direction is not in the horizontal plane for the upright head, thus the cross-sectional area perpendicular to the flow vector is not in the coronal plane. To obtain such a cross-sectional area it is appropriate to use the sound reflection method described by Hilberg et al. (1989), which gives flow cross section but no indication of surface area or passage shape.

MODEL CONSTRUCTION

The replicate nasal-passage model was made from "sandwich" elements of clear methacrylate plastic, each element having a thickness equal to the slice interval of the MR images, in this case, 3 mm. The elements were 9 cm square, large enough to accommodate any section of the nasal passages. In order to make a smooth transition from one image to the next, the image plane chosen was the junction plane of two adjacent elements. The images were spatially indexed to the plastic elements, and adjacent images were drawn on the front and back of each element; only the boundary of the air/mucosa surface was drawn.

The elements were then hand-carved with a high-speed cutting tool (Dremel Model 395, Racine, WI), using a cutting bit appropriate to the passage thickness. Because the elements were transparent and both boundaries were visible from either side, it was relatively easy to carve the contours from front to rear of the element. After initial cutting, adjacent elements were aligned to check for smooth transitions, and touch-up carving was employed where transitions seemed too abrupt to be physiologically realistic.

The front elements representing the external and internal nares were cut off just below their inferior boundaries and milled flat to give a planar surface. The nares were then drilled through to the passages in elliptical cross sections. A connecting tube of 2 cm i.d. was fitted to a gasketed rectangular flange which could be clamped against the nostril plane to provide convenient flow access to the passage. The flange had a cutout large enough to permit flow access from the tube to the nares entrances. Similar milling and a flanged connecting tube were provided at the inferior end of the nasopharynx so that the replicate model could be connected to any exposure system in an airtight fashion. Two holes were drilled through the elements, which were drawn tight with wing nuts on threaded stock. Air leakage between the elements was eliminated by a thin coating of vacuum grease around the edges of the elements.

FLUID MECHANICAL CHARACTERIZATION

The model thus constructed was evaluated fluid-mechanically by measuring the pressure gradients at different constant inspiratory and expiratory flow rates. The initial measurements were made with the entire model, measuring static pressure at the entrance and exit flange locations. By removing successive elements from the rear of the model and providing an exit element with an attached tube in which static pressure could be measured, the pressure drop could be determined as a function of distance through the model. Plots of pressure drop vs. flow rate for the entire model and for truncated anterior sections during inspiratory flow are shown in Figure 4.

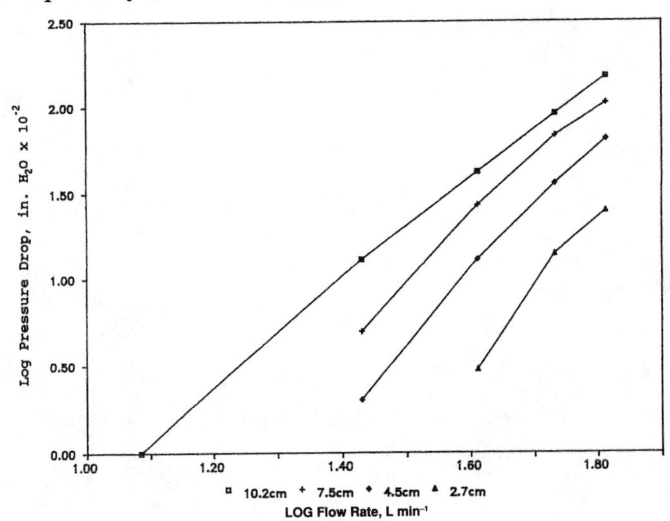

Figure 4. Plots of pressure drop vs. flow rate through nasal passage of upper-airway model.

No observations of streamline direction and turbulence were made in the current model with liquid flow at equivalent Reynolds number conditions. These could be obtained, however, provided that proper allowance was made for refractive index matching between the plastic and liquid. It would be valuable to compare flow and turbulence in this model of the in vivo nasal passages with Swift and Proctor's (1977) study which was conducted with models made from postmortem specimens. Such a comparison could reveal artifacts caused by postmorten dimensional changes.

MODEL USE IN DEPOSITION STUDIES

A photo of the MR-derived, 3-mm sandwich nasal models of subject DLS, assembled with connecting tubes, is shown in Figure 5. This replicate model can be integrated into various exposure systems so that aerosols can be passed through the model to determine the overall and/or local deposition characteristics of the particles.

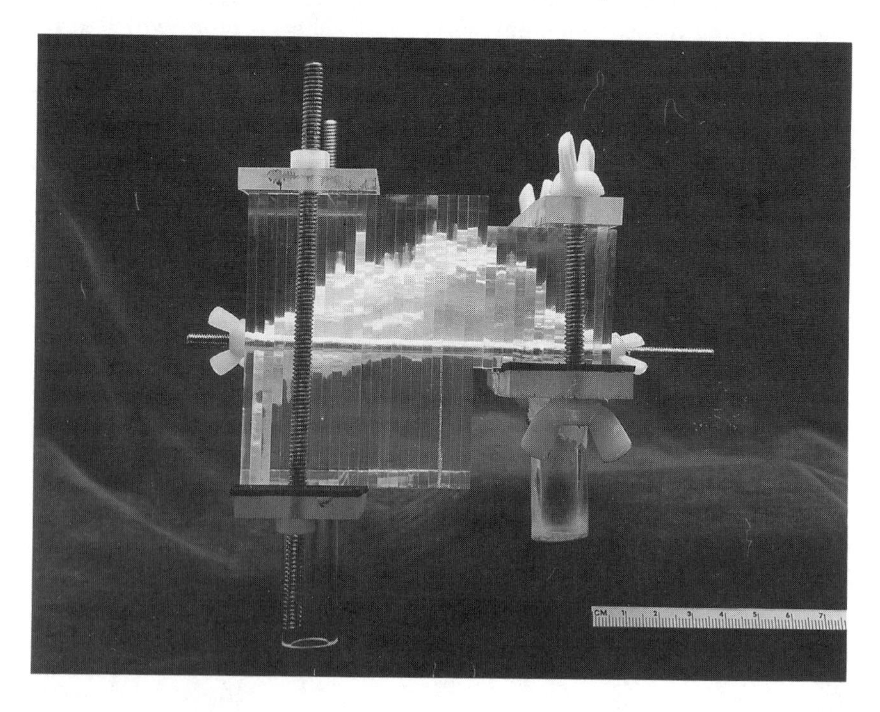

Figure 5. Assembled "sandwich" nasal model.

One such system, which has been used with cast upper-airway models by Yamada et al. (1988), is shown in Figure 6. Monodisperse aerosols ranging from 5 to 200 nm in diameter can be produced by evaporation condensation of a pure substance, followed by electrostatic classification and discharge to produce particles of the desired size. These sizes represent the range of radon progeny aerosols, from large unattached particles to the predominant size of attached radon progeny, in the normal indoor environment. Using the MR image sandwich model, we are now attempting to determine overall deposition at constant flow (either inspiratory or expiratory) by measuring entering and exit aerosol concentration with a condensation nucleus counter (TSI, model 2020, St. Paul, MN). The aerosols used for these studies are NaCl or metallic silver, and the assumption made is that radon progeny of the same size behave similarly with respect to diffusion-controlled deposition.

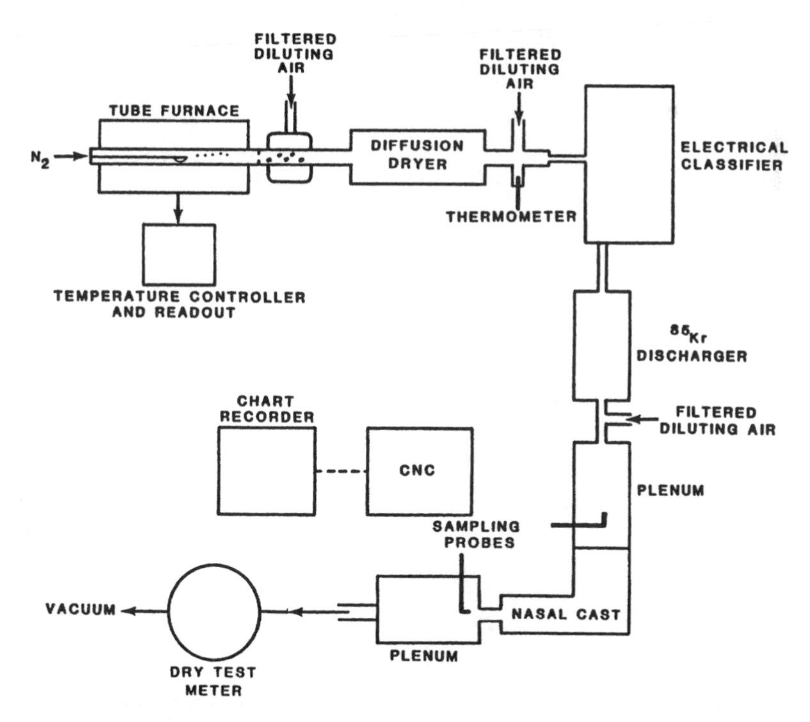

Figure 6. Aerosol exposure system for NaCl/Ag aerosols of particles 5 to 200 nm in diameter, integrated with magnetic-resonance-derived, "sandwich" nasal models. Aerosol concentration measured with condensation nucleus counter (CNC).

With such a system, it is only possible to measure overall deposition. However, studies are also being undertaken in which the sandwich replicate model is incorporated into a thoron progeny generation and exposure system at Inhalation Toxicology Research Institute, Albuquerque, New Mexico. This system, whose characteristics are described by Su et al. (1989), produces aerosols whose mean size ranges from 1 to 2 nm, as determined by screen diffusion batteries. A primary advantage of this system is that both the overall and local deposition characteristics can be determined. The overall deposition percentage is determined by filter-collection samples of aerosol entering and exiting the model. The local deposition of aerosol as a function of distance from the nose tip is determined by gamma-counting ^{212}Pb deposited on each segment. A plot of such a local deposition measurement of a 1-nm aerosol normalized for lateral surface area is shown in Figure 7. This kind of information can be combined with models of nasal clearance patterns (Swift and Proctor, 1988) to develop dosimetric models of nasally deposited radioaerosols and, by extension, to other upper-airway surfaces that are models for dose to the entire NOPL region.

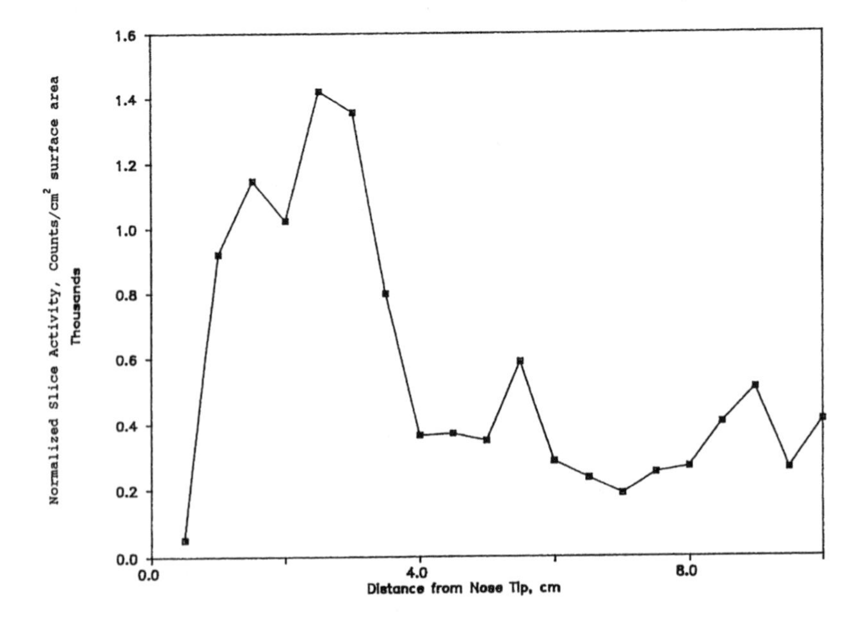

Figure 7. Plot of normalized local deposition of 1-nm aerosol measured in nasal model.

The sandwich airway models can also be employed in studies of aerosols in the inertial range, 0.5 to 10 μm. In preliminary studies of deposition in the model, recently reported by Swift (1991), a polydisperse aerosol of diethyl hexyl sebacate (DEHS) produced by a nebulizer flowed through the model at constant rates ranging from 13 to 90 L min^{-1}. The number distribution of this aerosol has a modal value of 1.2 μm, but numbers of aerosol particles at sizes from 0.8 to 10 μm are sufficient to determine the ratio of outlet concentration to inlet concentration using an aerodynamic particle size analyzer (TSI, model 3000). If aerosols in this size range contained a radioisotopic substance or a chemical tracer (e.g., sodium fluorescein), the local deposition characteristics could also be measured.

One drawback to the use of such replicate models is the difference in surface character between the actual organ and the model. Most of the nasal and other NOPL surfaces are covered with a liquid, while the model is dry. Is deposition on such dry surfaces similar to that in vivo? For diffusional deposition, the surface does not play a role in determining the particle flux. For inertial sizes, the use of liquid aerosols minimizes the possibility that bounce on the dry surface alters the deposition patterns. For unattached progeny, the question of particle growth in a progressively more humid inspired air volume is not important. However, the chemistry of ambient particles to which radon progeny attach and that of particles in the inertial size range may lead to hygroscopic growth. Further study should be devoted to the possibility that the evaporative flux of water vapor into the inspired air may somewhat impede diffusional movement of ultrafine particles from the air to the airway surface.

DIMENSIONAL CHANGES AND SCALING STUDIES

It is well established that the dimensional characteristics of the nasal passage of a particular individual are subject to "natural" and induced variability (Hilberg et al., 1989; Kennedy et al., 1988). It is possible with the sandwich replicate model to approximate these changes and study their effect on aerosol deposition. Some preliminary studies conducted in normal human subjects whose nasal passage patency was increased by topical application of xylometazolin are reported by Rasmussen et al. (1990). The decongestion resulted in enlargement of the anterior nasal cross sections. However, in general, this was accompanied by an increase in nasal deposition. Intersubject variability in nasal deposition was not simply correlated to a single cross-sectional dimension such as that determined by the acoustic reflection method. These questions deserve further in vivo study.

The sandwich replicate models can also be valuable in allowing the effects on deposition of systematic changes in dimensions of specified regions of the NOPL airways to be studied in detail with respect to aerosol particle size. A preliminary study of the effect of area changes at the nasal valve was reported by Swift (1991). In this instance, three sandwich elements of the nasal valve area were made, representing the normal, congested, and decongested state. Deposition of inertial-size particles at a single flow rate was found to be dependent on the cross-sectional area in a manner in which the Stokes inertial parameter for each instance was an appropriate scaling factor. That is, when the deposition percentage was plotted as a function of nasal-valve Stokes number, the deposition curves for the three cases collapsed to a single function.

The replicate model technique also lends itself to construction of scaled-up models of NOPL airways by factors which permit studies that are difficult to perform in life-size models; e.g., velocity measurements with anemometers. According to fluid-mechanical principles, dimensional similarity using Reynolds number scaling should preserve the flow features from life-size to scaled-up models; this can readily be checked by measurements of velocity distribution and turbulence.

FUTURE STUDIES WITH REPLICATE MODELS

There are a number of interesting scientific question with practical implications for radon dosimetry and exposure to other environmental aerosols which can be addressed with a combination of in vivo and replicate sandwich NOPL airway models. One such investigation should match a NOPL airway sandwich model to an upper-bronchial replicate model to determine whether changes from nasal to oral breathing or dimensional changes in either airway significantly influence bronchial distribution of radon-progeny aerosols. Further studies are also needed on the effect of the nasal cycle and changes induced by environmental agents on the overall and local deposition of radon progeny aerosols in the nasal airway.

Preliminary studies of deposition of such aerosols as a function of age, ranging from newborns to adults, have been carried out by Swift (1991). A more detailed investigation of age effects is needed, in which several "representative" airway models are employed. The effect of age beyond the normal working-age range, 20 to 65 y, needs to be studied as interest develops in radon exposure as well as exposure to other environmental aerosols.

With very few exceptions, both in vivo and model studies of the NOPL airways have involved healthy, nonsmoking, Caucasian adult males, a very selective slice of the population. Unanswered is the question of the influence on NOPL aerosol behavior of sex, ethnic background, health history, health status, and environmental exposure history. Replicate model studies of the NOPL airways also allow comparative studies of humans and animal species, either those widely used in exposure studies or species being considered for such studies, in order to better understand species variability and determine the limitations of species extrapolation.

ACKNOWLEDGMENTS

We obtained the MR images used to construct the model described in the text with the cooperation of Raymond A. Guilmette of ITRI, Albuquerque, New Mexico and Jeffrey D. Wicks of the University of New Mexico Center for Noninvasive Diagnosis, Albuquerque, New Mexico. The morphometric measurement, construction, resistance measurements, and inertial particle studies were carried out by Jana Nadarajah, The Johns Hopkins University, Baltimore, Maryland.

REFERENCES

Guilmette, RA, JD Wicks, and RK Wolff. **1989**. Morphometry of nasal human airways in vivo using magnetic resonance imaging. J Aerosol Med 2:365-377.

Hilberg, O, AC Jackson, DL Swift, and OF Pedersen. **1989**. Acoustic rhinometry: Evaluation of nasal cavity geometry by acoustic reflection. J Appl Physiol 66:295-303.

Itoh, H, GC Smaldone, and DL Swift. **1985**. Mechanism of aerosol deposition in a nasal model. J Aerosol Sci 16:529-534.

Kennedy, DW, JJ Zinreich, AE Rosenbaum, AJ Kumar, and ME Johns. **1988**. Physiologic mucosal changes within the nose and ethmoid sinus: Imaging of the nasal cycle by MRI. Laryngoscope 98:928-933.

Rasmussen, TR, DL Swift, O Hilberg, and OF Pedersen. **1990**. Influence of nasal passage geometry on aerosol particle deposition in the nose. J Aerosol Med 3:15-25.

Su, YF, YS Cheng, GJ Newton, and HC Yeh. **1989**. *Activity Size Distribution of ^{220}Rn Progeny in Ultra High Purity Air*, Technical Report LMF-126. Inhalation Toxicology Research Institute, Albuquerque, NM.

Swift, DL. **1991.** Inspiratory inertial deposition of aerosols in human nasal airway replicate casts: Implication for the proposed NCRP lung model. Radiat Protect Dosim 38:29-34.

Swift, DL and DF Proctor. **1977**. Access of air to the respiratory tract, pp. 905-911. In: *Respiratory Defense Mechanisms*, Part 1, JD Brain, DF Proctor, and L Reid (eds.). Marcel Dekker, New York.

Swift, DL and DF Proctor. **1988**. A dosimetric model for particles in the respiratory tract above the trachea, pp. 1035-1044. In: *Inhaled Particles VI*, J Dodgson, RI McCallum, MR Bailey, and DR Fisher (eds.). Pergamon Press, Oxford, UK.

Yamada, Y, YS Cheng, HC Yeh, and DL Swift. **1988**. Inspiratory and expiratory deposition of ultrafine particles in a human nasal cast. Inhal Toxicol 1:1-11.

DEPOSITION OF "UNATTACHED" RADON DAUGHTERS IN MODELS OF HUMAN NASAL AND ORAL AIRWAYS

J. C. Strong[1,2] and D. L. Swift[3]

[1]National Radiological Protection Board, Chilton, Didcot, United Kingdom

[2]Present address: AEA Technology, Harwell Laboratory, Didcot, United Kingdom

[3]The Johns Hopkins University, Baltimore, Maryland

Key words: *Nasal deposition, radon*

ABSTRACT

In order to estimate accurately an effective dose equivalent for exposures to radon daughters, knowledge of their deposition in the lung is required. However, the nose and mouth are effective filters for removing aerosol particles, especially in the range of sizes of "unattached" radon daughters. Therefore, it is equally important to have reliable data on deposition in this region of the respiratory tract.

We will describe our work in studying nasal and oral deposition of "unattached" radon daughters in casts of these airways. Several hollow casts of adult and child nasal and oral airways were fabricated at The Johns Hopkins University from layers of Perspex™ (an acrylic plastic). The shapes of the airway passages were obtained from nuclear magnetic resonance sectional images of healthy subjects. The casts were exposed to radon gas and daughters produced by flushing filtered air through a commercially available ^{226}Ra source. The gas stream was drawn through a 1.4-L cylindrical tube to allow measurable growth of ^{218}Po activity before it was passed through casts of both nasal passages or the oral cavity. The deposition of "unattached" ^{218}Po was measured by comparing the activity collected on filters mounted in series and in parallel with a cast. Measurements were made at various flow rates (Q; 4 to 20 L min^{-1}). The diffusion coefficient (D) of ^{218}Po was measured each time the flow rate was changed, by replacing the cast with a stainless steel gauze screen and measuring the activity penetrating the screen. The measured diffusion coefficient ranged from 0.02 to 0.05 cm^2 s^{-1} and was found to vary with the residence time of ^{218}Po in the growth tube.

The deposition efficiency (η) of ^{218}Po measured in these casts ranged from 50 to 70%, and was similar to values we found previously, using casts of nasal and oral airways from cadavers.

INTRODUCTION

Radon-daughter aerosols in domestic and occupational environments can be characterized by two components: the so-called "unattached" and attached fractions. A large proportion of the activity is attached to an aerosol having an activity median diameter (AMD) in the range 100 to 250 nm. The "unattached" fraction is associated with an aerosol size range from 0.5 to 5 nm. For a given exposure to radon daughters, radiation dose to the lung will be reduced if either aerosol component is deposited substantially in the nasal or oral airways. Our earlier work (Strong and Swift, 1987) indicated that penetration through a nasal cast increased with particle size and flow rate and that, under all flow conditions, particles larger than about 20 nm diameter would penetrate the nose without significant loss. The aim of this work was to examine experimentally the magnitude of nasal and oral deposition for aerosols in the size range of "unattached" radon daughters at various physiological flow rates.

METHOD

Freshly formed ^{218}Po was used to investigate the penetration of "unattached" radon daughters through plastic casts of human upper airways. Five models were used: one was a cast obtained at autopsy from the nasal cavity of an adult which was used in our original work, whereas the others were fabricated from sheets of Perspex™, the airway outlines being obtained from healthy human subjects using magnetic resonance imaging (MRI).

The experimental set-up is shown in Figure 1. Radon gas, produced by flushing air through two commercially available ^{226}Ra sources (each of nominal activity, 80 kBq, Pylon) was drawn through a cylindrical tube of 1.4-L volume, to allow measurable growth of ^{218}Po activity. Air leaving the growth tube was split: half was drawn through a 47-mm-diameter glass fiber filter (Grade GF/A, Whatman) used as the reference filter; the other half was drawn through the nasal cast, then through another 47-mm-diameter filter at the exit. Airflows were monitored by rotameters and vacuum gauges, drawn through sealed pumps, then passed back through the ^{226}Ra sources. The apparatus worked in a closed loop to give a constant rate of supply of ^{218}Po. A filter (HEPA cartridge, Gelman) placed before the growth tube ensured the removal of any particles formed in the loop that might provide nuclei for attachment of ^{218}Po. The concentration of particles in the air leaving the growth tube was measured with a condensation nucleus counter (Model 3020, TSI Inc.) and shown to be less than 20 cm^{-3}. The probability of ^{218}Po attaching to particles was therefore negligible.

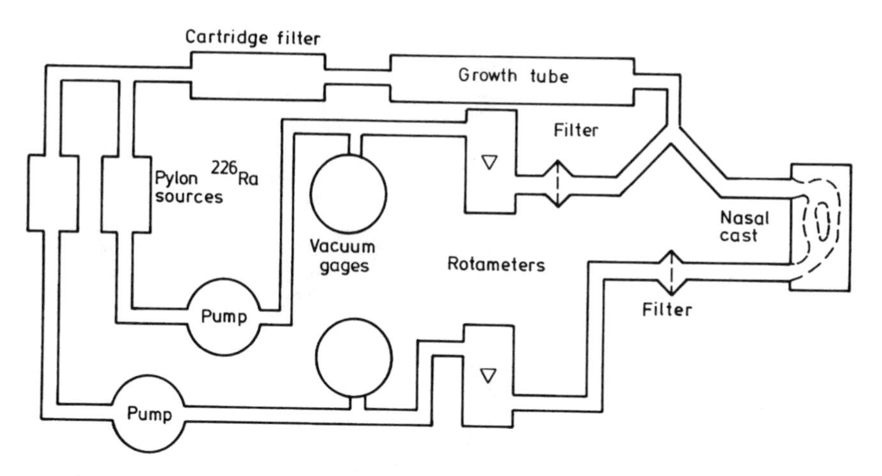

Figure 1. Apparatus used to measure penetration of "unattached" ^{218}Po through nasal casts.

In order to measure the diffusion coefficient of ^{218}Po for each experiment, the cast and backup filter were removed and replaced with a modified in-line filter holder which held a 200-mesh stainless steel wire screen and filter, separated by a 5-mm spacer. Alpha activity on the filter was compared with that collected on the reference filter in the other limb of the apparatus to obtain activity penetration through the wire screen. The diffusion coefficient of the "unattached" ^{218}Po activity was then calculated from the measured penetration and the theoretical penetration (Cheng and Yeh, 1980; Cheng et al., 1985; Ramamurthi and Hopke, 1989). Cast penetration data were obtained by measuring the alpha activity on the cast backup filter and comparing it with that on the reference filter.

RESULTS

Penetration data were obtained for the three adult nasal casts and two casts from children (ages 2.5 and 6 y); one of the adult casts included an oral cavity. The results are shown in Table 1 with the flow rates at which the tests were made, together with the measured particle diameters. The measured values of diffusion coefficient ranged from 0.02 to 0.05 cm^2 s^{-1} and were found to vary inversely with flow rate, suggesting that particle growth was indeed occurring in the 1.4-L reservoir.

Table 1. Measured penetration of "unattached" ^{218}Po through nasal casts, with estimates of the aerosol particle diameter.

Cast/Model	Cast flow rate, L min^{-1}	Diameter, nm	Penetration
2.5-y-old child Both passages	4.9 9 18	1.25 0.95 0.85	0.25 0.24 0.28
6-y-old child Both passages	6.4 9	1.04 0.95	0.42 0.37
Adult model A Both passages	4 10 20	1.30 0.96 0.82	0.30 0.37 0.40
Adult model B Both passages	4 10 20	1.30 0.92 0.80	0.30 0.29 0.29
Adult model B Right passage	19	0.92	0.37
Adult model B Mouth	4 10 20	1.28 0.92 0.79	0.50 0.31 0.50
Adult molded cast Single passage	5 11 18	1.40 1.10 1.10	0.36 0.35 0.39

DISCUSSION

Aerosol deposition in the range of particle sizes considered here is dominated by diffusional mechanisms which depend on the flow conditions and particle size. In the nasal airways, the flow has been shown to be turbulent (Proctor and Swift, 1971). Thus, the mechanism responsible for deposition of particles with diameters less than 20 nm in this region of the respiratory tract is likely to be controlled by a turbulent diffusion process. A simple model to describe the deposition of aerosols in a nasal cavity based on turbulent diffusion was suggested by Cheng et al. (1988):

$$\eta = 1 - \exp(-bQ^{-1/8}D^{2/3}), \tag{1}$$

where Q is the total volumetric flow rate (L min^{-1}) through both passages, D is the diffusion coefficient of the particles (cm^2 s^{-1}), b is a constant, and η is the fractional deposition in the cavity. However, this model was originally based on data for particles in the range from 5 to 200 nm. It has recently been found to overestimate nasal deposition in the size range of "unattached" radon daughters and has therefore been modified (Cheng et al., this volume):

$$\eta = 1 - \exp(-bQ^{-1/8}d^{1/2}). \quad (2)$$

Figure 2 shows deposition data for the 6-y-old and the adult nasal casts plotted against the deposition parameter $Q^{-1/8}D^{-1/2}$, as shown in Figure 2. Where only one nasal passage was tested, values for flow rate were doubled to give the total flow, Q. Figure 2 also shows the original set of data for NaCl and silver aerosols with particle diameters in the range from 5 to 15 nm (Strong and Swift, 1987). A function of the form suggested above was fitted to the data, and this is also shown in Figure 2. The value for the constant b was found to be +7.4.

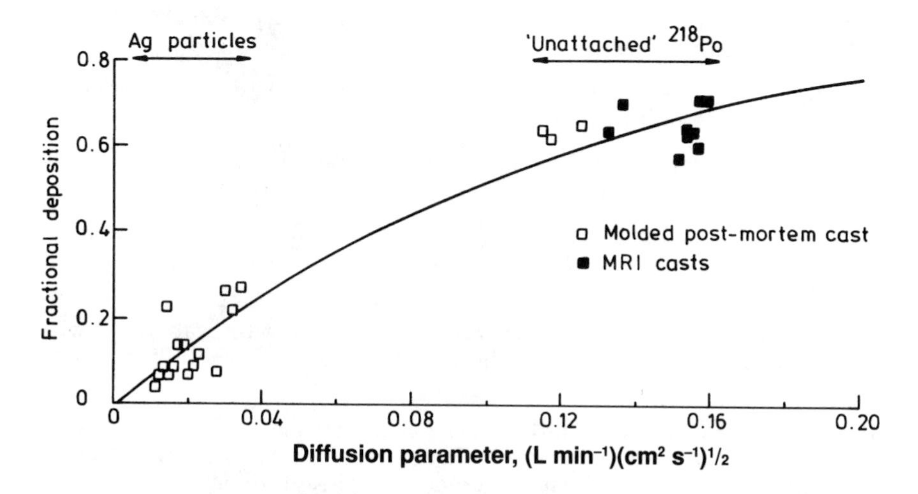

Figure 2. Aerosol deposition as a function of diffusion parameter ($Q^{-1/8}D^{1/2}$). The fitted function $\eta = 1 - \exp(-7.4Q^{-1/8}D^{1/2})$ is also shown. MRI = magnetic resonance imaging.

CONCLUSION

The limited deposition data obtained with the adult cast containing an oral cavity (Model B) are not significantly different from those obtained with the nasal casts (Models A and B). In the case of casts from children, the 2.5-y-old deposition data indicate higher values than those obtained from the 6-y-old and adult casts.

The results given here fit the model

$$\eta = 1 - \exp(-7.4Q^{-1/8}D^{1/2}). \quad (3)$$

This model is quite similar to that suggested by other investigators (Cheng et al., in press); however, the value presented here for b (–7.4) is less than that suggested by Cheng. The data presented here indicate substantial deposition of "unattached" radon daughters in the human nasal and oral airways, although the deposition in the oral region is slightly less. Thus, deposition in these regions of the respiratory tract should be taken into account when estimating radon-daughter doses.

REFERENCES

Cheng, YS and HC Yeh. **1980**. Theory of a screen type diffusion battery. J Aerosol Sci 11:313-320.

Cheng, YS, HC Yeh, and KJ Brinsko. **1985**. Use of wire screens as a fan model filter. Aerosol Sci Technol 4:165.

Cheng, YS, Y Yamada, HC Yeh, and DL Swift. **1988**. Diffusional deposition of ultrafine aerosols in a human nasal cast. J Aerosol Sci 19:741-751.

Proctor, DF and DL Swift. **1971** The nose—A defence against the atmospheric environment, pp. 59-69. In: *Inhaled Particles III*, WH Walton (ed.). Unwin, Surrey.

Ramamurthi, M and PK Hopke. **1989**. On improving the validity of wire screen "unattached" fraction Rn daughter measurements. Health Phys 56:189-194.

Strong, JC and DL Swift. **1987**. Deposition of ultrafine particles in a human nasal cast, pp. 109-112. In: *Aerosols: Their Generation, Behavior and Applications*. The Aerosol Society First Conference, Loughborough, UK.

QUESTIONS AND ANSWERS

Q: Park, PNL. Do you attempt to modify humidity in working with these models or control the humidity of the air?

A: Strong, NRPB, UK. No, not at this stage. We're just using lab air, which was at about 40% relative humidity.

Q: Park. In relating the results of these type of studies to people, do you have to estimate what the influence of humidity would be in the nose or in the aerosol, as humidification during inspiration?

A: Strong. Yes, I think that is the next step, and I'm hoping that we can do that, that we can actually humidify the airway and look to see what deposition is there. Because it's sure to change the particle size growth. That's the next step.

Q: Cross, PNL. When you say young children are protected in comparison with adults, you are obviously referring only to lung effects. Nobody's ever addressed the possible health effects of nasal dose and, maybe, carcinoma development.

A: Strong. Yes.

Q: Hopke, Clarkson University. With regard to the humidity effects on size: If we're looking at a cluster equilibrium phenomenon between water and the polonium, then there probably is a limited range of thermodynamic stability, even over the range of humidities. And so there probably is not going to be as much growth as you might think of in terms of normal particles. I think the size is going to stay fairly constant, but we'll be doing measurements of that during the next year.

Q: Hopke. I was wondering if you had enough data on these children's casts to get any idea as to what the coefficient would be, for the children?

A: Strong. Well, as you saw, we only had three points to put in the curve, and I suppose we could arbitrate and put a number on there. But no, we haven't done that yet, because I thought we didn't have enough information at the other end.

Q: Cheng, ITRI. If I understand correctly, your final equation is a combined result of a single- pass cast and an MRI laminated cast, and for the radon progeny those two casts gave similar results. [Yes.] We have also done some experiments with David Swift, using a similar cast that you have. We found that for radon progeny, the laminated cast gave a higher deposition than the single-pass cast, even though the result of the single-pass cast agrees with your data. But our result on the laminated cast has higher deposition; probably you have a 70% deposition, and ours is 80%, so this is probably an area that needs more experimentation.

A: Strong: By the way, with the single-pass model that we used, I've actually modified the flow to make it look like a double pass.

Q: Knutson, EML. John, you said that you measured the size of the unattached progeny. How were the values 0.8, 0.9, and 1.3 measured, and how do you explain the change in size? It was a function of flow rate, I believe; right?

A: Strong. Yes, we measured it with a single screen. We just looked at the penetration through a single screen compared with the reference filter, and then compared it with Cheng's work and also Phil Hopke's latest addition to the screen penetration work. I assume that, as the flow rate is going up, then if there is any growth. . . . The radium A is being produced, obviously, as ions, but as Phil has pointed out, you do get them growing with whatever is left in the system, and I think it's just in that tube, which has 1.4 liters volume, you're varying this sort of period, giving them the chance to grow to their normal size, perhaps.

Q: Knutson. How long was the residence time in that tube?

A: Strong. I can't remember offhand.

Q: Swift. It varies, depending on the flow rate, doesn't it? A fixed volume at a variable flow rate?

A: Strong. Yes, that's right. It's seconds, I think.

Q: Knutson. One additional question: Does that change in particle size from 0.8 to 1.3 really affect deposition noticeably, or . . . ?

A: Strong: No, it doesn't seem to, no.

DEPOSITION OF RADON PROGENY IN NONHUMAN PRIMATE NASAL AIRWAYS

H. C. Yeh[1], Y. S. Cheng[1], Y. F. Su[1] and K. T. Morgan[2]

[1]Inhalation Toxicology Research Institute, Albuquerque, New Mexico

[2]Chemical Industry Institute of Toxicology, Research Triangle Park, North Carolina

Key words: *Nasal deposition, ultrafine particles, radon progeny, rhesus monkey*

ABSTRACT

Radon progeny are usually associated with ultrafine particles ranging in diameter from 0.001 to 0.005 μm for "unattached" progeny and from 0.005 to 0.2 μm for those attached to indoor aerosols. To assess the health effects of inhaling indoor radon progeny, it is necessary to study the regional deposition of these inhaled ultrafine particles. Laboratory animals are often used in studies of the toxicity of inhaled particles and vapors. Information on the deposition of particles larger than 0.2 μm in the nasal passages of laboratory animals is available; however, there is little information on the deposition of particles smaller than 0.2 μm. In this report, we describe the use of nasal casts of a rhesus monkey to measure total deposition of ultrafine aerosols, including unattached ^{220}Rn progeny, in a unidirectional-flow inhalation exposure system. Deposition data were obtained for monodisperse silver aerosols with particle sizes ranging from 0.005 to 0.2 μm, at several inspiratory and expiratory flow rates that represented normal breathing as well as hypo- and hyperventilation. In addition, we studied the deposition of unattached ^{220}Rn progeny, at particle sizes from 0.001 to 0.003 μm. The deposition efficiency decreased with increasing particle size, indicating that diffusion was the dominant deposition mechanism. The effect of flow rate was essentially negligible. Based on assumptions that turbulent flow and complete mixing of aerosols occur in the nasal airways, a general equation,

$$E = 1-\exp(-a\,D^{b}Q^{c}) \qquad \text{for } d_p \leq 0.2\ \mu\text{m}, \tag{1}$$

was derived, where E is the deposition efficiency, d_p is the particle diameter, D is the diffusion coefficient, and Q is the flow rate. Constants a, b, and c are estimated from experimental data, for either inspiration or expiration. This mathematical expression will be useful for making modifications to both deposition and dosimetry models.

INTRODUCTION

Information on aerosol deposition patterns in the human respiratory tract is needed to improve human health risk estimates for exposure to airborne radon progeny. The respiratory tract is commonly divided into nasopharyngeal (or head airway), tracheobronchial, and pulmonary regions (Task Group on Lung Dynamics, 1966). Deposition of inhaled particles in these regions depends on physical properties of the particles (e.g., size, density, shape), airway morphometry, and breathing pattern (Yeh et al., 1976). The head airways, representing the first line of defense, are often the site of insult from inhaled particles. Those that deposit in the head airways will not be able to reach either the tracheobronchial or the pulmonary regions. Thus, information about inhaled particle deposition in the head airways is important for better understanding the potential risk to the lung from inhaled particles.

There is a significant body of information on the deposition of relatively large inhaled particles in the respiratory tracts of people and laboratory animals (Lippmann, 1977; Lippmann and Schlesinger, 1984; Schlesinger, 1985; Stuart, 1984). The most complete data are for people who inhaled particles in the range of 0.5 to 10 μm aerodynamic diameter (Lippmann, 1977; Stahlhofen et al., 1980, 1981, 1983; Schreider, 1984). There are few data for particle sizes beyond this range, although Heyder et al. (1986) reported on total deposition of aerosols as small as 0.005 μm. However, they did not provide information on the sites of deposition. With respect to inhaled radon progeny, the most important particle sizes are those between 0.001 and 0.005 μm, which is the effective size of unattached progeny, and between 0.005 and 0.5 μm, the most common sizes reported for attached radon progeny in mines and homes (NCRP, 1984; James et al., 1981).

The results of recent morphologic and morphometric animal studies indicate that, of the commonly used laboratory animals, nonhuman primates have nasal structures that most closely resemble those of humans (Harkema et al., 1987). Thus, the purpose of our study was to determine the total deposition efficiency of ultrafine particles, including those of radon progeny, in a monkey nasal cast. We used both monodisperse particles that ranged from 0.005 to 0.2 μm and unattached ^{220}Rn progeny in this study.

MATERIAL AND METHODS

One clear replica cast was made of the nasal airways of a rhesus monkey that weighed 8.5 kg; the cast included the nose, pharynx, and larynx (Figure 1). Details of the casting method can be found elsewhere (Patra et al., 1986; Morgan et al., 1989). Pressure drops across the nasal passage between the nares and the trachea were measured with an incline manometer at constant flow rates of between 1.5 and 7.0 L min^{-1}, for both inspiration and expiration. Figure 2 shows the experimental setup, which was similar to that described by Cheng et al. (1988), except that, in the case of ^{220}Rn progeny, a different aerosol generation system was used. Briefly, monodisperse silver particles were generated by placing silver wool in a quartz boat inside a tube furnace operated at 900-1050°C. A regulated N_2 carrier gas stream (2 L min^{-1}) carried the vaporized materials into a glass condensation chamber where cool diluting air was introduced. Singly charged monodisperse particles of known size were extracted by using an electrostatic classifier (TSI Model 3071, St. Paul, MN). After neutralizing the aerosol by passing it through a ^{85}Kr discharger, the monodisperse aerosol was drawn through the monkey nasal cast under constant flow conditions. Aerosol concentrations at the inlet (C_{in}) and outlet (C_{out}) of the cast were measured by using a continuous-flow condensation nucleus counter (TSI Model 3020, St. Paul, MN), which was operated in the count mode. The coincidence count errors of the aerosol concentrations were corrected. Deposition efficiencies (E) were calculated according to the expression

$$E = 1 - (C_{out}/C_{in}). \tag{2}$$

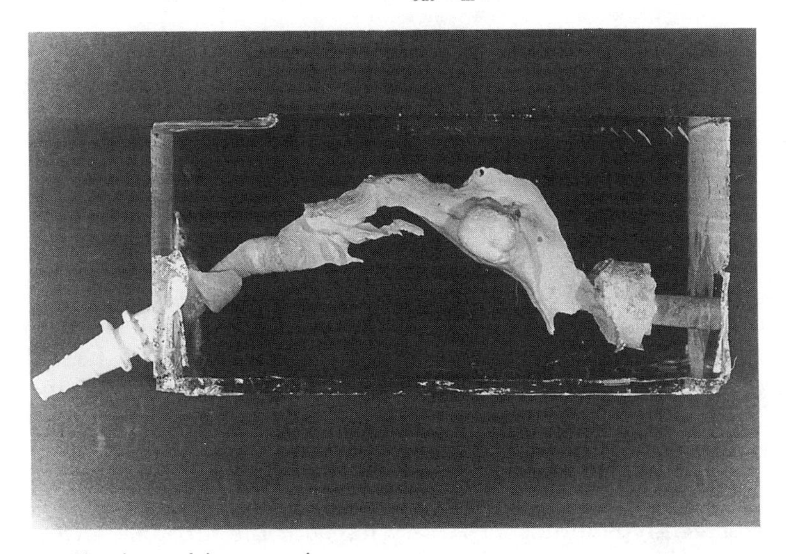

Figure 1. Nasal cast of rhesus monkey.

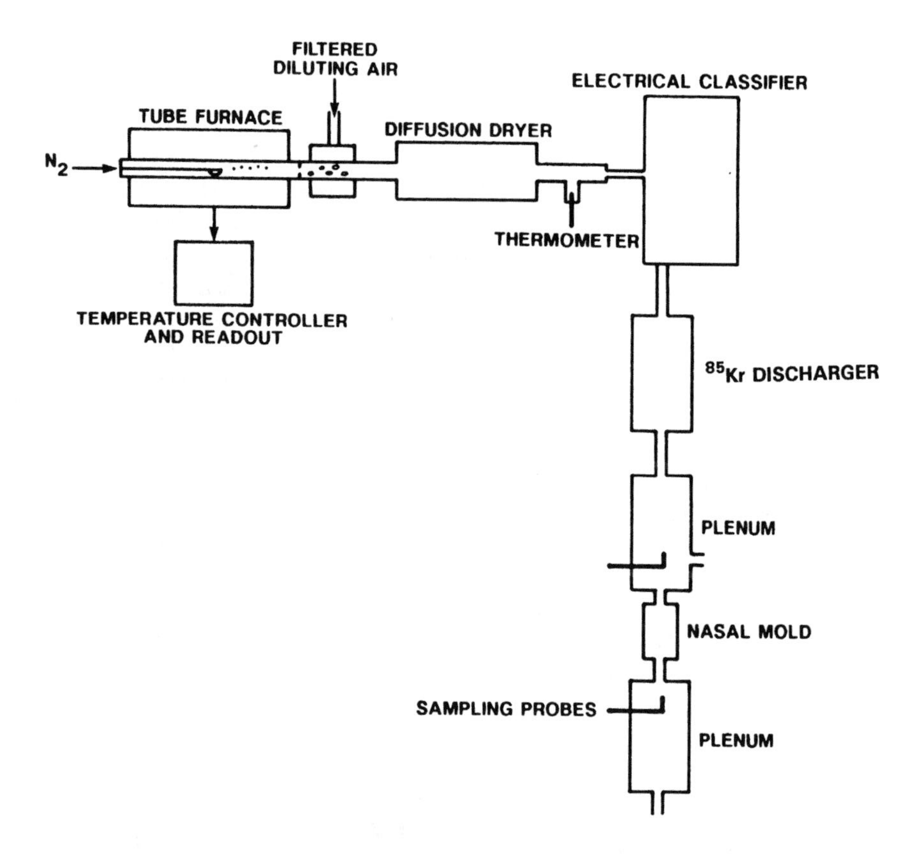

Figure 2. Experimental setup for ultrafine particle deposition study.

In the case of ^{220}Rn progeny, the experimental setup was modified (Figure 3). The aging chamber was designed so that the average residence time of the flow would be four to five times the half-life of thoron gas; thus, the effect of thoron gas on deposition would be negligible. The activity concentration measurements with and without the cast were used to calculate the deposition efficiency.

The monodisperse particle sizes were measured independently by using a screen-type diffusion battery for both the ultrafine silver particles and the unattached ^{220}Rn progeny.

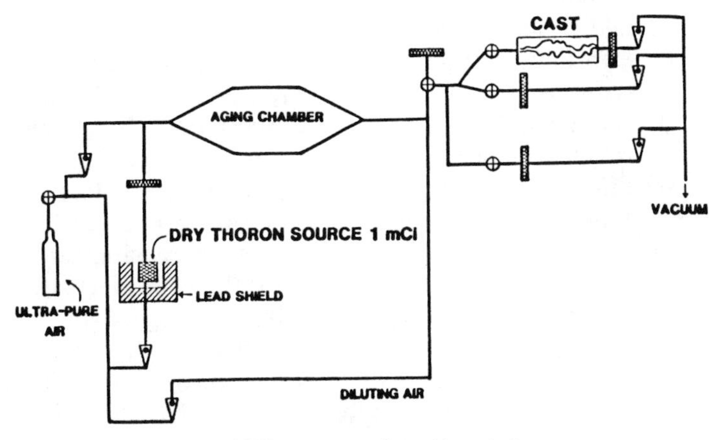

Figure 3. Experimental setup for ^{220}Rn progeny deposition study.

RESULTS AND DISCUSSION

Figure 4 shows the measured pressure drop for both inspiratory and expiratory flows across the nasal cast. The nonlinearity of the pressure-drop/flow-rate relationship suggests that the flow within the monkey nasal cast is turbulent and is similar to the nasal flow previously reported for a human nasal cast (Yamada et al., 1988). Like the results of the human nasal cast study, the expiratory flow pressure drop, on average, was slightly higher than that of the inspiratory flow, suggesting that the expiratory flow patterns may be somewhat different from the inspiratory flow patterns.

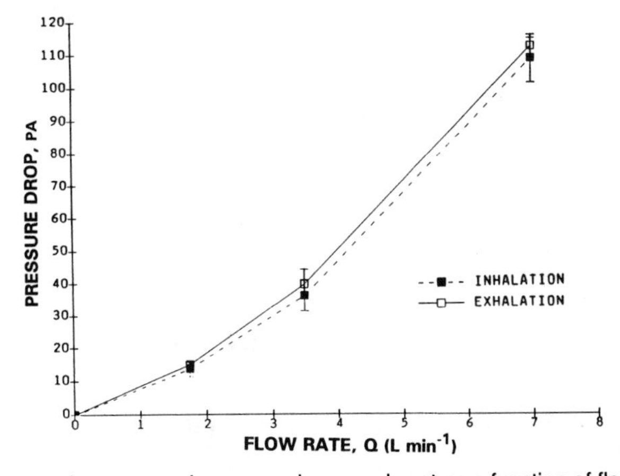

Figure 4. Pressure drop across rhesus monkey nasal cast as a function of flow rate.

Figures 5 and 6 show aerosol deposition efficiencies for inspiratory and expiratory flow, respectively, as a function of particle size, with flow rate as a parameter. Flow rates of 1.75, 3.5, and 7.0 L min^{-1} were used, with 3.5 L min^{-1} corresponding to the normal breathing rate (2 times the minute respiratory volume) for the rhesus monkey. The deposition efficiency increased with decreasing particle size for the size range studied, and the effect of flow rate was essentially negligible. This can be explained by the fact that increasing the flow rate decreases residence time, thus reducing diffusional deposition; however, it also increases turbulent diffusivity, thus increasing turbulent diffusion deposition. Therefore, the net effect, in this case, is very small. The same phenomenon was observed in the human nasal cast study (Yamada et al., 1988). Figure 7 shows both inspiratory and expiratory deposition at a normal breathing flow rate of 3.5 L min^{-1}. Similar deposition efficiency curves were obtained for two other cases at Q = 1.75 and 7.0 L min^{-1}

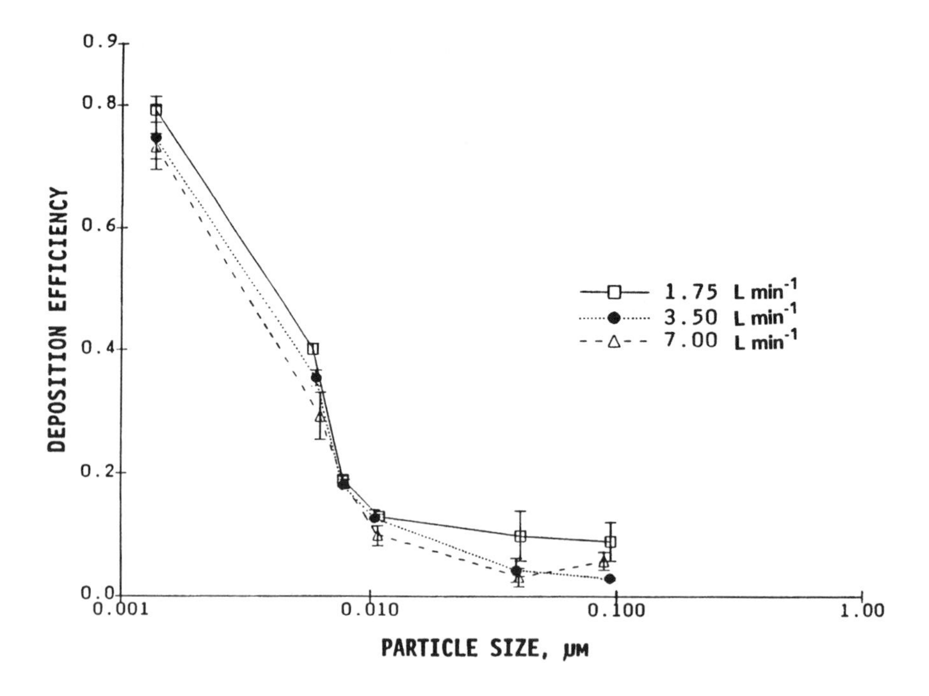

Figure 5. Deposition efficiency of ultrafine particles in rhesus monkey nasal cast at various inspiratory flow rates.

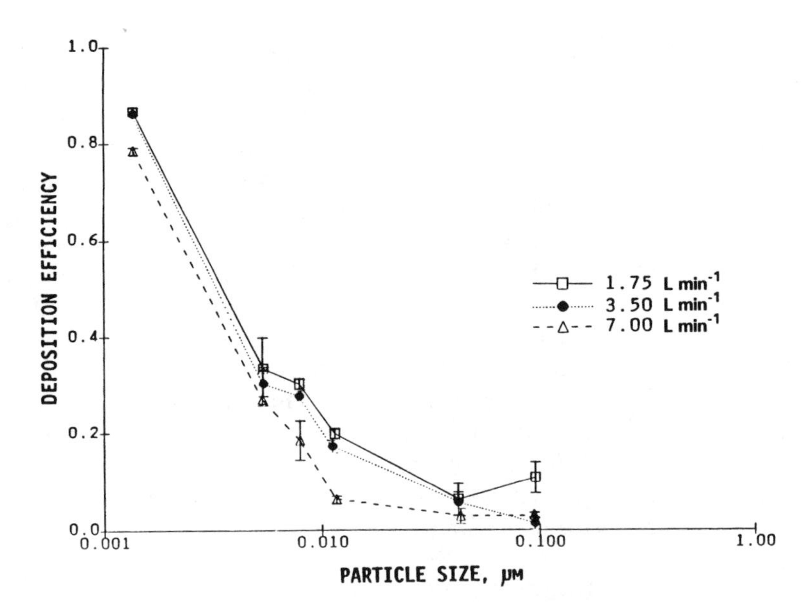

Figure 6. Deposition efficiency of ultrafine particles in rhesus monkey nasal cast at various expiratory flow rates.

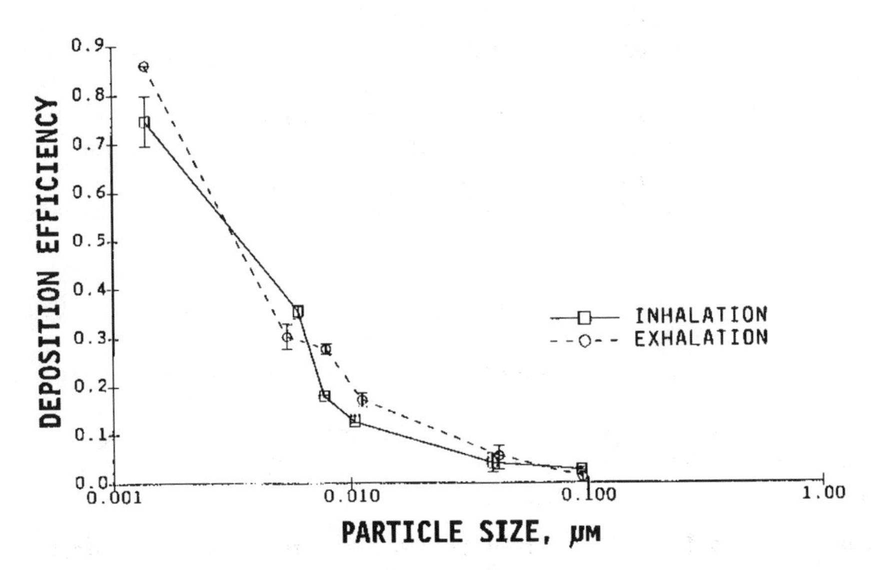

Figure 7. Comparison of ultrafine particle deposition efficiencies in rhesus monkey nasal cast for inhalation and exhalation at flow rate of 3.5 L min^{-1}.

Because of the lack of data for ultrafine particle deposition, the International Commission on Radiological Protection Task Group on Lung Dynamics (1966) assumed zero deposition in the nasal region for particles less than about 0.2 μm in diameter. This and a previous study (Cheng et al., 1988; Yamada et al., 1988), however, indicate that substantial deposition of ultrafine particles occurs in the nasal airways for both human and monkey nasal casts. It is well established that inertial deposition is the dominant deposition mechanism for particles larger than about 0.5 μm in diameter. Like aerosol filtration, the diffusional deposition mechanism is important for smaller particles. The data reported here for monkey nasal airways are similar to the previously reported data obtained for a human nasal cast (Cheng et al., 1988; Yamada et al., 1988). In earlier studies, it was suggested that the turbulent diffusion mechanism could explain the nasal deposition of ultrafine particles. A general equation for deposition efficiency, E, can be written as (Cheng et al., 1988)

$$E = 1 - \exp(-a\, D^b Q^c) \qquad \text{for } d_p \le 0.2\ \mu\text{m}, \qquad (3)$$

where a, b, and c are constants, D is the particle diffusion coefficient ($cm^2\ s^{-1}$), Q is the flow rate ($L\ min^{-1}$), and d_p is the particle diameter. Constants a, b, and c can be estimated, from experimental data, for either inspiration or expiration by estimating b and c independently for each particle size and flow rate, respectively, from equation (3). The mean values of b and c were then calculated. Finally, all of the respective data were used to estimate the final values of a, for inspiratory and expiratory flow (Cheng et al., 1991). Table 1 lists the estimated values of a, b, and c. With these constants, equation (3) can be rewritten as:

$$E = 1 - \exp(-13.3\, D^{0.543} Q^{-0.219}) \qquad \text{for inspiratory flow} \qquad (4)$$

$$\text{and } E = 1 - \exp(-14.6\, D^{0.543} Q^{-0.219}) \qquad \text{for expiratory flow.} \qquad (5)$$

Figure 8 shows the curves based on equations (4) and (5) along with the experimental data for particle sizes ranging from 0.001 to 0.1μm and for the three flow rates used.

Also listed in Table 1 are those constants obtained previously from the human nasal cast study (Yamada et al., 1988). More recently, using human nasal casts, Cheng et al. (in press) indicated that, when deposition data for ^{220}Rn progeny were included, the constant b could be estimated as 0.5, which agrees well with our study. Also, the absolute value of the constant c is relatively small, indicating that the flow dependence is very small or negligible. This phenomenon is different from the case with larger particles, where the inertial deposition mechanism is dominant, and the deposition efficiency is linearly proportional to flow rate on a semilogarithmic scale (Yu et al., 1981). Figure 9 shows monkey nasal cast deposition data, along with data previously obtained from studies of ultrafine aerosol deposition in a human nasal cast (Cheng et al., 1988; Yamada et al., 1988). Very comparable deposition efficiencies were observed in the monkey and human casts for the same particle sizes. These data indicate that the primary deposition mechanism for ultrafine particles in airways of both species is turbulent diffusion.

Table 1. Values of fitted constants (a,b,c) in equation [$E = 1\text{-}\exp(-aD^bQ^c)$] describing nasal deposition in the monkey and human for particles in the range of 0.0017 to 0.2 μm diameter.

Species	Breathing mode	a	b	c
Monkey	Inspiration	13.3	0.543	–0.219
	Expiration	14.6	0.543	–0.219
Human	Inspiration	40.3	0.667	–0.125
	Expiration	48.7	0.667	–0.125

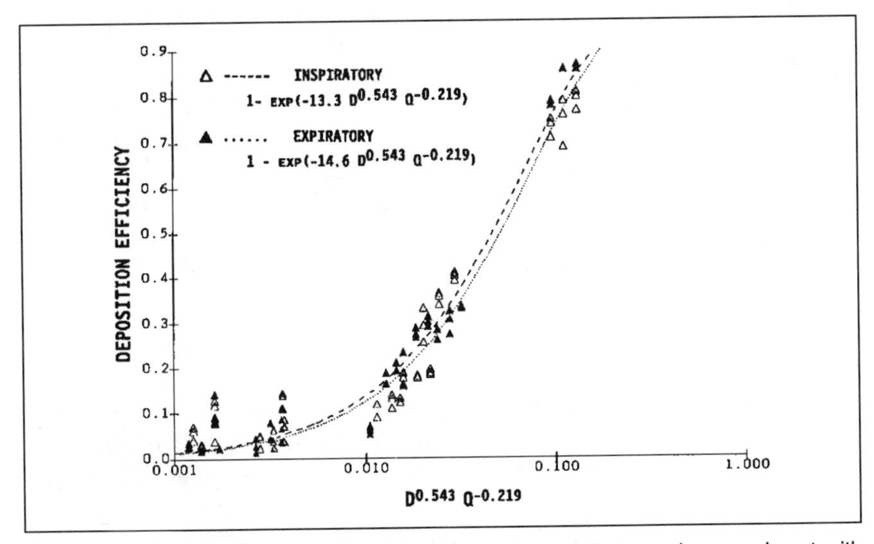

Figure 8. Comparison of deposition data obtained from rhesus monkey nasal cast with predictions for inspiration and expiratory flows based on empirical equations.

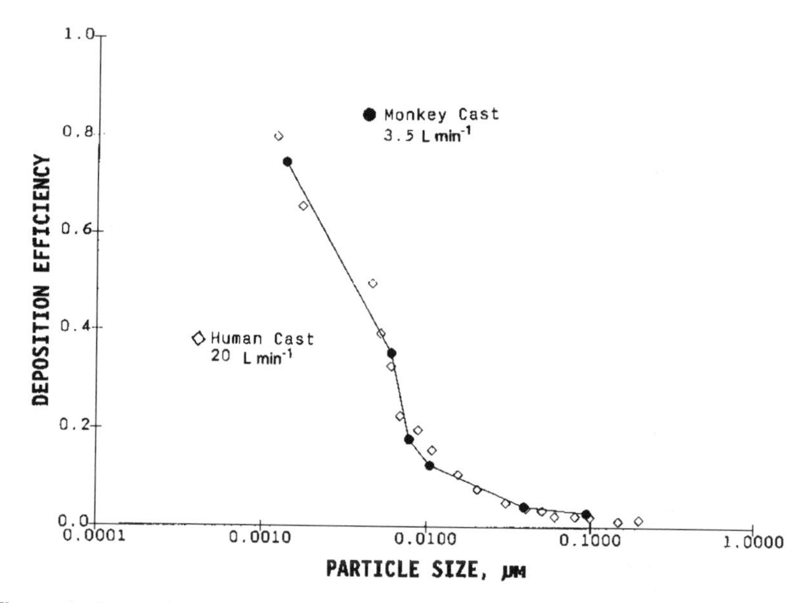

Figure 9. Comparison of rhesus monkey and human nasal cast deposition of ultrafine particles.

In summary, we found that a substantial fraction of inhaled ultrafine particles was deposited in the monkey nasal airway cast of monkeys. The deposition efficiency, which increased with decreasing particle size, was as high as about 80% for unattached ^{220}Rn progeny of 0.0017 μm diameter. The effect of breathing flow rate on deposition efficiency was minimal for the ultrafine particles. Deposition efficiencies for particles less than 0.1 μm in monkey and human nasal casts were similar, and an equation based on turbulent diffusion can be fitted to these data for either inspirational or expirational flow. These mathematical expressions will be useful for making modifications to both the inhaled particle deposition and dosimetry models.

ACKNOWLEDGMENT

The authors are indebted to our colleagues at Inhalation Toxicology Research Institute for their critical review, to Mr. R. D. Brodbeck and Mr. Thomas D. Holmes for their excellent technical help, and to Dr. T. A. Coons for editing. This research was supported by the U. S. Department of Energy, Office of Health and Environmental Research under Contract No. DE-ACO4-76EV01013.

REFERENCES

Cheng, YS, Y Yamada, HC Yeh, and DL Swift. **1988**. Diffusional deposition of ultrafine aerosols in a human nasal cast. J Aerosol Sci 19:741-751.

Cheng, YS, HC Yeh and DL Swift. **1991**. Aerosol deposition in human head airways for particles 1 nm to 20 μm. Radiat Protect Dosim 38:41-47.

Harkema, JR, CG Plopper, DM Hyde, DW Wilson, JA St George, and VJ Wong. **1987**. Nonolfactory surface epithelium of the nasal cavity of the bonnet monkey: A morphologic and morphometric study of the transitional and respiratory epithelium. Am J Anat 180:266-279.

Heyder, J, J Gebhart, G Rudolf, CF Schiller, and W Stahlhofen. **1986**. Deposition of particles in the human respiratory tract in the size range 0.005-15 μm. J Aerosol Sci 17:811-825.

James, AC, W Jacobi, and F Steinhausler. **1981** Respiratory tract dosimetry of radon and thoron daughters: The state-of-the-art and implications for epidemiology and radiobiology. In: *Radiation Hazards in Mining: Control, Measurement, and Medical Aspects*, M. Gomez (ed.). Kingsport Press, Kingsport, TN.

Lippmann, M. **1977**. Regional deposition of particles in the human respiratory tract. In: *Handbook of Physiology*, DHK Lee (ed.). Section 9: Reactions to Environmental Agents. Williams and Wilkins, Baltimore, MD.

Lippmann, M and RB Schlesinger. **1984**. Interspecies comparisons of particle deposition and mucociliary clearance in tracheobronchial airways. J Toxicol Environ Health 13:441-469.

Morgan, KT, TM Monticello, A Fleishman, and AL Patra. **1989**. Preparation of rat nasal airway casts and their application to studies of nasal airflow, pp. 45-58. In: *Extrapolation of Dosimetric Relationships for Inhaled Particles and Gases*, JD Crapo, ED Smolko, FJ Miller, JA Graham, and AW Hayes, (eds.). Academic Press, Inc., San Diego, CA.

NCRP. **1984**. *Evaluation of Occupational and Environmental Exposures to Radon and Radon Daughters in the United States*, NCRP Report No. 78. National Council on Radiation Protection and Measurements, Bethesda, MD.

Patra, AL, A Gooya, and KT Morgan. **1986**. Airflow characteristics in a baboon nasal passage cast. J Appl Physiol 61:1959-1966.

Schlesinger, RB. **1985**. Comparative deposition of inhaled aerosols in experimental animals and humans: A review. J Toxicol Environ Health 15:197-214.

Schreider, JP. **1984**. Comparative anatomy and function of the nasal passage, pp. 1-25. In: *Toxicology of the Nasal Passage*, CS Barrow (ed.). Hemisphere, New York.

Stahlhofen, W, J Gebhart, and J Heyder. **1980**. Experimental determination of the regional deposition of aerosol particles in the human respiratory tract. Am Ind Hyg Assoc J 14:385-398a.

Stahlhofen, W, J Gebhart, and J Heyder. **1981**. Biological variability of regional deposition of aerosol particles in the human respiratory tract. Am Ind Hyg Assoc J 42:348-352.

Stahlhofen, W, J Gebhart, J Heyder, and G Scheuch. **1983**. New regional deposition data of the human respiratory tract. J Aerosol Sci 14:186-188.

Stuart, BO. **1984**. Deposition and clearance of inhaled particles. Environ Health Perspect 55:369-390.

Task Group on Lung Dynamics. 1966. Deposition and retention models for internal dosimetry of the human respiratory tract. Health Phys 12:173-207.

Yamada, Y, YS Cheng, HC Yeh, and DL Swift. **1988**. Inspiratory and expiratory deposition of ultrafine particles in a human nasal cast. Inhal Toxicol 1:1-11.

Yeh, HC, RF Phalen, and OG Raabe. **1976**. Factors influencing the deposition of inhaled particles. Environ Health Perspect 15:147-156.

Yu, CP, CK Diu, and TT Soong. **1981**. Statistical analysis of aerosol deposition in nose and mouth. Am Ind Hyg Assoc J 42:726-733.

QUESTIONS AND ANSWERS

Q: Cross. Are you planning to do something similar with the rat nasal casts?

A: Yeh, ITRI. For rat casts, we have done the ultrafine particle portion, but not the thoron progeny. The rat airway is so small, we have to figure out how to avoid the loss between the connecting lines; but we are planning that.

Q: Cross. What's your guess, that the deposition efficiencies might be similar to those you find with the human and the monkey?

A: Yeh. Well, for the ultrafine particles, excluding the radon progeny, the deposition curve for 0.2 μm down to about 0.005 μm is very similar to that of the human, but maybe about 20% higher.

Q: Cross. Would you care, then, to stick your neck out? If there are such similar deposition efficiencies, and we so readily see nasal carcinomas in rats exposed to high unattached fractions, would you might see these in humans?

A: Yeh. I would think the nasal deposition for radon progeny in rat casts would be pretty high, probably higher than the human.

Q: Cross. You said 20% difference for ultrafine particles, is that right?

A: Yeh. In human casts, we found, it is somewhere around 75%. So you multiply 75% times 1.2, so it's probably around 90% for the rats.

Q: Cross. But close enough that you might suspect that you could see nasal carcinomas in high unattached human exposures?

A: Yeh. . . . Yes, but also depending on the clearance and how high the exposure was for the human. The nasal clearance compared to the bronchial airway is much, much higher. Normally we are talking about a human nasal clearance half-time of about 10 minutes, and hours for the tracheobronchial.

C: Cross. But I'm talking only in the nasal passages . . .

A: Yeh. Nasal passage clearance is so fast; also, the clearance between rats and humans may be different.

Q: Cross. But because of the short-lived activity, the clearance difference may not make an awful lot of difference in dose or risk?

A: Yeh. Probably not.

Q: Knutson. How did you measure a 1.4-nm size particle for unattached radon daughters? Do you see any change in that size? If the size was, instead, 0.8, would it affect the deposition?

A: Yeh: The size for the ultrafine particle was given by the electrical classifier. We also confirm that with a diffusion battery and for the radon-220 progeny we use a diffusion battery to determine the size. Actually, we are using a device very similar to what you are using, a five-screen graded diffusion battery for size determination. From the trend shown in the graph, it looks like 0.8 and 1.4 will be different. The 0.8 will be higher.

Q: Knutson. Do you see any change in the 1.4 figure with repeated tests, or is that a pretty stable number?

A: Yeh. The dots on the viewgraph there, the data are, I think, from about nine measurements. And the size range is, I think, between 1.25 and 1.5 nm. The average is about 1.4 nm, so it's very, very comparable. I think those are within the experimental errors.

Q: Hopke. What do you do to control electrostatic effects? Do you coat the inside, or do you do something to potentially remove static charge that would build up on this plastic?

A: Yeh. For the plastic, we prime it, using antistatic spray.

Q: Swift. Do you think if you put this primate on a treadmill that you could induce him to make the fateful mistake of opening his mouth and breathing orally?

A: Yeh. I think I have to ask our veterinarian and see if that can be done; I don't know.

Q: Swift. Most smaller animals do not take to opening their mouths and breathing. I wonder if the primate will so behave?

A: Yeh. I don't know. It depends on how well you can train the monkey, I guess, and see if the monkey can tolerate the mask. In addition to the cast study, we also studied the deposition in live monkeys for comparison. In that study we have to anesthetize the monkey, so I don't know.

Q: Park, PNL. Have any of the experimenters using casts evaluated what the influence of nasal hair would be on deposition?

A: Yeh. The study we have done thus far is with David Swift's human cast. Your C cast or F cast? [C—*Swift*]. In that cast he has both hair and no hair. In that study there was essentially no effect.

A: Cheng. May I add to this point? We've measured the deposition of thoron progeny with the cast, with and without hair. Essentially, with the hair the deposition efficiency increased slightly; probably 2-3% higher than without hair; so it's a very minor difference.

Deposition and Fate of Inhaled Radon and Radon Decay Products

WHO INDOOR AIR QUALITY REPORT

Preceding these questions, there was a general discussion by Suess of WHO on an indoor air quality report on radon written by a WHO Working Group (1988).

Q: James, PNL. What level of risk do you assume to be reasonable presuming you set the 100 Bq m^{-3} on the basis of a view of a risk that's acceptable? You said that below that is safe; do you really mean that? Do you really think that the risk is zero below a hundred and finite above?

A: Suess, WHO. I cannot answer straight, but I can say that in this particular book I refer to a number of pollutants, including asbestos. And I think everybody knows we are not going to go and discover asbestos, which is also controversial in some ways. But we will not give a guideline for safe levels because we don't think there is one. We gave the risk assessment, that one in a million, or what have you. I don't have the number for radon, I don't think we gave it at the time because the approach was, perhaps, following a little later. Maybe somebody else has it. But since we have a background of, give or take, three and then average, we cannot prevent it. I think it's ideal to get to natural background indoors too, but it may not be practical. We should go by the ALARA principle, and I think that everybody will agree in principle too. We have to let at least the politicians know what is the risk; they have to make the calculation of how many of the general public they are willing to sacrifice.

Q: Cross, PNL. You just made the statement and answered your own question when you said you believe there is no safe level, so obviously it's all reality, there is no myth to it. We know that there is a divided audience, I don't know if it's evenly divided or not, but of course you will hear some papers where there is no correlation with some of these epidemiology studies. You will hear an awful lot of discussion Thursday on why a lot of these studies are not expected to show results one way or the other, because of the improper design. So it's probably premature to say. I can say, based on the animal studies, that we haven't seen any lower levels where there still isn't some risk of lung cancer. So if you think that animals parallel humans in some ways, then there's a partial answer for the question of risk. But in defense of the program, I had hoped to get several proffered papers directly bearing on that question, and not one came forth. People are very timid. It's probably premature to go away with that very definite answer, of "reality" regarding the risk of indoor radon, but I'm hoping that a lot of what we hear here this week will at least influence our opinion and perhaps change our minds. By Friday we may have a stronger answer.

C. Suess. I heard about your problem of not getting the papers with the appropriate title, and that's why I raised it; it was really in support of your morning point. I would like to mention that the paper on which my points here are based is available for distribution outside the registration office (WHO Working Group, 1988). And I would like to indicate that on the back of that paper you can see the list of the subgroup that dealt with the radon issue. I'm sure you know a lot of the names and some of them are sitting here. These people that did deal with the issue, whom I had the pleasure of bringing together, seem to have had an opinion of reality, and again I think the discussion should go on and maybe we can go more deeply into it, but I think the chairman wants to . . .

C. Cross. By the end of Thursday we should be much more secure on that position.

C. Johnson, PNL. I hope you're right, Fred, but I doubt it. I'm going to take the next couple of minutes to just cover the material that was in the cancelled paper on radon uptake in humans, to at least give you my version of it. I think it's an interesting topic, and it does serve somewhat as an introduction to the next paper.

REFERENCE

WHO Working Group. 1988. Indoor air quality: Radon—Report on a WHO Working Group. J Environ Radioactivity 8:73-91.

ON THE EXHALATION RATE OF RADON BY MAN[a]

J. Rundo, F. Markun, and N. J. Plondke

Argonne National Laboratory, Argonne, Illinois

Key words: *Radon, exhalation, postprandial effect*

ABSTRACT

This paper describes some aspects of the exhalation rate of radon by man which may be relevant to its internal dosimetry and, therefore, to possible radiobiological consequences. Prolonged exposure of a person to radon results in a reservoir of radon dissolved in body fat and fluids. If the person then moves to an environment with a lower radon concentration, there is a net exhalation of radon, and the initial exhalation rate depends on the radon concentration in the first environment. This is demonstrated for seven persons whose houses contained radon at concentrations varying from 10 Bq m^{-3} to almost 1000 Bq m^{-3}. About 1 h after leaving the house, the subjects' average exhalation rate of radon, expressed as the equivalent volume of house air per unit time, was 236 ml min^{-1}.

In general, the exhalation rate declined in a manner that seemed to be predictable from the integral of the equation that describes the retention of a single inhalation exposure to radon. However, the behavior is complicated by a major but short-lived postprandial increase in the exhalation rate of radon by persons whose only source of radon was in the air of their homes. The phenomenon was studied in seven subjects, who showed initial exhalation rates ranging from 3 mBq min^{-1} to 200 mBq min^{-1}. The excess radon exhaled amounted to ~4 L to ~15 L, when expressed as the equivalent volume of house air. This radon must have come from a reservoir somewhere in the body. The possible dosimetric and radiobiological consequences of this phenomenon are unknown.

INTRODUCTION

Current concern about possible health effects of indoor radon necessitates investigation of all aspects of radon in man, i.e., physical as well as biological factors. In this paper we present some observations on the rate of exhalation of radon inhaled in the indoor environment.

It has been known for many years that retention in the human body of radon that has been inhaled can be described mathematically by the sum of five exponential functions of time (Harley et al., 1951;

(a) Work supported by the U.S. Department of Energy, Office of Health and Environmental Research, under Contract No. W-31-109-ENG-38.

Lucas and Markun, 1972). Each function is considered to represent retention in a particular compartment in the body. Thus, the compartment that clears very rapidly (half-time, 23 s) presumably represents the lung; the slowest compartment (half-time, about 18 h) is identified as fat. The other compartments have half-times of about 4.5 min, 41 min, and 120 min, respectively. Consequently, prolonged inhalation of radon results in a pool in the body, the size of which is determined both by the radon concentration in the air, the constituents of the body (i.e., percent fat), and by the duration of exposure. In general, an equilibrium will be reached in exceptional cases only (i.e., cases of continuous exposure for several days).

When a person moves to an environment with a lower radon concentration, there is a net exhalation of radon, and the exhalation rate is determined by the same factors. If the parameters of the retention equation were fairly similar for all persons, there would be a linear relationship between the initial exhalation rate and the concentration in the house for exposures of similar durations.

Our observation of a postprandial peak in the rate of exhalation of radon by persons with long-standing burdens of radium (Rundo et al., 1978) raised the question of the origin of the "excess" radon exhaled during the period of digestion. If it were the result of a transient change in the fraction of radon lost from bone, the dosimetric consequences would be trivial. On the other hand, if it were the result of the flushing of a reservoir of radon dissolved in soft tissue, it could be important to identify the reservoir, which might be the recipient of chronic irradiation at levels greater than those of many soft tissues. This question could be resolved by the demonstration of a postprandial peak in the exhalation of radon, the only source of which was inhalation in the home environment.

We therefore decided to determine serially the exhalation rate of radon in the breath of ourselves and some colleagues, paying particular attention to the period following a meal. In this way we could explore the relationship between a "high" environmental radon concentration and the initial exhalation rate soon after leaving that environment, as well as seek evidence for a post-prandial peak in the exhalation rate.

EXPERIMENTAL METHODS

Samples of air were collected from the houses of seven subjects, and the concentrations of radon to which they had been exposed for periods of about 14 h were determined with an alpha-particle scintillation radon counter (Lucas, 1957). Data are shown in Table 1.

Table 1. Details of subjects for whom radon exhalation rates were measured and the concentrations of airborne radon in their houses.

Subject No.	Sex	Height, m	Weight, kg	Age, y	Rn concentration in house air, mBq L^{-1} [a]
50-002	M	1.68	78.2	51	57 ± 4
50-009[b]	M	1.78	77.3	52	59 ± 4
50-009[b]	M	1.78	77.3	52	51 ± 4
50-026	F	1.58	94.5	29	962 ± 21
50-070	M	1.83	82.3	32	10 ± 3
50-109	M	1.71	84.1	38	46 ± 2
50-148	M	1.74	69.5	52	153 ± 6
50-150	M	1.73	79.3	30	518 ± 5

[a] Means of pairs of consistent results on samples taken the night before and on the morning of the test, except for subjects 50-026 and 50-070, for whom house air samples were taken 2 d later.
[b] Subject 50-009 was tested on two occasions, a week apart.

As soon as possible after arriving at Argonne, five of the subjects brought a breakfast of choice to the underground laboratory ("vault"), which is ventilated with radon-free air. One or two 10-min samples of breath were taken for determining the initial exhalation rate of radon, and the subject then ate breakfast. Breath-sampling was started immediately after the end of the meal and at first was almost continuous: serial 10-min collections were made during a period of 1-2 h, with a 2- or 3-min interval between each collection while the radon trap (charcoal) was being changed. After this time, less frequent collections were made.

RESULTS

The exhalation rates for the first 10-min samples of breath are plotted as a function of the indoor radon concentrations in Figure 1. There is an altogether respectable correlation between the two, the straight line having a slope of 236 ml min^{-1}, when radioactivity is expressed in units of the equivalent volume of room air (EVRA; Lucas and Stehney, 1956). The relatively slight spread of the points (in an admittedly small sample) indicates that the retention equations for the seven subjects must have been similar.

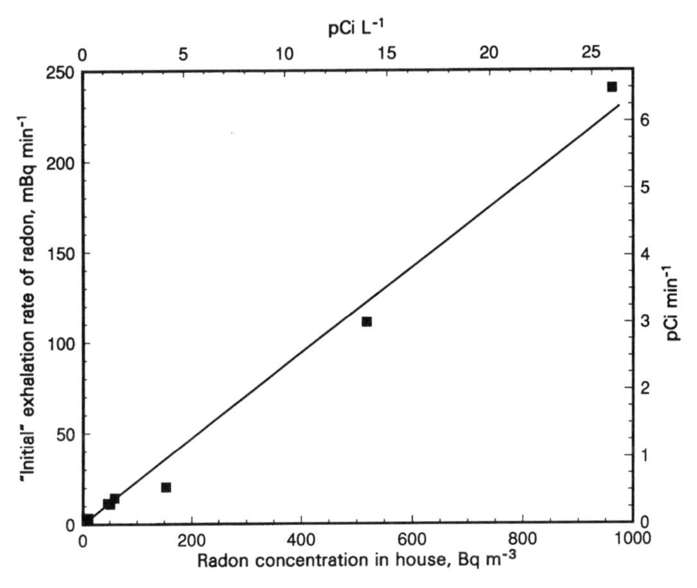

Figure 1. Exhalation rates of radon as a function of the indoor radon concentration for seven subjects.

Other factors being equal, one would expect the radon exhalation rate to decline monotonically. In Figure 2 the serial results for subject 50-150 (for whom the initial exhalation rate plotted in Figure 1 was just over 100 mBq min^{-1}) show that other factors are far from equal. The curve labeled "Integrated Retention Equation" is a plot of the retention function for a single intake of radon, integrated for continuous exposure for 14 h, then normalized to pass through the early experimental points. Statistical uncertainties are not indicated because, for all points, error bars ($\pm 1\sigma$) would be no greater than the heights of the symbols; the variability was therefore due to biological factors. After breakfast (eaten at the time indicated by the letter "B"), an initial decrease in the radon exhalation rate was followed by a substantial increase to a peak some 70-90 min after breakfast, then a gradual decline to a "normal" or "expected" level some 3 h after the meal. The subject also ate his lunch (at the time indicated by the letter "L") in the underground laboratory, and serial breath collections and analyses for radon were continued. The results suggested that a second postprandial peak was present, but the exhalation rate 3 h after lunch was considerably higher than expected on the basis of the smooth curve representing the integrated retention function.

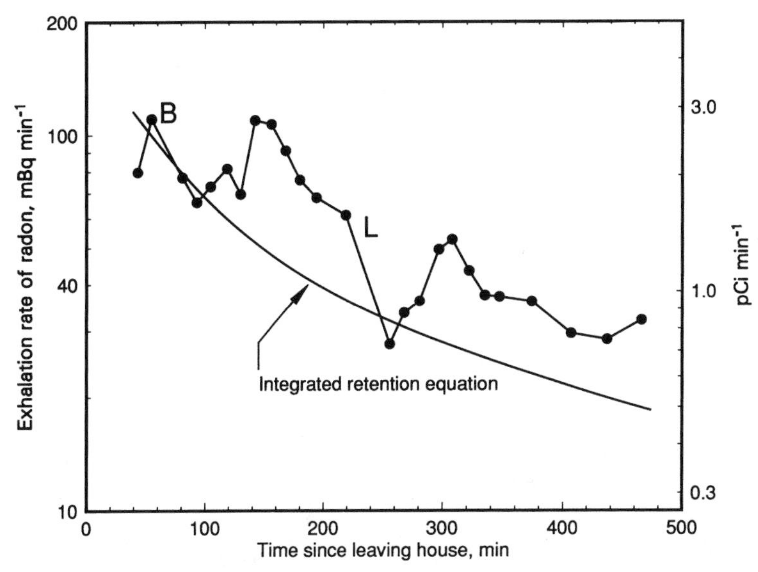

Figure 2. Exhalation rate of indoor radon by subject 50-150 (experimental points) as a function of time since inhalation. Rate may be compared with that predicted by multi-exponential retention function, integrated for continuous exposure for 14 h, and normalized by trial and error to early data (smooth curve). Letters indicate midpoints of meals (B = breakfast, L = lunch).

A similar series of breath collections and radon analyses was made for the two other subjects whose houses showed radon concentrations in excess of 150 Bq m^{-3}. The results are plotted in Figure 3 together with the results for subject 50-150.

In all three cases, there was a pronounced postprandial peak after breakfast, although the increase started sooner after the meal for subject 50-148. The dashed lines drawn under the first peaks represent assumed exponential baselines to permit calculation of the "excess" radon exhaled during the period of the peak. The subjective nature of this procedure is self-evident. The evidence for a peak after lunch was less conclusive for subjects 50-026 and 50-148 than for subject 50-150, although it is still suggestive.

We wished to determine if the postprandial peak could be deferred until after lunch. Accordingly, two subjects ate no breakfast; the exhalation rate of radon was determined at intervals during the morning and continued more frequently after lunch, which was eaten in the underground vault. The results are plotted in Figure 4; error bars ($\pm 1\sigma$) are again within the heights of the symbols. The postprandial peaks are quite unambiguous in both cases, but the shapes are very different. For subject 50-109, the dashed line, which represents the extrapolation of a least-squares fit of an exponential function to the data from 143 to 295 min, was used as the baseline for the peak. This procedure was not possible for subject 50-070 because the data showed no well-defined decrease before lunch; the results also suggest that sampling of the breath was discontinued before the exhalation rate of radon had recovered from the elevated values.

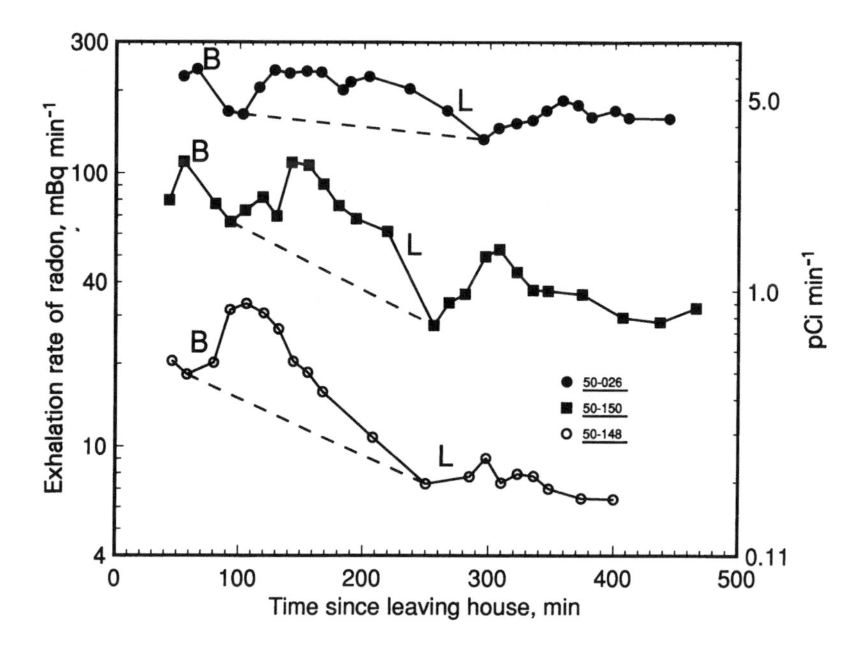

Figure 3. Exhalation rate of radon (logarithmic scale) for three subjects as a function of time since leaving home. Data plotted in Figure 2 are included. Letters indicate midpoints of meals (B = breakfast, L = lunch).

The total amount of radon exhaled during the period of the peak was calculated for each subject. The amount that was exhaled during the time between the collections of successive pairs of samples was assumed to be at a rate equal to the average of the rates before and after. These quantities were calculated and added to the sum of the amounts found in the sampling periods. With one exception, the estimated amount of exhaled radon which was not collected ranged among the subjects from 30 to 41% of the total. The exception was the first test on subject 50-009, where it was 69% of the total; this was an exploratory test, and gamma-ray measurements were also being made in vivo during the course of the morning.

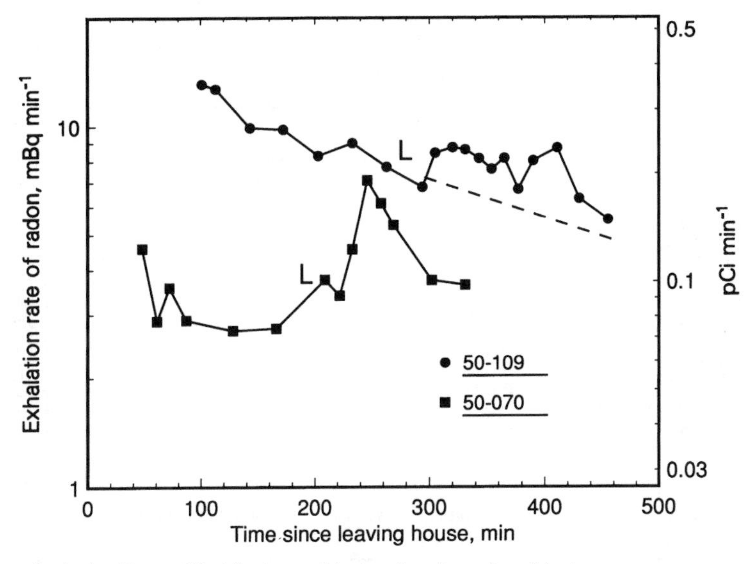

Figure 4. As for Figure 3 but for two subjects who ate no breakfast.

For each subject, we assumed that had there been no postprandial peak, the rate of exhalation of radon would have followed an exponential decline, as exemplified by the dashed lines in Figures 3 and 4. The total area under the baseline was determined by integration of the exponential function between appropriate limits. This quantity was then subtracted from the total radon exhaled during the period of the peak (Table 2). The uncertainties shown are based on statistical errors of counting; however, in calculating the error on the total amount exhaled, the estimated error on the amount exhaled between successive pairs of samples was doubled before propagation with those for the samples, to allow for some biological variability. For each subject, the uncertainty on the area under the baseline was derived from the error on the ratio of the exhalation rates at the beginning and end of the period of the peak.

In order to compare the data for subjects who had been exposed to widely differing concentrations of radon (but for comparable periods of about 14 to 16 h), the net area in each peak was converted from activity to the equivalent volume of room air by dividing by the appropriate concentration in the house air (column labeled "EVRA,L,"

Table 2). The range of values for the peak areas is reduced from about 70:1 to about 4:1. As a measure of the relative effect, the ratio of the net peak area to the area under the baseline was calculated, and the results (Table 2) show a range of about 3:1.

Table 2. Total, baseline, and net excess amounts of radon exhaled by seven subjects during postprandial peak.

Subject No.	Total radon exhaled during postprandial peak, Bq	Area under baseline, Bq	Net peak area		Peak area / Baseline area
			Bq	EVRA, L[a]	
50-002	1.63±0.01	1.25±0.04	0.38±0.04	6.7±0.8	0.31
50-009	1.87±0.03	1.59±0.04	0.27±0.05	4.7±1.0	0.17
50-009	1.12±0.01	0.93±0.03	0.19±0.03	3.7±0.7	0.20
50-026	43.6±0.3	32.9±0.7	10.7±0.7	11.1±0.8	0.33
50-070[b]	0.57–0.63	0.43–0.44	0.14–0.19	13.6–17.8	0.33–0.42
50-109	1.32±0.01	1.03±0.05	0.28±0.05	6.2±1.2	0.28
50-148	3.32±0.03	2.05±0.05	1.27±0.06	8.3±0.5	0.62
50-150	12.3±0.1	7.2±0.2	5.1±0.2	9.8±0.4	0.71

[a] Equivalent Volume of Room Air obtained by dividing the net peak area in mBq by the concentration in house air in mBq L^{-1} (Table 1).
[b] All values depend on starting and ending times assumed for peak and on how baseline is drawn. See Figure 4.

DISCUSSION

Observation of the postprandial peak in the exhalation of radon from the environment proves that the source of the similar peak in the exhalation of radon produced in vivo is a reservoir of radon dissolved in body fluids or soft tissue rather than a change in the fraction released from radium in bone. We had hoped that the area of the peak, when expressed as the equivalent volume of room air, might give some clue as to the identity of the reservoir. However, there is no obvious correlation between the area of the peak and any of the variables in Table 1. One possibility is the capillary bed; another, perhaps more likely, is the omentum, which may contain much adipose tissue. The greater solubility of radon in fatty media than in aqueous ones is well known. More data are needed before any conclusion can be drawn concerning the dosimetric or radiobiological consequences.

The detailed behavior of the exhalation rate during the period of the peak merits comment. For subjects 50-026 (Figure 3) and 50-109 (Figure 4), there is a suggestion of a splitting or doubling of the peak. When first observed (in subject 50-109), this was thought to be merely biological variation, but its observation in subject 50-026 tended to confirm it as a real phenomenon, even though its existence depends on the validity of one low point in each case. The peaks for subjects 50-148 and 50-150 (Figures 2 and 3) do not show a doubling, but there is marked asymmetry of the peak in each case. Because of variations in respiratory minute volume of subject 50-148, there was actually a small increase in the concentration of radon in the breath at 158 min. Similarly, the results for subject 50-070 (Figure 4) suggest that there may have been a "step" in the declining portion of the peak which was not observed because breath sampling was discontinued too early. These observations warrant further study, perhaps with a stable isotope of another of the noble gases, the behavior of which may be similar to that of radon. It is possible that this noninvasive technique may offer a tool to physiologists for the study of some aspects of the digestive process not otherwise amenable to investigation.

REFERENCES

Harley, JH, E Jetter, and N Nelson. **1951** *Elimination of Radon from the Body*, Report No. 3. Analytical Branch, Health and Safety Division, New York Operations Office, U.S. Atomic Energy Commission, Washington, DC. Reprinted as US AEC Report HASL-32 (1958).

Lucas, HF, Jr. **1957**. Improved low-level alpha scintillation counter for radon. Rev Sci Instrum 28:680-683.

Lucas, HF and F Markun. **1972**. Radon Breath Measurements: Estimates of Corrections for Previously Inhaled Radon, pp. 136-140. In: *Radiological and Environmental Research Division Annual Report, Center for Human Radiobiology,* July 1971- June 1972, ANL-7960, Part II. Argonne National Laboratory, Argonne, IL.

Lucas, HF and AF Stehney. **1956**. Radon Contamination in the Measurement of Low Levels of Expired Radon, pp. 6-11. In: *Argonne National Laboratory, Radiological Physics Division Semiannual Report, January - June 1956*, ANL-5596, Part II. Argonne National Laboratory, Argonne, IL.

Rundo, J, F Markun, and JY Sha. **1978**. Postprandial changes in the exhalation rate of radon produced in vivo. Science 199:1211-1212.

QUESTIONS AND ANSWERS

Q. (Speaker unidentified.) One possible source of exposure that we haven't considered in this conference yet is for radon mitigators to check pressure differential when working in a home where one of the things that they do is to drill through the concrete slab. What might occur quite commonly is that, drilling through the floor, where the pressure under the slab may be greater than inside the building, they could get a puff of radon amounting to 10,000, 50,000 pCi/L when they make that initial breakthrough. I wonder what your thoughts might be about the dosimetric significance of that short-duration, high-level radon exposure. Also, what are your thoughts about how to measure that, as far as dosimetry?

A. Rundo, Argonne National Laboratory. I think that would be extremely difficult to assess. Of course one could sit down with pencil and paper, or pocket calculator, and do all sorts of figuring. I don't know whether this would be a worthwhile exercise, but clearly you could do something about making calculations. The question is, what reliance do you place on those calculations, and how many people are going to be involved in this kind of thing?

SMOKING PRODUCED MUCUS AND CLEARANCE OF PARTICULATES IN THE LUNG

T. D. Sterling[1] and T. M. Poland[2]

[1]School of Computing Science, Simon Fraser University, Burnaby, British Columbia, Canada

[2]Theodor D. Sterling and Associates Ltd., Vancouver, British Columbia

Key words: *Radon, smoking, lung cancer, mucus, particle clearance*

ABSTRACT

Some studies of miners have shown a lesser relative lung-cancer risk for smokers than for nonsmokers. For example, experiments by Cross and associates with dogs have shown an apparent protective effect of cigarette smoke against radon-daughter and dust exposure. One reason for these changes may be the thickened mucus layer in the tracheobronchial region of smokers. Physiological changes in the lung due to smoking may decrease the effects of radioactive particles on cancers in the bronchial region by apparently promoting faster clearance, in that region, of radioactive particles and by decreasing the radiation dose through reduced penetration to the sensitive basal epithelial cells. Because of the short half-life of radon daughters, even if there is possible tobacco-related delay of particle clearance from the alveolar region it cannot affect radon clearance. Therefore, the possible mitigating effect of tobacco on radon-produced cancer appears to be limited to the tracheobronchial region. It would be of value to a number of occupations if the same changes in the lungs due to smoking could be produced in exposed workers in the absence of cigarette-smoking. Beta-carotene and vitamin A, which affect maintenance and secretion of the mucosal lining, appear to thicken mucus, thereby providing protection against radon-induced lung cancers that is similar to smoking-related changes in the lung.

INTRODUCTION

Estimates of risk associated with exposure to radon in residences and other indoor environments are mainly based on models which extrapolate from human occupational or experimental animal studies. In general, models predict that indoor levels of radon will be associated with significantly increased risk of lung cancer. However, these models are unclear as to the effect of radon when combined with cigarette-smoking.

Numerous studies have demonstrated that underground miners exposed to radon and radon daughters are at increased risk for lung cancer. Early studies of Colorado plateau miners attributed much of the excess risk to smoking (Wagoner et al., 1965). After extended follow up, however, Lundin et al. (1969) found that relative risks for lung cancer in uranium miners were the same for exposed smokers compared to unexposed smokers and for exposed nonsmokers compared to unexposed nonsmokers. More recent studies of Colorado plateau miners by Saccomanno et al. (1988) calculate only absolute risks for smoking and nonsmoking miners, using the white U.S. general population as referents for both smokers and nonsmokers. Therefore, their published results do not allow a comparison of relative risks computed for smokers and nonsmokers separately.

A number of studies of Swedish miners show that radiation exposure was the primary cause of lung cancer and that smoking miners actually had a lower relative risk of lung cancer than nonsmoking miners (Dahlgren, 1979; Damber and Larson, 1985; Radford and St. Clair Renard, 1984). One study even found that the absolute risk of lung cancer was significantly less in smokers than in nonsmokers (Axelson and Sundell, 1978). The results of these studies are summarized in Table 1.

Table 1. Lung-cancer risks for smokers and nonsmokers occupationally exposed to radon (from Reinstein et al., 1981).

	Risk	
Reference	Smokers	Nonsmokers
Axelson (1978)*	14.9	58.1
Dahlgren (1979)	0.16	0.4
Damber (1985)		
All ages:	4.7	5.4
Ages > 80:	2.9	3.9
Lundin (1969)	4.0	4.0
Radford (1984)	2.9	10.0

* Risks computed are absolute risks (i.e., exposed smokers and nonsmokers were compared to the same referent group—the general population). All other risks presented in this table are relative risks (i.e., exposed smokers were compared to unexposed smokers and exposed nonsmokers were compared to unexposed nonsmokers).

Axelson and Sundell (1978) and Sterling (1983, 1989) interpreted the difference between smokers' and nonsmokers' risks as the result of smoking-related changes in the lung, especially because of increased mucus thickness.

A study by Cross et al. (1982) on beagle dogs supports the interpretation of a possible mitigating effect of smoking. One group of dogs was exposed to radon, radon daughters, and uranium ore dust. A second group of dogs was exposed to the same particulates as well as cigarette smoke. In the first group of dogs, 8 out of 19 developed respiratory tract cancer; in the second group, only 2 of 19 dogs developed the lesions.

Results of both animal experiments and human occupational studies suggest that the effects of smoking are complex and may have a mitigating effect on radon-produced tumors. We review here the possible effects of smoking-induced changes in the lung on tumor production in order to explain the results of animal and human studies.

PHYSIOLOGICAL CHANGES IN THE LUNG DUE TO SMOKING

Many studies have found hypersecretion of mucus in the lungs as a result of smoking (Agnew et al., 1987; Ashford et al., 1970; Comstock et al., 1972; deHamel and O'Donnell, 1972; El-Sewefy, 1970; Evans et al., 1969; Higgins, 1971; Holland et al., 1969; Howard, 1970; Radsel and Kambic, 1978; Seely et al., 1971; Weiss et al., 1983; Woolf and Suero, 1971; Woolf, 1974). These findings are supported by studies that found hypersecretion of mucus in animals exposed to cigarette smoke: in dogs (Auerbach et al., 1967; Battista and Kensleer, 1970; Frasca et al., 1968; King et al., 1989); sheep (Mawdesley-Thomas, 1973); and rats (Hayashi et al., 1979; Jones et al., 1973; Lamb and Reid, 1969; Smith et al., 1978). Other studies also found smoking-induced thickening of bronchial epithelium (Frasca et al., 1968; Auerbach et al., 1961; Radsel and Kambic, 1978). One study found decreased mucus viscosity in dogs as a result of smoking (King et al., 1989). Cross et al. (1982) did not examine mucus content of the lung in the dogs used in their study.

EFFECT OF MUCUS CONTENT ON MIGRATION, CLEARANCE, AND DEPOSITION OF PARTICLES IN THE LUNG

The increased quantity and decreased viscosity of lung mucus among smokers has been found to facilitate migration of particles. In an important study by Cohen et al. (1979), smoking and nonsmoking subjects inhaled dust consisting of small magnetic particles and then were immediately subjected to high magnetic flux. When the remanent field was measured, the magnetic particles were inhomogenously distributed in the lungs of smokers but evenly distributed in the lungs of nonsmokers. The Cohen data can be interpreted to mean that clusters formed in the lungs of smokers but not of nonsmokers when subjected to magnetic flux. This difference in distribution suggests that particles migrate more readily in smokers' than in nonsmokers' lungs because of increased mucus production. This difference was confirmed experimentally by Reinstein et al. (1981), who studied migration of magnetite in a mucus-like medium. They found that particles migrated more readily in less viscous lung equivalent medium (Figure 1). While the Cohen experiment is important, it was designed with a regrettable disregard of studies which demonstrated that instillation of FeO_2 (the magnetite used by Cohen) could cause cancer of the lung in animals (Sterling, 1981).

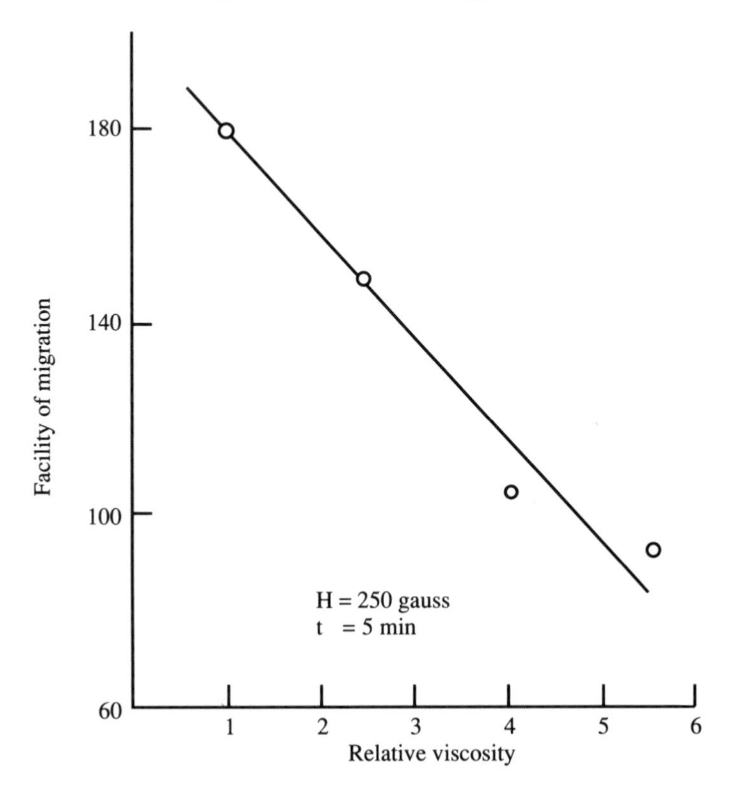

Figure 1. Migratory facility (measured thickness of clusters formed in 5 min under 250 gauss) as a function of viscosity of mucus-equivalent medium. (From Reinstein et al. 1981)

One would expect that increased mucus, facilitated migration, and productive cough would affect particle clearance from the lungs. However, it is difficult to draw any clearcut conclusions about the difference between smokers' and nonsmokers' particle clearance.

Results are conflicting in a number of inhalation studies in which subjects inhaled radioactively labeled aerosol particles, and gamma cameras or scintillation probes were used to measure particle retention and epithelial transport. Three studies found no difference in clearance rates between smokers and nonsmokers (Thomson and Pavia, 1973; Pavia et al., 1970; Lourenco et al., 1971). Six studies found decreased clearance due to smoking (Bohning et al., 1982; Foster et al., 1985; Green, 1970; Svartengren, 1986; Weiss et al., 1983, 1984). The study by Bohning et al. (1982) found decreased alveolar clearance. The study by Green (1970) found decreased mucociliary activity in vitro. The remaining four studies found slower tracheobronchial mucociliary clearance in smokers compared to nonsmokers. Finally, 10 studies found increased mucociliary clearance in smokers compared to nonsmokers, particularly in peripheral areas of the tracheobronchial tree (Agnew et al., 1982, 1987; Camner and Philipson, 1974; Groth et al., 1990; Huchon et al., 1984; Jones et al., 1980; Luchsinger et al., 1968; Mason et al., 1983; Sanchis et al., 1971).

In summary, 19 studies examined particle clearance from the lungs of smokers and nonsmokers. Three found no difference in clearance between smokers and nonsmokers, 6 found decreased clearance due to smoking, and 10 found increased clearance in smokers compared to nonsmokers.

Again, these results are consistent with animal research. King et al. (1989), using dogs, and Minty and Royston (1985), using rats, found increased clearance in animals exposed to cigarette smoke. On the other hand, two studies found reduced clearance in guinea pigs exposed to cigarette smoke (McFadden et al., 1986; Rylander, 1971).

The most reasonable conclusion to be drawn from these contradictory results is that smoking-related changes in the lung may increase peripheral clearance, but that cannot be concluded with certainty.

The effect of smoking-induced changes in mucus on particle deposition is important. A number of studies examined radioactively labeled aerosol deposition in the lungs of smokers and nonsmokers. One investigator found that deposition of particles was shifted to more proximal regions of the tracheobronchial tree in smokers (Lippman and Altshuler, 1976). Three investigators found no difference in deposition patterns between smokers and nonsmokers (Foster et al., 1985; Huchon et al., 1984; Lourenco et al., 1971).

EFFECTS OF SMOKING-INDUCED CHANGES IN THE LUNG ON PENETRATION OF RADIOACTIVE PARTICLES

Increased amounts of mucus and possible thickening of the pleura and bronchial epithelium found in smokers may reduce the radiation dose to the sensitive basal cells of the bronchial epithelium. Cancers tend to be rare in the trachea and main bronchi, where epithelial distances are too great for alpha particles to penetrate to the basal cells (Axelson, 1984). In peripheral areas of the bronchial tree, the distance from the mucus surface to the sensitive epithelial basal cells is critical because it is approximately the same as the range of alpha-particle penetration. Penetration by radon daughters polonium-218 and -214 is reported as 47 μm and 71 μm, respectively (Walsh, 1970).

The minimal distance from the surface of the mucus layer to the basal cells in the critical region of the bronchi is estimated as at least 36 μm (Altshuler et al., 1964). The thickness of the mucus sheath in the critical region is estimated to be as much as 20 μm or more (Jacobi, 1964).

Because smokers have increased mucus production, the mucus sheath covering their bronchial surfaces is thicker than in nonsmokers. In addition, because of decreased viscosity of mucus and productive cough, particles to which radon daughters adhere may be removed more readily from the tracheobronchial region in smokers' than nonsmokers' lungs. Therefore, the dose of radiation to the tracheobronchial region is correspondingly reduced. Walsh (1970) evaluated alpha-radiation dose to the critical basal cells and estimated that increasing mucus considerably reduced the effect of radon-daughter alpha particles. For instance, he estimated that increasing the mucus sheath by even as little as 10 μm would reduce the radiation dose by half (Axelson and Sundell, 1978).

Alveolar clearance times are much slower than tracheobronchial clearance times: on the order of several days, or even months. The half-lives of radon daughters polonium-214 and -218, which attach to particles and may lodge in the lungs, are 3.11 min and 1.5×10^{-4} s, respectively (WHO, 1988). Therefore, damaging alpha particles would be emitted long before alveolar clearance could take place. Any differences in alveolar clearance times between smokers and nonsmokers would not affect radiation dose in alveolar regions.

With respect to the influence of smoking-related changes in the lung on risk of cancer from exposure to radon and radon daughters, four factors emerge:

1. Because of lengthy alveolar clearance times and the short half-life of radioactive particles, effects on alveolar clearance times related to smoking are irrelevant. Therefore, any effect of smoking is limited to radon-induced lung cancer arising in bronchial regions.
2. Mucus hypersecretion and productive cough among smokers may enhance mucociliary clearance, particularly in peripheral areas of the tracheobronchial tree, where alpha particles penetrate to the sensitive basal epithelial cells.
3. Deposition of particles in the lungs of smokers may be in more proximal regions than for nonsmokers; there, epithelial distances are greater, and alpha particles cannot penetrate to the sensitive basal cells.
4. Enhanced mucus production increases the thickness of the mucus layer covering the bronchi in the lungs of smokers.

CONCLUSION

Physiological changes in the lung due to smoking may decrease effects of radioactive particles on cancers in the tracheobronchial region by faster clearance of radioactive particles and by decreasing the dose of radiation through reduced penetration to the sensitive basal epithelial cells. Because of the short half-life of radon daughters, possible smoking-related delay of particle clearance from the alveolar region probably does not affect radon-induced changes. Therefore, the mitigating effect of smoking on radon-induced cancer appears to be limited to the tracheobronchial region. It would be of value to a number of occupations if the changes in the lungs due to smoking could be produced in exposed workers in the absence of cigarette smoking. It is encouraging to find that beta-carotene and vitamin A, which affect maintenance and secretion of the mucosal lining, appear to thicken mucus and thereby provide protection similar to that of smoking-related changes in the lung from radon-induced lung cancers (Seifter et al., this volume).

ACKNOWLEDGMENT

We thank Dr. Olaf Axelson for critical comments.

REFERENCES

Agnew, J, F Little, D Pavia, and S Clarke. **1982**. Mucus clearance from the airways in chronic bronchitis—smokers and ex-smokers. Clin Respir Physiol 18:473-484.

Agnew, J, D Pavia, M Lopez-Vidriero, and S Clarke. **1987**. Mucus clearance from peripheral airways. Eur J Respir Dis Suppl W3:1580-1585.

Altshuler, B, N Nelson, and M Kuschner. **1964**. Estimation of lung tissue dose from inhalation of radon and daughters. Health Phys 10:1137-1161.

Ashford, J, D Morgan, S Rae, and R Sowden. **1970**. Respiratory symptoms in British coal miners. Am Rev Respir Dis 102:370.

Auerbach, O, A Stout, E Hammond, and L Garfinkel. **1961**. Changes in bronchial epithelium in relation to cigarette smoking and in relation to lung cancer. N Engl J Med 265:253-267.

Auerbach, O, E Hammond, D Kirman, L Garfinkel, and A Stout. **1967**. Histologic changes in bronchial tubes of cigarette smoking dogs. Cancer 20:2055-2066.

Axelson, O. **1984**. Room for a role for radon in lung cancer association? Med Hypoth 13:51-61.

Axelson, O and L Sundell. **1978**. Mining, lung cancer and smoking. Scand J Work Environ Health 4:46-52.

Battista, SP and CJ Kensleer. **1970**. Mucus production and ciliary transport activity. Arch Environ Health 20:326-338.

Bohning, DE, HL Atkins, and SH Cohn. **1982**. Long-term particle clearance in man: Normal and impaired. Ann. Occup. Hyg. 26:259-271, 1982.

Camner, P and K Philipson. **1974**. Mucociliary clearance. Scand J Respir Dis (Suppl 90):45-48.

Cohen, D, S Arai, and J Brain. **1979**. Smoking impairs long term dust clearance from the lung. Science 204:514.

Comstock, G, W Brownlow, R Stone, and P Sartwell. **1972**. Cigarette smoking and change in respiratory findings. Arch Environ Health 21:50.

Cross, F, R Palmer, R Filipy, G Dagle, and B Stuart. **1982**. Carcinogenic effects of radon daughters, uranium ore dust and cigarette smoke in beagle dogs. Health Phys 42:33-52.

Dahlgren, E. **1979**. Lung Cancer, Cardiovascular disease and smoking in a group of mine workers. Lakartidningen 76:4811-4814.

Damber, L and L Larson. **1985**. Underground mining, smoking and lung cancer: A case-control study in the iron ore municipalities in Northern Sweden. J Natl Cancer Inst 74:1207-1213, 1985.

deHamel, E and T O'Donnell. **1972**. Smoking habits and respiratory symptoms in British coal miners. N Z Med J 76:1.

El-Sewefy, A. **1970**. Chest symptomatology in a Sheffield steel works. J Egypt Med Assoc 52:578.

Evans, S, E Wilkes, and D Dalrymple-Smith. **1969**. Presymptomatic diagnosis: A study in country practice. J R Coll Gen Pract 17:2337.

Foster, W, E Langenback, and E Bergofsky. **1985**. Disassociation in the mucociliary function of central and peripheral airways of symptomatic smokers. Am Rev Respir Dis 132:633-639.

Frasca, JM, O Auerbach, VR Perks, and JD Jamieson. **1968**. Electron microscopic observations of the bronchial epithelium of dogs. Exp Pathol 9:380-384.

Green, G. **1970**. The J Burns Amberson Lecture—In Defense of the Lung. Am Rev Respir Dis 102:691-703.

Groth, S, F Hermansen, and N Rossing. **1990**. Pulmonary clearance of inhaled $^{99}Tc^{m}$-DTPA: Significance of site of aerosol deposition. Clin Physiol 10:85-98.

Hayashi, M, G Sornberger, and G Huber. **1979**. Morphometric analyses of tracheal gland secretion and hypertrophy in male and female rats after experimental exposure to tobacco smoke. Am Rev Respir Dis 119:67-73.

Higgins, I. **1971**. Chronic respiratory disease: Findings in surveys carried out in 1957 and 1966 in Stavely. Chest 59(Suppl):345.

Holland, W, H Kasap, J Colley, and W Cormack. **1969**. Respiratory symptoms and ventilatory function: A family study. Br J Prev Soc Med 23:77.

Howard, P. **1970**. A long term follow up of respiratory symptoms in a group of working men. Br J Ind Med 27:326.

Huchon, G, J Russel, L Barritault, A Liparsky, and J Murray. **1984**. Chronic airflow limitation does not increase respiratory epithelial permeability assessed by aerosolized solute but smoking does. Am Rev Respir Dis 130:451-460.

Jacobi, W. **1964**. The dose to the human respiratory tract by inhalation of shortened Rn and Rn decay products. Health Phys 10:1163-1174.

Jones, R, P Bolduc, and L Reid. **1973**. Goblet cells, glycoprotein and tracheal gland hypertrophy in rat airways. Br J Exp Pathol 54:229-232.

Jones, J, P Lawler, J Crawley, B Minty, C Hulands, and N Veall. **1980**. Increased alveolar epithelial permeability in cigarette smokers. Lancet 1:66-68.

King, M, A Wight, G DeSanctis, J El-Azab, D Philps, G Angus, and M Cosio. **1989**. Mucus hypersecretion and viscoelasticity changes in cigarette smoking dogs. Exp Lung Res 15:375-389.

Lamb, D and L Reid. **1969**. Goblet cell increase in rat bronchial epithelium after exposure to cigarette and cigar tobacco smoke. Br Med J 1:33-35.

Lippman, M and B Altshuler. **1976**. Regional deposition of aerosols, pp. 25-48. In: *Air Pollution and the Lung*, EF Aharonson, A Ben-David, MA Klinberg, (eds.). Halsted Press, John-Wiley, Jerusalem.

Lourenco, R, M Klimek, and C Borowski. **1971**. Deposition and clearance of 2-micron particles in the tracheobronchial tree of normal subjects—smokers and nonsmokers. J Clin Invest 50:1411-1420.

Luchsinger, PC, B La Garde, and JE Kilfeather. **1968**. Particle clearance from the human tracheobronchial tree. Am Rev Respir Dis 97:1046.

Lundin F, J Lloyd, and E Smith. **1969**. Mortality of uranium miners in relation to radiation exposure, hard-rock mining and cigarette smoking—1950 through September 1967. Health Phys 16:571-578.

Mason, G, J Uszler, R Effros, and E Reid. **1983**. Rapidly reversible alterations of pulmonary epithelial permeability induced by smoking. Chest 83:6-11.

Mawdesley-Thomas, L, P Healy, and D Barry. **1973**. Experimental bronchitis in animals due to sulphur dioxide and cigarette smoke: An automated quantitative study, pp. 509-525. In: *Inhaled Particles III*, HW Walton (ed.). Gresham Press, Old Woking, Surrey, United Kingdom.

McFadden, D, J Wright, B Wiggs, and A Churg. **1986**. Smoking inhibits asbestos clearance. Am Rev Respir Dis 133:372-374.

Minty, B, and D Royston. **1985**. Cigarette smoke induces changes in rat pulmonary clearance of ^{99m}Tc-DTPA: A comparison of particulate and gas phases. Am Rev Respir Dis 132:1170-1173.

Pavia, D, M Short, and M Thomson. **1970**. No demonstrable long term effects of cigarette smoking on the mucociliary mechanism of the lung. Nature 226:1228-1231.

Radford E and K St Clair Renard. **1984**. Lung cancer in Swedish iron miners exposed to low doses of radon daughters. N Engl J Med 310:1W5-1494.

Radsel, Z and V Kambic. **1978**. The influence of cigarette smoke on the pharyngeal mucosa. Acta Otolaryngol 85:128-134.

Reinstein, L, L Robinson, and A Glicksman. **1981** Formation of iron clusters in a mucus-like medium. J Appl Physiol 52:2572.

Rylander, R. **1971** Lung clearance of particles and bacteria: Effects of cigarette smoke exposure. Arch Environ Health 23:321-326.

Saccomanno, G, G Huth, O Auerbach, and M Kuschner. **1988**. Relationship of radioactive radon daughters and cigarette smoking in the genesis of lung cancer in uranium miners. Cancer 62:1402-1408.

Sanchis, J, M Dolovich, R Chalmers, and M. T. Newhouse. **1971** Regional distribution and lung clearance mechanisms in smokers and nonsmokers, pp. 55-61. In: *Inhaled Particles*, WH Walton (ed.). Unwin, Old Woking, Surrey, United Kingdom.

Seely, J, E Zuskin, and A Bouhuys. **1971** Cigarette smoking: Objective evidence for lung damage among teenagers. Science 172:741.

Smith, G, LV Wilton, and R Binns. **1978**. Sequential changes in the structure of rat respiratory system during and after exposure to cigarette smoke. Toxicol Appl Pharmacol 46:579-591.

Sterling, T. **1981** Possible risks to human lungs from magnetometric dust clearance experiments. J Appl Physiol 52:2575-2577.

Sterling, T. **1983**. Possible effects on occupational lung cancer from smoking related changes in the mucus content of the lung. J Chron Dis 36:669-676.

Sterling, T. **1989**. Theory (or model) of the joint influence of occupational exposure to carcinogenic dust and to cigarette smoke and occupational lung cancer. Exp Pathol 37:181-185.

Svartengren, M, E Hassler, and K Philipson. **1986**. Spirometric data and penetration of particles to the alveoli. Br J Ind Med 43:188-1991.

Thomson, M and D Pavia. **1973**. Long term tobacco smoking and mucociliary clearance from the human lung in health and respiratory impairment. Arch Environ Health 26:86-89.

Wagoner, J, V Archer, and F Lundin. **1965**. Radiation as the cause of lung cancer among uranium. N Engl J Med 273:181-188.

Walsh, P. **1970**. Radiation dose to the respiratory tract of uranium miners: A review of the literature. Environ Res 3:14-36.

Weiss, T, P Dorow, and R Felix. **1983**. Regional mucociliary removal of inhaled particles in smokers with small airways disease. Respiration 44:338-345.

Weiss, T, P Dorow, and R Felix. **1984**. Continuous aerosol inhalation scintigraphy in the evaluation of early and advanced airways obstruction. Eur J Nucl Med 9:62-67.

WHO (World Health Organization). **1988**. Indoor air quality and radon—Report of a WHO working group. J Environ Radioactivity 8:73-91.

Woolf, C. **1974** Clinical findings, sputum examinations and pulmonary function tests related to the smoking habits of 500 women. Chest 66:652.

Woolf, C and J Suero. **1971** The respiratory effects of regular cigarette smoking in women. Am Rev Respir Dis 103:26.

QUESTIONS AND ANSWERS

C. Poland, Sterling & Associates, Ltd., Canada. Yes, we're suggesting that smoking may induce changes, and it would be beneficial if we could induce similar changes without smoking.

Q. Guilmette, ITRI. How do you reconcile your ideas with the BEIR-IV conclusions that, in fact, radon exposure and cigarette-smoke exposure are synergistic. That is, that they fit a multiplicative or submultiplicative model?

A. Poland. Most of the studies that we are aware of on uranium miners have found an increased risk of lung cancer in nonsmokers that are occupationally exposed over that in the smokers. Those are the studies that we presented in the first table.

Q. Guilmette. Why, then, are 96% of the cancers that have been recorded in the Colorado Plateau miners in smokers?

A. Poland. Those were in earlier studies, and we believe there's a more recent study by Lundin et al. in the Colorado Plateau workers that found the risks in smokers and nonsmokers were the same; the relative risk, that is.

C. Guilmette. The numbers I'm giving you are from Gene Saccomanno, who spoke to us this last April. I don't call those too old.

C. Johnson, PNL. I think, Ray, that what you're saying is a relative risk, not the absolute risk, of course.

C. Poland. I think that Dr. Sterling, the senior author, would like to address that question.

A. Sterling, Simon Fraser University. I think there is a reality here, and we point out this reality. For instance, it has been known since 1955 in relation to coal miners' pneumoconiosis, that there seems to be some smoking-produced changes in the lungs which seem to be beneficial to the worker. This doesn't mean that the worker should go out and smoke. But it should mean that we should take cognizance of such data and that, as Poland has pointed out. Study after study of uranium miners shows that the relative risk of exposed smokers compared to nonexposed smokers is smaller than the relative risk of exposed nonsmokers compared to nonexposed nonsmokers. It's data, and I think models have to fit their data.

C. Johnson. Are there further questions?

C. Harley, New York University. I think there is some confusion about relative risk here. The relative risk in nonsmokers is higher than in smokers. But the baseline is so much higher in smokers, that the absolute risk is higher in smokers. There is no question about that in all the data sets.

Now, in one data set, the Swedish miners, the absolute risks are about the same. Which means that your relative risk in nonsmokers will be higher than that in smokers. But, in all probability, in this data set that's true because the Swedes do not smoke very much. They are light smokers compared to the conventional heavy smokers in the United States and, say, Czech populations. But, clearly, the risk to smokers is greater than to nonsmokers from radon exposure. That's not questioned.

C. Cross, PNL. I'd like to clarify a slightly different point: what we saw in the earlier dog study was probably repeated in the study in mice, that came out of Harwell recently, where the promotional effect of cigarette smoke may have been obscured because the radiation doses were so high, causing some cell sterilization. And you tend to see more of a promotional effect when you have a combined exposure to radiation and cigarette smoke if the radiation exposures are at lower levels. So what we see in the underground miner populations can sometimes be confusing, depending on how high those exposure levels were.

Stochastic Dosimetry Methods

ALPHA PARTICLE SPECTRA AND MICRODOSIMETRY OF RADON DAUGHTERS

R. S. Caswell and J. J. Coyne

National Institute of Standards and Technology, Gaithersburg, Maryland

Key words: *Radon, alpha-particle spectra, microdosimetry, radiation transport*

ABSTRACT

We are interested in understanding the physics of the process by which radon-daughter alpha particles irradiate cells, leading to the induction of cancer. We are focusing initially on two aspects: the alpha spectra incident upon cells, which are needed for input to biophysical models of cancer induction; and microdosimetric spectra and parameters which give information on radiation quality. Adapting an analytical method previously developed for neutron radiation, we have calculated the alpha-particle slowing-down spectra (the spectra incident upon cells) and, subsequently, the microdosimetric spectra and parameters for various cell nuclei or site diameters. Results will be presented from three modes of program operation. MODE 1 is for the thin, plane source of radon-daughter activity adjacent to the epithelium. MODE 2 is for the thick source layer (the mucous-serous layer) adjacent to the epithelium. MODE 4 is for cylindrical airways of various radii, lined by the mucous-serous layer. MODE 1 is most useful for understanding the problem; MODE 4 is most anatomically relevant. MODE 3 is not discussed in this paper.

Alpha-particle spectra and microdosimetric spectra and parameters are studied as a function of cell depth, ^{218}Po/^{214}Po ratio, airway radius, and cell nucleus or site size. Also available from the calculation is mean dose as a function of depth below the airway surface. The results described here are available on personal computer diskettes. We are beginning to compare our studies with the calculations of other workers and plan to extend the calculations to the nanometer target level.

INTRODUCTION

We are studying the physics of the process by which radon-daughter alpha particles irradiate cells, leading to the induction of cancer. The calculation focuses on two aspects: the alpha spectra, "slowing-down spectra," which are incident upon the cells at risk; and microdosimetric event-size spectra, which give information on radiation quality. Adapting an analytical method previously developed for neutron radiation

(Caswell, 1966), we have calculated the alpha-particle slowing-down spectra and, subsequently, the microdosimetric spectra and parameters for various cell nuclei or site diameters. Calculation is done in the continuous slowing-down approximation which assumes that the alpha-particle energy loss is given by the stopping power, or LET_{∞}.

Results are presented for three calculational modes. MODE 1 is for the thin, plane source of radon-daughter activity adjacent to the epithelium. MODE 2 is for a thick source layer, the mucous-serous layer of otherwise similar geometry to that of MODE 1. MODE 4 is for cylindrical airways of various radii, lined by the mucous-serous layer. MODE 3 is not discussed in this paper. Alpha-particle and microdosimetric spectra and parameters are studied as a function of cell depth, $^{218}Po/^{214}Po$ ratio, airway diameter, and cell nucleus or site size.

ALPHA-PARTICLE SLOWING-DOWN SPECTRA

Figure 1 presents the alpha-particle (slowing-down) spectrum for a 1-Bq-cm^{-2} thin source layer of ^{218}Po; Figure 2 presents a similar source layer of ^{214}Po, both at a cell depth of 25 μm. Note the characteristic peak at high energy due to regions of the source layer near the cells, a tapering off of the spectrum at lower energies; and the high tail as one approaches zero energy, which occurs because the slowing-down spectrum is proportional to the inverse of the stopping power. The ^{214}Po spectrum is more energetic because the alpha particles are emitted with 7.7 MeV, in contrast to 6.0 MeV for ^{218}Po. Figure 3 shows the combined spectrum at the same cell depth for a source with a $^{218}Po/^{214}Po$ ratio of 1/3. We now consider the radioactivity distributed uniformly through a 15-μm mucous-serous layer (rather than in a thin layer). The slowing-down spectrum (Figure 4) is smoothed out somewhat by the thickness of the source and is slightly lower in energy because of the greater average distance from the thick radioactive source layer to the cells.

In the more realistic case of a cylindrical airway lined with a mucous-serous layer, we use the geometry of Figure 5, assuming an airway diameter of 0.61 cm. At the same cell depth, 25 μm, the alpha spectrum has what is essentially an additional component: the alpha particles from the opposite side of the airway, which must cross the airway to reach the cells in question. These produce a lower-energy, broad peak (Figure 6) at about 2.5 to 3 MeV. The energy is lower because of the extra slowing-down of the alpha particles in the air and in the additional source layer. Typical alpha-particle slowing-down spectra for ^{218}Po for cell depths of 10 μm and 30 μm are given in Figure 7, and for ^{214}Po in Figure 8. Clearly seen are the energy degradation with depth, as well as the loss of the broad peak near the middle of the spectrum, which has been attenuated by the air gap and additional material at the 30-μm depth. Figure 9 shows the slowing-down spectra for an airway diameter of 0.61 cm for cell depths between 10 and 55 μm for a source layer with $^{218}Po/^{214}Po$ ratio of 1/3.

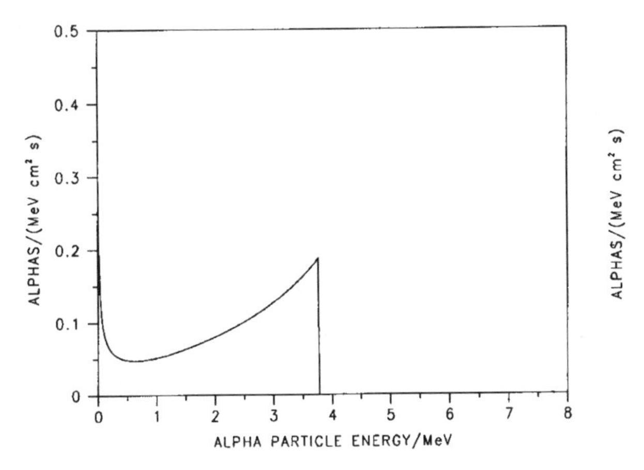

Figure 1. Slowing-down spectrum of alpha particles for plane, thin source (MODE 1) of 1 Bq cm^{-2} ^{218}Po at cell depth of 25 μm.

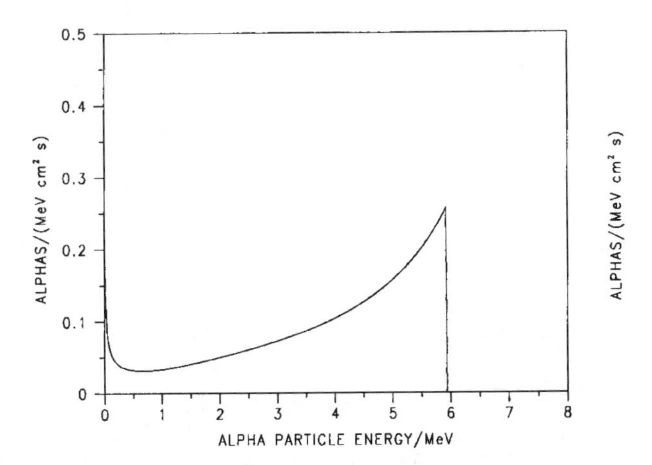

Figure 2. Slowing-down spectrum of alpha particles for plane, thin source of 3 Bq cm^{-2} ^{214}Po at cell depth of 25 μm.

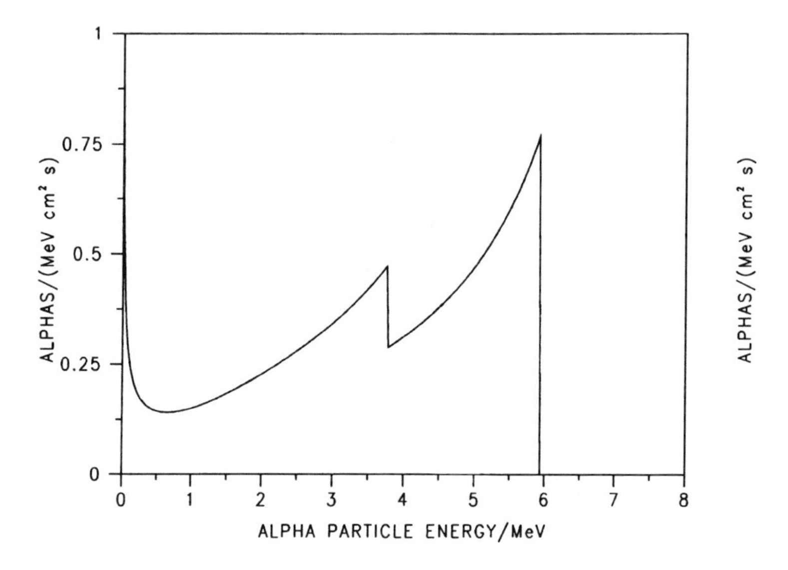

Figure 3. Slowing-down spectrum of alpha particles for plane, thin source with combined activity of 1 Bq cm^{-2} ^{218}Po and 3 Bq cm^{-2} ^{214}Po; cell depth, 25 μm.

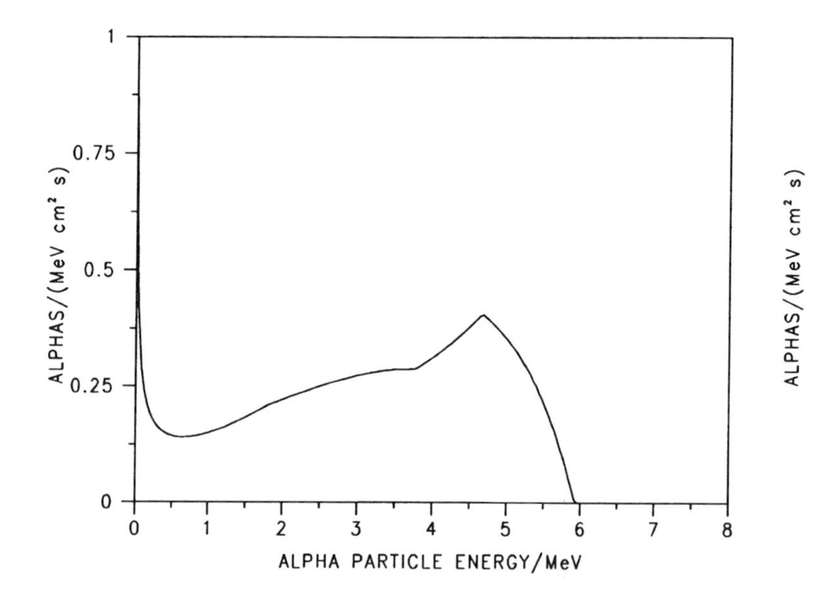

Figure 4. Alpha-particle slowing-down spectrum for source layer of 15-μm thickness with ^{218}Po/^{214}Po activity ratio of 1/3 (MODE 2) and cells at a depth of 25 cm, measured from bottom of source layer.

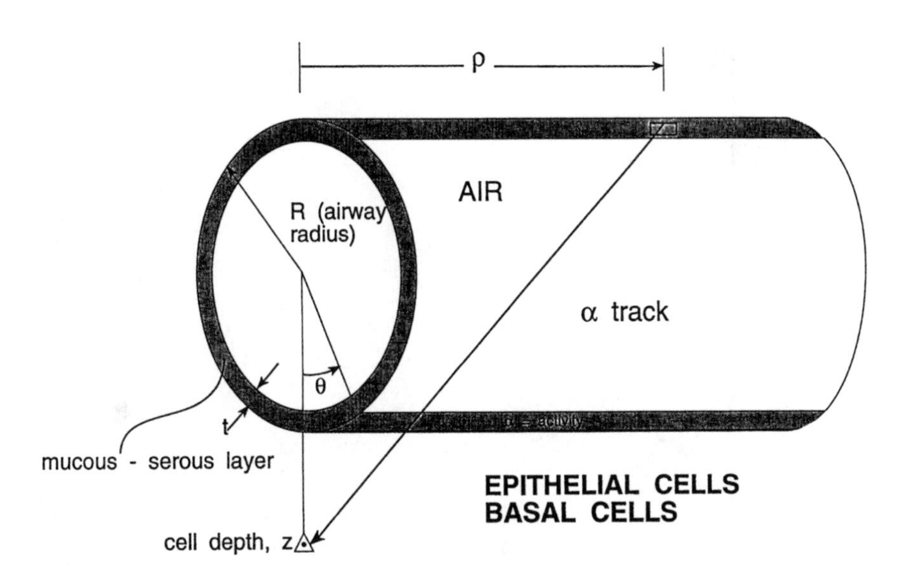

Figure 5. Geometry of cylindrical airway source calculation (MODE 4).

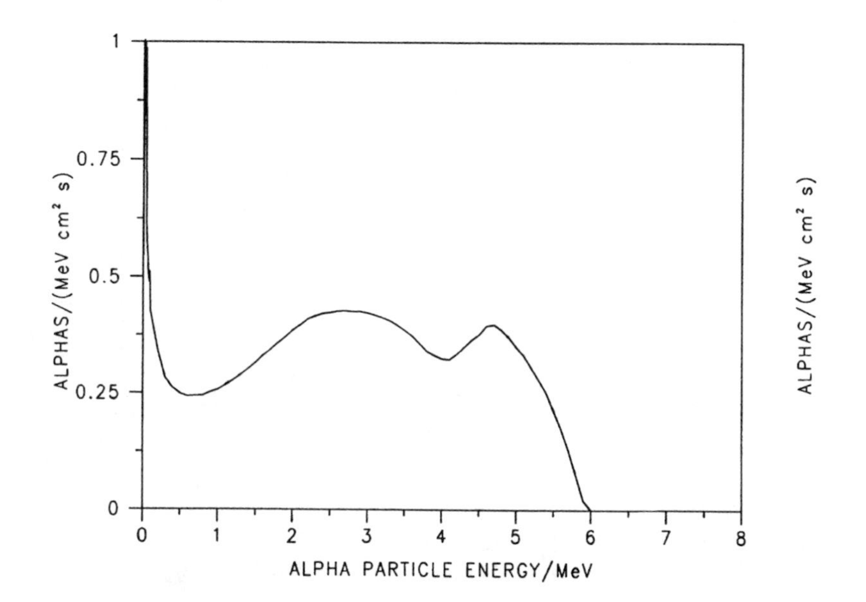

Figure 6. Alpha-particle slowing-down spectrum for cylindrical airway of 0.61-cm diameter for a 15-μm-thick source layer with combined radioactivity of 1 Bq cm^{-2} ^{218}Po and 3 Bq cm^{-2} ^{214}Po; cell depth, 25 μm.

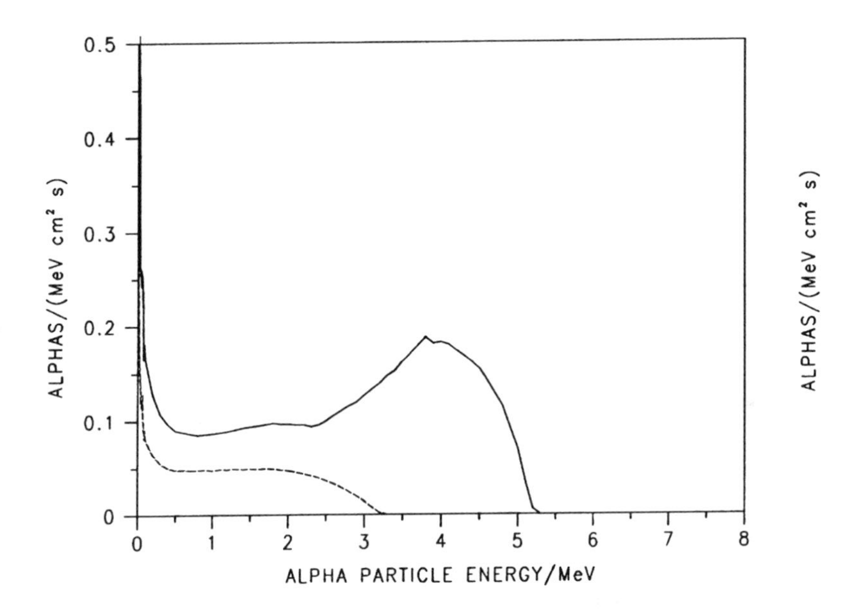

Figure 7. Alpha-particle slowing-down spectra for a cylindrical airway of 1.13 cm diameter for a 1-Bq-cm^{-2} ^{218}Po source layer of 15-μm thickness for cell depths 10 μm (solid curve) and 30 μm (dashed curve).

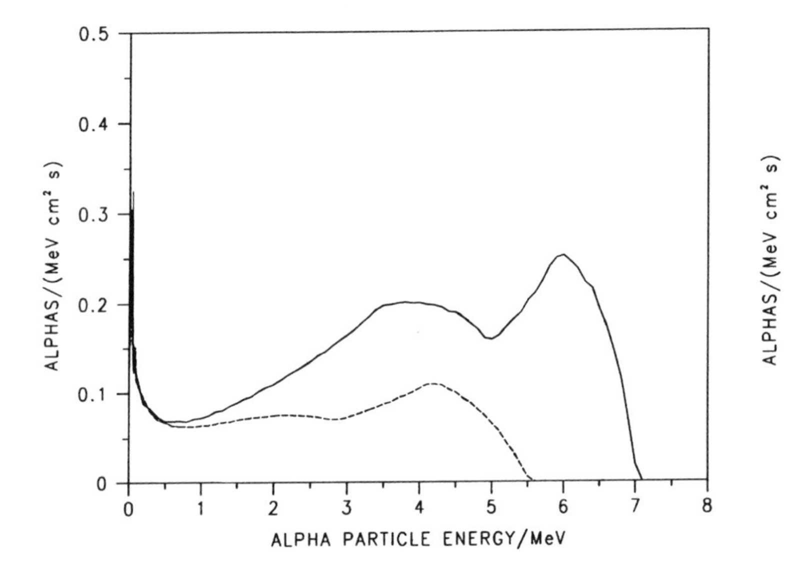

Figure 8. Alpha-particle slowing-down spectra for a cylindrical airway of 1.13-cm diameter for a 1-Bq-cm^{-2} ^{214}Po source layer of 15-μm thickness for cell depths 10 μm (solid curve) and 30 μm (dashed curve).

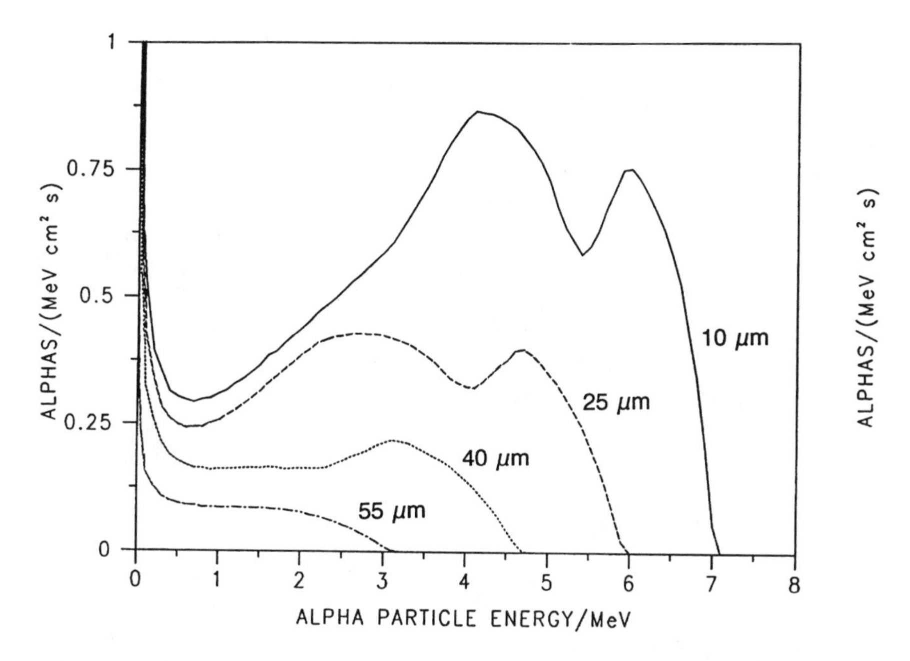

Figure 9. Alpha-particle slowing-down spectra for an airway of 0.61 cm diameter for cell depths between 10 and 55 μm for a 15-μm source layer with $^{218}Po/^{214}Po$ activity ratio of 1/3.

MICRODOSIMETRIC SPECTRA

One application of slowing-down spectra is to use them as input to calculations of microdosimetric spectra in lineal energy, y, or specific energy, z. In microdosimetry one is concerned with energy deposition in a spherical sensitive "site" in a cell, which is frequently taken to correspond to the cell nucleus. Figure 10 shows the y spectrum for a cell depth of 20 μm for a 15-μm-thick source layer containing 1 Bq cm^{-2} ^{218}Po for sensitive site (cell nucleus) diameters of 5 and 10 μm, respectively, and an airway diameter of 1.13 cm. Obviously, the spectra are not very sensitive to site diameter. In Figure 11 we compare the y spectra for ^{218}Po and ^{214}Po for an airway diameter of 1.13 cm at the same depth, 20 μm, for a 5-μm site diameter. These spectra are not normalized to unit area but are calculated on an absolute basis. Since the area under the curves is the absorbed dose rate, the greater dose at the given depth is for the more energetic alpha ray, ^{214}Po. The change of y spectra with depth for a mixed source of 1 Bq cm^{-2}

^{218}Po and 3 Bq cm^{-2} ^{214}Po is shown in Figure 12 for cell depths of 10 μm, 25 μm, and 40 μm. If it is legitimate to use y spectra as measures of radiation quality, we conclude that, although the dose changes rapidly with depth, the change in radiation quality is only moderate. As the alpha particles slow down at greater depth, LET increases as the particles approach the energy of the Bragg peak. This is shown in the increasing y values of the spectra with depth. Table 1 gives microdosimetric parameters and quality factors at various depths for a ^{218}Po/^{214}Po ratio of 1/3 for a cylindrical airway of 0.61 cm diameter. The values of the quality factor depend more on the prescription of quality factor used than on variations in quality with depth of alpha-particle radiation.

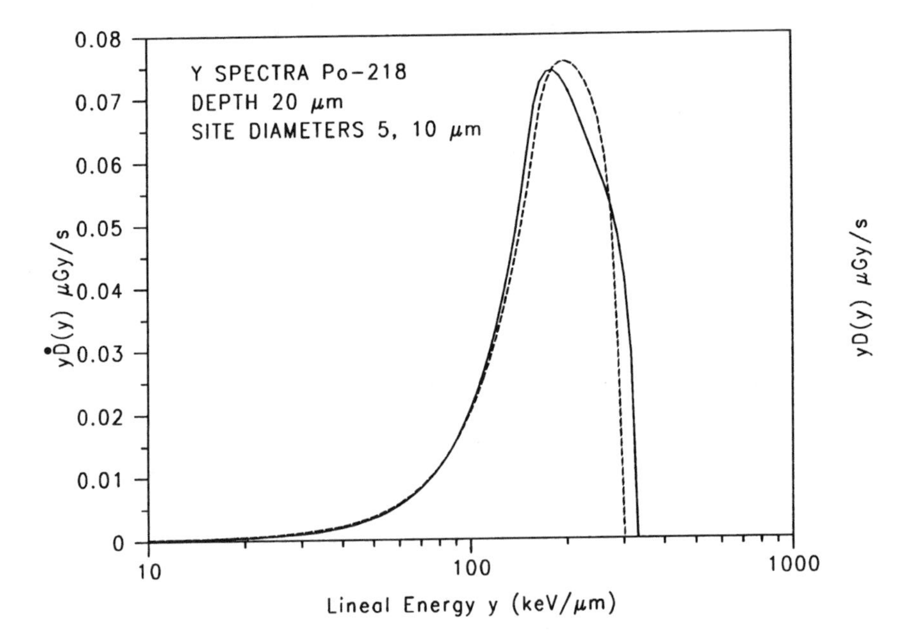

Figure 10. Lineal energy (y) spectra at 20-μm cell depth for a 15-μm source layer containing 1 Bq cm^{-2} ^{218}Po; airway diameter, 1.13 cm. Cell nucleus diameters: 5 μm (solid curve) and 10 μm (dashed curve). Units of y are keV/μm, of $\dot{D}(y)$ are Gy/(keV μm^{-1} s), thus units of $y\dot{D}(y)$ are Gy s^{-1}.

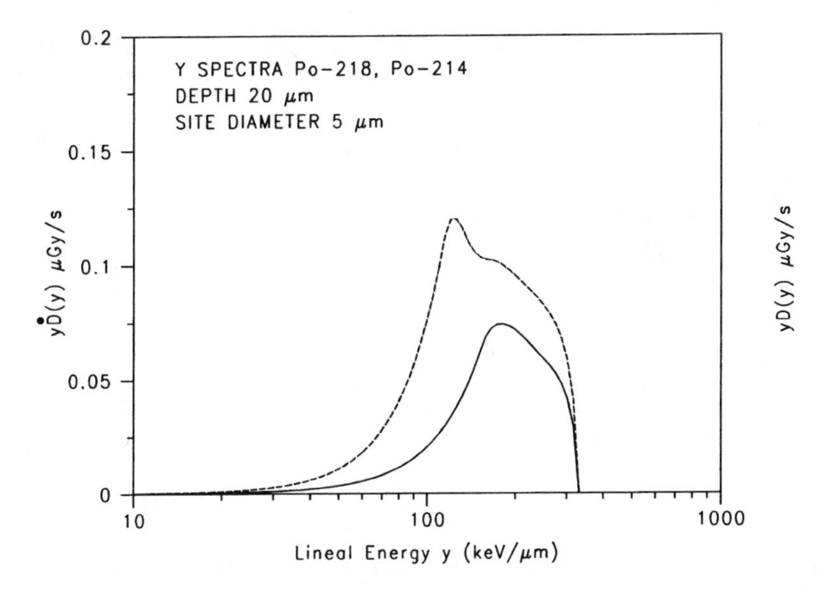

Figure 11. Comparison of y spectra at 20-μm cell depth for ^{218}Po (solid curve) and ^{214}Po (dashed curve). Sensitive-site diameter, 5 μm; airway diameter, 1.13 cm; source-layer thickness, 15 μm.

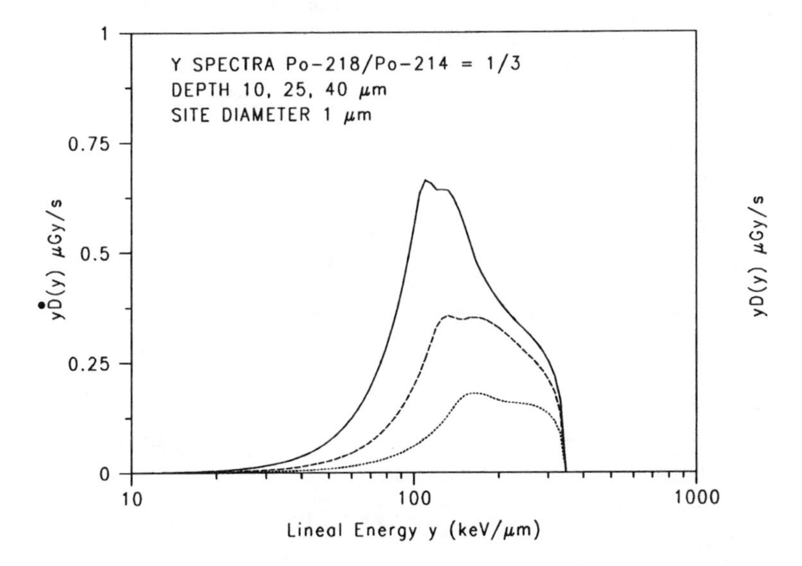

Figure 12. Change of y spectra with cell depth for 15-μm source layer of 1 Bq cm^{-2} ^{218}Po and 3 Bq cm^{-2} ^{214}Po; site diameter, 1 μm. Spectrum for 10-μm depth, solid line; for 25-μm, dashed line; for 40-μm, dotted line. Airway diameter, 0.61 cm.

Table 1. Microdosimetric parameters and quality factors of alpha-particle spectra ($^{218}Po/^{214}Po$ ratio, 1/3) for a cylindrical airway of 0.61 cm diameter. Site diameter, 1 μm.

	Microdosimetric Parameters (keV/μm)				
Cell depth (μm)	$\bar{y}_F$	$\bar{y}_D$	y^*	Q (ICRP, 1966)[a]	Q (ICRU, 1986)[b]
10	115	148	70	16.2	26.0
25	134	169	70	17.2	26.0
40	147	184	69	17.8	25.6
55	168	209	66	18.6	24.2

[a] Quality factor prescription of ICRP Report 9, with the formula of Kellerer and Hahn (1988) adapted for the conversion of D(y) to D(L).
[b] Quality factor prescription given by the ICRU-ICRP Task Group (1986) in ICRP Report 40.

SUMMARY

This analytical method provides a convenient method of calculating alpha-particle slowing-down energy spectra, microdosimetric spectra, and radiation quality parameters for radon-daughter alpha rays in the airways of the bronchial epithelium. In part, these results are comparable to the Monte-Carlo calculations of Brenner (1990) and the inverse Fourier transform calculations of Hui et al. (1990). We have begun a collaboration with Professor Werner Hofmann of the University of Salzburg, Austria, to study several biophysical cancer induction models which use these spectra as input.

ACKNOWLEDGMENT

This work has been supported by the Office of Health and Environmental Research, U.S. Department of Energy.

REFERENCES

Brenner, DJ. **1990**. The microdosimetry of radon daughters and its significance. Radiat Protect Dosim 31:399-403.

Caswell, RS. **1966**. Deposition of energy by neutrons in spherical cavities. Radiat Res 27:92-107.

Hui, TE, JW Poston, and DR Fisher. **1990**. The microdosimetry of radon decay products in the respiratory tract. Radiat Protect Dosim 31:405-411.

ICRP. 1966. *Recommendations of the International Commission on Radiological Protection 1965*, Publication 9. International Commission on Radiological Protection, Oxford, United Kingdom.

ICRU. 1986. *The Quality Factor in Radiation Protection*, Report 40. International Commission on Radiation Units and Measurements, Bethesda, MD.

Kellerer, AM and K Hahn. **1988.** The quality factor for neutrons in radiation protection: Physical parameters. Radiat Protect Dosim 23:73-78.

QUESTIONS AND ANSWERS

Q. Borak, Colorado State University. The airways aren't necessarily right-circular cylinders, and what happens is, a very slight irregularity of a few microns, which could perhaps change it. But maybe it all washes out. Have you given some thought to this irregularity system?

A. Caswell, National Institute of Standards and Technology. We haven't done any calculations. We've discussed this with Dr. Hofmann, University of Salzburg, who thinks that these effects will not be large. One of the other things is that we assumed essentially an infinite cylinder, but of course that's cut off in any calculation by the alpha particle range. So it isn't infinite, it has some finite length. But then, if you can calculate what the doses are in the middle, which is what we would get for a segment or a generation, then, as you move toward the edge, you're getting a contribution from the next one either higher or lower. I think it might be that you have to go to a Monto Carlo calculation to do that.

Q: Borak. This morning, or earlier this afternoon, we also saw this possibility of hot spots, particularly at a bifurcation. Have you tried to simulate some of the LET distributions that you get at these bifurcations?

A. Caswell. We haven't, but we should be able to do hot spots very easily. But, basically, we would add a hot spot term to our source term, so that should be very easy.

Q. James, PNL. Does the LET distribution depend too much on dose rate? You show that it varied very slightly with depth below the epithelial surface. The hot spot would be just twice the surface density.

A. Caswell. Basically, it would give you a source point, somewhere in this layer. The way we get our spectrum is to add up the elements of the slowing-down spectrum produced by each volume element in the source. The hot spot, therefore, would be some small volume with a lot of activity in it. I think that would be very easy to put in.

C. James. But we've got to remember Naomi's comment, which is very valid. You might have a hot spot in deposition but that's washed out with subsequent clearance.

A. Caswell. Yes, but what we do at this point is to calculate for a moment in time. If the source changes by being washed out, then we could integrate over time. I think it's certainly doable, but we haven't done it.

Q. Curtis, Lawrence Berkeley Laboratory. Yes, I'd like to make a comment on just this last one. It seems to me that as far as hot spots are concerned, you don't know whether or not the clearance is at the same rate over the hot spot as that in the bifurcation. It might be measured in a non-hot-spot situation, so that there might be some accumulation that wouldn't be cleared as fast in the bifurcation.

But what I want to suggest to you, or ask you, is, have you calculated the dose? You can, of course, integrate under those curves and come up with a dose. Have you done that and compared it with the much earlier calculations of dose that are already in the literature? As a function of depth, that is.

A. Caswell. We do calculate the dose as part of the calculation, and we have made one or two comparisons, but we haven't done a careful job of it yet. I think that, basically, the cases we've looked at are in rough agreement.

NANODOSIMETRY OF RADON ALPHA PARTICLES

M. Zaider[1] and M. N. Varma[2]

[1]Center for Radiological Research, Columbia University, New York, New York

[2]U.S. Department of Energy, Washington, DC

Key words: *Microdosimetry, nanodosimetry, radon alpha particles, associated volume, associated surface*

ABSTRACT

It is currently accepted that energy deposition at the nanometer level (rather than conventional microdosimetry) determines the biological effects of ionizing radiation. Many previously established experimental techniques (e.g., the Rossi proportional counter) or theoretical methods (e.g., simplified calculations using the continuous slowing-down approximation (CSDA) are inapplicable to the study of nanodosimetry. The peculiarities of the geometry of exposure to radon progeny further complicate the problem. This is because the conditions under which several "classical" models of radiation action are obtained (e.g., the alpha-beta formulation of the Theory of Dual Radiation Action, which is built on microdosimetry) are no longer valid. It thus becomes clear that not only new techniques but new concepts are required to describe the effects of radon alpha particles.

In this paper we discuss a number of computational aspects specific to radon nanodosimetry. In particular, we describe the novel concept of "associated surface" (AS) which is necessary for efficiently converting Monte-Carlo-generated particle tracks to nanodosimetric spectra. The AS is the analog of Lea's associated volume, applied to radiation sources subject to the geometrical restrictions of internal exposure.

We systematically analyze factors affecting the nanodosimetry of radon progeny, such as the distance between the radioactive source and the sensitive volume, the size of the sensitive volume, and CSDA versus full Monte-Carlo track generation.

INTRODUCTION

Practical assessment of the risk incurred by humans exposed to radon involves extrapolation. One extrapolates epidemiological results obtained at high doses to the range of doses typical for the average population. In addition to some intrinsic difficulties (poor retrospective dosimetry, confusing factors such as smoking, etc.) one has the basic problem of selecting an extrapolation procedure, and consequently (whether overtly acknowledged or not), a model of radiation action.

Even such seemingly "obvious" procedures as linear or quadratic extrapolation may result in one or two order-of-magnitude differences between predictions, an unsatisfactory situation.

One also extrapolates results obtained with other radiations and/or systems. As opposed to epidemiological results where absolute risks are evaluated, one is concerned here with the rather more modest goal of determining the risks of radon radiation relative to another, perhaps better understood, system or type of radiation. The device used for this purpose is the quality factor, a number which should be a single-value function of a physical quantity (field descriptor) characterizing the radiation. Neither linear energy transfer (LET)—the field descriptor currently in use—nor the newly proposed lineal energy (y) in a 1-μm sphere rigorously satisfies this criterion. There are notorious examples of fields with similar LET values or microdosimetric spectra that have different biological effects. This situation arises because of the attempt to select a "model-independent" field descriptor, an unlikely proposition at best.

It appears, then, that in estimating radon risks (or those of any other radiation), the problem of determining a model of radiation action (field descriptor plus probability of effect) cannot be avoided if accurate predictions are desired. In this paper we examine conceptual and practical problems related to the possibility of using nanodosimetric distributions of specific energy in order to understand the biological effects of radon-associated alpha particles. The relevance of this approach stems directly from the currently accepted tenet that energy deposition at the nanometer level (rather than conventional microdosimetry) determines the biological effects of radiation.

The use of nanodosimetry as field descriptor has a number of important consequences. The methodology based on simplified representation of charged-particle tracks [e.g., continuous slowing down approximation (CSDA), tracks consisting of straight-line segments with no radial dose distribution] is inadequate. One needs to use, almost exclusively, event-by-event Monte-Carlo transport methods (Paretzke et al., 1974; Turner et al., 1980; Zaider et al., 1983).

Conventional methods used to obtain microdosimetric distributions are very inefficient when applied to nanometer-sized volumes and new, nanodosimetric-specific techniques are needed. One such approach is based on the concept of associated volume (Lea, 1962). A novel technique using the associated surface of the track (see below), should be used for radon nanodosimetry.

Because direct experimental nanodosimetric techniques have yet to be demonstrated, calculations can be verified only against measured radial dose distributions (Varma et al., 1977) or variance-covariance results (Kellerer and Rossi, 1984). Other parameters used to check the codes, such as LET or w-values, provide only limited sensitivity as testing devices.

There is a need for radon-specific, nanodosimetry-based models of radiation action, that is, models that invoke (and use predictively) energy deposition in sites with dimensions comparable with sensitive cellular substructures. Multihit, multitarget models such as Katz's track structure theory (TST; Katz et al., 1972) or the Theory of Dual Radiation Action (TDRA; Kellerer and Rossi, 1978) could serve for this purpose. The former needs to be reformulated in nanodosimetric terms. The generalized version of TDRA cannot be applied in its present form to radon nanodosimetry because of the peculiar geometry of radon-progeny exposure (discussed further below).

To the extent that nanodosimetric field descriptors are successful, the actual shape (morphology) of the target cells containing the sensitive substructures becomes largely irrelevant. This is particularly pertinent in radon research, where significant efforts are being made toward characterizing the geometry of the presumably cancer-prone cells.

GEOMETRY OF CELLS AT RISK

The transport of, and energy deposition by, alpha particles with energies of several MeV are reasonably well understood. Likewise, microdosimetric spectra for these alpha particles have been determined repeatedly, both experimentally and theoretically. [Microdosimetry was actually "invented" using alpha particles (Rossi, 1967). Their spectra are routinely used to calibrate Rossi proportional counters.] What makes the radon microdosimetry different is: (1) Sensitive cells (those at risk) are positioned at fixed locations relative to alpha-particle sources; and (2) the short (relative to cellular dimensions) range in tissue of alpha particles emitted by radon and thoron progeny (see Table 1). It is thus apparent that energy deposited in nanometer-sized volumes might change significantly as a function of such factors as incidence angle or source-site distance (when comparable with particle range).

Table 1. The range (R) and energy of the alpha particles emitted by radon progeny.

	R/μm	Energy/MeV
Rn^{222}(radon)	41	5.5
Po^{218}(RaA)	48	6.0
Po^{214}(RaC')	71	7.7
Rn^{220}(thoron)	52	6.3
Po^{216}(ThA)	58	6.8
Bi^{212}(ThC)	49	6.1
	89	8.8

The exact type of cells involved in the development of lung cancer is not known. The mucosa (bronchial tissue) consists of ciliated epithelial cells (Figure 1), goblet cells and mucous (serous) cells supported, in part, by triangular-shaped basal cells (Jeffery and Reid, 1977). Of these, the first two kinds are fully differentiated cells and are therefore unlikely to be involved in tumor development. A number of dosimetric calculations have used the basal cells as targets, although there is evidence that other cells (serous or Clara, for instance) are linked to specific tumors. Arguments against the basal cells being the primary targets have also used the fact that the peripheral epithelium has a very low density of basal cells to start with. We adopt the position that sensitive cells extend over the whole thickness of the epithelium (James, 1988).

Radon progeny can rest, in principle, at any position in the mucous layer supported by the ciliated cells, although perhaps with higher probability on the surface of the mucosa. Epithelial thickness has been found to vary from 80 ± 6 μm in the main bronchi to 15 ± 5 μm in the bronchioles (Gastineau et al., 1972). Mucosal thickness is usually assumed to be about 7 μm (James, 1988).

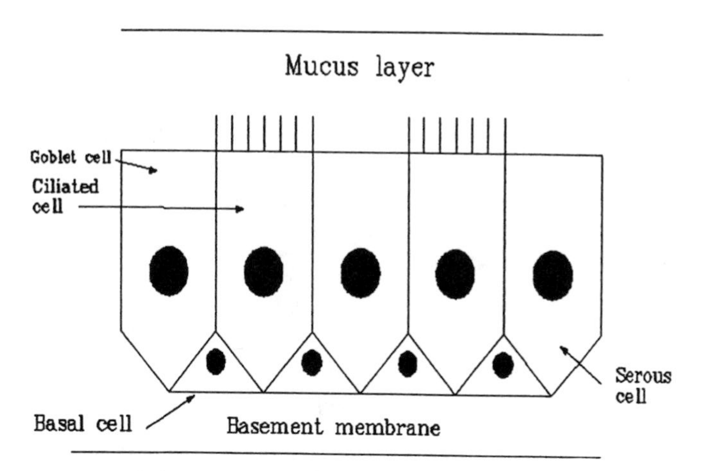

Figure 1. Graphic representation of ciliated mucosa.

MONTE-CARLO TRANSPORT OF CHARGED PARTICLES

The Monte-Carlo method is the basic tool for performing track calculations. In this approach, particles are transported on an "event-by-event" basis, simulating their transport in water vapor. A brief description of our code follows; for further details see Zaider et al. (1983).

The necessary input consists of cross sections covering particles and energies of interest as well as angular distributions. Much of the experimental data on cross sections in water vapor have been incorporated in the code. Where experimental data were not available, theoretical calculations have been used.

The basic unit in the particle transport codes is an electron code (DELTA), which can transport electrons from relativistic down to thermal energies. All other particle transport codes (e.g., for ions) ultimately use the results of this code.

Electrons interacting with a water molecule can either elastically scatter, excite, or ionize the molecule. The two latter processes lead to dissociation of the molecule. The total cross sections for elastic scattering have been characterized experimentally in the past few years, and are described by Zaider et al. (1983). The low-energy, elastic angular-distributions are particularly important in determining the spatial distribution of ionizations and excitations. Earlier works used either the "screened Rutherford" scattering approach of Bethe (1953) or the "phase-shift" approach of Mott (1929). At low energies both these models are very inaccurate (Brenner and Zaider, 1984). In light of the considerable body of experimental data on elastic scattering angular-distributions that recently became available, particularly that of Nishimura (1979) from 30 to 200 eV, we have used a semiempirical approach, fitting the available data to a distribution particularly suitable for Monte-Carlo transport codes.

We have used five different energy levels in the water molecule. Experimental data from Schutten et al. (1966) and Mark and Egger (1976) were used for the total and partial cross sections, with the secondary electron distribution represented analytically.

The code DELTA transports electrons down to thermal energies. This is important in order to determine the position of the electron before solvation. As electron energy decreases below about 5 eV, vibrational interactions form the dominant mode of energy loss. These cross sections are taken from the experimental work of Seng and Linder (1976). Below about 0.5 eV, rotational interactions dominate. Here, few experimental data are available; these cross sections were therefore calculated using the plane-wave Born approximation. Above 5 eV, a variety of states can be excited. The 10 most important are individually incorporated in the code, for the most part with experimentally based cross sections.

For proton and heavy-ion transport, the three main processes are ionization, excitation, and charge-exchange. In terms of ionization, two sets of experimental data are available: that of Toburen and Wilson (1977) from 300 to 3000 keV, and that of Bolorizadeh and Rudd (1986) from 20 to 200 keV. Both data sets include secondary electron cross sections, differential in angle and energy, and have been parameterized in the code. Below a few hundred keV, charge exchange to the continuum, where the proton gains or loses as electron, is also included in the code, based on the experimental data of Dagnac et al. (1970) and Toburen et al. (1968).

Heavy-ion transport, which is clearly of significance for radon studies, presents peculiar problems, mainly because the double-differential ionization cross sections of heavy ions have been only sparsely measured. In our treatment of alpha particles, we scale the cross sections for equal-velocity protons by Z^2 to obtain cross sections for an ion of charge Z. This procedure is rigorously valid whenever the ion is fully ionized; the criterion is that $\beta = v/c \geq 0.025\ Z^{2/3}$. Alpha particles with energies larger than 2.8 MeV fulfill this condition. For lower energy, we use (for lack of a better alternative) an effective charge, Z^*:

$$Z^* = Z[1 - \exp(-125\beta Z^{2/3})]. \tag{1}$$

CALCULATION OF NANODOSIMETRIC SPECTRA

The net result of transporting a charged particle event-by-event is a simulated track, consisting of the geometrical positions of all energy deposition events as well as the amount of energy locally deposited at each interaction point. Calculation of a nanodosimetric (or microdosimetric) spectrum uses, again, the Monte Carlo technique. While in an actual experiment the energy deposited in a fixed volume (the counter) by traversing particles is recorded, calculation adopts the opposite procedure: on one (or at most several) tracks, a very large (i.e., sufficient to produce statistically significant results) number of sampling volumes are placed, and the total energy deposited is stored each time. In this paper, a spherical volume will be used because it is convenient and does not contradict any biological evidence. The techniques described are, of course, valid for volumes of any shape.

The straightforward approach for producing a spectrum of specific energy is to “throw” a random sphere in a volume (we shall call it, generically, a box) completely enclosing the track. The efficiency of this procedure, measured as the ratio of successes (spheres containing at least one energy-transfer event) to the total number of spheres sampled, is generally less than one. “Missing” is, in fact, rather costly in terms of computing time since verifying the content of each sphere involves a loop over a large number of transfer points. It is therefore desirable to bring the efficiency as close to one as possible. (This is particularly important for nanodosimetry in general, and critical for radon nanodosimetry.)

The volume around a track which has a sampling efficiency of exactly one is called the associated volume of that track (Lea., 1962; and Figure 2). Conceptually, it can be built as follows: Centered at each transfer point, place a sampling sphere; the union of all these spheres is the associated volume. This volume has the required property, since by its construction any sphere placed with its center in the volume will be within, at most, one radius from at least one transfer point.

Let v_A be the volume of the associated volume (AV) and v_B the volume of the box. Then, for box-sampling, the efficiency, ϵ, is v_A/v_B. This ratio depends on the radius of the sphere, R:

$$\lim_{R \to 0} \frac{v_A}{v_B} = 0 \qquad (v_B \text{ fixed}). \tag{2}$$

For alpha particles and R = 10 nm, efficiencies as low as 1% are not uncommon.

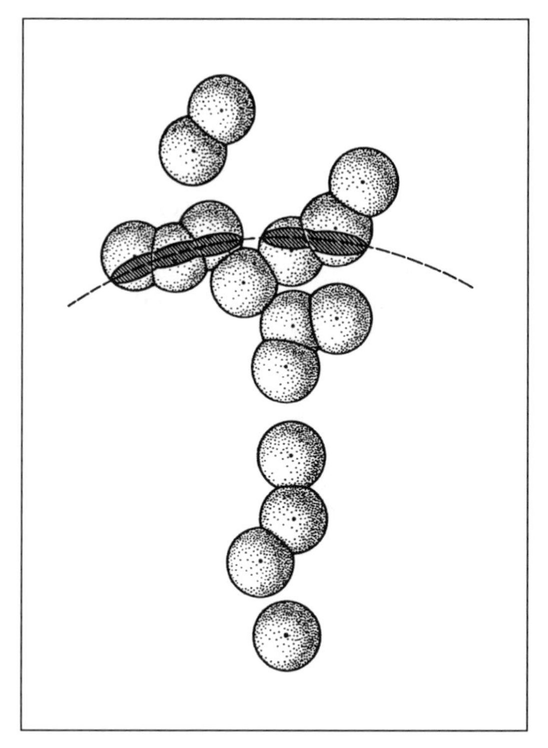

Figure 2. Graphic representation of concepts of associated volume and associated surface (dashed area).

We have devised a computational scheme for directly simulating the AV, that is, for selecting spheres in the AV only: (1) Randomly select a transfer point; (2) within a distance, R, of this point, randomly place a sphere of radius R; (3) calculate the total energy in this sphere. The event thus obtained should be scored in the spectrum with a weight inversely proportional to the number of events (n) in the sphere, since this method of selecting points in the AV is biased toward regions of high density of transfer points (straight sampling should be spatially uniform).

A key element in this method (and also in the definition of the AV) is the fact that the tracks intersecting the sphere originate uniformly from random points in space and are isotropically distributed. This has been called μ-randomness (Kellerer, 1980). In the calculation, μ-randomness translates into selecting spheres homogeneously. Also, many models of radiation action, in particular TDRA, are built implicitly (and occasionally tacitly) on the assumption of μ-randomness. It is for this reason that the TDRA, for instance, cannot be applied, as currently formulated, to radon exposure which is not characterized by μ-randomness.

Exposure by radon progeny of bronchial cells occurs under what might be called modified surface randomness (designated here by σ-randomness): In its most basic configuration a cell is exposed to an isotropic distribution of particles originating from a point at a fixed distance, ℓ, relative to the center of the cell (or site). Actual distributions are linear combinations of such elementary spectra. The unit-efficiency locus is now the intersection between the AV and a sphere of radius ℓ, centered at the origin of the track. We call this the associated surface (AS) of the track (see Figure 2). The simulation of the AS is similar to that of the AV. (The geometry of the problem is, however, more complicated.) The main difference is in calculating the bias introduced by this method: The selection of spheres on the associated surface is (a) proportional to the number of transfer points between $1 + R$ and $1 - R$ from origin; and (b) inversely proportional to the AS zone contained in the sphere. A number of examples are given below.

FACTORS AFFECTING THE NANODOSIMETRY OF RADON DAUGHTERS

Distance Between Radioactive Source and Sensitive Volume

Assuming that cells at risk are located throughout the epithelium, it is necessary to calculate nanodosimetric spectra for a relatively wide range of distances, ℓ, between the source and the center of the sphere. For this purpose it is instructive to examine first a highly simplified situation: a sphere intersected by infinite, straight lines of constant LET (=L) and distributed under σ-randomness (Figure 3). Within a scale factor, L, the nanodosimetric spectrum is identical with the distribution of chords, x, due to tracks intersecting the sphere. This probability distribution function can be readily obtained:

$$p_\sigma(x)dx = \frac{1}{2l(1-\cos\theta_M)} \quad \frac{x}{(x^2-4R^2+4l^2)^{1/2}} \; dx \qquad (3)$$

$$\theta_M = \sin^{-1}(R/l) \qquad x \leq 2R.$$

For a radioactive source on the surface of the sphere, all chords are equiprobable:

$$\lim_{l\to R} p_\sigma(x) = \frac{1}{2R} \qquad x \leq 2R. \qquad (4)$$

For l >> R the distribution becomes triangular, corresponding to μ-randomness, as expected:

$$\lim_{l\to\infty} p_\sigma(x) = \frac{2x}{(2R)^2} \qquad x \leq 2R. \qquad (5)$$

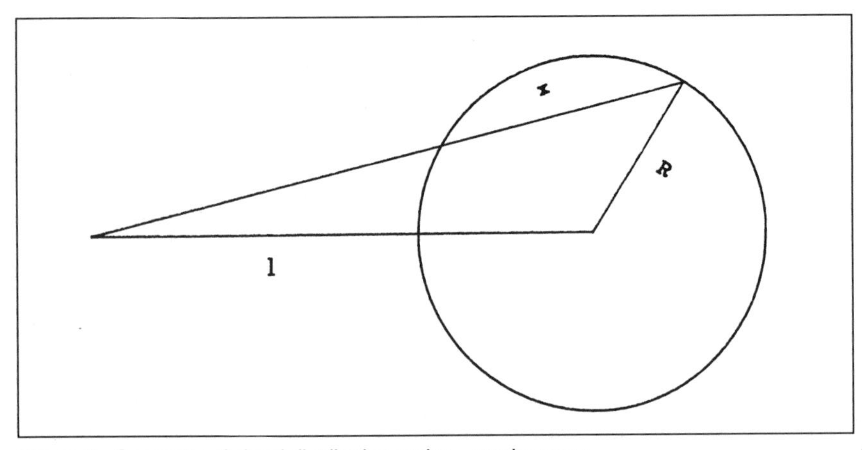

Figure 3. Geometry of chord distribution under σ-randomness.

Equations 4 and 5 are the limiting cases dictated by chord distribution only (Figure 4). Some microdosimetric calculations use CSDA: The LET is allowed to change along the chord and, consequently, tracks have finite ranges. With the exception of track-end effects, the spectra continue to be dominated by the chord distribution, (equation 3) and, implicitly, by the dimensionless parameter ℓ/R (Figure 5).

The full Monte-Carlo calculations show a similar trend (i.e., insensitivity to ℓ when the chord-length distribution has been factored out). However, in nanometer volumes, the spectra look quite different from those obtained with any of the approximations discussed above.

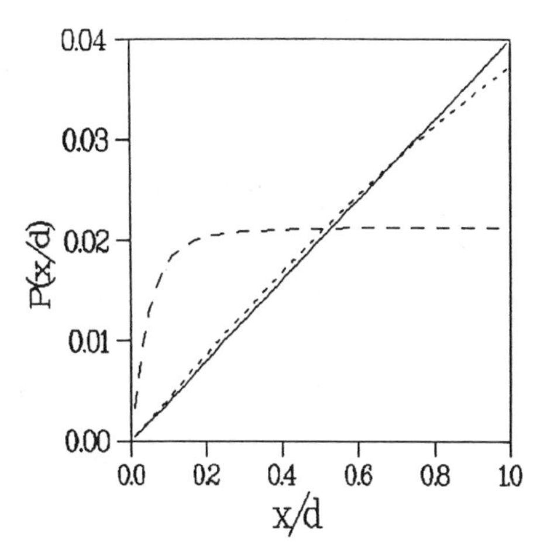

Figure 4. Probability distribution of chord length, x, in a sphere of diameter d under σ-randomness. Solid line, l/δ = 40; dotted line, l/δ = 1; dashed line, l/δ = 0.501.

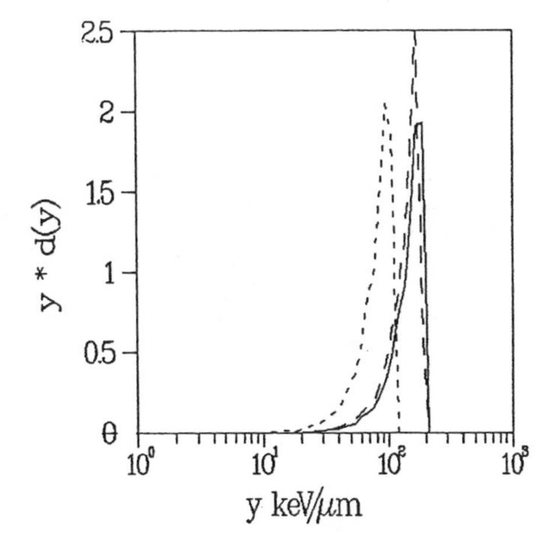

Figure 5. Microdosimetric spectrum for a 7-MeV alpha particle in a 50-nm- diameter sphere. Calculation uses continuous slowing-down approximation (CSDA). Distance source-sphere is ℓ = 40 μm (dash), ℓ = 50 nm (short dash) and ℓ = 60 μm (solid).

Size of Sensitive Volume

Within the CSDA there are no qualitative differences between nano- and microdosimetric distributions (Figure 6). This can be understood in the context of the discussion above. It is, of course, the result of neglecting the radial extent of the track which, indeed, dominates the nanodosimetry. Figure 7 shows spectra at d = 1000 nm and d = 10 nm (d = diameter), calculated with the Monte-Carlo technique. Somewhat loosely, one can say that the spectra show, respectively, the result of a track sampling a 1-μm sphere or the result of a 10-nm-diameter sphere sampling the core radial extent (penumbra) of the track.

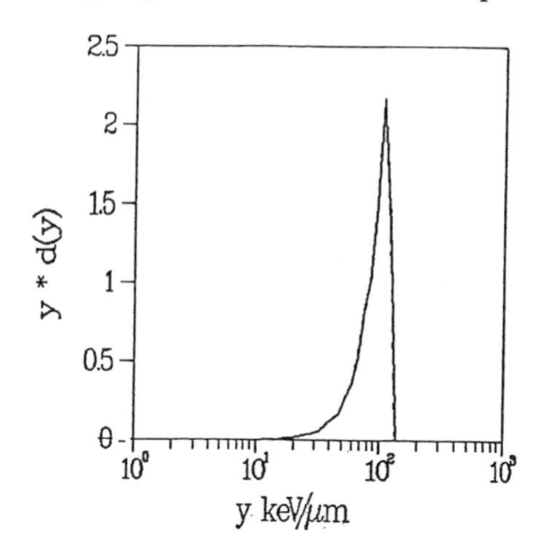

Figure 6. Microdosimetric spectrum for 7-MeV alpha particles at distance l = 3 μm from spheres of diameters d = 1 μm (solid line) and d = 10 nm. Using continuous slowing-down approximation (CSDA), the two spectra are indistinguishable.

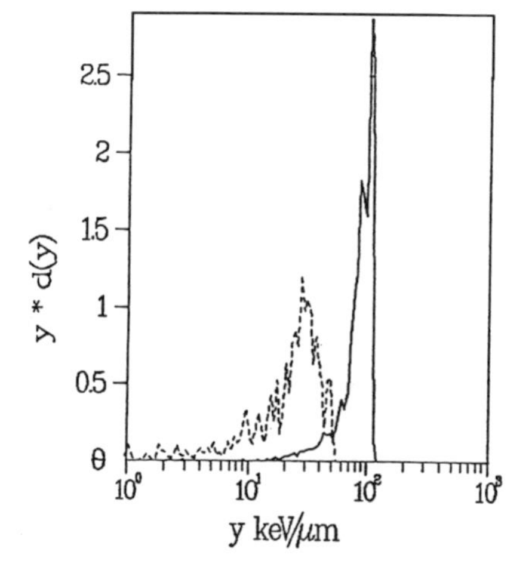

Figure 7. Full Monte-Carlo calculation of microdosimetric spectrum of a 7-MeV alpha particle at a distance ℓ = 3 μm from a sphere of diameter d = 1 μm (solid line) or d = 10 nm (dashed line). Concept of associated surface used in this calculation.

DISCUSSION

A critical question in performing these calculations is the extent to which CSDA can be used as a substitute for the full Monte-Carlo calculation. The answer depends on the ratio r_p/R (r_p is the penumbra radius, that is, the maximum radial extension of the track). At $r_p/R << 1$, CSDA provides an adequate description of the track. Whenever $r_p/R \geq 1$ a full Monte-Carlo treatment is necessary. For radon-type alpha particles ($E_\alpha \approx 7$ MeV), the limits given above correspond to micro- and nanodosimetric distributions, respectively.

As a next approximation, one may consider using radial dose distributions (which can be accurately measured) to obtain, via analytical methods, the microdosimetric spectra. This problem has been discussed in a recent publication (Varma and Zaider, 1990); by using radial dose distributions the criterion $r_p/R << 1$ may be significantly relaxed.

Finally, it appears imperative to assess the precise size of the nanodosimetric volume to be used. The problem is far from trivial, as neither the TDRA formulation (based on micron-randomness of the track relative to the sensitive volume) nor multihit, multitarget models (based on Poissonian distributions of "hits") are valid for radon exposures. The techniques presented here allow calculation of nanodosimetric spectra. In order to make further progress, it will be necessary to express models of radiation action exclusively in terms of such distributions.

ACKNOWLEDGMENTS

This investigation was supported in part by grants DE-FG02-88ER60631 and DE-FG02-90ER61022 from the Department of Energy, and by grants CA12536 and CA15307 awarded by NCI, DHHS, to the Center for Radiological Research, Columbia University, New York.

REFERENCES

Bethe, H. **1953**. Moliere's theory of multiple scattering. Phys Rev 89:1256-1266.

Bolorizadeh, M and ME Rudd. **1986**. Angular and energy dependence of cross sections for ejection of electrons from water vapor. II. 15-150-keV proton impact. Phys Rev A33:888-892.

Brenner, DJ and M Zaider. **1984**. A computationally parameterization of experimental angular distributions of low energy electrons elastically scattered of water vapor. Phys Med Biol 29:443-447.

Dagnac, J, D Blanc, and D Molina. 1970. A study on the collision of hydrogen ions H^+, H_2^+, H^+ with a water vapor target. J Phys B3:1239-1251.

Gastineau, RM, PJ Walsh, and N Underwood. **1972**. Thickness of bronchial epithelium with relation to exposure to radon. Health Phys 23:857-860.

James, AC. **1988**. Lung dosimetry, pp. 259-309. In: *Radon and Its Decay Products in Indoor Air*. John Wiley Sons, New York.

Jeffery, PK and LM Reid. **1977**. The respiratory mucus membrane, pp. 193-215. In: *Respiratory Defense Mechanisms*, Part I. JD Brain, DF Proctor, and LM Reid (eds.). Dekker, New York.

Katz, R, SC Sharma, and M Homayoonfar. **1972**. The structure of particle tracks, pp. 317-383. In: *Topics in Radiation Dosimetry*, F. H. Attix (ed.). Academic Press, New York.

Kellerer, AM. **1980**. Concepts of geometrical probability relevant to microdosimetry and dosimetry, pp. 1049-1062. In: *Proceedings of the 7th Symposium on Microdosimetry*, J Booz, HG Ebert, and HD Hartfiel (eds.). Oxford, UK.

Kellerer, AM and HH Rossi. **1978**. A generalized formulation of dual radiation action. Radiat Res 75:471-488.

Kellerer, AM and HH Rossi. **1984**. On the determination of microdosimetric parameters in time-varying radiation fields: The variance-covariance method. Radiat Res 97:237-245.

Lea, DE. **1962**. *Action of Radiation on Living Cells*. Cambridge University Press, Cambridge, UK.

Mark, TD and F Egger. **1976**. Cross section for single ionization of H_20 and D_20 by electron impact from threshold up to 170 eV. Int J Mass Spectrom Ion Phys 20:89-99.

Mott, NF. **1929**. The scattering of fast electrons by atomic nuclei. Proc R Soc London Ser A123:425-442.

Nishimura, H. **1979**. Elastic scattering cross sections of H_20 by low-energy electrons, pp. 314-318. In: *Electronic and Atomic Collisions*, Vol. II, K Takayanagi and N Oda (eds.). Society for Atomic Collision Research, Kyoto, Japan.

Paretzke, HG, G Leuthold, G Burger, and W Jacobi. **1974**. Approaches to physical track structure calculation, pp. 123-140. In: *Fourth Symposium on Microdosimetry*, J Booz, HG Ebert, R Eickel, and A Wacker (eds.). CEC, Luxembourg.

Rossi, HH. **1967**. Microscopic energy distribution in irradiated matter, pp. 43-92. In: *Radiation Dosimetry*, Vol. I, FH Attix and WC Roesh (eds.). Academic Press, New York.

Schutten, J, FJ de Heer, HR Moustafa, AJH Boerboom, and J Kistemaker. **1966**. Gross- and partial-ionization cross sections for electrons on water vapor in the energy range 0.1 -20 keV. J Chem Phys 44:3924-3928.

Seng, G and F Linder. **1976**. Vibrational excitation of polar molecules by electron impact. II. Direct and resonant excitation in H_2O. J Phys B9:2539-2551.

Toburen, LH, MY Nakai, and RA Langley. **1968**. Measurement of high-energy charge transfer cross sections for incident protons and atomic hydrogen in various gases. Phys Rev A 171:114-122.

Toburen, LH and WE Wilson. **1977.** Energy and angular distributions of electrons ejected from water vapor by 0.3-1.5 MeV protons. J Chem Phys 66:5202-5213.

Turner, JE, RN Hamm, HA Wright, JT Modolo, and GMAA Sardi. **1980**. Monte Carlo calculations of initial energies of Compton electrons and photoelectrons in water irradiated by photons with energies up to 2 MeV. Health Phys 39:49-55.

Varma, MN, JW Baum, and AV Kushner. **1977.** Radial dose, LET and W for 0^{18} ions in N_2 and tissue equivalent gases. Radiat Res 70:511-518.

Varma, MN and M Zaider. **1990**. The radial dose distribution as a microdosimetric tool. Radiat Protect Dosim 31:155-160.

Zaider, M, DJ Brenner, and WE Wilson. **1983**. The application of track calculations to radiobiology. 1. Monte Carlo simulation of proton tracks. Radiat Res 95:231-247.

QUESTIONS AND ANSWERS

Q. (Speaker not identified.) As you know, in this business, when we do track structure work, there are two approaches: site-specific modeling, such as the Y distribution and the Z distribution, and also things pioneered by Kellerer and Rossi looking at track specifics—modeling of the events and looking at things like proximity functions. Do you think there is any extrapolation or use of the proximity function in problems such as these?

A. Zaider, Columbia University. No, because the notion of proximity function is based on a randomness which they call μ-randomness, a totally uniform overlap between the track and sensitive volume. Only then can you apply this notion of proximity function. In this case, radon, you cannot use it, and I think we'll have to invent something new.

Q. Morgan, PNL. When you talk about biological modeling, what scale are you talking about? Cells, or DNA in solution? What specifically do you mean when you say that these assumptions will have to be tested with biological models?

A. Zaider. When I talk about models of radiation action, I mean models that use micro- dosimetric information. An example of such a model is the Kellerer/Rossi dual-radiation action, which was initially built on a micrometer size. That turned out not to be good, because soft x-ray data, as an example, could not be predicted by that model. Then it went down to what they call a distance model, involving proximity functions and so on; that was all right. But it cannot be applied to the radon problem. So the need was to go to a model which would invoke size of the order of, say, 10 nm, so really, a DNA level. Another model would be the so-called multihit, multitarget model, which also has to have such an assumption. Katz has a series of models along these lines; most of them do invoke nanometer-size targets. But what we need to do is basically take all these models, one after another, and try to discover, based on biological data, which size will fit, if any. Then go back and do all our stochastic calculations for that particular size.

Q. Morgan. Are you still talking about using cell-survival studies to test these models?

A. Zaider. No it doesn't have to be cell-survival. These models are not, fortunately or unfortunately, specific for a particular end point. In fact, they were invented for chromosome aberrations, as it happens.

Q. Caswell. You seem to say it's either/or: nanodosimetry or microdosimetry. There's a model by Feinendegen, Bond, Sondhaus and others which says it's both. Would you comment on that?

A. Zaider. Yes, you're absolutely right. Let me answer this way, making my answer more precise, and thank you for the question. The fact that, for instance, in survival curves we get bending survival curves for x rays leads, in a model-independent way, to the conclusion that the overall sensitive site has to be on the order of micrometers. There is no doubt about that. However, it appears that in this big gross volume you have to have what is called flocculi. Often, nanometer size is used to describe this. So, yes, I think it will turn out to be a combination of both. This is not unreasonable.

MICRODOSIMETRY OF RADON PROGENY: APPLICATION TO RISK ASSESSMENT[(a)]

D. R. Fisher,[1] T. E. Hui,[1] V. P. Bond,[2] and A. C. James[1]

[1]Pacific Northwest Laboratory, Richland, Washington

[2]Brookhaven National Laboratory, Upton, Long Island, New York

Key words: *Risk assessment, microdosimetry, radon, radon progeny*

ABSTRACT

We developed methods for calculating radiation doses to individual cells and cell nuclei of human bronchial epithelium from radon and progeny. These methods were for specified levels of exposure, breathing rates, equilibrium factors, unattached fraction of progeny, and other factors that are important in radon dosimetry. If we also know which cells are likely precursors for cancer, and we know their locations in the respiratory tract, we may then calculate the statistical probability of certain events: that these cells are irradiated by alpha particles, the number of single alpha-particle hits, and the spectrum of doses delivered (as a probability density in specific energy).

As we continue to study the relationship between microdosimetry and biological effects, we hypothesize that the corresponding probability of lung cancer is related to the specific energy imparted to nuclei of single cells in bronchial epithelium by radon and its progeny. The mathematical relationship between specific energy distribution and the probability of important biological effects may be determined experimentally from results of irradiating cultured bronchial epithelial cells and exposing laboratory animals to radon and progeny. The concept of "hit-size effectiveness function" proposed by Bond and Varma is useful for interpreting calculated specific energy spectra. When applied to the lung, this concept implies that to unfold the cell-specific transformation probability needed for predicting the dose-response function, new data are needed. Epithelial cells must be exposed to radiations of different quality and different absorbed dose levels. Factors that remain to be quantified are the time course of irradiation, dose rate, linearity of the response at low levels of exposure, and relative impact of cofactors (initiators and promoters) which, in addition to the radiation dose, are important in cancer induction.

(a) Work performed for the U.S. Department of Energy under Contract No. DE-AC06-76RLO 1830.

INTRODUCTION

One of the most important aspects of the study of health risks associated with environmental exposure to radon and progeny is the relationship between the radiation dose delivered to single cells and the probability of lung cancer. An understanding of the health effects of radon is particularly important because low-level radon progeny are prevalent, to varying degrees, in the air of homes and work places. Analyses of lung-cancer risks from exposure to radon and daughter products have proceeded from the epidemiological data base obtained from studying several different populations of underground miners. Laboratory animal studies have confirmed the causative relationship between radon exposure and lung cancer. In vitro irradiations of single cells by alpha particles have shown that the observed radiation damage includes chromosomal aberration, mutation, and transformation. These studies have resulted in some important general concepts:

- Cell transformation leading to lung cancer is a rare event.
- The excess incidence of lung tumors is associated with radiation exposure.
- Critical targets are most likely nuclei of epithelial cells of the respiratory tract.
- There are enormous uncertainties in individual exposure levels, wide variations in the dose to bronchial epithelium, and numerous interwoven factors that influence the distribution of dose to individual cell nuclei.
- The exposure-risk relationship is approximately linear in the range of about 50 to 500 WLM[a] cumulative lifetime exposure to radon and daughter products.

There is a tendency to try to relate lung-cancer risk with exposure to radon progeny in air rather than with actual dose at the cellular level. However, a direct association of risk with dosimetry at the cellular level should be useful for improving the quality and accuracy

(a) Working level month (WLM) is a common unit of exposure to radon progeny in air. It is defined as the exposure resulting from inhalation air with a concentration of 1 working level (WL) of radon progeny for 170 working hours. One working level (WL) is any combination of the daughter products in 1 liter of air that will result in the eventual emission of 1.3×10^5 MeV potential alpha energy.

of risk assessments. The purpose of our study was to more closely evaluate dose distribution at the cellular level and extend the concepts of radon microdosimetry to better understand the long-term probability of lung cancer. Of particular interest were methods for predicting the biological effects of exposure to alpha particles from exposure to low-level environmental radon and progeny. An approach for extending microdosimetric concepts in radon dosimetry to risk assessment is presented in this paper.

ESTIMATED RISK OF LOW-LEVEL EXPOSURES TO RADON AND PROGENY

As many as 20 separate follow-up studies of underground mining populations show an excess of lung cancer related to cumulative exposure to radon progeny, generally at levels much greater than typically found in dwellings (NRC, 1988; Harley, 1989). However, the risk of cancer or other harmful effects from exposure to low-level radon and progeny is not known.

The U.S. Environmental Protection Agency (USEPA) and others estimated that indoor exposure to radon progeny could account for 5,000 to 25,000 lung-cancer deaths annually in the United States (Lubin, 1989; USEPA, 1987; NCRP, 1984). Publication 50 of the International Commission on Radiological Protection (ICRP, 1987) reviewed current data from epidemiological studies on radon-exposed uranium miners and proposed estimates of individual risk of lung cancer from chronic indoor exposure to natural levels of radon and progeny. The ICRP considered both relative-risk and absolute-risk projection models and obtained lifetime risks per 1 WLM exposure (chronic) in each year of 0.007 to 0.016, and 0.004 for nonsmokers only. Each radon-induced lung cancer was estimated to reduce life expectancy by about 15 y.

The functional relationship between exposure to radon-daughter radiation and long-term probability of lung-cancer mortality includes complicating factors such as age at exposure, prolongation of exposure, time since exposure, age at exposure, and compounding factors such as smoking history, gender, and exposure to other environmental carcinogens. The National Academy of Sciences (NRC, 1988) adopted a relative-risk model to describe the rate of lung-cancer mortality per person-year at risk, using data from four cohorts of underground miners:

$$\tau(\text{age, period, dose history}) = \tau_0(\text{age})\,[1 + 0.025\gamma\,(\text{age})(W_1 + 0.5\,W_2)\,], \quad (1)$$

where τ is the age-specific risk or chance of dying of lung cancer in 1 y at a certain age, given that a person is still alive at that age;

τ_0(age) is the age-specific background risk of lung cancer from all causative agents;

the dependence of risk on age γ(age) is 1.2 for age < 55 y, 1.0 for ages 55-64 y, and 0.4 for age > 65 y;

W_1 is the cumulative exposure in working level months (WLM) incurred between 5 and 15 y before the age when cancer develops;

and W_2 is the exposure in WLM incurred 15 y or more before this age (NRC, 1988).

The WLM unit of exposure to radon progeny is a convenient term for expressing an estimate of the time-integrated exposure of workers to potential alpha-particle energy from radon progeny in air. It can be calculated from airborne concentration measurements and estimates of the time spent by a worker in an area where he breathes air having that concentration. However, estimates of radiation dose to sensitive tissues of the respiratory tract are dependent on additional factors, and the dose/WLM conversion factor may range over an order of magnitude or greater (James, 1988). These additional factors pertain to breathing rate and whether the subject inhaled by mouth or nose; the unattached fraction of potential alpha energy; particle-size distribution, and other characteristics of the aerosol that affect deposition and retention; the geometrical structure of the respiratory tract; progeny equilibrium; the thickness of the mucus and sol gel layers in which the radon progeny are deposited and transported; the transfer of radon progeny in mucus to epithelial tissues and uptake by blood; and the spatial distribution of target cell nuclei relative to the source distribution within the epithelium. These factors can lead to wide variations in radiation dose to the sensitive cells of the respiratory tract epithelium for constant levels of exposure in working level months. A correspondingly wide variation in lung-cancer risk would be expected. We therefore consider evaluation of radiation doses to sensitive tissues to be an important step toward understanding the risks of exposure to radon and progeny. Radiation absorbed doses may then be calculated directly, given certain assumptions about these factors.

DOSIMETRY OF RADON AND PROGENY

Doses to epithelial tissues of the respiratory tract from inhaled radon and progeny cannot be measured directly. However, they may be calculated if sufficient information is known about the deposition and spatial distribution, residence times and translocation rates of the radionuclides, and their proximity to target tissues. Deposition is determined by the fraction of progeny attached to inhaled particles, particle size, electrical charge, breathing rate, and airway geometry and morphometry. Clearance, on the other hand, is determined by mucociliary transport, absorption into tissue and blood, and physical decay rates. The relative distances between alpha-emitting sources and target nuclei are dependent on the distribution of sources in mucus and tissue, the size of airways, thickness of the mucus layer, and the distributions of basal and secretory cell nuclei within the respiratory tract epithelium. The dose is dependent on the number of radioactive transformations and decay energies. The number of atoms available for decay, therefore, depends on the airborne concentration and the degree of equilibrium of radon with progeny.

Numerous investigators have calculated "average" doses to the respiratory tract from inhaled radon and its alpha-emitting decay products ^{218}Po and ^{214}Po. These calculations were made for a number of different exposure parameters, as reviewed in the Experts Group Report (NEA, 1983), in NCRP Report No. 78 (1984), and in the BEIR-IV Report (NRC, 1988). Although it is not our purpose to review those methods in this paper, we emphasize that the estimated dose to lung tissue is highly variable and depends on assumptions for each of the terms used in the calculation. Values of the dose conversion factor given in NCRP Report No. 78 for various studies ranged from 0.002 to 14 rad WLM^{-1} (or 0.00002 to 0.14 Gy WLM^{-1}; NCRP, 1984), with the most common factors cited ranging from 0.2 to 2 rad WLM^{-1}. The average dose varies among the different generations of the respiratory tract, with generations 2 to 6 receiving generally higher doses than the smaller airways of generations 7 to 15. Within a single airway, the differences in local absorbed dose may also vary because of nonuniform distributions of progeny. Doses may also be significantly higher at bronchial bifurcations, particularly at carinal ridges, due to the combined effects of enhanced radionuclide deposition and relatively slow clearance (Hofmann and Martonen, 1988). Although the average absorbed dose per unit exposure was about 0.15 Gy WLM^{-1}, Hui et al. (1990) showed that a large fraction of individual cells remained unirradiated and that the doses to individual cells with energy imparted ranged from 0.01 to more than 300 Gy. Thus, the "average" dose is clearly inadequate for describing energy depositions among different generations, within a region of a single generation, or to single cells within a region.

STATISTICAL VARIATIONS IN DOSE AND RESPONSE

Although we know that biological effects result from discrete energy-deposition events at the cellular and subcellular level, it is difficult to establish the dose-response relationship at low levels of exposure. This is because there are large stochastic variations in both the ionization densities within cell nuclei and in the types of biological response that are possible for any amount of energy imparted. A variety of different biological responses are possible at constant absorbed dose when the factors affecting hit probability at the cellular level are modified. This means that two outcomes are possible: (1) The same absorbed dose could result in different biological effects. (2) The same biological effect could result at two different absorbed dose levels.

To analyze the dose-response relationship, two statistical variations need to be dealt with simultaneously: First, for a given average (absorbed) dose, the energy imparted to cell nuclei will be highly variable, depending on whether the cells are hit or missed by alpha particles and the ionization density (which is determined by the length of tracks and the number of tracks). The distribution of "doses" imparted to microscopic targets may be represented by a probability density in specific energy, $f(z)$. Second, for a given specific energy distribution there will be a variable biological response because of the large variety of biochemical changes that may result when radiolysis products (free radicals and ions) randomly interact with molecular DNA. Accordingly, the dose-effect relationship was interpreted by Morstin et al. (1989) as an integral convolution (or inner product) of these two separate functions (see also Bond et al., 1985; and Sondhaus et al., 1990). Thus,

$$E(D) = \int f(z,D) \cdot \epsilon\,[z, f(z,D)]\,dz, \tag{2}$$

where $E(D)$ is the biological response at an absorbed dose D, $f(z,D)$ is the probability density in specific energy (ICRU, 1980, 1983), and $\epsilon[z,f(z,D)]$ is the corresponding cellular response function. Evaluation of the microdosimetric function is described in the next section.

MICRODOSIMETRIC METHODS FOR CELL-SPECIFIC DOSIMETRY

General methods for calculating f(z) for irradiation of basal and secretory cells of the respiratory tract epithelium by inhaled radon and progeny are given in Hui et al. (1990) and in Fisher et al. (1990). These methods also provide the absorbed dose (D); the probability that cell nuclei are hit once, twice, or n times; and the probability that cell nuclei are completely missed by alpha particles and secondary electrons. As with conventional radon dosimetry, these methods account for the many factors that influence the probability of alpha-particle energy depositing in cell nuclei.

Monte-Carlo techniques were applied to determine chord-length distributions for distances between alpha-emitting sources and nuclear targets. The mathematical methods of Roesch (1977) for internal emitters were then applied to determine probability densities in specific energy for any region of the respiratory tract and any set of assumptions with regard to these variables.

The product of a microdosimetry calculation is a probability density in specific energy; an example is shown in Figure 1, a statistical distribution of doses f(z,D) to individual target cell nuclei from alpha-particle irradiations. It includes contributions from both ^{218}Po and ^{214}Po; the cell nuclei were assumed to be spherical and to have a known radius. The distribution includes a high probability of zero specific energy (0.27), meaning that 27% of the targets received no energy deposition. Thus, 73% of the targets received at least one "hit" by an alpha particle. The mean of the distribution (or the absorbed dose D) was 0.7 Gy, and the mode, 0.5 Gy, corresponds to the dose mean for single hits. Most targets received less than 2 Gy, but a small fraction received 2 to 3.5 Gy. The probability of a site receiving 1, 2, 3, or n hits may also be determined from the calculation.

A computational model for evaluating radiation dose distributions to cell nuclei from exposure to radon progeny throughout the human bronchial tree was described by Fisher et al. (1990). The model incorporated current information on respiratory tract geometry, nasal and oral filtration efficiencies for unattached radon progeny, characteristics of bronchial deposition by diffusive and inertial processes, mucus clearance, transfer of progeny into airway epithelium, locations of secretory and basal cell nuclei, and other factors important for assessing the probability of radiation interactions with cell nuclei. In addition, this model was used with microdosimetric theory to determine:

(1) probability densities in specific energy, (2) mean absorbed doses, (3) fraction of cell nuclei receiving no hits and therefore no radiation energy, and (4) hit probabilities for nuclei with single and multiple hits. The model was used to compare these values for different total exposures (WLM) in homes or underground mines. Both soluble (transportable in epithelial tissue) and insoluble progeny were considered. The results (Fisher et al., 1990) are summarized in Table 1. The exposure levels chosen for comparison were 0.15 and 1.0 cumulative WLM (typical of exposures in home atmospheres) and 10 and 100 cumulative WLM (typical of exposures in underground mines).

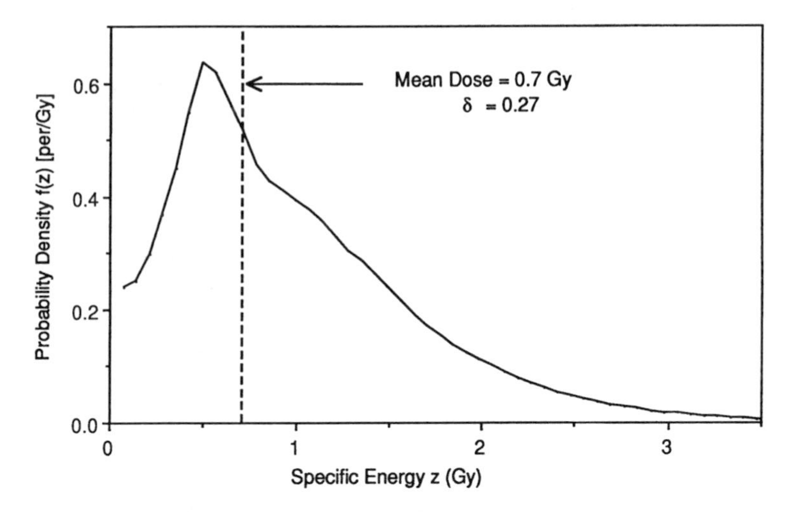

Figure 1. Example of a probability density in specific energy calculated for 6.5-μm cell nuclei irradiated by 6-MeV and 7.69-MeV alpha particles from uniformly distributed sources (^{218}Po and ^{214}Po). Mean absorbed dose is 0.7 Gy; fraction of unirradiated sites (δ) is 0.27.

Table 1. Calculated values of absorbed dose and alpha-particle hit probabilities for secretory cell nuclei.

Location	Solubility	Exposure (WLM)	Mean Specific Energy (Gy)	Hit probabilities 0	1	2	>2
Indoors in home	Soluble	0.15	0.0034	0.996	0.004		
	Insoluble	0.15	0.0033	0.996	0.004		
	Soluble	1.0	0.023	0.974	0.026		
	Insoluble	1.0	0.022	0.977	0.023		
Underground mine	Soluble	10	0.31	0.69	0.26	0.05	
	Insoluble	10	0.28	0.75	0.22	0.03	
	Soluble	100	3.1	0.029	0.087	0.16	0.72
	Insoluble	100	2.8	0.053	0.156	0.23	0.56

These results show that single-hit interactions predominated at low exposure levels, as expected. However, we also found that more than 99% of potential target nuclei were completely uninvolved at the 0.15-WLM cumulative exposure level. The hit probability ratios and dose distributions also changed as the exposure level increased. Multiple-hit interactions were more frequent at higher levels of exposure. The resulting dose conversion factors (Gy/WLM) obtained from Table 1 ranged from 0.008 to 0.033 Gy WLM^{-1} (0.8 to 3.3 rad WLM^{-1}).

EVALUATION OF MICRODOSIMETRIC INFORMATION FOR RISK ASSESSMENT

One question, for many years, has been, "What do we do with a probability density in specific energy $f(z,D)$, and how can this spectrum help us predict the biological effects of radiation (such as that received by the respiratory tract epithelium from radon and progeny)?" We have already suggested one possible answer to this question by presenting the concept of hit-size effectiveness function. The probability density $f(z,D)$ may be useful for evaluating the biological effectiveness of radiation, using the relationship given in equation (2) if sufficient data are available for determining the cellular response function $\epsilon[z,f(z,D)]$. The cellular response function is not yet available for the irradiation of epithelial cell nuclei by alpha particles. We can predict, however, some characteristics of this function from other information that is available.

Because radiation effects begin at the cellular level, we believe that the specific energy to the nucleus is directly related to the probability of biological effects. The transformation of cells from their normal state to a malignant state is generally believed to result from low-specific-energy interactions in the cell nucleus. These are necessarily nonlethal energy depositions, because inactivated cells cannot reproduce. The transformation response to exposure is probably linear, with increasing probability of single hits. As the probability of multiple hits increases, there may be increased probability of cell death and somewhat reduced probability of transformation per unit exposure. Therefore, a nonlinear response may be anticipated with increasing levels of exposure.

Kellerer and Rossi (1972, 1978) proposed a theory of dual radiation action in which the biological response of a cell resulted from an interaction of two sublesions in a sensitive volume. The number of sublesions is proportional to the specific energy, z, in the target volume, and the expectation value for the number of lesions is proportional to the square of the specific energy:

$$\mathrm{E}(z) = Kz^2. \tag{3}$$

The mean number of lesions as a function of the absorbed dose $E(D)$, obtained by averaging over $E(z)$ for all sensitive volumes in a population, is

$$E(D) = \int_0^\infty Kz^2\, \mathrm{f}(z,D)dz = \mathrm{K}\bar{z}_\mathrm{D} + D^2, \tag{4}$$

where $\bar{z}_\mathrm{D}$ is the single-event dose mean specific energy (ICRU, 1983). The average diameter of the sensitive volume over which sublesions combined was approximately 1 μm. However, the sensitive volume may, in fact, be the entire inhomogeneous nucleus, having a diameter of 4 to 8 μm and a complex substructure of various sensitive sites. We therefore look to other possible interpretations of the microdosimetric spectrum.

Figure 2 is a representation of a probability density in specific energy for a hypothetical population of cell nuclei irradiated by a 5-MeV alpha particle. We may assume that for every irradiated cell, there exists a value of specific energy z_0, above which the imparting of that amount of energy within the nucleus will be lethal to the cell, or below which the cell may survive and continue division. If we integrate this "partial area," then the surviving fraction, the cells available for transformation, is proportional to the integrated area (indicated with diagonal lines). This is an idea we have not tested experimentally because the value of z_0 is not a constant or a step function but is more likely to be a continuous function s(z), which is the survival probability of a cell receiving a specific energy z. The survival probability s(z) may be related to an exponential of z. Therefore, the probability of a biological end point for the surviving cells is an integral that incorporates s(z), f(z), and λ(z) (Hui, 1989):

$$E(S) = \mathrm{k}\int_0^\infty \mathrm{s}(z)\, \lambda(z)\, \mathrm{f}(z,D)dz, \tag{5}$$

where s(z) is the probability of a cell surviving receipt of a specific energy z,

λ(z) is the probability of a cell transforming or having another biological change from z, and

k is a constant of proportionality.

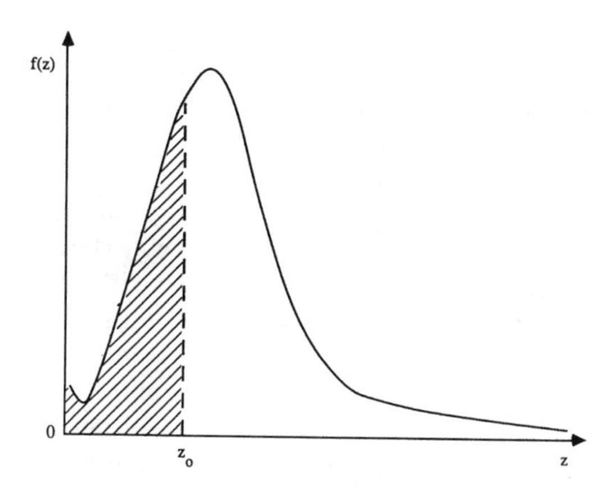

Figure 2. Schematic representation of a probability density distribution for a single cell, indicating a value of z_o, below which the specific energy imparted is nonlethal to the cell.

If we assume no effect at zero dose and a definitive effect at high values of z, then $\lambda(z)$ may be represented by the function shown in Figure 3, although the function has yet to be determined experimentally. Further evaluation shows that equations (5) and (2) are mathematically equivalent and that the function described in Figure 3 is the hit-size effectiveness function. The expectation values of s(z) and $\lambda(z)$ may be determined experimentally by irradiating cells at different absorbed dose levels, whereas f(z,D) may be calculated for different absorbed dose levels.

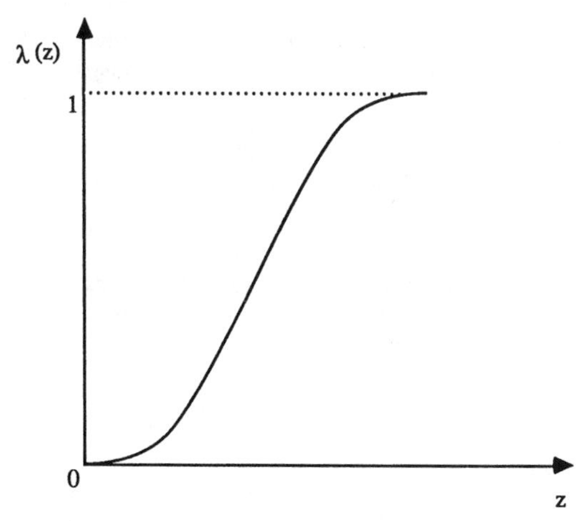

Figure 3. Schematic representation of the probability of a biological end point as a function of specific energy imparted to the cell nucleus.

SUMMARY

Alpha particles from radon progeny impart a specific energy to epithelial cell nuclei; the probability density in specific energy may be calculated from the dosimetric model and assumptions describing the physical properties of the aerosol and its deposition and clearance in the lung. Only cells having a nonlethal energy deposition event may eventually transform, and transformation is an extremely low-probability process. A hit-size effectiveness function may be unfolded from experimental data to describe the probability of a transformation for a discrete value of specific energy. This function may enable one to understand the relationship between microdosimetric results and biological effects, and should be useful for risk assessment. The application of this work will allow prediction of the effects of radon progeny for low-level environmental exposures. More information is needed about the response of cell populations to alpha-particle radiation to evaluate the dose-response relationship. Additional factors that remain to be quantified are the time course of irradiation, dose rate, linearity of the response at low levels of exposure, and relative impact of cofactors (initiators and promoters) which, in addition to the radiation dose, are important in cancer induction.

REFERENCES

Bond, VP, MN Varma, CA Sondhaus, and LE Feinendegen. **1985**. An alternative to absorbed dose, quality and RBE at low exposures. Radiat Res 104 (Suppl. 8):S552-S557.

Fisher, DR, TE Hui, and AC James. **1990**. Model for assessing radiation dose to epithelial cells of the human respiratory tract from radon progeny. Radiat Protect Dosim (in press).

Harley, N. **1989**. Overview and history of radon and lung cancer, pp. 5-11. In: *Proceedings No. 10, 24th Annual Meeting of the National Council on Radiation Protection and Measurements*, March 30-31, 1988, Washington, DC. National Council on Radiation Protection and Measurements, Bethesda, MD.

Hofmann, W and TB Martonen. **1988**. The significance of enhanced radionuclide deposition at bronchial bifurcations. Ann Occup Hyg 32:1055-1065.

Hui, TE. **1989**. *The Microdosimetry of Radon and Its Daughter Products in the Human Respiratory Tract*, doctoral dissertation. Texas A&M University, College Station, TX.

Hui, TE, JW Poston, and DR Fisher. **1990**. The microdosimetry of radon decay products in the respiratory tract. Radiat Protect Dosim 31:405-411.

ICRP. **1987.** *Lung Cancer Risk from Indoor Exposure to Radon Daughters*, ICRP Publication 50. International Commission on Radiological Protection, Oxford.

ICRU. **1980**. *Radiation Quantities and Units*, ICRU Report 33. International Commission on Radiation Units and Measurements, Bethesda, MD.

ICRU. **1983**. *Microdosimetry*, ICRU Report 36. International Commission on Radiation Units and Measurements, Bethesda, MD.

James, AC. **1988**. Lung dosimetry, pp. 259-309, Chapter 7. In: *Radon and Its Decay Products in Indoor Air*, AV Nero (ed.). John Wiley, New York.

Kellerer, AM and HH Rossi. **1972**. The theory of dual radiation action. Curr Top Radiat Res Q 8:85-158.

Kellerer, AM and HH Rossi. **1978**. A generalized formulation of dual radiation action. Radiat Res 75:471-488.

Lubin, J. **1989**. On the BEIR IV lung cancer risk projection model for radon exposure, pp. 86-114. In: *Proceedings No. 10, 24th Annual Meeting of the National Council on Radiation Protection and Measurements*, March 30-31, 1988, Washington, DC. National Council on Radiation Protection and Measurements, Bethesda, MD.

Morstin, K, VP Bond, and JW Baum. **1989**. Probabilistic approach to obtain hit-size effectiveness functions which relate microdosimetry and radiobiology. Radiat Res 120:383-402.

NCRP. **1984**. *Evaluation of Occupational and Environmental Exposures to Radon and Radon Daughters in the United States*, NCRP Report No. 78. National Council on Radiation Protection and Measurements, Bethesda, MD.

NRC. **1988**. *Health Risks of Radon and Other Internally Deposited Alpha-Emitters* (BEIR IV). National Research Council, National Academy Press, Washington, DC.

NEA. **1983**. *Dosimetry Aspects of Exposure to Radon and Thoron Daughter Products*. Nuclear Energy Agency, Organization for Economic Co-operation and Development, Paris.

Roesch, WC. **1977**. Microdosimetry of internal sources. Radiat Res 70:494-510.

Sondhaus, CA, VP Bond, and LE Feinendegen. **1990**. Cell-oriented alternatives to dose, quality factor and dose equivalent for low-level radiation. Health Phys 59:35-48.

USEPA. **1987**. *Radon Reference Manual*, EPA 520/1-87-21. U.S. Environmental Protection Agency, Office of Radiation Programs, Washington, DC.

QUESTIONS AND ANSWERS

C. Morgan, PNL. I might suggest that in the absence of having enough information about transformation, that you look to the mutation literature. We assume that transformation is a result of DNA mutations caused by the incident radiation. There is a substantial body of evidence on a variety of radiations with mutations—far more than there is for transformation. The second thing I'd like to point out is that it's also becoming clear, from the same literature, that different genetic loci are going to respond differently to the same kind of radiation. So, for example, when you have the APRT gene irradiated by low-LET radiation, you have a dose-effect curve. When you go to a different locus, the TK locus in Helen Evans's system, you get a 10-fold or a 100-fold increase in the yield of mutations for the same dose. What I'm suggesting is that it's very locus-specific, that the target is different, depending on what locus you're talking about. The hit-size events spectrum is going to be different for every locus; that it could be potentially transforming. And while your data fit well, for the point mutations, that's only a single locus. If you were to change locus, go to a different genetic locus, where the nearest dominant lethal is farther away or closer, you might find a different fit to the data.

A. Fisher, PNL. I don't know if I can remember all of your points or respond to them. But as to whether or not the mutation data are good models for transformation, I don't know. But I think that, in many cases, it is possible to do both. And, I would also like to add, it is very important when doing these experiments to do good dosimetry and microdosimetry so that those comparisons can be made.

In regard to the comment you made on the specific locus, remember that what we calculate in terms of dosimetry is the dose to the cell nucleus as a whole. Dr. Zaider mentioned the need to do this also at the nanometer level, but for the purposes of our work we calculate the dose to the cell nucleus for an assumed volume. So I believe that this approach would be equally applicable to many different biological end points and qualities of radiations for that purpose. I hope I covered the excellent points that you mentioned.

C. Curtis, Lawrence Berkeley Laboratory. I want to congratulate you on your calculations. I think these are some very nice calculations that you have made. I want to emphasize the points that were made just at the end of the last talk, that there may well be two different volumes of relevance here, and I agree with the last comment that there are two different ones. One is the size of the nucleus of the cell, and the other is the size of, say, a few nanometers, where you have the action in the DNA. Now I think that microdosimetry may well apply to the type you (Fisher) are talking about, the size of the nucleus, which is in the micrometer range. But I would like to underscore very much the comments of Dr. Zaider. If you really want to talk about what's going on in transforming a cell, you are going to have to go to the mechanisms of what, on the molecular level, is happening to produce a transformed cell. Now, looking at what happens to a 10T 1/2 cell is going to be very different from what happens, for instance, to a human lung epithelial cell. We already know this, that the frequency of transformation is very, very different. And that there are a number of studies already to look at the transformation from a 10T 1/2 cell, for instance, for alpha particles and other types of high- and low-LET radiation. I think we need to emphasize the cells that are important, namely the human lung epithelial cells, and I think that is a much more difficult problem. I do think that we need to keep these two volumes in mind when we make these calculations. And the hit-size effectiveness function, I think, is a very phenomenological way of looking at it. I think that we need to get down to the molecular biology of what's going on in order to really understand what happens when there are a bunch of ionization clusters nearby, let's say a strand of DNA.

Fisher, PNL. Thanks, Stan. I'm in complete agreement with those comments and also what Marco Zaider said earlier. A lot of our work has been to characterize the surface equilibrium concentrations of radon progeny, and then to go from that step to the dosimetry at the cellular level. I also agree that the basic mechanisms of some of these biological end points are dependent on nanometer-size volumes. So we have more work to do.

RADON TRANSPORT IN SOILS AND INTO STRUCTURES

MEASUREMENT AND MODELING OF RADON INFILTRATION INTO A TEST DWELLING

P. Stoop, R. J. de Meijer, and L. W. Put

Rijksuniversiteit, Groningen, The Netherlands

Key words: *Indoor radon, infiltration, modeling, test dwelling*

ABSTRACT

To understand radon transport in Dutch residences, we made a detailed study of its characteristics in a test house. Information on the dynamic aspects of radon transport was obtained from leakage parameters and from continuously monitoring the concentration in the dwelling and crawl space and the pressure differences between compartments. Concentrations and dynamic variables were used as input for a multicompartment model to derive the radon production rate for the two compartments.

Depressurization and pressurization of the crawl space led to substantial changes in the radon concentration in the crawl space but had less effect in the living room. The results suggest that pressure-driven flow of radon through the soil is an important source for radon in the crawl space. The production rate for the dwelling cannot be explained in terms of known exhalation rates and known air flows; therefore the remainder, which appeared to be pressure-dependent, might be due to radon-rich air entering the dwelling via the cavity wall.

INTRODUCTION

Put et al. (1985) reported that the median and the geometric standard deviation of the distribution of indoor radon concentrations in The Netherlands are 24 Bq m^{-3} and 1.6, respectively. Because rock is almost totally absent near the surface in our country, the percentage of "high-radon" dwellings is small compared to those in Sweden and the United States, for example. Nevertheless, for the average Dutch citizen, radon contributes more than 50% of the radiation dose from natural sources. Improving our knowledge of radon entry and transport in a "typical" dwelling may lead to measures which will reduce exposure to radon.

We present part of the results of an extensive study of the static and dynamic aspects of radon concentrations in a test dwelling. Exhalation rates were measured, and quasi-continuous measurements of radon concentrations and of pressure differences between the dwelling and the outside air were made. These data were entered in models for the static and dynamic behavior of radon concentrations.

The main conclusions from our previous investigations (Stoop et al., 1990a) were that the concentrations in the crawl space and living room were greater than those calculated from measured static source strengths. We believe that the difference for the crawl space may be accounted for by pressure-driven flow of soil gas into the crawl space. Furthermore, depressurization of the crawl space had a clearly reducing effect on radon concentration in the crawl space but not on the concentration in the living room. The latter remained higher than expected on the basis of the concentration in the outside air and the exhalation by surfaces; therefore an unexpected entry route for radon was suspected. To investigate this in more detail, radon source strengths in the dwelling and in the crawl space were determined as a function of time at three pressure regimes maintained by a fan.

TEST DWELLING

The test house is located in Roden, southwest of the city of Groningen in the northern part of The Netherlands. It is on the edge of the "Drents Plateau," whose soil consists of silty fine sand with intercalations of boulder clay on "pot clay," the top of which lies, in general, 1 to 2 m below grade. The impermeability of the pot clay causes large and rapid variations in the groundwater level.

The single-family house, built in 1973, has a rectangular floor plan, with a wall extending from the crawl space to the roof, dividing the house into a northern and southern part. The southern part consists of a crawl space, a high-ceilinged living room, and an "open" kitchen that has no internal ceiling. The northern part of the dwelling contains, on the ground floor, a scullery, a bedroom, a bathroom, a study, and a toilet. An open staircase connects the ground floor with the first floor, where there are two bedrooms, a bathroom, and three storage areas.

The ceiling of the crawl space consists of prefabricated hollow concrete segments which are covered by a few centimeters of cement; in the living room the cement is covered with ceramic tiles. The outside walls are cavity walls of brick masonry (about 10 cm thick). In 1976 the cavity was filled with ureaformaldehyde foam for insulation. The bottom of the crawl space is uncovered sand and is about 40 cm below the level of the surrounding soil.

Around the house, at a distance of 1 to 1.5 m, a ring-type drain was installed, in February, 1989, 70 to 90 cm below the surface. The dividing wall in the crawl space has two openings in addition to feed-throughs for central heating (hot water) and utilities. In October, 1989, a 16-cm-diameter ventilation duct was installed from the crawl space to the roof; in this duct a fan was mounted to depressurize the crawl space. The crawl space has shafts to the outside for ventilation; in principle, these shafts should have no connection with the cavity wall. During the investigation, these shafts were sealed at the outside wall.

The term "dwelling" will be used for the total space in the house minus the crawl space; i.e., the living room plus the bedroom part. The volumes of the crawl space and of the dwelling were 61 ± 5 m^3 and 374 ± 30 m^3, respectively.

EXPERIMENTAL TECHNIQUES

Radon exhalation rates were measured in situ with a device described by Aldenkamp et al. (1990). For measuring, the device is placed on the surface to be investigated; soft-rubber rings at the ends of the two coaxial cylinders are designed to provide an air-tight fit to the surface. For rough surfaces, caulk is applied to the surface and to the outer cylinder. For measurements in soil, a 30-cm-long cylinder with a flat, smooth top surface is driven into the soil, and the device is mounted on this surface. Prior to each measurement, the device is flushed with dry nitrogen. The exhalation rate is determined by fitting the growth curve for the radon concentration in the device to a (1-exp[-at]) function.

Radon concentrations were measured continuously with cylindrical, 30-cm-long, Lucas cells with a diameter of 13 cm, segmented into nine longitudinal sections. The inside of each cell and the segment walls are covered by ZnS(Ag), except for the side where the cell is mounted on a 13-cm-diameter, low-background photomultiplier tube (PMT). Because of the geometry, a relatively high efficiency for detecting the decay of ^{222}Rn of about 40% is obtained. The efficiency for detecting alpha particles from the decay of ^{218}Po and ^{214}Po is about 50%. Mainly because of Cerenkov light induced by cosmic radiation in the window of the PMT, the background is about 1 cpm. Compared to the 1.2-L cell described by Busigin et al. (1979), for which the efficiencies for counting radon and decay products were 30 and 24%, respectively, our 3-L instrument is three to four times more sensitive. With this detector, concentrations of 10 Bq m^{-3} can be measured in 30 min with a statistical uncertainty of 50%. The count rate of the detector is converted to radon concentration by means of an algorithm based on the count rates for the decay series (Busigin et al., 1979).

The system is operated in a quasi-continuous mode. A sample is taken by flushing the cell over a filter and a drying column for 5 min at a rate of 3 L min^{-1}. Subsequently, the sample is counted for 25 min, and the number of counts is stored in the memory of a data logger. Simultaneously, the values of the following parameters, averaged over 25 min, are recorded: pressure differences between crawl space and outside air at the crawl-space ventilation openings (about 30 cm above grade); pressure difference between living room and crawl space; pressure difference between living room and outside of the roof; temperature and relative humidity of outside air, air in living room, and air in crawl space; barometric pressure and groundwater level in crawl space. Air leakage parameters between the dwelling and the crawl space, and between the dwelling and the outside air were determined using helium as a tracer gas. For details, see Stoop et al. (1990b).

MEASUREMENTS

Measurements were carried out from summer, 1989, to August, 1990. The first results were published by Stoop et al. (1990a). In this paper we present the results obtained from June 10 to 25, 1990. During this period all outside doors and windows were kept closed; all inside doors were open. The living area, including bedrooms, bathrooms and study, were regarded as a single compartment; the crawl space was considered the other compartment of the house. To ensure that the radon concentration in the crawl space was the same at each measuring position, the crawl-space air was mixed using three fans in the crawl space. Measurements were made at three different settings of the fan in the duct between the crawl space and the outside: "exhausting," with air being exhausted from the crawl space; "supplying," with outside air supplied to the crawl space; and "fan off."

RESULTS AND ANALYSIS

Radon Concentrations

Figure 1 shows the radon concentrations measured by radon meters in the crawl space, the living room, the hall, and outside. During the exhausting phase (June 10 to 13), the concentration in the crawl space was on the order of 22 Bq m^{-3}, which is a factor of about two higher than the concentrations in the dwelling (10 Bq m^{-3}), and a factor of three higher than the average concentration outside (6.6 Bq m^{-3}). The concentrations in the living room and the hall are about equal. During this period, the main air flow was from the dwelling to the crawl space and from outside to the crawl space. This means that there is a radon supply to the crawl space which is not accounted for by these air flows.

When the fan was not operating (June 13 to 16), the radon concentration in the crawl space increased to about 80 Bq m^{-3}. The concentration in the living room was slightly higher than in the hall, averaging about 22 Bq m^{-3}. In the second "fan-off" period (June 21 to 25), the concentration in the crawl space was about 75 Bq m^{-3}, and in the living room it was again somewhat higher than in the hall, with an average for the dwelling of 16 Bq m^{-3}.

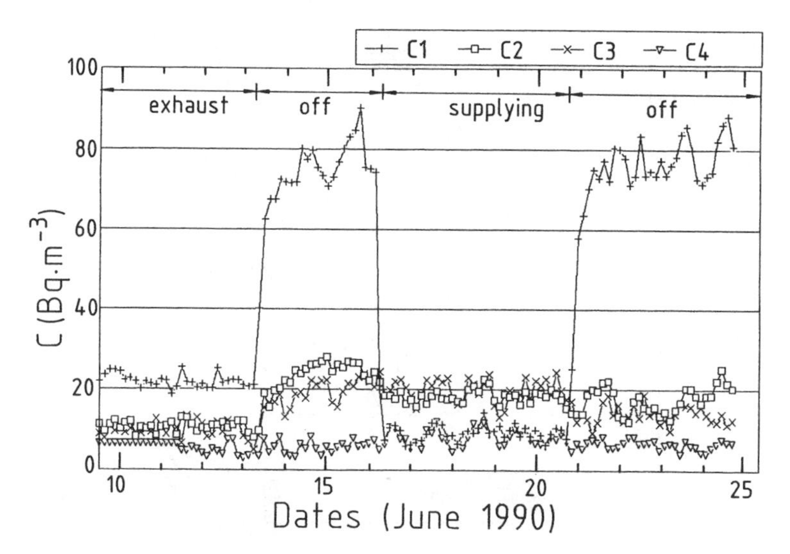

Figure 1. Radon concentrations in crawl space (C1), living room (C2), hall (C3), and outside (C4) in a test residence from June 10 to 25, 1990. Half-hour measurements averaged over 3 h. Divisions along abscissa correspond with 12 o'clock noon.

During "supplying" (June 16 to 21), the radon concentration in the crawl space was equal to the radon concentration outside. Moreover, the fluctuations in concentration in the crawl space and outdoors were identical. We therefore concluded that the fluctuations were real and were not due to uncertainties in determining the concentrations.

Air Flows

Figure 2 shows schematically the dwelling, the duct between the crawl space and the roof, and the air flows. Plus and minus signs indicate the possible direction of the flows, with "plus" corresponding to the direction of the arrows. Q_1 is the flow through the duct, calculated from the conservation of air mass in the crawl space. Q_2 is the total flow between the outside and the crawl space via the walls. Q_3 is the flow between dwelling and crawl space; Q_4 is the unidirectional flow from outside into the dwelling via the four walls and/or roof; Q_5 is the unidirectional flow from the dwelling to the outside via walls, roof and chimney; and Q_6 is a rest flow in order to balance the air mass in the dwelling.

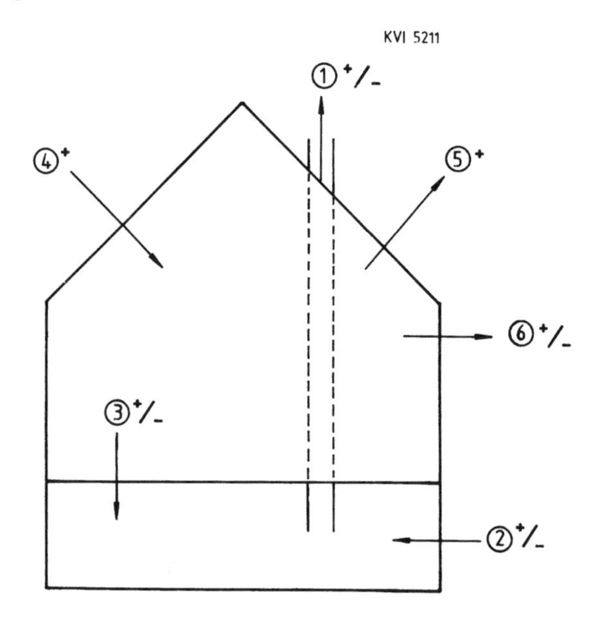

Figure 2. Diagram of test dwelling, with air flows indicated by arrows. The + and – indicate possible direction of air flows, with + corresponding to direction of arrows.

Flows were calculated using the equation

$$Q = T \left[\frac{\Delta p}{1\ \text{Pa}}\right]^{1/n}, \tag{1}$$

with Q being the air flow through the barrier, and Δp the pressure difference over the barrier. The quantities T and n are the leakage parameters representing the air transparency coefficient and flow exponent, respectively, for that particular barrier. The inaccuracy in the magnitude of the air flows varies from about 10% at larger pressure differences to more than 25% for ambient conditions.

Measured pressure differences and leak parameters published by Stoop et al. (1990b,c) or by Phaff et al. (1983) were inserted in equation 1. They are listed in the second column of Table 1. Air flows calculated with these parameters and the measured pressure differences indicated that Q_6 became unacceptably large. This observation most likely indicates that the pressure differences measured at one location on each wall and on the roof are not fully representative for the actual situation. The rest current could be reduced effectively by distributing part of the leak transparency coefficient of the roof over the walls. The adjusted values are listed in the last column of Table 1; these values were used to calculate the air flows presented in this section and for further analysis.

Table 1. Transparency coefficients (T) obtained from leakage parameter measurements and adjusted values (T_{adj}) needed to reduce rest current in dwelling (see text). n = 1.5 unless otherwise indicated.

Surface	T ($m^3\ h^{-1}$)	T_{adj} ($m^3\ h^{-1}$)
Crawl Space		
North	0[a,b]	0
East	34[a]	34
South	11.5[a]	11.5
West	2.5[a]	2.5
Dwelling		
Floor	31 (n = 1.9)[c]	31 (n = 1.9)
North	10[d]	20
East	44[d]	50
South	6.5[d]	20
West	37[d]	50
Roof	172[d]	130

[a] Values estimated from transparency coefficients for southern part of crawl space (Stoop et al., 1990c).
[b] Value set at zero because most of northern wall of crawl space is adjacent to garage; remainder of wall has no ventilation shafts.
[c] Values from Stoop et al., (1990b).
[d] Calculated from values given in Phaff et al. (1983).

Table 2. Radon production rates, S (kBq h^{-1}), and pressure differences between outside air and crawl space, $\Delta P = P_{oa} - P_{cs}$ (Pa), averaged over periods indicated; uncertanties represent standard deviations.

Period	Fan Mode	S_{cs} [a]	ΔP	S_{cs}^{calc} [b]	S_{dw} [c]
6/10–6/13	Exhausting	4.9 ± 0.7	9.5 ± 1.7	42	1.0 ± 0.4
6/13–6/16	Off	4.0 ± 1.0	0.7 ± 0.3	4.5	3.1 ± 0.6
6/16–6/21	Supplying	0.4 ± 0.5	-8.4 ± 0.2	0.6	4.0 ± 0.4
6/21–6/25	Off	3.6 ± 0.9	1 ± 4	7.0	2.5 ± 1.0

[a] For crawl space, measured
[b] For crawl space, calculated
[c] For dwelling, measured

Figure 3 shows the air currents Q_1–Q_6 as schematically indicated in Figure 2. The values were calculated as the average over six half-hour measuring periods. For each period we used the average pressure difference over that period and the adjusted transparency coefficients of Table 1. From Figure 3 it is clear that during "exhausting," Q_5, the flow exiting the dwelling, becomes smaller. Similarly, during "supplying," Q_4, the flow entering through the walls from the outside, becomes smaller. Moreover, Q_4 and Q_5 show variations with a 24-h repetition time; these variations are often in phase, indicating that cross-ventilation occurs in the dwelling.

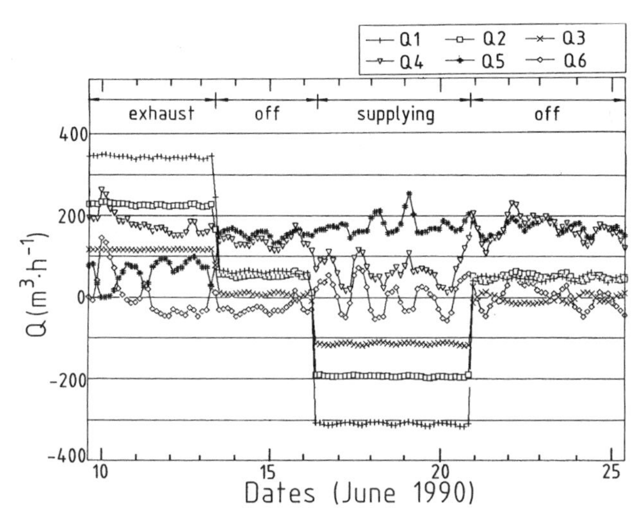

Figure 3. Air flows (Q) in crawl space and dwelling from June 10 to 25, 1990, as derived from measured pressure differences and adjusted transparency coefficients. Each data point represents average of six half-hour measurements. Flows are identified in Figure 2.

These observations reflect the fact that pressure differences, especially over the walls and roof, fluctuated rapidly and often showed diurnal oscillations. These oscillations were, surprisingly enough, in phase with the oscillations in the radon concentration, both in the dwelling and outside. The maxima occurred during the night, when temperature differences were greatest between inside and outside air and mixing in the outside air was minimal. This indicates that the diurnal oscillations in Q_5 were due to the "stack effect" and that part of the influx of the air into the room was from outside via the wall.

Radon Production Rates

With known air flows and radon concentrations, the production rate, S, can be calculated from a radon mass balance. For compartment i, with concentration C_i, and all compartments j, with radon concentrations C_j, that can directly exchange air with compartment i by means of an air flow Q_{ij} flowing from i to j, the radon production rate is given by

$$S_i(t) = \frac{V_i}{\Delta t}\left[C_i(t + \Delta t) - C_i(t)\right] + C_i(t)\sum_j Q_{ij}(t) - \sum_j C_j(t)\,Q_{ji}(t). \qquad (2)$$

In this equation, the left side is the production rate during the current period. The first term of the right side is the growth of radon activity per unit of time during the current period. The second term is the radon activity flow out of compartment i, and the third term is the radon activity flow into compartment i. Loss of radon due to nuclear decay was ignored.

Figure 4 shows the radon production rates in the crawl space and dwelling for the period June 10 to 25, 1990. From the figure, one notices that the production rates are strongly dependent on the ventilation situation in the crawl space. While "exhausting," the production rate in the living room is clearly lower than during the other periods and, similarly, the production rate in the crawl space is clearly lower during "supplying." Variations in the source strength for each fan status reflect inaccuracy in determining the value of S from equation 2.

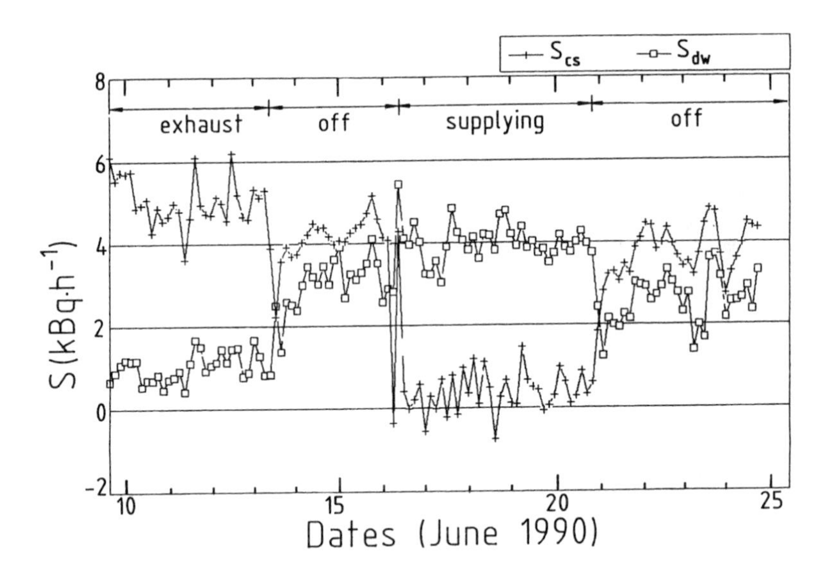

Figure 4. Radon production rate (S) for crawl space (cs) and test dwelling (dw) from June 10 to 25, 1990. Each data point represents average of six half-hour measurements.

To allow a more quantitative comparison, the production rates were averaged over periods with a constant fan mode (Table 2); the uncertainties represent standard deviations. The radon production rate for the crawl space during "supplying" was much smaller than during "fan-off" or "exhausting." The value during "exhausting" seems to be slightly higher than during "fan-off." These results are consistent with the contribution of a pressure-driven flow through the soil. The static radon production rate due to exhalation of soil and building materials may be estimated from the values given by Stoop et al. (1990a). The static radon production rate for the crawl space is estimated at 0.6 kBq h^{-1}. This value is consistent with the production rate obtained while "supplying." The higher production rate during "fan-off" is expected, since during this period the crawl space is at lower pressure than the outside (see Table 2), allowing radon-rich soil gas to enter the crawl space.

The slightly higher value during "exhausting" (with about an order-of-magnitude larger pressure difference) compared to the "fan-off" value indicates that the radon transport through the soil is limited by the "ventilation" rate in the soil. Increasing the flow above a certain value would not lead to a proportionally higher production rate. This implies that the pressure-driven flow component through the soil is no longer proportional to the pressure differences over the walls, as was assumed in our previous model (Stoop et al., 1990a).

From analysis of the data for week 24, 1989 (see Stoop et al., 1990a), the pressure-driven flow source term for the total crawl space is 3.2 kBq h^{-1} Pa^{-1}, whereas the static source term is 0.6 kBq h^{-1} (see above). These values, combined with the average pressure difference during the four intervals in the period June 10 to 25 lead to S_{cs}^{calc} values listed in Table 2. A comparison of these estimated values and the values derived from this investigation shows that except for the first interval, the values agree reasonably well. The large deviation when "exhausting" supports the theory that the production rate is limited by the "ventilation" rate of the soil.

For the dwelling, the production rate is clearly lowest during "exhausting." The values during the two "fan-off" periods are the same within the uncertainties, but the weighted average (3.0 ± 0.5 kBq h^{-1}) is 1.0 ± 0.6 kBq h^{-1} smaller than the value during "supplying"; it is 2.0 ± 0.6 kBq h^{-1} higher than the value during "exhausting." These results indicate that the production rate in the dwelling has a pressure-driven flow component. Since the flow through the floor has already been accounted for, this component must enter via a different route. Radon production can occur only in soil or in building materials. Therefore, two pressure-driven flow sources are possible. A flow from the crawl space to the dwelling via the cavity wall may pick up radon from the soil or from the masonry at the inner part of the cavity wall. Or, a flow through the pores in the concrete floor may push the radon in the floor into the dwelling. The latter, however, is unlikely, because during "fan off" the flow through the floor (Q_3) is, on average, from dwelling to crawl space (see Figure 3).

The static component for the dwelling was estimated from the measured exhalation rates for walls and floors as 0.75 ± 0.06 kBq h^{-1}. This value would be consistent with the production rate calculated from radon concentrations and air flows during "exhausting." From results for the radon concentration in the cavity wall at the three fan modes, the variations in the production rate for the dwelling can be qualitatively understood. A quantitative explanation, however, cannot yet be made because we do not know how representative the measured concentrations are for the whole foam-filled cavity wall. From the measured values, the flow from cavity wall to dwelling is calculated to be about 40 m^3 h^{-1} or less, which does not seem unrealistic in view of the porosity of the wall.

CONCLUSIONS

This work confirms the existence of a yet unaccounted-for radon production in the dwelling. This production is pressure-dependent, and it is absent when the crawl space is being exhausted. Furthermore, it is not caused by a direct flow from crawl space to dwelling. It may be caused by transport of radon into the dwelling via the porous cavity wall. The origin of the radon might be either the soil or the brick of the cavity wall.

ACKNOWLEDGMENT

The authors thank the occupants of the dwelling for their cooperation; especially for accepting the inconveniences from installation of equipment and the noise of pumps and ventilators. We acknowledge Ir. L.E.J.J. Schaap for his contribution in determining the leakage parameters.

This investigation was financially supported by the Dutch Ministry of Housing, Planning and Environmental Hygiene as part of the Dutch National Research Programme "RENA." The work was carried out as part of the program "Environmental Radioactivity Research" of the Kernfysisch Versneller Instituut (KVI) of the University of Groningen.

REFERENCES

Aldenkamp, FJ, RJ de Meijer, LW Put, and P Stoop. **1990**. *An Assessment of a Method for in-situ Radon Exhalation Measurements*, KVI internal report R-12. Kernfysisch Versneller Instituut, Groningen, The Netherlands.

Busigin, A, AW van der Vooren, and CR Phillips. **1979**. Interpretation of the response of continuous radon monitors to transient radon concentrations. Health Phys 37:659-667.

Phaff, JC, WF de Gids, and B Knoll. **1983**. *Ventilatie van Gebouwen, Metingen van de Luchtlekken en Voorspelling van de Ventilatie van een Woning in Roden*, IMG-TNO Technical Report C535 (in Dutch). TNO, Delft.

Put, LW, RJ de Meijer, and B Hogeweg. **1985**. Survey of radon concentrations in Dutch dwellings. Sci Tot Environ 45:441-448.

Stoop, P, FJ Aldenkamp, EJT Loos, RJ de Meijer, and LW Put. **1990a**. Measurements and modelling of radon infiltration into a dwelling, pp. 45-53. In: *1990 International Symposium on Radon and Radon Reduction Technology*, Atlanta, Georgia, February 19-23, 1990, Vol. 2. EPA Office of Air and Radiation, Cincinnati, OH.

Stoop, P, RJ de Meijer, LW Put, LEJJ Schaap, and ITM Vermeer. **1990b.** *Ventilation and Radon Measurements in a Dwelling*, KVI internal report R-17. Kernfysisch Versneller Instituut, Groningen, The Netherlands.

Stoop, P, EJT Loos, RJ de Meijer and LW Put. **1990c**. *Measurements on Modelling and Control of Infiltration of Radon into Dwellings*, KVI internal report R-13. Kernfysisch Versneller Instituut, Groningen, The Netherlands.

QUESTIONS AND ANSWERS

Q: Reimer, U.S. Geologic Survey. If the supply air vents not to the crawl space but to the living area, what do you estimate the source term to be?

A: Put, The Netherlands (presenter). I think the source term would be only the exhalation. The definition we used for source term (production rate) is the total exhalation rate plus unknown flows. It does not include known flows into a compartment. Supplying outdoor air to the living area results in an overpressure, thus suppressing the observed pressure-driven flow component for the dwelling. It certainly would be worthwhile to try it. I think we'll find that the concentration is just like the outdoor concentration.

A SIMPLE MODEL FOR DESCRIBING RADON MIGRATION AND ENTRY INTO HOUSES

R. B. Mosley

U.S. Environmental Protection Agency, Research Triangle Park, North Carolina

Key words: *Radon transport model, radon entry, indoor radon, convective transport in soils*

ABSTRACT

While it is possible to formulate a fairly rigorous mathematical model to describe radon transport through soil, such a model requires rather complex numerical solutions that are time-consuming to evaluate. Numerical solutions are also cumbersome for evaluating both the relative importance of individual mechanisms and the appropriateness of alternative boundary conditions. Analytical solutions, even if only approximate, are much more informative for understanding the relative importance of specific physical mechanisms.

A simplified model for gas transport through soil surrounding the substructure of a house is discussed. Simplifying assumptions about the distribution of entry routes and driving forces are used to relate indoor radon levels to soil characteristics and to dynamics within the house. Preliminary validation of the model consists of demonstrating reasonable predicted values for indoor radon concentrations.

Model calculations indicate that the soil permeability is the most important single parameter that influences indoor radon concentrations. According to the model, most of the soil gas flow occurs near the house, with 90% being within six multiples of the basement depth. For a 0.002-m crack, it was found that the total flow increased by only a factor of two when the crack width (cylinder diameter) increased by a factor of 50.

INTRODUCTION

Elevated indoor radon levels are now recognized as a major environmental health problem around the world. New, low-cost methods for reducing exposure to radon in indoor air are urgently needed. The development of new reduction strategies, including modifications of construction practices, requires a complete understanding of the mechanisms by which radon migrates through the soil and enters a building. One way to enhance our understanding of these physical processes is to construct mathematical models of them and test predictions using the models against measured parameters.

It is commonly agreed that elevated indoor radon concentrations most often result from radon in soil gas entering the house. For houses with basements and slab-on-grade houses, the entry route mentioned most frequently is the perimeter crack between the wall and slab, whose width expands and contracts. Our discussion will consider only houses with basements.

In most such houses, large openings to the soil have been avoided to reduce the risks of moisture problems. Consequently, the most prevalent entry routes from soil into a basement are the openings that are most difficult to avoid, such as cracks and openings around utility penetrations, or openings that result from such basic construction practices as the crack between the concrete slab and the wall. Furthermore, a common practice in some localities is to leave a substantial wall/floor crack as part of the water drainage system. Another construction feature, which is intended to control potential water problems, contributes significantly to radon entry. That is, connecting perimeter drainage tiles to an open basement sump to allow water near the foundation to drain into the sump.

This practice effectively couples a long, porous tube buried in the ground directly with the interior of the basement. Since basement air tends to be at lower pressure than ambient air, pressure-driven air flows through the soil into the buried tube. For discussion purposes, we will assume that soil gas entering the buried tube can readily enter the basement with negligible pressure drop relative to the basement. Consequently, it is assumed that virtually all the pressure drop between the interior of the basement and the ambient air occurs within the soil. It is also assumed that the effects of cracks in the slab can be simulated by flow into an appropriately sized tube. This representation of cracks is similar to descriptions given by Mowris (1986), Nazaroff (1988), and Nazaroff and Sextro (1989). Our model could be considered an extension of these works, using an alternative mathematical formulation.

To obtain closed-form solutions, an idealized representation of the building will be used. Its physical structure will be ignored, and the complete entry process will be represented by the interaction of ambient air and the buried tube (Figure 1). Because of the reduced pressure in the tube, air will flow from the atmosphere through the soil and into the tube. Radon, generated uniformly in the soil, will be transported through the soil both by convective flow and by diffusion resulting from the concentration gradients.

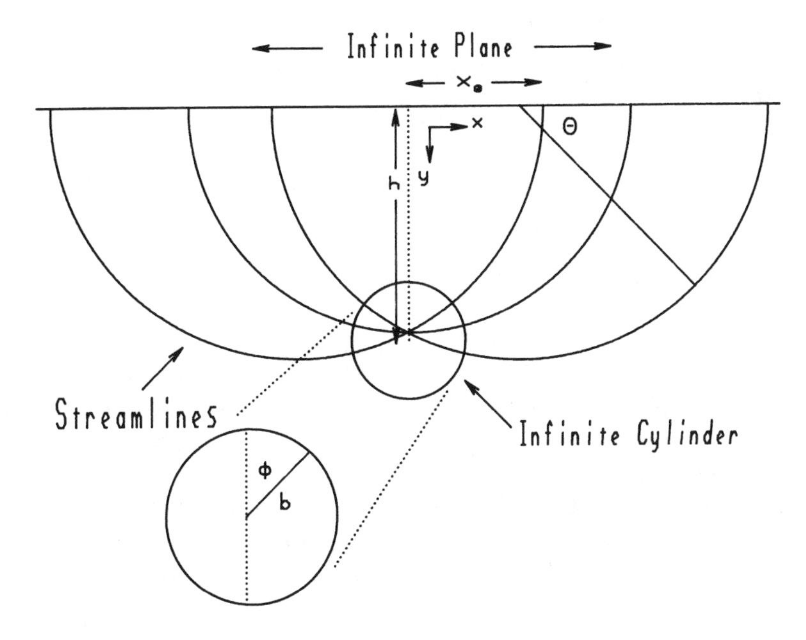

Figure 1. Schematic of the idealized geometry (infinite plane and cylinder) used to solve Poisson's equation.

Representative flow lines are illustrated in Figure 1. The radon entry rate into the tube will be calculated using the convective activity flux across the surface of the tube. It is assumed that no barrier exists between the soil and the hollow tube.

DEVELOPMENT OF THE EQUATIONS

In describing the flow of air through soil, we assume that the moisture is sufficiently low that migration of radon within the liquid phase can be neglected. The radon gas then moves through the soil gas within the soil pores. Radon movement can result either from the convective movement of the soil gas driven by a pressure difference, or from molecular diffusion in the soil gas. Because the pressure differences that produce convective flow are much smaller than absolute atmospheric pressure, the flow properties of the gas will be considered incompressible. Darcy's approximation to flow through a porous medium will be assumed.

In the steady state, the governing equations become

$$\nabla^2 P = 0, \tag{1}$$

$$\bar{v} = -\frac{k}{\mu} \nabla \bar{P}, \tag{2}$$

$$D_e \nabla^2 C - \frac{1}{\epsilon} \bar{v} . \nabla \bar{C} + G - \lambda\, C = 0, \tag{3}$$

where ∇ is the gradient operator, P is the pressure difference driving the flow, $\vec{v}$ is the superficial velocity, k is the permeability of the soil, μ is the dynamic viscosity of the gas, D_e is the effective diffusion coefficient of radon in the soil, C is the activity concentration of radon, ϵ is the porosity of the soil, G is the generation rate of radon activity, and λ is the radon decay constant. For simplicity of solution, the soil properties of permeability, porosity, and generation rate are taken to be uniform and isotropic.

Equation (1) can be solved to yield the pressure profiles within the soil. By differentiation, equation (2) yields the velocity streamlines for convective flow. Solution of equation (3) then yields the activity concentration at any point in the soil. The radon entry rate into the house will be estimated by computing the activity flux through specific openings into the structure. This approximation neglects the contribution of diffusion within the opening in the barrier, although the contribution of diffusion to transport in the soil is included in equation (3). Such an approximation is justified as long as the convective flow rate through the opening is large compared with the diffusive flux; that is,

$$vC >> D_0 \frac{C}{T}, \tag{4}$$

or

$$v >> \frac{D_0}{T}, \tag{5}$$

where D_0 is the diffusion coefficient in air, and T is the thickness of the barrier containing the opening. For relatively small leakage areas, such as cracks in concrete slabs, this condition will usually be met.

As with most transport problems, the complexity of the solution is largely determined by the boundary conditions, which depend on the locations and geometries of the entry openings in the substructure of the building. In this case, the boundary conditions consist of specifying the pressure at the air/soil interface and at the tube/soil interface. Also specified are the activity concentration at the air/soil interface and at great depth in the soil, where the concentration depends only on the generation rate and the decay rate. The pressure difference will be zero at the soil surface and P_c at the surface of the cylinder. The activity concentration will be zero at the soil surface and a maximum ($C_\infty = G/\lambda$) at infinite depth.

Equation (1) is the classical Laplace equation, which can be solved by a number of methods. Numerous sources on applied mathematics and physics describe coordinate systems in which Laplace's equation is separable (Arfken, 1968). The advantage of using coordinate transformations to solve boundary-value problems is most apparent when the surfaces corresponding to constant values of the coordinates match the surfaces on which the boundary conditions are to be specified in the original problem. For instance, bipolar coordinates contain parallel cylinders with a plane of symmetry.

Laplace's equation for constant potential (or pressure) on an infinitely long cylinder parallel to a plane held at constant potential (or pressure) is readily solved in bipolar coordinates since the solution is proportional to the radial coordinate (Morse and Feshbach, 1953). When transformed to Cartesian coordinates, the solution for the pressure can be written

$$P(x,y) = P_c \frac{\ln\left(\dfrac{h^2 - b^2 + x^2 + y^2 + 2y\sqrt{h^2 - b^2}}{h^2 - b^2 + x^2 + y^2 - 2y\sqrt{h^2 - b^2}}\right)}{\ln\left(\dfrac{h + \sqrt{h^2 - b^2}}{h - \sqrt{h^2 - b^2}}\right)}, \tag{6}$$

where P(x,y) represents the pressure at the point (x,y) relative to the pressure at the air/soil interface, P_c is the pressure at the surface of the cylinder relative to the pressure at the air/soil interface, and ln is the natural logarithm. As can be seen in Figure 1, h represents the depth of the cylinder below the surface of the soil, b is the radius of the cylinder, and x and y represent the horizontal and vertical coordinates, respectively. Note that y is positive when measured in the downward direction.

Surfaces of constant pressure are cylinders with centers located on the positive y-axis. Flow streamlines follow cylindrical surfaces with centers located on the positive x-axis at $[x_0 - (h^2 - b^2)/x_0]/2$, where x_0 is the intersection of the streamline with the positive x-axis. The value of x_0 defines a particular streamline. These surfaces are illustrated in Figure 2.

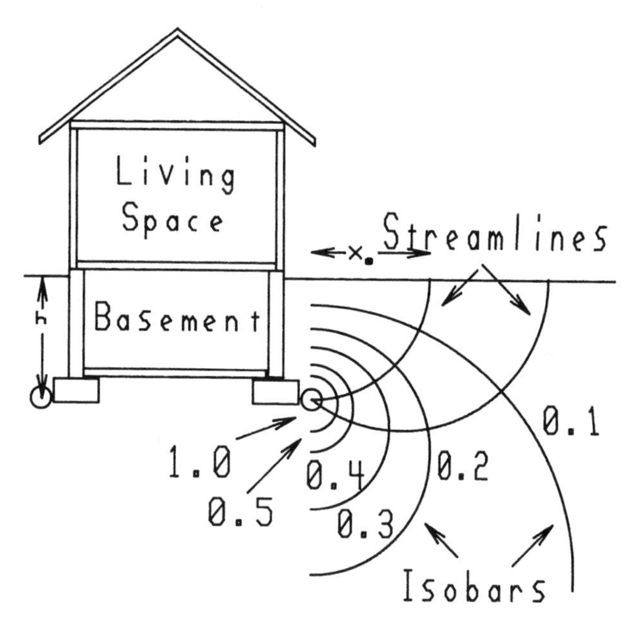

Figure 2. Schematic of a house, showing the basement surrounded by soil, with constant pressure surfaces and flow streamlines illustrated.

By applying equation (2) to equation (6), the components of velocity can be written as

$$v_x = \frac{4kP_c\sqrt{h^2 - b^2}}{\mu \ln\left(\dfrac{h + \sqrt{h^2 - b^2}}{h - \sqrt{h^2 - b^2}}\right)} \left(\frac{2xy}{(h^2 - b^2 + x^2 + y^2)^2 - 4y^2\,(h^2 - b^2)}\right) \tag{7}$$

and

$$v_y = \frac{-4kP_c\sqrt{h^2 - b^2}}{\mu \ln\left(\frac{h + \sqrt{h^2 - b^2}}{h - \sqrt{h^2 - b^2}}\right)} \left(\frac{h^2 - b^2 + x^2 + y^2}{(h^2 - b^2 + x^2 + y^2)^2 - 4y^2(h^2 - b^2)}\right). \quad (8)$$

Note that P_c is negative when the flow is into the cylinder.

The magnitude of the velocity along a streamline can be written in rectangular coordinates as

$$v = \frac{4k|P_c|\sqrt{h^2 - b^2}}{\mu \ln\left(\frac{h + \sqrt{h^2 - b^2}}{h - \sqrt{h^2 - b^2}}\right)} \left(\frac{1}{\sqrt{(h^2 - b^2 + x^2 + y^2)^2 - 4y^2(h^2 - b^2)}}\right), \quad (9)$$

where $|P_c|$ in the absolute value of P_c. In polar coordinates, the velocity along a streamline is

$$v = \frac{4k|P_c|\sqrt{h^2 - b^2}}{\mu \ln\left(\frac{h + \sqrt{h^2 - b^2}}{h - \sqrt{h^2 - b^2}}\right)} \left(\frac{2x_0^2}{x_0^2 + h^2 - b^2}\right)\left(\frac{1}{x_0^2 - (h^2 - b^2) + (x_0^2 + h^2 - b^2)\cos\theta}\right), \quad (10)$$

where θ is the polar angle, with the origin located on the x-axis at the center of the arc forming the streamline (see Figure 1). The velocity can be integrated either along the x-axis from 0 to ∞ or over the surface of the cylinder from 0 to π. Both integrations yield

$$Q_T = \frac{2\pi L k|P_c|}{\mu \ln\left(\frac{h + \sqrt{h^2 - b^2}}{h - \sqrt{h^2 - b^2}}\right)}, \quad (11)$$

where Q_T is the total flow rate of soil gas into the tube, and L is the length of the tube.

Integrating along the x-axis to a distance x_0 yields

$$Q(x_0) = \frac{4Lk|P_c|}{\mu \ln\left(\frac{h + \sqrt{h^2 - b^2}}{h - \sqrt{h^2 - b^2}}\right)} \tan^{-1}\left(\frac{x_0}{\sqrt{h^2 - b^2}}\right). \quad (12)$$

$Q(x_0)$ is the total air flow per unit time that passes through the surface of the soil between x = 0 and x = x_0. Note the fraction of total flow that passes through this surface area. From the last two equations,

$$\frac{Q(x_0)}{Q_T} = \frac{2}{\pi} \tan^{-1}\left(\frac{x_0}{\sqrt{h^2 - b^2}}\right). \quad (13)$$

Inverting equation (13) yields

$$x_0 = \sqrt{h^2 - b^2} \tan\left(\frac{\pi}{2} \frac{Q(x_0)}{Q_T}\right) \approx h \tan\left(\frac{\pi}{2} \frac{Q(x_0)}{Q_T}\right). \quad (14)$$

This expression shows the fraction of the total flow that passes through the soil surface within distance x_0 of the house. It is independent of the soil properties and the dimensions of the house except for basement depth. This expression allows determination of the linear expanse of soil, measured from the house, that must be considered to incorporate a prescribed fraction of the total flow.

The time of flight along a streamline can be computed by integrating the reciprocal of the velocity along the streamlines. This integration yields t =

$$\frac{\mu h^3}{16k|P_c|\sqrt{h^2-b^2}} \ln\left(\frac{h+\sqrt{h^2-b^2}}{h-\sqrt{h^2-b^2}}\right)\left(\frac{(\xi^2+1)^2}{\xi^3}\right)[2\xi+(\xi^2-1)\,(\frac{\pi}{2} + \cos^{-1}\left(\frac{2\xi}{\xi^2+1}\right))], \quad (15)$$

where $\xi = x_0/h$ is the distance (in multiples of h) from the house at which the streamline intersects the x-axis. The time of flight of an air parcel along a streamline will be compared with the average lifetime, τ, of a radon atom.

Assume that the first term in equation (3) could be neglected. In this case, transport would be entirely due to convective flow. Equation (3) could then be written as

$$\frac{dC}{ds} + \frac{\lambda}{\epsilon v} C = \frac{G}{\epsilon v}, \quad (16)$$

where s is a distance measured along a streamline. The transport equation has been reduced to an ordinary differential equation that is readily solved. Using an integrating factor, the solution is

$$C(s) = \frac{G}{\lambda}\ [1 - \exp\left(- \lambda\epsilon \int \frac{ds}{v}\right)]. \tag{17}$$

Using the velocity from equation (10), the integral can be evaluated. When h is much larger than b, the concentration on a streamline becomes

$$C(y)= \frac{G}{\lambda}[1-\exp\left(\frac{-\lambda\epsilon\mu h^2 \ln(2h/b)}{8k|P_c|}\ \frac{(\xi^2+1)^2}{\xi^3}\left[\frac{2\xi y}{h}+(\xi^2-1)\sin^{-1}\left(\frac{2\xi y/h}{\xi^2+1}\right)\right]\right)]. \tag{18}$$

Since the value of ξ identifies a particular streamline, a unique value of x (on the streamline) corresponds to each value of y. If this expression is evaluated on the surface of the cylinder, it becomes

$$C(\phi) = \frac{G}{\lambda}\ [1 - \exp\left(\frac{-\lambda\epsilon\mu h^2 \ln(2h/b)}{k|P_c|}\left[\frac{1-\phi\cot\phi}{\sin^2\phi}\right]\right)], \tag{19}$$

where ϕ is the polar angle with the origin at the center of the cylinder (see Figure 1).

The radon entry rate is given by

$$E = \int Cvda. \tag{20}$$

The integration is over the surface of the cylinder. The velocity evaluated at the surface of the cylinder can be written

$$v = \frac{k|P_c|h}{b\ \mu\ \ln(2h/b)}\left(\frac{1}{h - b\cos\phi}\right). \tag{21}$$

Equation (20) can be integrated numerically, using equations (19) and (21), to yield the entry rate. The steady-state indoor radon concentration is given by

$$C_i = \frac{E}{V(\lambda + \lambda_v)}, \tag{22}$$

where V is the volume of the house, and λ_v is the air exchange rate. This expression assumes the house can be treated as a single, well-mixed zone. It also assumes zero radon concentration in the ambient air. Equation (20) was evaluated using the Adams-Bashforth method.

RESULTS

In order to evaluate this model and its potential for describing radon entry into buildings, it is necessary to specify numerical values for the parameters occurring in the equations that identify a particular house and soil combination. However, since this is the initial development of the model, it is appropriate to demonstrate the model's response to a typical range of parameters. The more detailed validation step should wait for a more complete solution of the transport equation using both convection and diffusion.

A baseline set of parameters was used as a reference case. Individual parameters were then varied over typical ranges in a limited sensitivity study to observe the effect of varying a few key parameters. Table 1 shows the baseline set of parameters and the range of variation for key parameters.

Table 1. Parameters used to evaluate equations in model.

Parameter	Baseline Value	Range of Variation
Pc	5 Pa	0 – 25 Pa
μ	1.7×10^{-5} kg $m^{-1}s^{-1}$	
k	$1\times10^{-10}m^2$	$1\times10^{-11} - 1\times10^{-8}$ m^2
D_e	1.0×10^{-6} m^2 s^{-1}	
ϵ	0.5	
G	0.0334 Bq m^{-3} s^{-1}	
λ	2.11×10^{-6} s^{-1}	
D_0	1.2×10^{-5} m^2 s^{-1}	
T	0.1 m	
b	0.0508 m	0.001 – 0.0508 m
h	2.0 m	
τ	4.75×10^5 s	
L	40 m	
V	400 m^3	
λ_v	1.39×10^{-4} vol s^{-1}	

The baseline set of values corresponds to a house with a basement 2 m below grade, with 40 m of perimeter drain pipe connected to the open basement sump. The house has a volume of 400 m^3 and an air exchange rate of 0.5 air changes per hour. The soil is relatively dry, with modest permeability (10^{-10} m^2) and typical radium content, yielding a typical radon production rate of 0.0334 Bq m^{-3} s^{-1}.

Soil Gas Flow

The solutions to equations (1) and (2) are illustrated in Figure 2. The curves labeled "Isobars" represent surfaces of constant pressure. These curves have been normalized to be zero at the soil surface and 1 on the surface of the cylinder. The curves labeled "Streamlines" represent the solution to the velocity equation. The shape and spacing of the streamlines indicate that the velocity decreases at greater distances from the building. However, the gradient of pressure (and consequently the velocity) approaches the same value as the surface of the cylinder is approached along any of the streamlines.

The integrated flow of soil gas per unit time into the cylinder can be seen in Figure 3. Total flow of soil gas per unit time into the cylinder is shown as a function of the radius of the cylinder and the applied pressure difference. Note that the total flow through a 0.051-m (2-in.) radius drainage pipe is only about twice as great as that through a 0.001-m radius tube (corresponding to a 0.002-m-wide crack). This reflects the assumption that essentially all the pressure drop occurs in the soil rather than across the soil/tube interface.

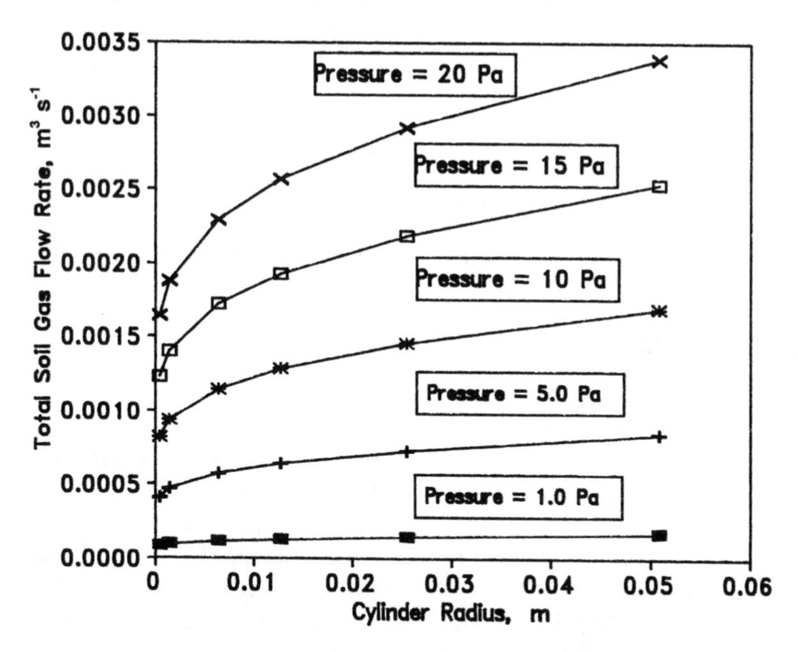

Figure 3. Soil-gas flow rate as a function of the cylinder radius for five different values of applied pressure.

The fraction of flow occurring within a specified distance of the house is shown in Figure 4 as a function of the normalized distance from the building. The unit of normalization is the basement depth, h. Approximately 90% of the flow occurs within six multiples of the basement depth from the building. For the baseline case the depth is about 2.0 m. In this case, 90% of the flow would occur within about 12 m of the building.

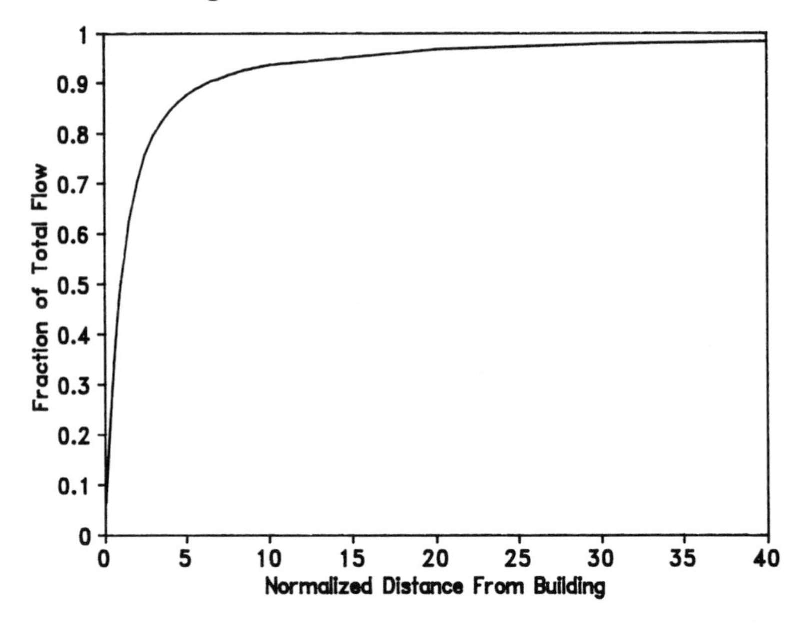

Figure 4. Fraction of total air flow as a function of the normalized distance from the building.

Radon Transport

Soil gas flow and radon transport are not necessarily equivalent. The previous discussion indicated that 10% of the soil gas entering the cylinder originates at distances greater than six multiples of the basement depth. As we shall see, radon does not usually travel such great distances. Figure 5 shows the normalized time of flight of an air parcel along a flow streamline as a function of the normalized distance of the streamline from the building. The unit of normalization is the basement depth. The various curves represent different values of soil permeability, spanning the typical range of values. The time of flight is normalized relative to the average lifetime of radon atoms.

This figure shows that convective flow is an important transport mechanism for permeabilities of 10^{-10} m^2 and higher.

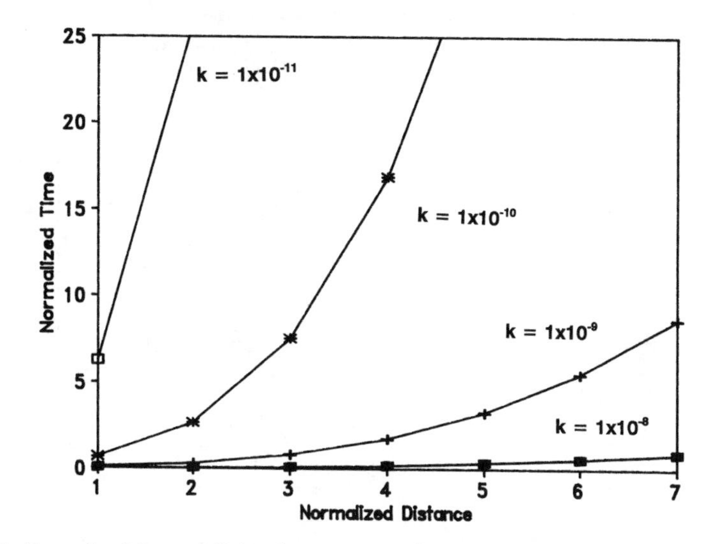

Figure 5. Normalized time of flight along a streamline as a function of the normalized distance of the streamline from the building for four values of soil permeability.

Figure 6 shows normalized time of flight as a function of normalized distance (along the x-axis) of the streamline from the building for three different-sized cylinders. The time of flight is normalized relative to the average lifetime of radon atoms. The unit of normalization for distance is the basement depth. Note that for permeability of 10^{-10} m^2 radon does not live long enough to travel very far on the long streamlines.

By computing the average velocity on a streamline, the distance that radon could travel in an average lifetime can be estimated. Figure 7 shows the distance traveled during one radon lifetime as a function of the normalized distance to the particular streamline (x_0/h) that the motion follows. The three curves correspond to three values of soil permeability. For a permeability of 10^{-11} m^2, air parcels travel only a fraction of a meter during one radon lifetime. Under these conditions, radon could migrate only for distances less than 1 m through the convective flow mechanism.

In marked contrast, the curve corresponding to a permeability of 10^{-9} m^2 indicates that convective flow of radon could occur over distances up to 50 m along the streamline with $x_0/h = 2$, and over distances of 20 m along streamlines with $x_0/h = 6$. Recall that 90% of the total flow occurs on streamlines with $x_0/h < 6$. For permeability of 10^{-10} m^2, convective flow occurs over distances of about 5 m along streamlines with $x_0/h = 2$ and over distances of about 2 m along streamlines with $x_0/h = 7$.

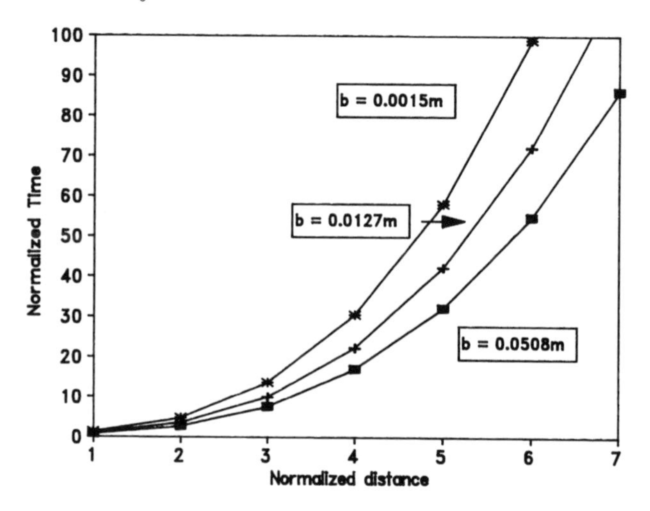

Figure 6. Normalized time of flight along a streamline as a function of the normalized distance of the streamline from the building for three different-sized cylinders.

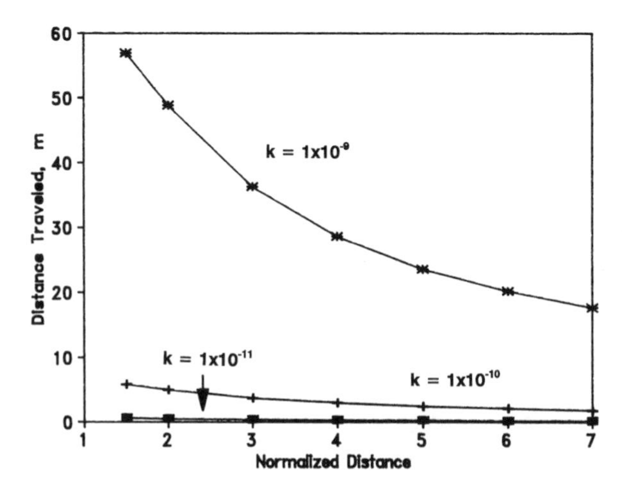

Figure 7. Distance radon travels along a streamline as a function of the normalized distance of the streamline from the building for three different soil permeabilities.

These observations have significant implications in terms of the effect of the volume of soil surrounding a basement on indoor radon concentrations. The 10^{-10} m^2 permeability case is further illustrated in Figure 8, which shows four streamlines that are spaced about one basement depth (h) apart. The dashed curve delineates the estimated volume of soil surrounding the basement that contributes to the indoor radon concentration through the convective flow process. This result, of course, depends on assumed values of other parameters, such as an applied pressure difference of 5 Pa. The values of the parameters that have been assumed are given in Table 1.

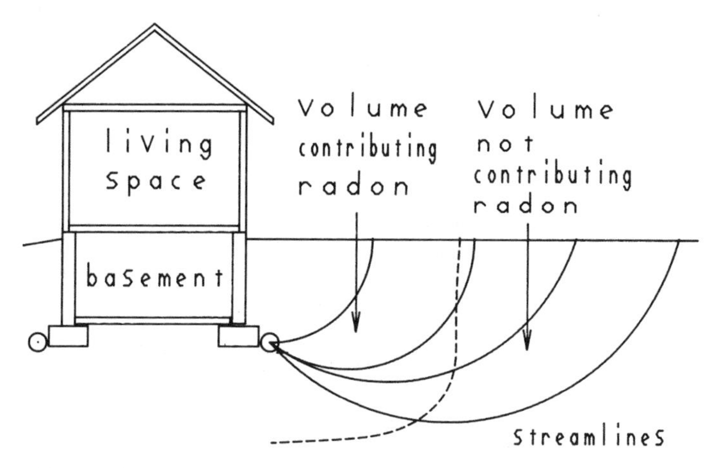

Figure 8. Schematic of building, soil, and flow streamline system, indicating volume of soil that contributes radon to indoor air by pressure-driven flow under baseline set of conditions.

Figure 8 also shows that the volume of soil beyond about two basement depths (h) from the house does not contribute to indoor radon levels because the radon decays before it reaches the basement. However, Figure 4 shows that about 30% of the total air flow arises from the volume that does not contribute radon. This observation suggests that a house surrounded by a finite thickness layer of soil of low radium concentration would be shielded from strong radon sources outside that layer. The lower the permeability, therefore, the thinner the layer required to shield the house.

Figure 9 shows the indoor radon concentration as a function of applied pressure for different values of soil permeability. Note the nonlinear dependence on pressure, which indicates the fact that the radon concentration near the entry cylinder varies with the pressure.

Primarily, this variation reflects a dilution effect from unsaturated air flowing from the atmosphere.

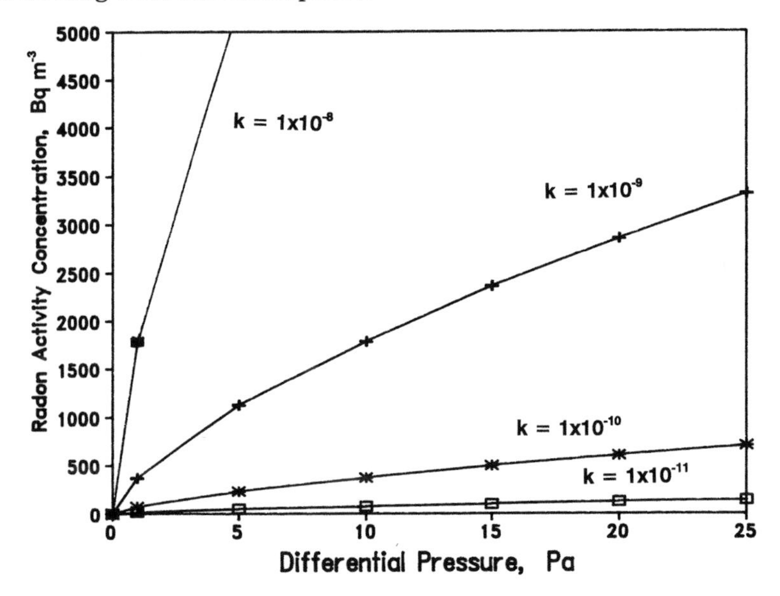

Figure 9. Indoor radon activity concentration as a function of applied pressure for four values of soil permeability.

The radon concentration can vary by several orders of magnitude, depending on the value of soil permeability. As suggested by Figure 9, in this paper, soil permeability is the most important single factor in determining indoor radon concentration. This is because soil permeability often varies by 5 or 6 orders of magnitude with commonly occurring variations in soil properties and moisture content.

SUMMARY

To the extent that soil-gas leakage into a basement can be simulated by pressure-driven flow through the soil into a long hollow tube, the pressure and velocity profiles have been solved in closed form. To the limit that diffusion can be neglected, radon transport and entry into the building can be computed. These approximations appear to be quite reasonable for soil permeabilities of 10^{-10} m^2 and higher. The neglect of diffusion probably is not justified for permeabilities of 10^{-12} m^2 and lower.

The total soil-gas flow rates were observed to vary by a factor of only 2 when the radius of the tube (width of the crack) varied by a factor of 50. Also, 90% of the total soil-gas flow occurs within six basement depths of the house. A large fraction of the total flow occurs along streamlines for which the time of flight is large compared to average lifetime of radon atoms. Consequently, only a finite volume of soil immediately surrounding the basement contributes radon to the interior of the building. For a permeability of 10^{-10} m^2, this volume of soil extends about two basement depths from the house.

For permeabilities less than 10^{-11} m^2, this volume probably extends only a small fraction of a meter from the house. However, this last statement should be viewed with caution, since the thickness of this thin soil layer may be less than the diffusion length. In this case, contributions from diffusion may be particularly significant. For a relative comparison, the estimated effective diffusion length ($\sqrt{De/\lambda}$) can be as small as 0.01 m in wet mud and as large as 1.2 m in clayey sand.

Our model yielded realistic predictions of indoor radon for permeabilities higher than 10^{-11} m^2, the range for which the assumptions of the model are expected to apply. The next important step in advancing the model is a more rigorous solution of the transport equation with both transport mechanisms operative.

REFERENCES

Arfken, G. **1968**. Coordinate systems, pp. 87-92. In: *Mathematical Methods for Physicists*. Academic Press, New York.

Morse, PM and H Feshbach. **1953**. Solutions of Laplace's and Poisson's equations, p. 210. In: *Methods of Theoretical Physics*, Vol. 2. McGraw-Hill, New York.

Mowris, RJ. **1986**. Analytical and Numerical Models for Estimating the Effect of Exhaust Ventilation on Radon Entry in Houses with Basements or Crawl Spaces, LBL-22067, M.S. Thesis. Lawrence Berkeley Laboratory, Berkeley, CA.

Nazaroff, WW. **1988**. Predicting the rate of radon-222 entry from soil into the basement of a dwelling due to pressure-driven air flow. Radiat Protect Dosim 24:199-202.

Nazaroff, WW and RG Sextro. **1989**. Techniques for measuring the indoor ^{222}Rn source potential of soil. Environ Sci Technol 23:451-458.

QUESTIONS AND ANSWERS

Q: Riddle, Cavalier Engineering, Spokane. I am curious how we could justify such a model when we seldom see a, b, and c horizons of similar permeability.

A: Mosley, EPA. Frankly, I guess if I had a little more time I would talk a little bit about the conclusions and results. I would say it is obviously very sensitive to permeability. If I were trying to fit this model to a particular house, the first thing I would have to do would be to have fairly high permeabilities to say it applies. The original intent, of course, was to show some comparison with the data. But I concluded that, in the final analysis, I didn't have enough data on any houses up in that high permeability range that would allow me to do it. I could fit them, because I have enough uncertainty in the parameters that I could come pretty doggone close to the radon concentrations. But, exactly for that reason, there's enough uncertainty in the measurement of permeability across the dimensions of the space (a factor of 3 or 4 would be a good limit on the variation of permeability) and that alone, with the uncertainty in the radium concentration in the soil (which never got pointed out there), but was a key component in the coefficient of the concentration profile. There is probably enough uncertainty in those two parameters to allow you to give a reasonable fit to a lot of data. So the problem is, there are enough parameters in the model that you couldn't do a very rigorous fit at this point.

RADON TRANSPORT PROPERTIES OF SOIL CLASSES FOR ESTIMATING INDOOR RADON ENTRY

K. K. Nielson and V. C. Rogers

Rogers and Associates Engineering Corporation, Salt Lake City, Utah

Key words: *Radon entry, diffusion, permeability, soil texture, moisture*

ABSTRACT

Radon diffusion coefficients and air permeabilities of soils are required for modeling indoor radon entry, for interpreting data on soil radon sources, and for designing and evaluating potential radon mitigation or containment systems. Diffusion and permeability coefficients depend on soil moistures, particle sizes, and compactions. We present a systematic method for estimating these transport parameters for the 12 common Soil Conservation Service soil textural classes for predicting radon entry.

Diffusion coefficients and air permeabilities are estimated from predictive correlations with soil particle-size distribution, porosity, and water matric potential. The matric potential and particle sizes are preferred for estimating soil moistures when site-specific data are unavailable because long-term average matric potentials near structures are relatively constant. They provide a useful basis to compare radon transport properties of different soil textural classes in a given location.

Predicted diffusion coefficients range from about 10^{-9} m^2 s^{-1} for the silty clay class to 4 x 10^{-6} m^2 s^{-1} for the sand class. Air permeabilities range from 8 x 10^{-18} m^2 for the silty clay class to 2 x 10^{-11} m^2 for the sand class. Comparison of measured diffusion and permeability coefficients with predicted values are within geometric standard deviations of 1.4 and 1.7, respectively.

The method provides generic diffusion and permeability coefficients as functions of soil textural class and matric potential. Because long-term average matric potentials near structures are relatively constant (10 to 30 kPa), specifying the soil textural class characterizes the soil radon transport properties sufficiently for estimating indoor radon entry.

INTRODUCTION

Elevated indoor radon (^{222}Rn) occurrences, which pose significant health risks, result mainly from radon gas emanating from foundation soils (USEPA, 1986). Geographic patterns in occurrences (Nero et al., 1986; Gundersen et al., 1988; Otton et al., 1988) suggest the possible location of problem areas from soil properties. Properties that affect radon availability, besides parent ^{266}Ra concentrations and emanation fractions, are soil air permeabilities (permitting advective transport by pressure-driven air flow) and radon diffusion coefficients (permitting diffusive transport) (Eaton and Scott, 1984; Nazaroff and Sextro, 1989; Rogers and Nielson, 1984, 1990). Both air permeability and diffusion coefficients have been described recently by simple correlations with soil water content, particle sizes, and porosities (Rogers and Nielson, 1991); however quantitative values have not been previously associated with standard soil classifications.

This paper characterizes air permeabilities and radon diffusion coefficients of soils as classified by the 12 textural classes of the U.S. Soil Conservation Service (SCS) (USDA, 1975). These classes are used as the framework for comparing trends in diffusion, permeability, and related water-content effects because they are simple and systematic, are widely used and understood, and can be visually identified in many cases. For more general use, however, the methods presented here for estimating radon diffusion coefficients and air permeabilities are not limited to the average properties of the SCS soil textural classes, or to a particular classification, but apply to the continuum of soil types because of the diversity of materials on which the methods are based.

The water-content variations of soils and their dominant effects on permeability and diffusion are transposed to soil matric potentials to provide a more constant, fundamental basis for comparing different soils. Because the matric potential is the environmental tension or suction affecting a soil, the soil's water retention curve (matric potential vs. water content) best defines its water content in a given environment. This permits comparisons of radon transport properties and related performance of different soils for particular applications. The use of these methods for predicting radon entry rates into dwellings is illustrated by sample calculations.

BACKGROUND AND THEORY

Radon diffusion coefficients depend mainly on soil water content, but also on soil porosity and other parameters (Tanner, 1964). Several empirical functions have been used to predict radon diffusion coefficients from commonly measured soil properties (Rogers et al., 1980; NRC, 1989). These provide a simple, common basis for estimating diffusion coefficients over a broad range of soil and moisture conditions. The most recent and broadly applicable function, which was used in this study, has the following form (Rogers and Nielson, 1991):

$$D = p\,D_o \exp(-\,6pm - 6m^{14p}), \qquad (1)$$

where D = radon diffusion coefficient of the soil pore space ($m^2\ s^{-1}$),
p = soil porosity (dimensionless),
D_o= diffusion coefficient of radon in air ($1.1 \times 10^{-5}\ m^2\ s^{-1}$), and
m = volume fraction of moisture saturation (dimensionless).

Soil air permeabilities depend strongly on water content and also on soil pore sizes and porosities. They have also been described by a new, simple, empirical function of commonly measured parameters (Rogers and Nielson, 1991) as:

$$K = (p/500)^2\, d^{4/3} \exp(-\,12m^4), \qquad (2)$$

where K = bulk soil air permeability coefficient (m^2), and
d = arithmetic mean soil particle diameter (m).

Soil water content depends strongly on soil texture as well as other hydrogeological parameters. However, for many conditions near dwellings, long-term average soil water contents can be estimated from the soil water-retention characteristics. Despite hysteresis (between draining vs. wetting cycles), soil water-retention curves give useful estimates of long-term average water contents for different soils in a given environment. Although measurements of the water-retention curve generally are tedious, several empirical predictive correlations and models are available. Some involve multiple regressions of water contents at selected matric potentials on soil texture, density, and organic matter (Salter and Williams, 1967; Gupta and Larson, 1979; Rawls and Brakensiek, 1982; Puckett et al., 1985; Meng et al., 1987). Others propose a general power curve with fitting constants to be adjusted empirically (Brooks and Corey, 1964; Clapp and Hornberger, 1978). A physico-empirical model has been proposed (Arya and Paris, 1981) to relate soil particle sizes to pore sizes, from which the water-retention characteristic is computed. The Arya and Paris model was chosen for primary use here because of its potentially more detailed description of soils when particle-size analyses are available. Water contents computed from the Gupta and Larson (1979) and Rawls and Brakensiek (1982) regressions were also computed for comparisons with the Arya and Paris model.

To implement the Arya and Paris model for water-retention calculations, their equation relating pore size to particle size was combined with the equation of capillarity that relates pore size (cylindrical equivalent) to matric potential. Eliminating the pore-size parameter, the resulting equation simplifies to

$$\psi_i = c\, W_i^{(\alpha-1)/2}\, R_i^{(1-3\alpha)/2}, \tag{3}$$

where ψ_i = soil matric potential (–Pa),
W_i = mass fraction of particles in size class i,
R_i = arithmetic mean particle radius of size class i (cm),
α = empirical constant from Arya and Paris Model,

$$c = \frac{2\gamma \cos\theta}{0.01\,\rho_w g}\left[\frac{2}{3}\,\frac{p}{1-p}\left(\frac{3}{4\pi\rho_p}\right)^{1-\alpha}\right]^{-1/2}$$

γ = surface tension of water (0.072 N/m @ 25°C),
θ = contact angle (assumed 0°),
ρ_w = water density (997 kg m^{-3} @ 25°C),
g = acceleration due to gravity (9.8 m s^{-2}),
ρ_p = particle density (2.7 g cm^{-3}),
0.01 = cm H_2O per Pa unit conversion.

The empirical constant α originally was thought to range from 1.35 to 1.40. Later laboratory comparisons suggested a broader range, with possible systematic dependence on texture and matric potential (Schuh et al., 1988). Since the range in α causes relatively small variations in m, and the intent of this work is to demonstrate relative trends in diffusion and permeability with soil texture, the constant value of $\alpha = 1.38$ was used throughout this work.

At a given matric potential, all pores with radii less than those corresponding to the potential ψ_i are considered filled with water, as also assumed by Arya and Paris. This leads to the following expression for the degree of water saturation associated with each ψ_i,

$$m_i = \sum_{j=1}^{i} W_j, \tag{4}$$

where the summation over j starts with the small particle sizes. Water contents were computed as fractions of saturation instead of the more traditional volume fraction because the saturation fraction is a more fundamental unit for defining radon diffusion coefficients and air permeabilities (Rogers and Nielson, 1991).

PARTICLE SIZE DISTRIBUTION AND MOISTURE CALCULATIONS

A particle size distribution was defined for each SCS soil textural class from the three-parameter mathematical centroid calculated from its classification area (Figure 1). The resulting percentages of clay, silt and sand (Table 1) then were used to define detailed particle-size distributions. For each of the three size fractions, a gaussian distribution of particle size was defined to be geometrically centered in its size-range window. The resulting diameter distributions were centered at 0.35 μm for clay (0.06 - 2 μm range), 11 μm for silt (2 - 60 μm range), and 350 μm for sand (60 - 2000 μm range). Size range windows utilized the MIT definitions, which are commonly used with the SCS textural classes (Dunn et al., 1980). Distribution widths were chosen as GSD = 2.5 to minimize particle frequency discontinuities that occur at the window boundaries with very large distribution widths and to avoid particle frequency gaps that occur with very narrow widths. Distributions were truncated at the size-window boundaries, and normalized in height to correspond to the desired weight percent for each of the three components. The resulting particle-size distributions were defined at regular size intervals corresponding to standard sieve openings.

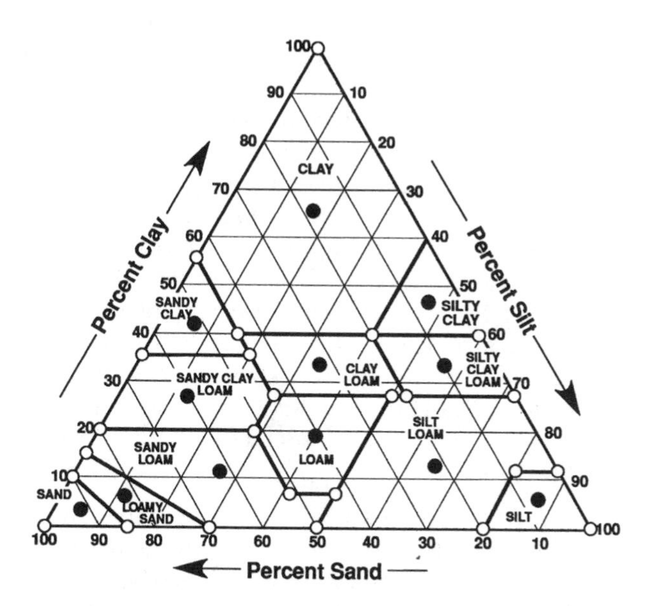

Figure 1. U.S. Soil Conservation Service classification chart showing the centroid compositions (solid circles) and vertices (open circles) for each soil textural class.

Table 1. Centroid compositions and mean particle sizes of the 12 U.S. Soil Conservation Service soil textural classes.

Textural Class	% Clay	% Silt	% Sand	Arithmetic Mean Particle Diameter (m)
Sand	3.33	5.00	91.67	4.4×10^{-4}
Loamy sand	6.25	11.25	82.50	4.0×10^{-4}
Sandy loam	10.81	27.22	61.97	3.0×10^{-4}
Sandy clay loam	26.73	12.56	60.71	2.9×10^{-4}
Sandy clay	41.67	6.67	51.66	2.5×10^{-4}
Loam	18.83	41.01	40.16	2.0×10^{-4}
Clay loam	33.50	34.00	32.50	1.6×10^{-4}
Silt loam	12.57	65.69	21.74	1.1×10^{-4}
Clay	64.83	16.55	18.62	9.2×10^{-5}
Silty clay loam	33.50	56.50	10.00	5.6×10^{-5}
Silt	6.00	87.00	7.00	4.6×10^{-5}
Silty clay	46.67	46.67	6.66	3.9×10^{-5}

Additional particle-size distributions were similarly defined for the material compositions at each of the 26 vertices that mark the extremities of the 12 texture classes (open circles, Figure 1). All vertices contiguous with a given classification area were used to define the range of variation associated with a given textural class.

The soil air permeability, computed from equation 2, depends on the mass-weighted arithmetic mean soil particle diameter, which correlates better with permeability than does the geometric mean or median diameter (Rogers and Nielson, 1991). The arithmetic mean emphasizes large-diameter particle sizes more than the median. This is illustrated by the arithmetic mean particle sizes that were computed for the 12 SCS texture classes, which are listed in Table 1. The mean diameter for SCS clay is significantly larger than that for SCS silty clay loam, silt, and silty clay. This is because the centroid composition for SCS clay (Figure 1) includes a significant fraction of particles in the sand size range, thus causing the average SCS clay diameter to exceed those of the low-sand classes such as SCS silty clay loam, silt, and silty clay. Similarly, the entire list of soil particle diameters in Table 1 follows the order of the mass fractions of sand for each classification.

Soil water-saturation fractions were computed using equations 3 and 4 for each of the 12 texture classes at varying densities. The water-saturation fractions were computed for each particle-size distribution at 43 different matric potentials spaced logarithmically over the range -10^2 to -10^8 Pa ($1 - 10^6$ cm H_2O). These are plotted in Figure 2 for the

1,600-kg m^{-3} density. The ranges of variation of each matric potential curve were determined from corresponding computations of water contents from the particle-size distributions that resulted from the vertices of each soil classification. The respective maxima and minima of the water contents obtained from all vertex calculations for each soil class were used to estimate the ranges of variation in the matric potential curves. These are shown by the shaded regions in Figure 2.

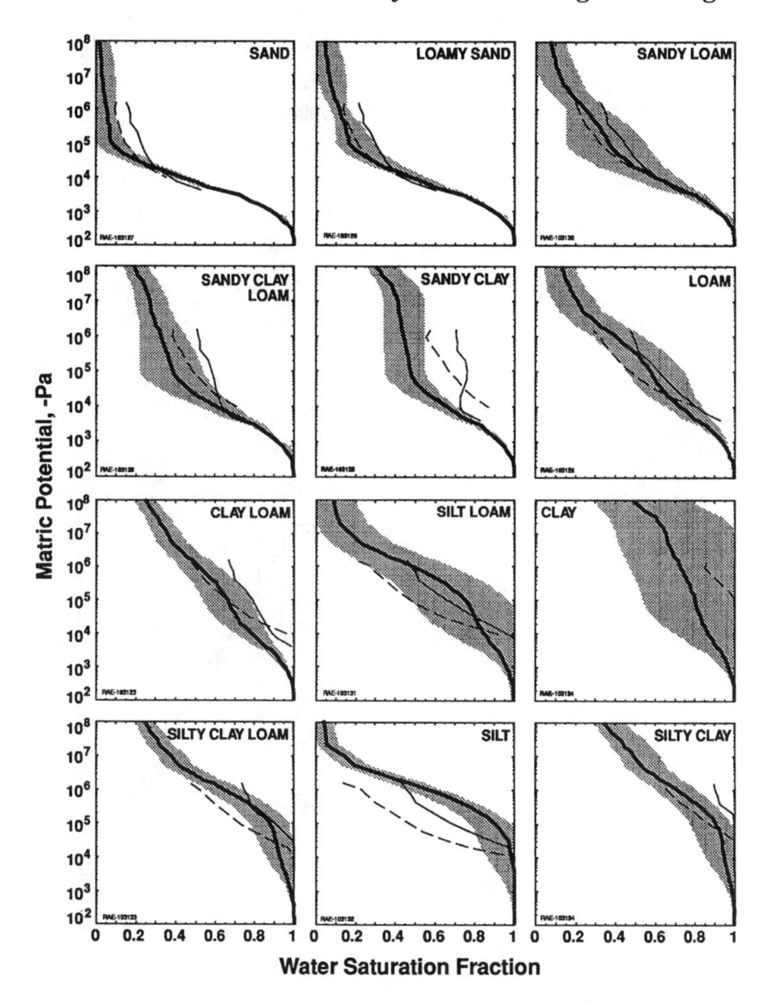

Figure 2. Soil water-retention curves (thick solid lines) and ranges (shaded areas) calculated from centroid and extremity compositions of the U.S. Soil Conservation Service soil textural classes, with comparisons to regressions of Gupta and Larson, 1979 (thin solid lines) and Rawls and Brakensiek, 1982 (dashed lines).

Comparative soil water-retention curves were also computed from two different regression correlations (Gupta and Larson, 1979; Rawls and Brakensiek, 1982) for comparison with the predictions of the Arya and Paris model. These were based only on the sand, silt, and clay percentages in Table 1 and the bulk soil density of 1600 kg m^{-3}. Organic fractions were assumed to be zero. The resulting values also are plotted in Figure 2 and, except for sandy clay, at least partially fall within the shaded regions defined by the range of SCS class variability as estimated by the Arya and Paris model. Water-saturation fractions from the Gupta and Larson correlation nearly always exceeded those of the Rawls and Brakensiek correlations. Water-saturation fractions from both correlations generally exceeded those of the Arya and Paris centroid values (Figure 2) for low-silt materials (sand, loamy sand, sandy clay loam, sandy clay, and clay), and were less than the Arya and Paris centroid values for high-silt materials (silt, silt-loam).

The effect of varying compaction of the soils was examined by similar calculations in which the soil density was varied over the range 1200 kg m^{-3} to 1800 kg m^{-3}. The resulting matric potential curves, given in Figure 3, showed similar trends for each soil type, with high soil densities having a greater fraction of saturation at a given matric potential.

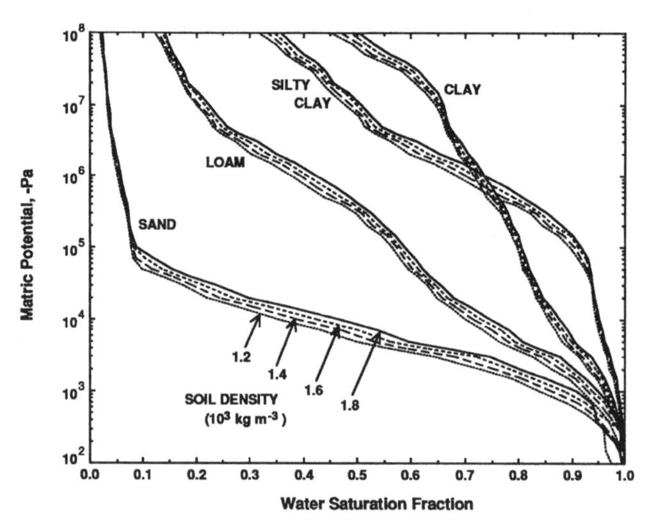

Figure 3. Effects of soil density on soil water-retention curves.

RADON DIFFUSION AND AIR PERMEABILITY CALCULATIONS

Each of the soil water-saturation fractions computed for the matric potential curves was used in equations 1 and 2 to compute corresponding radon diffusion coefficients and soil air permeabilities. Soil porosities of 0.407 (1600 kg m^{-3} bulk density) were used except in separate evaluations of the effects of varying soil density. The resulting radon diffusion coefficients and soil air permeabilities are plotted as a function of the soil matric potential in Figures 4 and 5, respectively. Both diffusion coefficients and permeabilities exhibit similar trends with matric potential, and they maintain identical ordering at matric potentials below 5×10^4 Pa. At higher matric potentials, the ordering of diffusion coefficients by soil type is more varied than that of permeabilities, owing to the explicit particle-size dependence in the permeability correlation (equation 2). Coarse-grained soils such as sand, loamy sand, sandy loam, sandy clay loam, and sandy clay have relatively uniform, dry-soil values up to 4×10^{-6} m^2 s^{-1} throughout the matric potential range above $\sim 5 \times 10^4$ Pa (-0.5 bar) for diffusion coefficients and up to 2×10^{-11} m^2 above $\sim 2 \times 10^4$ Pa (-0.2 bar) for permeabilities. The coefficients decrease rapidly at lower matric potentials. Diffusion coefficients and permeabilities decrease with decreasing matric potential for all of the finer-grained soils, particularly in the -10^4 to -10^6 Pa range encountered most commonly under field conditions. Silty soils (silt loam, silty clay loam, silt, and silty clay) exhibit the greatest variations in this range (factors of 20 to 500 for diffusion, and factors of 200 to 20,000 for permeability).

Soil compaction effects on radon diffusion coefficients and air permeabilities are determined by similarly computing the coefficients from the moistures and porosities of the more and less dense soils (Figure 3). Diffusion coefficients generally increase with decreasing density in the -10^{-4} to -10^{-6} Pa range of matric potentials. For loamy soils, the density effects are negligible in certain ranges of matric potential except for densities exceeding 1600 kg m^{-3}. For the other soils, however, a 50% density reduction increases diffusion coefficients by a factor of 2 to 3. Air permeabilities also increase with reduced density, but by factors of 3 to 6, and with more constant trends with variations in matric potential.

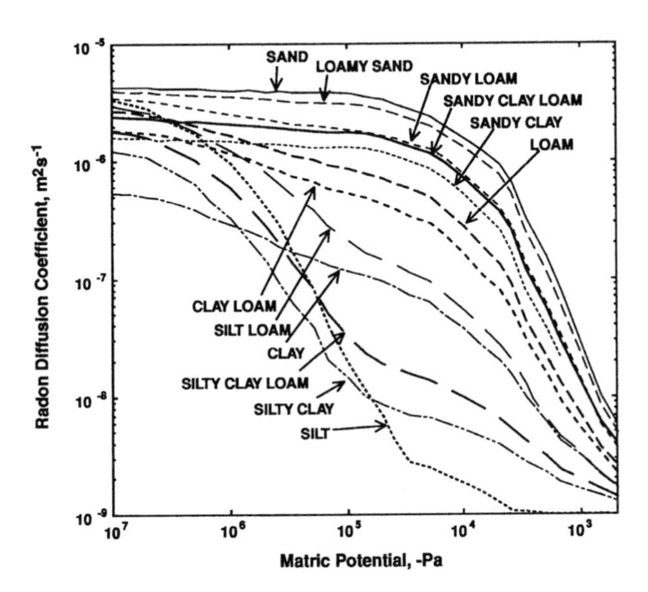

Figure 4. Radon diffusion coefficients of the U.S. Soil Conservation Service soil textural classes, calculated using equation 1.

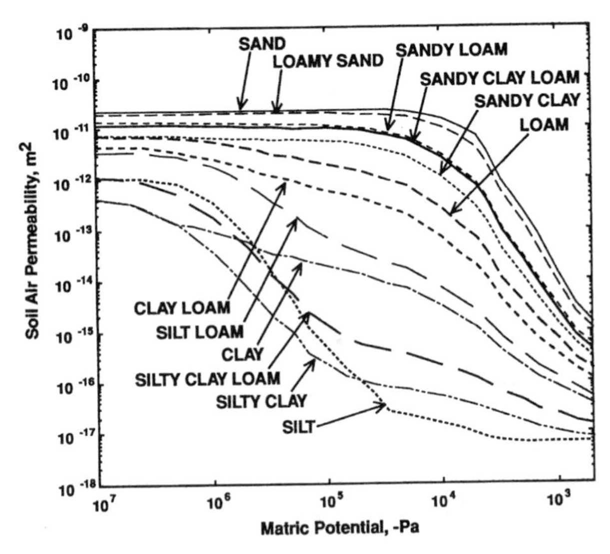

Figure 5. Soil air permeabilities of the U.S. Soil Conservation Service soil textural classes, calculated using equation 2.

APPLICATIONS

For some applications, the environment in which soils must control radon releases has already been defined in terms of soil matric potential. For example, earthen covers for uranium mill tailings impoundments are designed conservatively to control radon releases at water contents corresponding to 1.5 x 10^6 Pa matric potential (NRC, 1989). This represents the permanent wilting point of plants, beyond which plant roots become ineffective for further water extraction. The ranges of potential water-saturation fractions at this matric potential are illustrated in Figure 2 for the SCS soil texture classes. If the soil particle-size distribution is known, however, the uncertainty in its water content may be reduced from the widths of the shaded regions in Figure 2 to the uncertainty associated with the selected matric potential model. Although precisions of matric potential models are not well defined, the relative convergence of the comparisons for SCS sand, loamy sand, and sandy loam at low matric potentials (Figure 2) suggests that at least these water estimates should be more reliable than those for a more divergent case such as SCS clay or SCS silt at a moderate matric potential of -10^5 Pa. From given water-saturation fraction estimates, diffusion and permeability coefficients can be estimated within nominal factors of 2.0 and 2.3 GSD from equation 1 and equation 2, respectively.

For indoor radon predictions in Florida, recent field measurements on undisturbed soils (0.6 m depth) suggest ambient water contents corresponding to the –10 to –50 kPa matric potential range, and in-situ air permeabilities ranging from 3 x 10^{-16} to 3 x 10^{-11} m^2. The measurement precision of the air permeabilities was estimated to be GSD = 1.5 from 18 pairs of duplicate measurements taken at different sites. The radon diffusion measurements, made at field moistures and densities, used a procedure that has demonstrated a precision of about GSD = 1.4 (Nielson et al., 1982). The measured radon diffusion coefficients and air permeabilities are compared with correlation predictions in Figure 6. The 13 predicted diffusion coefficients exhibited a GSD of only 1.4 from corresponding laboratory diffusion measurements, equivalent to the diffusion measurement precision. The 37 predicted air permeabilities exhibited a GSD of 1.7 from in situ field permeability measurements, slightly exceeding the field permeability measurement precision.

The influence of foundation soils on radon entry into slab-on-grade houses was characterized by computer analyses of radon entry into a model house that was analyzed on each of the 12 SCS soil textural classes. The model house and its interaction with the underlying soils was characterized using a two-dimensional, three-phase, multiregion radon transport model called RAETRAD (Rogers and Nielson, 1990). RAETRAD explicitly considers radon advective transport by soil gas flow and diffusive transport in both the liquid and air phases of soil pores. The properties of the foundation soils were modeled at a constant soil water matric potential of –50 kPa to reflect the expected in-situ values of soil water content, permeability, and diffusion coefficient with the different soil textures. The model house parameters for the calculations are given in Table 2. The resulting radon entry rates are shown in Figure 7. The radon entry rates varied from nearly 0.12 Bq s^{-1} per Bq L^{-1} subslab (L s^{-1}) for the SCS sand class to about 0.005 L s^{-1} for SCS silty clay, in identical order to the soil permeabilities, diffusion coefficients, and moisture saturation fractions.

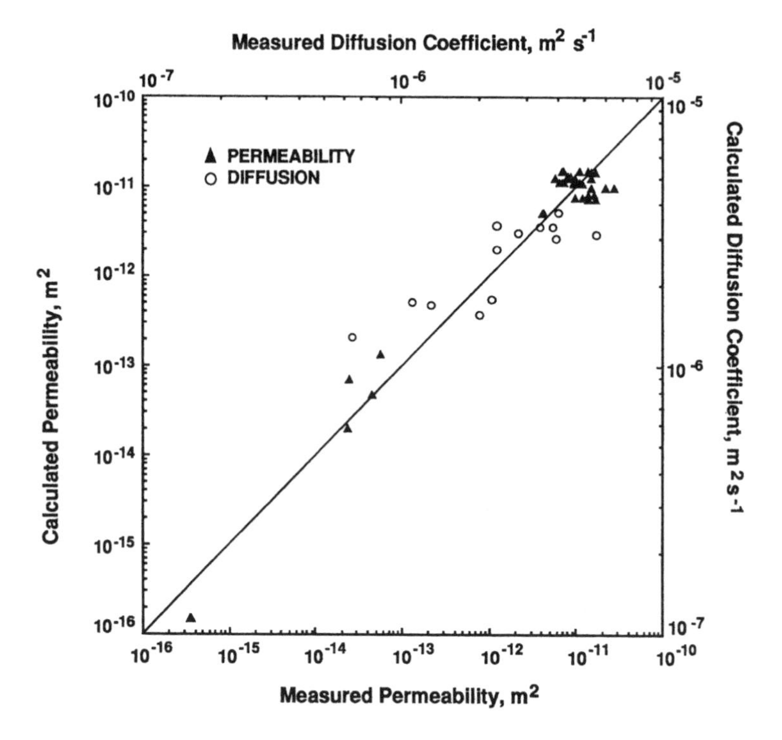

Figure 6. Comparison of measured and calculated soil air permeabilities (triangles) and radon diffusion coefficients (open circles) for Florida soils.

Table 2. Model house parameters for calculating indoor radon entry rates.

House radius	6.7 m
House volume	344 m^3
House air exchange	2.8×10^{-4} s^{-1} (1 h^{-1})
Indoor radon	74 Bq m^{-3} (2 pCi L^{-1})
House pressure	–2.4 Pa
Floor thickness	0.1 m
Concrete floor porosity	0.2
Concrete slab permeability	1×10^{-16} m^2
Concrete slab diffusion coefficient	5×10^{-8} m^2 s^{-1}
Concrete slab radium	11 Bq kg^{-1} (0.3 pCi g^{-1})
Outdoor radon	0 Bq m^{-3}
Floor crack width	0.01 m
Footer width	0.3 m
Footer depth	0.9 m
Footer permeability	1×10^{-15} m^2
Footer diffusion coefficient	5×10^{-8} m^2 s^{-1}
Footer porosity	0.407
Footer radium	11 Bq kg^{-1} (0.3 pCi g^{-1})
Soil matric potential	-5×10^4 Pa
Soil porosity	0.407
Soil radium	37 Bq kg^{-1} (1 pCi g^{-1})
Radon emanation	0.25
Fill soil thickness	0.3 m

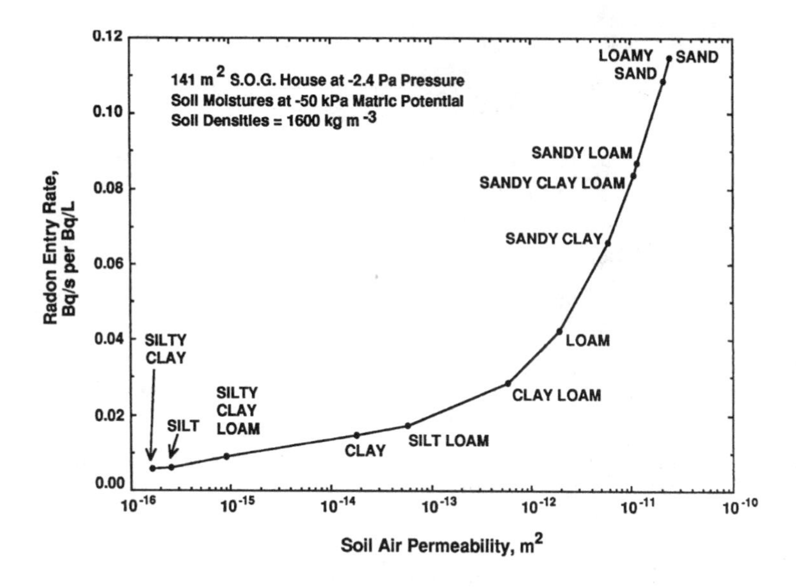

Figure 7. Radon entry rates computed by RAETRAD (see text) for a slab-on-grade structure.

These analyses show the large, water-related variations in radon transport properties that are dominated by the textural properties of soil. Even under identical environmental conditions, soils of different textural classes are estimated to cause very different radon entry rates and resulting indoor radon concentrations. Although soil water content traditionally requires site-specific measurement, values estimated from matric potentials may provide reliable alternatives for many modeling activities in which soil textural class or particle-size distribution is known.

ACKNOWLEDGMENT

This work is supported in part by U.S. Department of Energy Grant DE-FG02-88ER60664 and in part by U.S. Environmental Protection Agency Subcontract IAG-RWFL933783.

REFERENCES

Arya, LM and JF Paris. **1981**. A physicoempirical model to predict the soil moisture characteristic from particle-size distribution and bulk density data. Soil Sci Soc Am J 45:1023-1030.

Brooks, RH and AT Corey. **1964**. Hydraulic Properties of Porous Media, Hydrol Paper No. 3. Colorado State University, Ft. Collins, CO.

Clapp, RB and GM Hornberger. **1978**. Empirical equations for some soil hydraulic properties. Water Resour Res 14:601-604.

Dunn, IS, LR Anderson, and FW Kiefer. **1980**. *Fundamentals of Geotechnical Analysis*. Wiley & Sons, New York.

Eaton, RS and AG Scott. **1984**. Understanding radon transport into houses. Radiat Protect Dosim 7:251.

Gundersen, LCS, GM Reimer, CR Wiggs, and CA Rice. **1988**. *Map Showing Radon Potential of Rocks and Soils in Montgomery County, Maryland*, Map MF-2043. U.S. Geological Survey, Denver CO.

Gupta, SC and WE Larson. **1979**. Estimating soil water retention characteristics from particle size distribution, organic matter percent, and bulk density. Water Resour Res 15:1633-1635.

Meng, T. P., H. M. Taylor, D. W. Fryrear, and J. F. Gomez. **1987**. Models to predict water retention in semiarid sandy soils. Soil Sci Soc Am J 51:1563-1565.

Nazaroff, WW and RG Sextro. **1989**. Technique for measuring the indoor ^{222}Rn source potential of soil. Environ Sci Technol 23:451-458.

Nero, AV, MB Schwehr, WW Nazaroff, and KL Revzan. **1986**. Distribution of airborne radon-222 concentrations in U.S. homes. Science 234:992-997.

Nielson, KK, DC Rich, VC Rogers, and DR Kalkwarf. **1982**. *Comparison of Radon Diffusion Coefficients Measured by Transient-Diffusion and Steady-State Laboratory Methods*, NUREG/CR-2875. U.S. Nuclear Regulatory Commission, Washington, DC.

NRC. **1989**. *Calculation of Radon Flux Attenuation by Earthen Uranium Mill Tailings Covers*, Regulatory Guide 3.64. U.S. Nuclear Regulatory Commission, Washington, DC.

Otton, JK, RR Schumann, DE Owen, N Thurman, and JS Duval. **1988**. *Map Showing Radon Potential of Rocks and Soils in Fairfax County, Virginia*, Map MF-2047. U.S. Geological Survey, Denver, CO.

Puckett, WE, JH Dane, and BF Hajek. **1985**. Physical and mineralogical data to determine soil hydraulic properties. Soil Sci Soc Am J 49:831-836.

Rawls, WJ and DL Brakensiek. **1982**. Estimating soil water retention from soil properties. J Irrig Drain Div Am Soc Civil Eng 108(IR2):166-171.

Rogers, VC and KK Nielson. **1984**. *Radon Attenuation Handbook for Uranium Mill Tailings Cover Design*, NUREG/CR-3533. U.S. Nuclear Regulatory Commission, Washington, DC.

Rogers, VC and KK Nielson. **1990**. Benchmark and application of the RAETRAD model, pp. 1-10. In: *1990 International Symposium on Radon and Radon Reduction Technology*, EPA/600/9-90/005c, VI-1. U.S. Environmental Protection Agency, Washington, DC.

Rogers, VC and KK Nielson. **1991**. Correlations for predicting air permeabilities and ^{222}Rn diffusion coefficients of soils. Health Phys 61:225-230.

Rogers, VC, RF Overmyer, KM Putzig, CM Jensen, KK Nielson, and BW Sermon. **1980**. *Characterization of Uranium Tailings Cover Materials for Radon Flux Reduction*, NUREG/CR-1081. U.S. Nuclear Regulatory Commission, Washington, DC.

Salter, PJ and JB Williams. **1967**. The influence of texture on the moisture characteristics of soils, IV. A method of estimating the available water capacities of profiles in the field. J Soil Sci 18:174-181.

Schuh, WM, RL Cline, and MD Sweeney. **1988**. Comparison of a laboratory procedure and a textural model for predicting in situ soil water retention. Soil Sci Soc Am J 52:1218-1227.

Tanner, AB. **1964**. Radon Migration in the Ground: A Review, pp. 161-190. In: *The Natural Radiation Environment*, JAS Adams and WM Lowder (eds.). University of Chicago Press, IL.

USDA. **1975**. Soil taxonomy, pp. 469-474. In: *Agriculture Handbook 436*. U.S. Department of Agriculture, Soil Conservation Service, Washington, DC.

USEPA. 1986. *A Citizen's Guide to Radon*, EPA-86-004. U.S. Environmental Protection Agency, Washington DC.

QUESTIONS AND ANSWERS

Q: Holford, PNL. I was wondering about the alpha parameter that was used to fit the particle size distributions to the moisture characteristic curves. Did you find that to be a constant, or did it vary with the soil types?

A: Nielson, Salt Lake City. We held it constant based on the observation of Arya and Paris's original 1981 paper. Some of the later papers, where they've evaluated the comparisons with different soils, showed that there were some trends in the alpha parameter value with soil type. But the scatter in the data was large, and suggested inconsistent trends in a couple of cases. We first attempted to parameterize alpha with grain size, and we found that the uncertainty of doing that far exceeded the parameter we fitted it to, and so we fell back to using the constant that they first proposed.

Q: Holford, PNL. All right, one more thing. Many soil physicists predict permeabilities from the moisture characteristic curve using the Van Genuchten model, for instance. Do you think it is possible to predict the diffusion coefficients from the moisture characteristic curve?

A: Nielson. Yes, it is. We've done that, in fact, solely from theoretical bases; from first principles. You can calculate diffusion coefficients from the matric potential curve, using that to characterize the pore-size distribution. In the cases where we have done that, it's a fairly complicated calculation, and we found that for everyday use it's a little bit awkward and time-consuming, so for ordinary use we've chosen to use the correlations in place of it. But it can be done.

Q: Holford. Yes, because when you do that you avoid that alpha-fitting parameter in the Arya and Paris model.

A: Nielson. That's right. To date, we have not had enough confidence in the completely theoretical approach that we would choose it over the alpha-fitting parameter.

DETERMINATION AND MEASUREMENT OF SOIL PARAMETERS FOR CHARACTERIZING RADON HAZARD OF SOILS

T. E. Blue and M. S. Jarzemba

Ohio State University, Columbus, Ohio

Key words: *Soil, radon hazard*

ABSTRACT

There is little correlation between radon concentrations in soil and radon concentrations in homes. One explanation is that the soil radon concentration does not fully characterize the soil as a radon hazard. A mathematical model for the determination of important soil parameters for characterizing the flow of radon into a basement has been analyzed. We have identified important soil properties by mathematically modeling ventilated air enclosed in basement walls of thickness T (through which radon convects) and surrounded by soil of infinite extent (through which radon diffuses). The radon instantaneously mixes uniformly with the basement air and is lost from the basement air by ventilation (λ_v) and decay (λ). It was found that not only the soil pore gas radon concentration, C_S, but also the radon gas diffusion length, L_3, and the soil porosity, ϵ_3, are important to characterize the soil as a radon hazard.

A model for determining the parameters C_S, L_3, and ϵ_3 has also been analyzed. It was found that it is possible to measure in situ these important soil parameters by monitoring the radon gas concentration time history of two cavities of different radii formed in the same soil.

INTRODUCTION

It has been reported (Nason and Cohen, 1987) that in Pittsburgh, Pennsylvania, there was little correlation between radon-222 concentrations in the soil and radon concentrations in homes. One possible explanation, presented by the authors, for this low correlation is that the soil in which the radon concentration was measured was not representative of the soil surrounding the foundation of the home. This reasoning is supported by the fact that the radon concentration in the soil was measured at a depth of only 37 cm. However, another study (Akerblom et al., 1984) in Sweden, also found little correlation between radon concentrations in the soil and in adjacent homes, and in this study the soil was measured at a depth of 1 m. Other factors such as the construction of the homes, their heating systems, and the habits of their occupants affect the indoor radon concentration and may account for the lack of correlation. Studies (George and Breslin, 1982) of the radium-226 concentration in the soil and the radon-222 concentration in adjacent homes in New York and New Jersey support this idea. They found that the radium concentration in soil varied by 20% from its mean. However, the radon concentration in the homes varied from its mean by a factor of two. A third explanation is that the steady-state concentration of radon in the pore gas does not fully characterize the soil as a radon hazard. Other properties, such as the diffusion length for radon in the soil, may also be important.

The purpose of this paper is to explore the third explanation by: (1) identifying the soil properties important in characterizing radon transport into the basement of a home, and (2) by developing a model for analyzing dynamic measurements of the radon concentration of a cavity in the soil as a means for determining these properties in situ.

DETERMINING IMPORTANT SOIL PARAMETERS

Three-Region Model

A three-region model was used to determine the soil parameters of interest for characterizing radon gas flow into homes (Figure 1). In the three-region model, radon diffuses through soil of infinite extent and is transported by convection through concrete basement walls, where it instantaneously mixes with basement air. Radon is lost from the basement air by ventilation and decay.

The pore-gas radon concentration in the soil (C_3) is described by

$$\frac{\delta}{\delta t} C_3(x,t) = D_3 \nabla^2 C_3 - \lambda C_3 + S. \tag{1}$$

Since diffusion dominates pressure-driven transport in the soil (Akerblom et al., 1984), for simplicity only diffusion is recognized as a significant mode of transport in equation 1. Our neglect of pressure-driven transport in the soil (region 3) does not imply that, in our model, pressure-driven transport is unimportant as a mechanism for radon entry into a home. In fact, it is recognized in our model as being the dominant mechanism for radon migration through the basement wall. The pore-gas radon concentration in the concrete basement wall (C_2) is described by

$$\frac{\delta}{\delta t} C_2(x,t) = \frac{K}{\epsilon_2 \mu} \nabla P \cdot \nabla C_2 - \lambda C_2, \tag{2}$$

where ∇P is written as $\Delta P/T$, which is exact in one-dimensional planar geometry, since P satisfies Laplace's equation. Because convection dominates diffusion in the basement walls (Akerblom et al., 1984), diffusion has been ignored in equation 2.

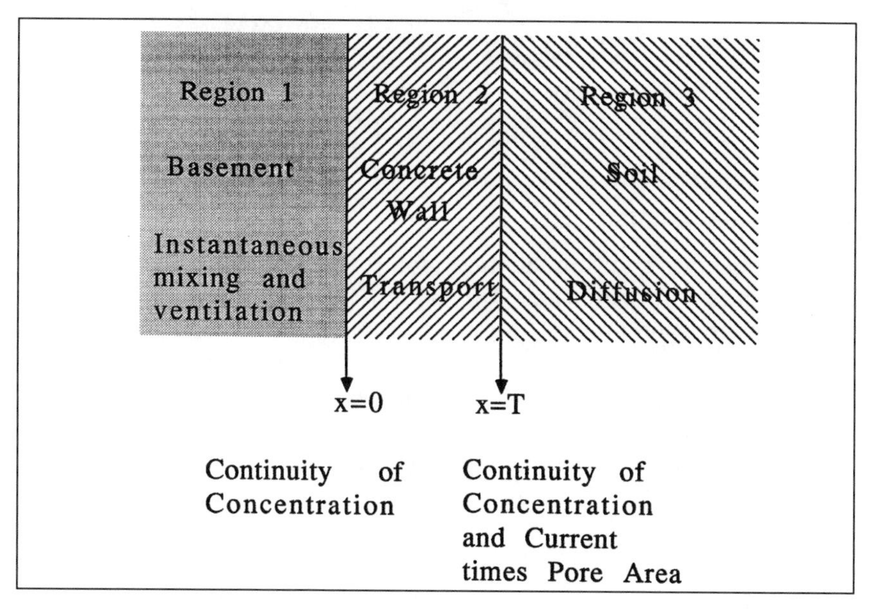

Figure 1. Three-region model for determining soil parameters for characterizing radon transport into houses.

The kinetic equation for the radon concentration in the basement air (C_1) is

$$\frac{\delta}{\delta t} C_1(t) = \frac{K \Delta P A}{\mu T V} C_2 (x = 0,t) - (\lambda + \lambda_v) C_1 + \lambda_v C_0. \tag{3}$$

In equations 1, 2, and 3, and their associated interface conditions, equations 4, 5, and 6, S is the soil pore-gas radon source strength, D_3

is the soil pore-gas radon diffusion coefficient, ϵ_3 is the soil porosity, ϵ_2 is the wall porosity, K is the wall's intrinsic permeability, μ is the viscosity of the air flowing through the basement wall, ΔP is the difference in air pressure across the basement wall, T is the thickness of the basement wall, A is the surface area of the basement wall, V is the volume of air within the basement, λ is the radon radioactive decay constant, and λ_v is the air exchange rate.

The boundary and interface conditions for this model are:

$$C_2(x = T,t) = C_3(x = T,t) \tag{4}$$

$$\epsilon_3 D_3 \nabla C_3(x,t)\big|_{x=T} = C_3(x = T,t)\ \frac{K\Delta P}{\mu T} \tag{5}$$

$$C_3(x=\infty,t) = S/\lambda. \tag{6}$$

Methods and Results

The steady-state solution for the radon concentration in the basement is calculated from equations 1-5 by first solving equation 1 to find a general expression for C_3:

$$C_3(x,\infty) = C_s(1 - B\exp(-x/L_3)\ ), \tag{7}$$

where $C_s = S/\lambda$, and the constant B is determined from equation 5 to be

$$B = \frac{\dfrac{K\Delta P}{\mu T}\ \exp(T/L_3)}{\left(D_3\ \dfrac{\epsilon_3}{L_3} + \dfrac{K\Delta P}{\mu T}\right)}\ , \tag{8}$$

where L_3 is the diffusion length for radon in the soil, and $L_3 = (D_3/\lambda)^{1/2}$. Then equation 2 is solved to find a general expression for C_2:

$$C_2(x,\infty) = E\ \exp\left(\frac{\lambda\epsilon_2\mu Tx}{K\Delta P}\right), \tag{9}$$

and the constant E is determined from equation 5 to be

$$C_s \left(1 - \frac{\dfrac{K\Delta P}{\mu T}}{\epsilon_3 \dfrac{D_3}{L_3} + \dfrac{K\Delta P}{\mu T}} \right) \exp\left(\frac{-\lambda \epsilon_2 \mu T^2}{K\Delta P} \right). \tag{10}$$

Finally, the solution for $C_2(x,\infty)$ is evaluated at x = 0, and equation 3 is solved for C_1, to find:

$$C_1(\infty) = C_s \frac{K\Delta P}{(\lambda + \lambda_v)\mu T} \; \frac{A}{V} \left(\frac{\epsilon_3 L_3}{\epsilon_3 L_3 + \dfrac{K\Delta P}{\mu T\lambda}} \right) \exp\left(\frac{-\lambda \mu \epsilon_2 T^2}{K\Delta P} \right). \tag{11}$$

Discussion of Results

From equation 11 and the definition of C_S, it is apparent that the indoor radon concentration in a basement is directly proportional to, but not solely dependent on, the radon source strength in the soil. A graph of the term in brackets in equation 11 versus realistic values of $\epsilon_3 L_3$ is shown in Figure 2, with typical values of $K\Delta P/(\mu T\lambda)$ as a parameter.

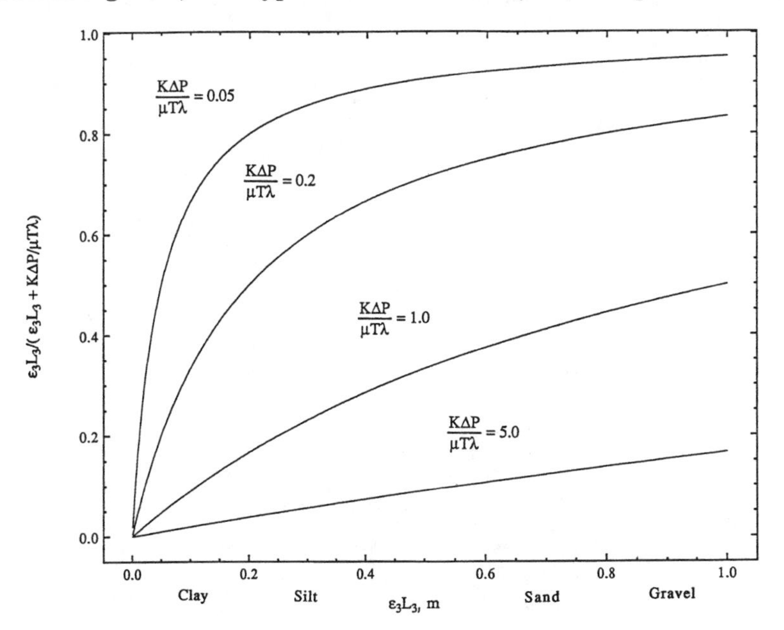

Figure 2. A graph of $\epsilon_3 L_3/(\epsilon_3 L_3 + K\Delta P/\mu T\lambda)$ vs. $\epsilon_3 L_3$ with $K\Delta P/\mu T\lambda$ as a parameter in units of meters.

One can see from equation 11 that for the extreme case of a very tight home [i.e., $K\Delta P/(\mu T\lambda)$ much less than $\epsilon_3 L_3$], the term in brackets is approximately equal to 1, and it is sufficient to measure C_s to characterize the radon hazard of the soil. On the other hand, for the extreme case of a very leaky home, [$K\Delta P/(\mu T\lambda)$ much greater than $\epsilon_3 L_3$], $\epsilon_3 L_3$ can be neglected in the denominator, and the product of ϵ_3, L_3, and C_s is a more appropriate parameter for assessing the radon hazard of the soil. In general, however, where neither of these two extreme cases exist (as can be seen in Figure 2), the expression for $C_1(\infty)$ is a more complicated function of C_s and $\epsilon_3 L_3$, and both values must be known to characterize the radon hazard of the soil.

A METHOD FOR MEASURING IMPORTANT SOIL PARAMETERS

Two-Region Model

As discussed previously, the soil parameters of interest are C_s, L_3 and ϵ_3. A two-region model is used to determine these parameters (Figure 3). For consistency with the model presented in the first part of this paper, the soil region is denoted by the subscript 3; in this case, there is no region 2.

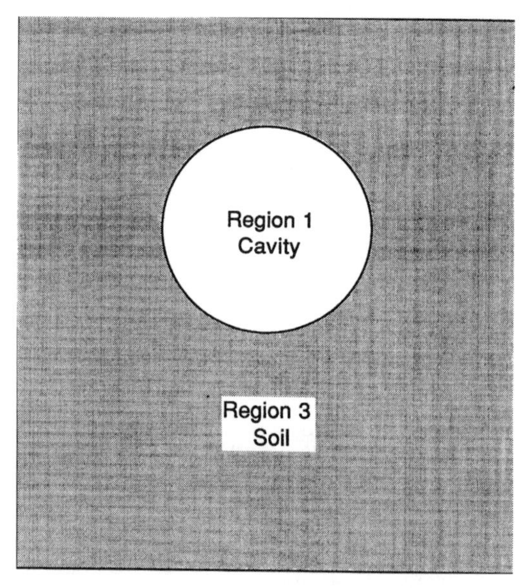

Figure 3. Two-region model for measuring important soil parameters.

The method, which is analyzed for measuring C_S, L_3 and ϵ_3, is to first form a spherical cavity in the soil. This cavity is left open to air, and the soil surrounding the cavity is allowed to attain its equilibrium radon concentration distribution. Then the cavity is closed, and the radon concentration within the cavity is measured as a function of time.

The determination of C_S, L_3, and ϵ_3 from the time history of the radon gas concentration in the cavity is based on a two-region diffusion model in spherical coordinates. The cavity, located at the origin of the coordinate system, has a radius R. Within the cavity (region 1, $0 < r < R$), the radon concentration (C_1) is uniform and is described by the equation

$$\frac{\delta C_1(t)}{\delta t} = \frac{3D_3\epsilon_3\nabla C_3(R,t)}{R} - \lambda C_1(t). \tag{12}$$

The pore-gas radon concentration in the soil (C_3) surrounding the cavity (region 3, $R < r < \infty$) is described by the equation

$$\frac{\delta C_3(r,t)}{\delta t} = D_3\nabla^2 C_3(r,t) - \lambda C_3(r,t) + S. \tag{13}$$

In equations 12 and 13, D_3 is the pore-gas radon diffusion coefficient in the soil, and λ is the radon radioactive decay constant. The inhomogeneous term in equation 12 is the product of the current of radon at R, the cavity surface area, and the soil porosity (ϵ_3), divided by the volume of the cavity.

Again defining $L_3 = (D_3/\lambda)^{1/2}$, equations 12 and 13 are solved with the interface condition that $C_3(R,t) = C_1(t)$, and with the boundary condition that $C_3(\infty,t) = C_S$. The initial condition on C_1 is $C_1(t{=}0) = 0$, and the initial condition on C_3 is

$$C_3(r,0) = C_S\left[1 - \frac{R}{r}\exp\left(\frac{-(r-R)}{L_3}\right)\right], \tag{14}$$

i.e., the cavity is left open to the atmosphere until the soil pore-gas radon concentration reaches its equilibrium distribution in the soil; the cavity is then closed.

Methods and Results

To determine the important soil properties C_s, L_3, and ϵ_3, it is necessary to Laplace-transform equations 12 and 13 to obtain expressions for $C_1(s)$ and $C_3(r,s)$. The ordinary differential equation in space for $C_3(r,s)$ is solved to find

$$C_3(r,s) = \frac{RC_1(s)\exp\left(-(r-R)/L_3\right)}{r} - C_s\frac{R}{rs}\exp\left(-(r-R)/L_3\right) + C_s\frac{1}{s}. \quad (15)$$

Next, equation 14 is differentiated with respect to r, evaluated for r=R, and substituted into the inhomogeneous term in transformed equation 12. The resulting equation is then solved for $C_1(s)$:

$$C_1(s) = \frac{C_s\epsilon_3\frac{1}{s}\left(\sqrt{a}+\sqrt{3\lambda}\right)}{\frac{1}{\sqrt{a}}(s+\lambda)+\epsilon_3\sqrt{a}+\epsilon_3\sqrt{3(s+\lambda)}}, \quad (16)$$

where $a = 3\lambda L^2/R^2$.

Discussion of Results

By monitoring the dynamic radon gas concentration in two cavities of different radii in the same soil, C_s, L_3, and ϵ_3 can be determined. This is accomplished by measuring for these cavities the slope of the radon concentration curves at time zero and the steady-state cavity radon gas concentrations.

Using the initial value theorem of Laplace transforms on $sC_1(s)$ [with $C_1(s)$ as given in equation 16] to solve for the slope of the cavity radon gas concentration at time zero, an expression for L_3/R (the ratio of soil diffusion length to cavity radius), can be written as

$$\frac{L_3}{R} = \sqrt{\frac{1}{4}+\frac{C_1'(0)}{3\lambda C_s\epsilon_3}} - \frac{1}{2}, \quad (17)$$

where $C_1'(0)$ is the slope of the cavity radon gas concentration at time zero.

Using the final value theorem of Laplace transforms on $C_1(s)$ [with $C_1(s)$ as given in equation 16] to solve for the steady-state cavity radon gas concentration, another expression for L_3/R is

$$\frac{L_3}{R} = \sqrt{\frac{1}{4} + \frac{C_1(\infty)}{3\epsilon_3(C_s - C_1(\infty))}} - \frac{1}{2}, \tag{18}$$

where $C_1(\infty)$ is the steady-state cavity radon gas concentration.

Because equations 17 and 18 form a system of two equations in three unknowns, it is necessary to obtain more information to be able to solve for C_s, L_3, and ϵ_3 uniquely. To accomplish this, it is necessary to measure $C_1'(0)$ and $C_1(\infty)$ for two cavities, A and B. This will yield a system of four equations in three unknowns, making it possible to solve the system completely.

Using measurements of $C_1(\infty)_A$, $C_1'(0)_A$, $C_1(\infty)_B$, and $C_1'(0)_B$, the steady-state and initial-slope values for cavities A and B respectively, expressions for C_s, L_3, and ϵ_3 are

$$C_s = C_1(\infty)_A \left[\frac{C_1'(0)_A}{C_1'(0)_A - \lambda C_1(\infty)_A}\right], \tag{19}$$

$$L_3 = \left[\frac{R_A R_B[R_A C_1'(0)_A - R_B C_1'(0)_B]}{R_B^2 C_1'(0)_B - R_A^2 C_1'(0)_A}\right], \tag{20}$$

and

$$\epsilon_3 = \frac{C_1'(0)_A}{3\lambda C_s\left(\frac{L_3^2}{R_A^2} + \frac{2L_3}{R_A}\right)}, \tag{21}$$

where R_A and R_B are the radii of cavities A and B, respectively. Of course, the assignment of cavity identification is arbitrary, and the analysis could be performed reversing the cavity identification. The results could then be averaged to find more accurate values for C_s, L_3, and ϵ_3.

Equations 19 through 21 show how to calculate C_s, L_3, and ϵ_3 from experimentally measured values. Thus we have an in-situ method for measuring C_s, L_3, and ϵ_3, the important soil parameters for characterizing radon gas transport into homes.

CONCLUSIONS

The solution for the steady-state radon concentration in the basement is:

$$C_1 = \frac{C_s K \Delta P}{(\lambda + \lambda)\mu T} \frac{A}{V} \left[\epsilon_3 L_3 / \left(\epsilon_3 L_3 + \frac{K \Delta P}{\mu T \lambda}\right)\right] \exp[-\lambda \mu \epsilon_2 T^2/(K \Delta P)]. \quad (22)$$

In equation 22, C_S, L_3, and ϵ_3 are the soil's radon concentration, radon diffusion length, and porosity; ϵ_2 and K are the wall's porosity and permeability; μ and ΔP are the air's viscosity and pressure difference across the basement wall; and V and A are the basement's volume and wall surface area. From equation 22, for a tight home (i.e., $K\Delta P/\mu T\lambda << \epsilon_3 L_3$), it is sufficient to simply measure C_S to characterize the radon hazard of a soil. If a home is very leaky (i.e., $K\Delta P/\mu T\lambda >> \epsilon_3 L_3$), then $C_S \epsilon_3 L_3$ is a more appropriate parameter for assessing the radon hazard.

REFERENCES

Akerblom, G, P Anderson, and B Clavensjo. **1984**. Soil gas radon—A source for indoor radon daughters. Radiat Protect Dosim 7:49-54.

George, AC and AJ Breslin. **1982**. The Distribution of Ambient Radon and Radon Daughters in Residential Buildings in the New Jersey-New York Area, p. 1272. In: *Natural Radiation Environment*, TF Gesell and WM Lawder (eds.). CONF-780422, NTIS, Springfield, VA.

Nason, R and B Cohen. **1987**. Correlation between radium-226 in soil, radon-222 in soil gas, and radon-222 inside adjacent houses. Health Phys 52:73-77.

QUESTIONS AND ANSWERS

C: Swedjemark, Swedish Radiation Protection Institute. You refer to a table in a Swedish report and the fact that you had no cross-correlations. That table is very simplified and is a summary of the basis for our classification. We have cross-correlations. We arranged courses for Swedish consultants who classified the soil in

Sweden. And the radon maps which we get after such classification is used mostly for planning our future building.

C: Durham, PNL (for Blue). Thank you. As I said, I was unaware of that.

STATISTICAL UNCERTAINTY ANALYSIS OF RADON TRANSPORT IN NONISOTHERMAL, UNSATURATED SOILS[a]

D. J. Holford, P. C. Owczarski, G. W. Gee, and H. D. Freeman

Pacific Northwest Laboratory, Richland, Washington

Key words: *Radon, transport, computer simulation, uncertainty analysis, unsaturated soils*

ABSTRACT

Transient radon flux from soil is affected by soil properties, as well as meteorological factors such as the infiltration of rainwater, air pressure, and temperature changes at the soil surface. Natural variations in meteorological factors and soil properties contribute to uncertainty in subsurface model predictions of radon flux. When coupled with a building transport model, these will add uncertainty to predictions of radon concentrations in homes. A statistical uncertainty analysis using the Rn3D finite-element numerical model was conducted to assess the relative importance of these meteorological factors and the properties of the soil affecting the transport of radon. The Rn3D model has been enhanced to simulate the nonisothermal transport of radon by diffusion and advection in both the water and the air phases. Each input parameter was treated as a random variable, with either a uniform, normal, or lognormal distribution. Monte Carlo simulations were run after using Latin hypercube sampling to select random combinations of the model input parameters for each simulation. The relative contribution of each input parameter to the variability in radon flux predictions was assessed using multiple linear regression. The input parameters that contribute the most uncertainty to model predictions should be measured with the highest degree of spatial and temporal accuracy in the field.

INTRODUCTION

To accurately predict radon fluxes from soils to the atmosphere, we must know more than the radium content of the soil. Radon flux from soil is affected not only by soil properties but also by meteorological factors such as air pressure and temperature changes at the soil surface, as well as the infiltration of rainwater. Natural variations in meteorological factors and soil properties contribute to uncertainty in

(a) This work is funded in part by the U.S. Department of Energy under Contract DE-AC06-76RLO 1830.

subsurface model predictions of radon flux. When coupled with a building transport model, these will also add uncertainty to predictions of radon concentrations in homes. A statistical uncertainty analysis using our Rn3D finite-element numerical model was conducted to assess the relative importance of these meteorological factors and the soil properties affecting radon transport.

RADON TRANSPORT IN NONISOTHERMAL, UNSATURATED SOILS

The Rn3D model has been enhanced to simulate the nonisothermal transport of radon by diffusion and advection in both the liquid (water) and the gas (air) phase. This three-dimensional, finite-element code was used to simulate the effect of air pressure, water pressure, and temperature gradients on radon concentration in partially saturated soil with parallel, partially penetrating cracks. For this uncertainty analysis, a two-dimensional, steady-state scenario was assumed (Figure 1).

Using Einstein's summation convention, steady-state, two-phase transport of radon in partially saturated soil is governed by

$$\frac{\partial}{\partial x_i}\left(D_{ij}\frac{\partial C_a}{\partial x_j}\right)-\frac{\partial}{\partial x_i}\left[\frac{v_{a_i}-v_{w_i}\kappa}{\epsilon(1-s+s\kappa)}C_a\right]-\lambda C_a+\frac{R\rho_b\lambda E/n}{1-s+s\kappa}=0, \quad (1)$$

where C_a = concentration of radon in air-filled pore space,
D = bulk diffusion coefficient of radon in partially saturated soil,
v_a = Darcy velocity of air,
v_w = Darcy velocity of water,
κ = solubility coefficient of radon in water relative to air,
ϵ = soil porosity,
s = percent saturation of the soil pore space with water,
λ = radon-222 decay coefficient,
R = radium content of the soil,
ρ_b = bulk density of the soil,
E = radon emanation coefficient, and
i and j = directional indices of the Cartesian coordinate system.

The Darcy velocity of air is given by

$$v_{aj}=-\frac{k_{aij}}{\mu_a}\left(\frac{\partial p_a}{\partial x_j}+\rho_a g\frac{\partial z}{\partial x_j}\right), \quad (2)$$

where k_a = permeability of the air phase (in units of length squared),
p_a = air phase pressure,
μ_a = dynamic viscosity of air, and
z = vertical distance above an arbitrary datum.

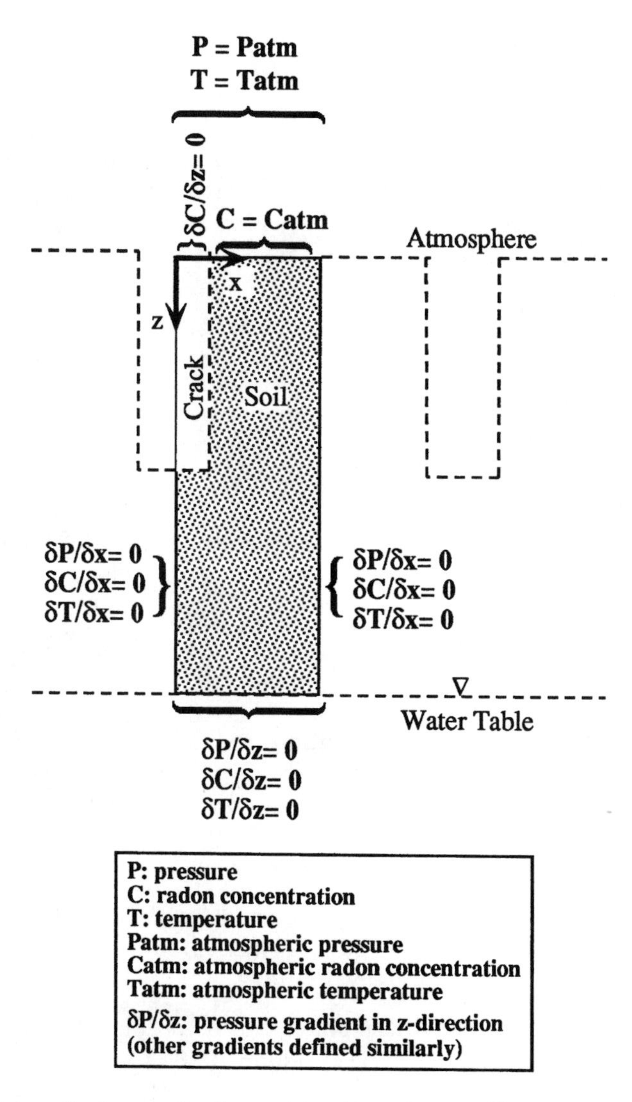

Figure 1. Schematic diagram showing model and boundary conditions.

Similarly, the Darcy velocity of water is given by

$$v_{w_i} = -\frac{k_{w_{ij}}}{\mu_w}\left(\frac{\partial p_w}{\partial x_j} + \rho_w g\,\frac{\partial z}{\partial x_j}\right), \tag{3}$$

where k_w = permeability of the water phase (in units of length squared),
p_w = water phase pressure, and
μ_w = dynamic viscosity of water.

Several of the model variables are dependent on soil pore saturation with water. The capillary pressure of partially saturated soil is defined as the difference between the water and the air phases, and is related to matric suction by the density of water and gravitational acceleration:

$$p_c = p_a - p_w = -\psi\rho_w g, \tag{4}$$

where p_c = capillary pressure,
ψ = matric suction,
ρ_w = density of water, and
g = gravitational acceleration.

The matric suction is related to the saturation by the following empirical relation (van Genuchten, 1980):

$$\frac{s - s_T}{s_w - s_T} = \left[\frac{1}{1 + (\alpha\psi)^n}\right]^m, \tag{5}$$

where S_r = residual water saturation,
S_w = maximum water saturation,
n = empirical constant correlated with pore size distribution,
m = 1-2/n, and
α = inverse of the air-entry pressure.

The values of n, m, and α were obtained for three different soils by fitting the van Genuchten model to measured water-retention data from a catalog of soils (Mualem, 1976a).

The permeability of the water phase can be predicted using the variables n and m from equation 5 (Mualem, 1976b):

$$k_w = k\, s^{\xi}[1 - (1 - s^{1/m})^m]^2, \tag{6}$$

where k = intrinsic permeability, and ξ = exponent commonly fixed at 0.5. The permeability of the air phase was assumed to be related to the water permeability by

$$k_a = k(1 - k_w)(1 - s)^2. \tag{7}$$

The bulk diffusion coefficient for radon in partially saturated soil was calculated as a function of water saturation, using

$$D_{ij} = \tau_{ij}\left\{\frac{D_a}{\left[1 + \frac{s\kappa}{(1-s)}\right]^{\chi}} + \frac{D_w}{1 + \frac{(1-s)}{s\kappa}}\right\}, \tag{8}$$

where τ = tortuosity factor for a soil,
D_a = diffusion coefficient of radon in pure air,
D_w = diffusion coefficient of radon in pure water, and
χ = an exponent set, in this case, to four.

If χ is set equal to 1, equation 8 represents the bulk diffusion coefficient of radon in partially saturated soil, assuming no pore blockage occurs until a pore fills with water (Nielson et al., 1984). In reality, pore blockage occurs before a pore is completely filled. Setting χ to four provides a better fit to data measured by Nielson et al. (1984).

The fraction of radon emanating from the soil was calculated from

$$E = \left\{\begin{array}{ll} E_w S/S^* + E_a(1 - s/s^*) & s < s^* \\ E_w & s > s^* \end{array}\right\}, \tag{9}$$

where E_w and E_a = emanation coefficients at saturation and at dryness, and S^* = minimum moisture on the plateau of an emanation-versus-moisture curve (Nielson et al., 1984).

The properties of water and air are treated as functions of temperature and pressure in Rn3D. The viscosity of water, viscosity of air, and density of water as functions of temperature were implemented tabular functions for temperatures between 0 and 100°C (Weast, 1982). The density of air as a function of temperature and pressure was calculated from the ideal gas law (Weast, 1982):

$$\rho_a = \left(\frac{1.293}{1 + 0.00367\,T}\right)\left(\frac{p}{1333.224 * 76}\right), \tag{10}$$

where ρ_a = density of dry air (kg m^{-3}),
T = temperature (°C), and
p = absolute pressure (Pa).

The diffusion of radon in pure water was calculated as a function of temperature (Bird et al., 1960; Hart, 1986):

$$D_w = 7.4 \times 10^{-12} \frac{(T + 273.15 \sqrt{\varphi_w M_w}}{(\mu_w) \left(M_{Rn}/\rho_{Rn} \right)^{0.6}} , \quad (11)$$

where D_w = diffusion of radon in pure water (m^2 s^{-1}),
T = temperature (°C),
ϕ = "association parameter" for water given as 2.6,
M_w = molecular weight of water,
μ_w = viscosity of water as a function of temperature (centipoise),
M_{Rn} = molecular weight of radon, and
ρ_{Rn} = density of radon at the normal boiling point given as $4.4 \times 10_3$ kg m^{-3} (Herreman, 1980).

The diffusion of radon in pure air was calculated as a function of temperature and pressure (Bird et al., 1960):

$$D_a = \frac{3.640 \times 10^{-8}}{1.01325 \times 10^7 \, p} \left(\frac{T+273.15}{\sqrt{T_{cRn} T_{ca}}} \right)^{2.334} (100 p_{cRnPca})^{1/3} (T_{cRN} T_{ca})^{5/12} \left(\frac{1}{M_{Rn}} + \frac{1}{M_a} \right), \quad (12)$$

where D_a = diffusion of radon in pure air (m^2 s^{-1}),
T = temperature (°C),
p = absolute pressure (Pa),
T_{cRn} = critical temperature of radon (°K),
T_{ca} = critical temperature of air (°K),
p_{cRn} = critical pressure of radon (Pa),
p_{ca} = critical pressure of air (Pa), and
M_a = is the molecular weight of air.

The diffusion coefficients calculated in this manner show good agreement with compiled data (Hart, 1986).

The solubility of radon in water/air at atmospheric pressure was implemented as a tabular function of temperatures between 0 and 100°C (Table 1).

Table 1. Solubility of radon in water/air at atmospheric pressure.

Temperature (°C)	Solubility coefficient
0	0.507
10	0.340
10	0.250
30	0.195
37	0.167
50	0.138
75	0.114
100	0.106

STATISTICAL UNCERTAINTY ANALYSIS

If some input variables are very influential on Rn3D's prediction of radon flux to the atmosphere, they should be measured at field sites with the highest degree of spatial and temporal accuracy. An uncertainty analysis attempts to quantify the uncertainty of model output, giving an indication of the probable range of possible outcomes. A schematic diagram (Figure 2) shows the statistical method used to perform this uncertainty analysis. Each input variable was treated as a random variable, with a certain range and a uniform, loguniform, normal, or lognormal distribution (Table 2).

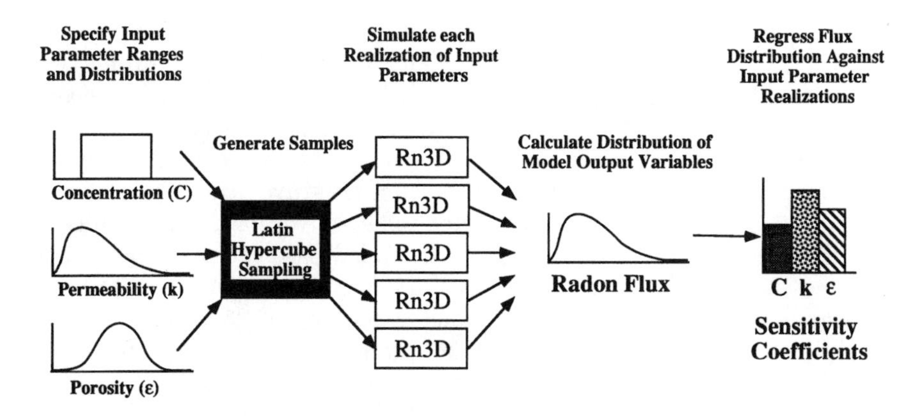

Figure 2. Schematic diagram showing Latin hypercube sampling (LHS), Monte Carlo simulation, and sensitivity analysis using multiple linear regression.

Table 2. Input variables that influence prediction of radon flux to atmosphere; their ranges and distributions.

Variable	Range	Units	Distribution
Atmospheric radon	0.1 to 100	Bq m^{-3}	Loguniform
Atmospheric pressure	8.5×10^4 to 10^5	Pa	Uniform
Capillary pressure	10^3 to 5×10^4	Pa	Loguniform
Gas pressure gradient	10^{-3} to 1	Pa m^{-1}	Loguniform
Liquid pressure gradient	10^{-1} to 10^2	Pa m^{-1}	Loguniform
Temperature gradient	–0.9 to 0.5	°C m^{-1}	Uniform
Intrinsic permeability	±1 order mag.	m^2	Lognormal
Porosity	±5%	—	Normal
Tortuosity	±10%	—	Normal
Radium content	10 to 200	Bq kg^{-1}	Loguniform
Crack width	10^{-4} to 10^{-2}	m	Loguniform
Crack depth	0.3 to 6	m	Loguniform
Crack spacing	0.6 to 20	m	Loguniform
Water table depth	–30 to –10	m	Uniform

Latin hypercube sampling (Iman and Shortencarier, 1984) was used to select random combinations of the model input variables within the ranges listed in Table 2. Each random combination of input variables was used for one run of Rn3D. A suite of such runs composes a Monte Carlo simulation (Hammersley and Handscomb, 1964). The advantage of Latin hypercube sampling is that, rather than running the model many times (as with the traditional Monte Carlo technique), the model must be run only two or three times the number of input variables.

Most of the input variables were given either uniform or loguniform distributions to evenly sample their entire ranges. The soil property's permeability, porosity, and tortuosity were assumed to be less uncertain than the rest of the input variables and were given normal or lognormal distributions. In addition, the moisture characteristic curves, relative permeability, diffusion-versus-moisture and emanation-versus-moisture curves were assumed to be known for three soil types. In these four curves (Figures 3, 4, 5, and 6, respectively), moisture content is defined as the product of soil saturation (s) and porosity (ϵ). Three Monte Carlo simulations, each consisting of 35 runs of Rn3D, were conducted for three soils (sand, loam, and clay) in order to assess uncertainties for different soil types. The base-case soil properties for each of the three Monte Carlo simulations are listed in Table 3.

Figure 7 is a probability plot of the results of the three Monte Carlo simulations. The predicted radon flux varies over five orders of magnitude. Interestingly, the variability between soil types is much less than the total variability for each soil type.

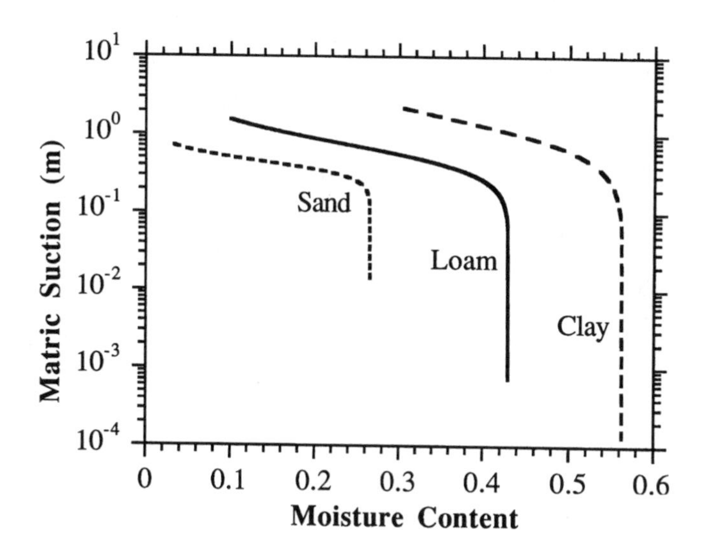

Figure 3. Matric suction versus moisture content for three soil types.

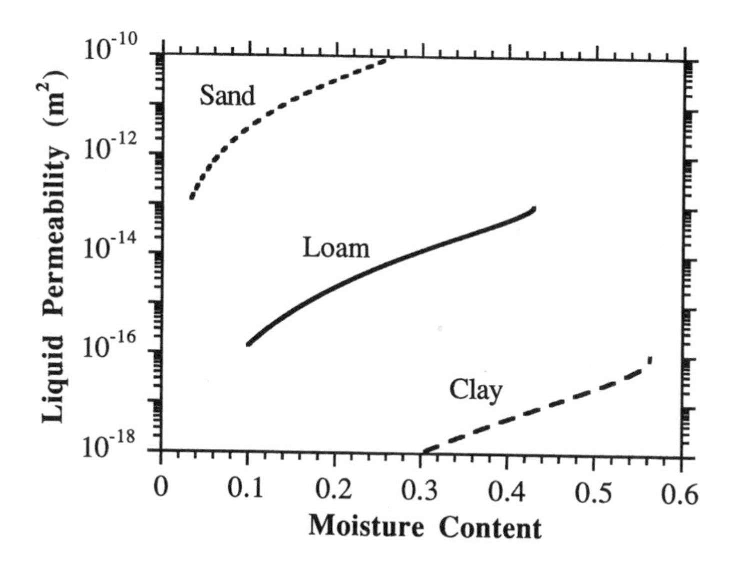

Figure 4. Liquid permeability versus moisture content for three soil types.

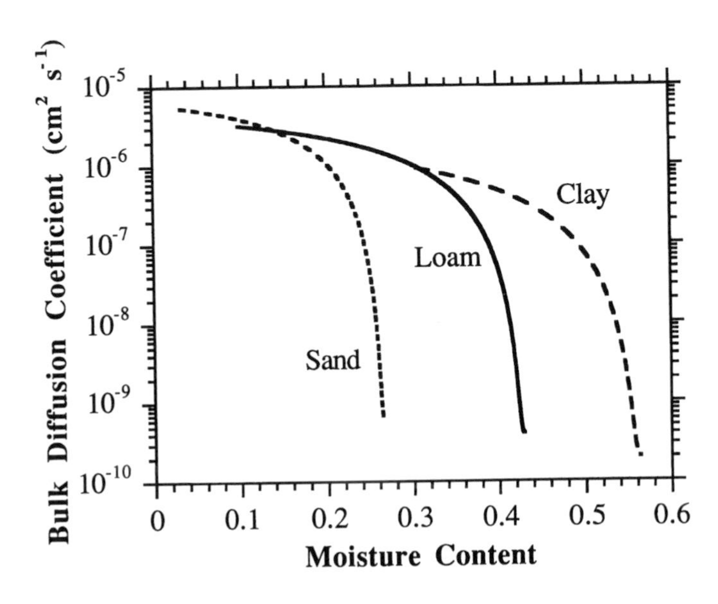

Figure 5. Bulk diffusion coefficient versus moisture content for three soil types.

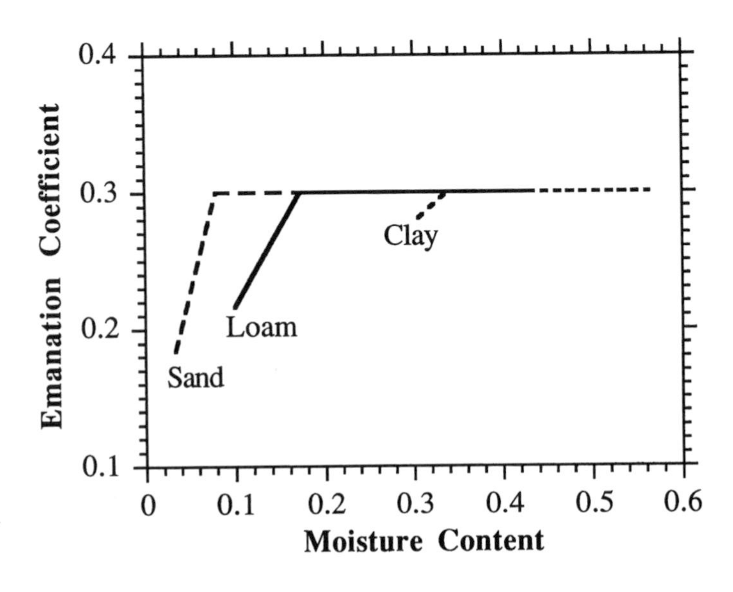

Figure 6. Emanation versus moisture content for three soil types.

Table 3. Base-case soil properties for three Monte Carlo simulations of radon flux transport to atmosphere.

Property	Sand	Loam	Clay
Permeability (m^2)	1×10^{-10}	1×10^{-13}	1×10^{-16}
Porosity	0.2526	0.4491	0.5631
Tortuosity	0.6	0.4	0.2
van Genuchten (Mualem, 1976a)			
Residual saturation	0.1229	0.2299	0.5409
n	4.886	2.502	1.8133
Alpha (m^{-1})	2.37	1.66	0.83
Emanation			
Wet	0.3	0.3	0.3
Dry	0.1	0.1	0.1
Minimum saturation	0.3	0.4	0.6

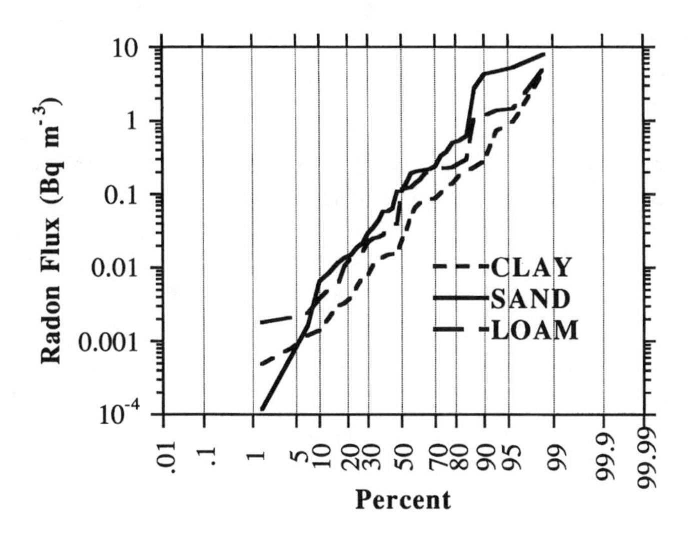

Figure 7. Normal probability plot of radon fluxes predicted by Monte Carlo simulations.

SENSITIVITY ANALYSIS

The results of the Monte Carlo simulations can be used to determine the sensitivity of the radon flux to each input variable, expressed as sensitivity coefficients. The sensitivity coefficient represents the percentage change in flux that results from a percentage change in an input variable:

$$a_i = \frac{dF/F}{dX_i/X_i} = \frac{d(\ln F)}{d(\ln X_i)}, \tag{13}$$

where a_i = sensitivity coefficient,
F = total radon flux from the soil to the atmosphere, and
X_i = i^{th} input variable.

The sensitivity coefficients can be estimated, using multiple linear regression, by assuming a first-order model for the relationship between the radon flux and the input variables:

$$\ln\hat{F} = \hat{a}_0 + \hat{a}_1 \ln X_1 + \hat{a}_2 \ln X_2 + \ldots + \hat{a}_k \ln X_k, \tag{14}$$

where $\hat{a}$ = i^{th} regression coefficient (estimated sensitivity coefficient), and k is the number of input variables.

The sensitivity coefficient can be standardized so that input variables with widely varying ranges can be compared:

$$\hat{a}_i^* = \hat{a}_i / \sigma_{\hat{a}_i}, \tag{15}$$

where $\hat{a}_i$ = standardized sensitivity coefficient, and $\sigma_{\hat{a}_i}$, = standard deviation of the sensitivity coefficient. This standardized sensitivity coefficient is regarded as significant if its absolute value is larger than 2 (representing variability larger than two standard deviations of the sensitivity coefficient). A negative coefficient indicates that the radon flux increases as the input variable's value decreases; a positive coefficient indicates the converse.

The standardized sensitivity coefficients for the three Monte Carlo simulations are shown in Figure 8. Input variables that have standardized sensitivity coefficients larger than 2 are:

- width of cracks (or other macropores),
- radium content,
- gas pressure gradient, and
- capillary pressure (soil saturation).

Variables with sensitivity coefficients between 1 and 2, which are therefore marginally significant, are:

- permeability,
- atmospheric pressure,
- depth to water table, and
- crack spacing.

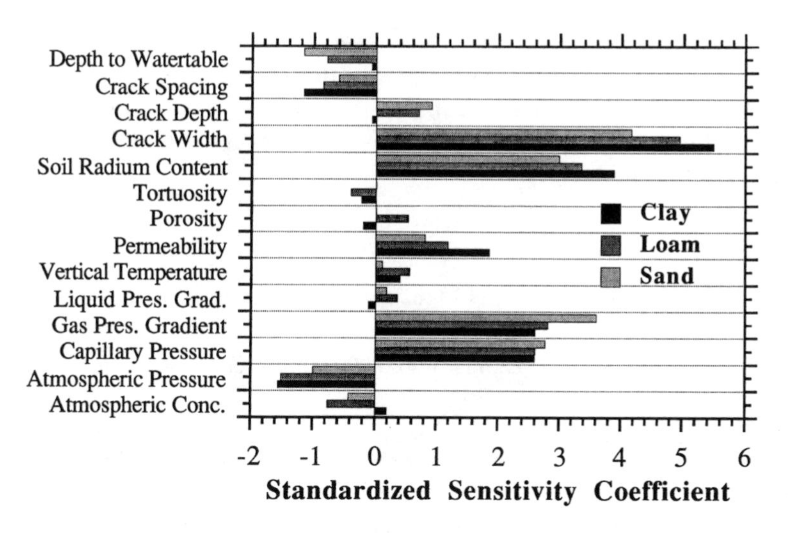

Figure 8. Standardized sensitivity coefficients for Monte Carlo simulations.

The differences in sensitivity coefficients calculated for the different soil types are not great, although some trends are apparent. For instance, the sensitivity coefficient for crack width is larger for a clay soil than for a sandy soil, because of the larger permeability contrast between cracks in a low-permeability clay than in a high-permeability sand. For the same reason, the clay soil scenario is more sensitive to variations in the soil permeability than is the sand scenario. Because of higher gas velocities, the gas pressure gradient is a more significant factor in the higher-permeability sand simulations.

CONCLUSIONS

This analysis indicates several input variables to the Rn3D model that should be carefully measured to accurately predict radon fluxes from soils. Obviously, nondestructive field measurements of crack width, spacing, and depth are not practical. But the sensitivity of radon flux to the presence of cracks in the soil suggests that using the laboratory-measured properties of a small soil sample to predict radon fluxes into a house is an adequate technique if cracks are, indeed, present. Field methods for measuring the bulk permeability and porosity of a large, in situ soil sample are necessary to accurately predict the radon gas transport properties of the soil.

Future work will include sensitivity and uncertainty analyses of more realistic scenarios with time-varying atmospheric pressure, temperature, and rainfall. In those cases, neither the pressure and temperature gradients, nor the soil saturation, will be assumed to be uniform in the soil (as in this paper).

REFERENCES

Bird, RB, WE Stewart, and EN Lightfoot. **1960**. *Transport Phenomena*. John Wiley & Sons, New York.

Hammersley, JM and DC Handscomb. **1964**. *Monte Carlo Methods*. Methuen, London, England.

Hart, KP. **1986**. *Radon Exhalation from Uranium Tailings*, Volume 1, Ph.D. Thesis. University of New South Wales, Kensington, New South Wales, Australia.

Herreman, W. **1980**. Calculation of the liquid density of radon. Cryogenics 20:133-134.

Iman, RL and MJ Shortencarier. **1984**. *A FORTRAN 77 Program and User's Guide for the Generation of Latin Hypercube and Random Samples for Use with Computer Models*, NUREG/CR-3624, SAND83-2365. Sandia National Laboratories, Albuquerque, NM.

Mualem, Y. **1976a**. *A Catalogue of the Hydraulic Properties of Soils*, Research Progress Report 442. Technion, Israel Institute of Technology, Haifa, Israel.

Mualem, Y. **1976b**. A new model for predicting the hydraulic conductivity of unsaturated porous media. Water Resour Res 12:513-522.

Nielson, KK, VC Rogers, and GW Gee. **1984**. Diffusion of radon through soils: A pore distribution model. Soil Sci Soc Am J 48:482-487.

van Genuchten, MT. **1980**. A closed-form equation for predicting the hydraulic conductivity of unsaturated soils. Soil Sci Soc Am J 44:892-898.

Weast, RC. **1982**. *CRC Handbook of Chemistry and Physics*. CRC Press, Inc., Boca Raton, FL.

QUESTIONS AND ANSWERS

Q: Reimer. The approach for measuring something like the width of a crack in the soil: I'd love to be able to do that. I'm just curious how you do it.

A: Holford. well, the point I was trying to make there (of course you can't go out in the field and measure the width of the crack), is that it's important not to just take a small soil core because you might be missing larger macropores in the soil, and that could be affecting the bulk permeability of the soil. So I think it's important to develop field methods for testing air permeability that measure a large volume of the soil.

RADON TRANSPORT FROM THE SUBSURFACE: THE ROLES OF CERTAIN BOUNDARY CONDITIONS AT SUBSURFACE/ENVIRONMENT BOUNDARIES

P. C. Owczarski, D. J. Holford, G. W. Gee, H. D. Freeman, and K. W. Burk

Pacific Northwest Laboratory, Richland, Washington

Key words: *Radon transport, soils, diffusion, advection, winds, boundary conditions*

ABSTRACT

The effects of wind on radon transport from its soil source to the environment are examined in two situations. In the first, removal of radon from the soil-air interface was found to be partially rate-limiting in conditions of nearly stagnant air (low wind). This gas-phase mass transfer limitation became especially important in the case of high-exhaling advective velocities or high-diffusion fluxes to the air. A detailed mathematical formulation for one-dimensional, steady-state radon transport from soil to air, using an air side mass transfer coefficient, was developed for this analysis. In the second situation, the Rn3D computer code was used to estimate radon concentration profiles in soils beneath a two-dimensional, slab-on-grade dwelling subjected to wind pressures. Of five generic dry, homogeneous soils studied (gravel, sand, silt, loam, and clay), only gravel showed significant changes in subslab concentrations as a result of wind pressures.

INTRODUCTION

This analysis is part of a larger study performed by the Pacific Northwest Laboratory (PNL) for the U.S. Department of Energy Office of Health and Environmental Research. The primary objectives of the larger study are to develop, evaluate, and apply a comprehensive model of radon transport within and from soils to the environment and structures.

In this paper, we examine two situations where atmospheric variables might affect the flux of radon from the subsurface to the environment. The atmospheric variables in question are surface winds and surface-induced pressure gradients, which can exist both vertically within the soil and horizontally along the soil/environment interface. The two situations can be expressed as questions: (1) Does wind affect the transport of radon across soil/air boundaries; for example, on open plains or in building crawl spaces? (2) Does wind affect the concentration profiles at soil/building interfaces, thus influencing the amount of radon available for transport into the building? These questions are answered quantitatively with a one-dimensional analysis, using analytical solutions to the soil radon-transport equation for the first answer and a two-dimensional analysis using the Rn3D code for the second. The answers to these questions should help prevent some modeling pitfalls and experimental oversights when studying radon transport from soils.

ONE-DIMENSIONAL TRANSPORT OF RADON ACROSS SOIL/AIR BOUNDARIES

Previous modeling studies at PNL (Holford et al., 1988, 1989; Owczarski et al., 1990) have examined the parameters controlling the flux of radon from soil to atmosphere. In these studies we have assumed that the soil/air boundary condition was type one, where the radon surface concentration is specified; we specified zero concentration of radon at the soil surface. Specifying the surface concentration ignores the resistance to radon transport in the gas phase above the surface. Technical details of the limitations of type-one boundary condition, where the effects of wind are ignored, are addressed here. The specific limitation of the type-one boundary condition is that the radon transported to the soil/air surface might build to a sufficiently high level to make the zero-concentration assumption invalid. Clearly, this buildup can happen only if the transport processes (e.g., turbulent eddies and molecular diffusion) on the air side cannot remove the radon as fast as it flows into the boundary region. Transport processes that remove radon are a combination of molecular diffusion and turbulent eddies in the air. To examine these processes, we constructed a simple, one-dimensional model of radon transport to the surface and subsequent removal of the radon by the wind, using a type-three boundary condition (defined below), where the radon transport resistance in the gas phase is included. We assumed steady state and ignored the dynamic nature of the wind, which, even when steady, produces rapid fluctuations in surface pressures that penetrate the soil and affect instantaneous surface fluxes. Our assumptions do not allow the wind to create pressure gradients within the soil.

The usual differential equation and boundary conditions for steady-state transport in homogeneous soils are:

$$\frac{\partial c}{\partial t} = D \frac{\partial^2 c}{\partial z^2} - \frac{V_z}{n} \frac{\partial c}{\partial z} - \lambda c + \phi = 0. \quad (1)$$

Boundary conditions:

a. $\frac{dc}{dz} \rightarrow 0$ as $z \rightarrow -\infty$

b. Type one or type three at $z = 0$,

where c = Rn concentration in soil gas, Bq m^{-3},
n = soil porosity,
λ = Rn decay constant, 2.1×10^{-6} s^{-1},
ϕ = Rn generation rate, Bq m^{-3} s^{-1},
D = Rn diffusivity in soil, m^2 s^{-1}
V_z = superficial soil gas velocity, m s^{-1},
t = time, s,
z = vertical distance into soil, m.

The type-one boundary condition is expressed as

$$c(z = 0) = c_0 = 0 \text{ (V) } c_a \text{ (ambient)}. \quad (2)$$

The type-three boundary condition as used in this paper is defined as

$$\textit{surface Rn flux} = k(c_0 - c_a),$$

where c_0 = surface Rn concentration,
c_a = ambient Rn concentration,
k = mass transfer coefficient, m s^{-1}.

The surface Rn flux is further defined as the sum of soil diffusive and advective fluxes at the surface, i.e.,

$$-nD \frac{dc}{dz} \Big|_{z=0} + V_z c_0 = k(c_0 - c_a), \quad (3)$$

where $V_z = -\frac{\kappa}{\mu} \frac{dP}{dz}$,

$\frac{dP}{dz}$ = soil pressure gradient, Pa m^{-1},

κ = soil permeability, m^2,

μ = soil gas viscosity, kg m s^{-1}.

The steady-state solutions of equation (1) for soil Rn concentration and surface flux were derived previously (Clements, 1974) for the type-one boundary condition:

$$\frac{c - c_a}{c_\infty - c_a} = 1 - \exp\left\{\left[\frac{V_z}{n} + \left(\frac{V_z^2}{n^2} + 4D\right)^{1/2}\right]\frac{z}{2d}\right\} = 1 - \exp(\alpha z). \quad (4)$$

$$Flux\ (z = 0,\ type\ 1) = n(c_\infty - c_a)\,\alpha D + V_z c_a, \quad (5)$$

where $c_\infty = \frac{\phi}{\lambda}$, and V_z is constant for all z. (Note that z is negative in the soil.) The corresponding solution for the type-three boundary condition is first presented here:

$$\frac{c - c_a}{c_a - c_a} = 1 - A\exp(\alpha z), \quad (6)$$

where

$$\alpha = \left[\frac{V_z}{n} + \left(\frac{V_z^2}{n^2} + 4\lambda D\right)^{1/2}\right] / 2D,$$

$$A = [k - V_z c_\infty / (c_\infty - c_a)] / (\alpha D n - V_z + k)$$

$$flux(z{=}0,\ type\ 3) = k(c_0 - c_a)(1 - A). \quad (7)$$

With careful algebra, equations (6) and (7) reduce to equations (4) and (5) as k approaches infinity. This reduction is expected, because a large mass-transfer coefficient corresponds to rapid removal of surface radon, and the resulting surface concentration is c_a. Thus we conclude that the type-three boundary condition is more general than the type one, where the latter ignores the resistance to radon transport in the gas phase above the surface.

The error in using the type-one equations when type three should apply can be expressed as

$$\%\ error = 100\left[\frac{flux\ (z = 0,\ type\ 1)}{flux\ (z - 0,\ type\ 3)} - 1\right] \quad (8)$$

$$= 100\left[\frac{\alpha Dn\,(c_\infty - c_a) + V_z C_a}{k(c_\infty - c_a)\,(1 - A)}\right]. \tag{9}$$

If the ambient concentration is $c_a = 0$, equation (9) reduces to

$$\%\,error = 100\,(-V_z + n\alpha D)/k. \tag{10}$$

We now proceed to evaluate equation (10) for a range of soil types, pressure gradients, and wind velocities.

First, we evaluate the mass transfer coefficient or escape velocity, k. A relationship for k can be derived from correlations for flow across a flat plate (Bird et al., 1960). Here, we use the expression

$$k = \bar{U}\frac{f}{2}(Sc)^{2/3}, \tag{11}$$

where the Schmidt number, Sc, = (ρDa/μ). Da is radon diffusivity in bulk air, ρ and μ are air density and viscosity, $\bar{U}$ is the mainstream wind velocity, and f is the friction factor. A rough surface friction factor of 0.01 is used here. (This f is the turbulent flow limit in large pipes.) For zero wind speed, k is diffusion-limited over a stagnant gas boundary layer, which we assume is 0.1 m thick, so k becomes Da/0.1 = 1×10^{-4} m s^{-1}. Table 1 shows values of k used in this paper for different wind speeds.

Table 1. Typical k values used in evaluating equation 10.

$\bar{U}$,[a] m s^{-1}	$\bar{U}$,[b] mph	k,[c] m s^{-1}
0	0	1×10^{-4}
1	2.24	4.38×10^{-3}
10	22.4	4.38×10^{-2}
25	55.9	0.11

[a] Local wind velocity.
[b] Wind velocity away from building or soil.
[c] Mass transfer coefficient (escape velocity).

Table 2 lists the pertinent soil properties used in evaluating equation (10). Five types of dry soils were used to limit the scope of this study.

Table 2. Dry soil properties used in evaluating equation 10 (Owczarski et al., 1990).

Soil Type	Porosity	Permeability, m^2	Diffusivity, $m^2\ s^{-1}$
Clay	0.5131	1.0×10^{-15}	1.98×10^{-8}
Silt	0.4026	1.5×10^{-14}	2.57×10^{-6}
Loam	0.4362	2.0×10^{-13}	6.89×10^{-6}
Sand	0.27	3.4×10^{-12}	7.10×10^{-6}
Gravel	0.519	1.9×10^{-9}	7.72×10^{-6}

Variations of the controlling parameters in equation (10) are represented in Figures 1 and 2 and Table 3. In Figure 1 we vary the vertical soil pressure gradient and wind speed for dry sand. The other soils show similar patterns, with the lower-permeability silt, clay, and loam showing lower percent errors, and higher-permeability gravel showing higher percent errors than sand. Surface flux values are always positive here (an overprediction that results from using the type-one boundary condition), and positive pressure gradients and low wind speeds are conducive to larger errors than negative-pressure gradients and high wind speeds. The overall result is that higher surface fluxes need higher escape velocities to maintain a low interface concentration as specified by a $c_a = 0$ boundary condition.

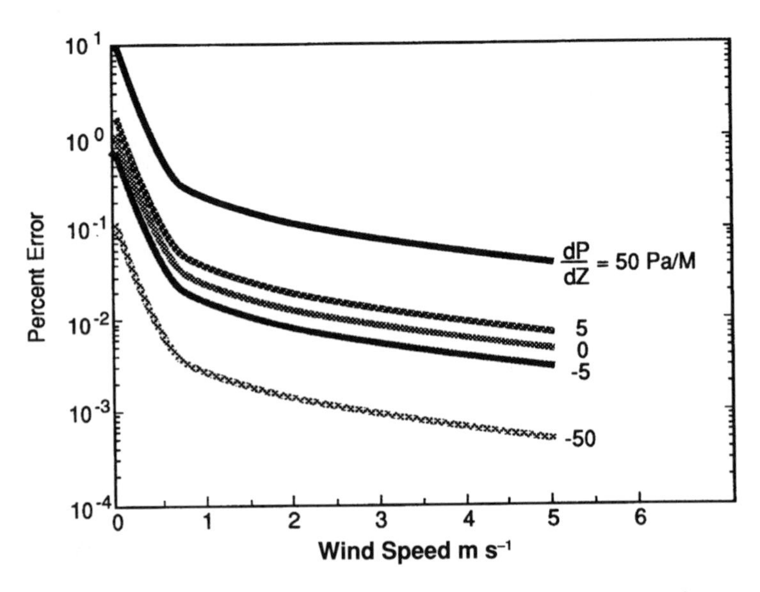

Figure 1. Percent error in radon surface flux for homogeneous dry sand, using type-one boundary condition. dP/dZ is soil gas pressure gradient.

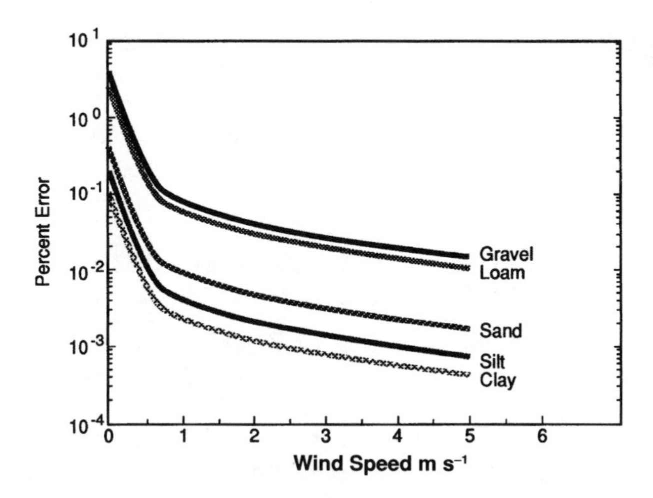

Figure 2. Percent error in radon surface flux for five dry soil types, using type-one boundary condition. No soil pressure gradients.

Table 3. Percent error in radon surface flux at zero wind.

	$\frac{dP}{dz}$ (Pa m^{-1})						
Soil type	50	5.0	0.5	0	–0.5	–5.0	–50
Gravel	5300	530	53	2.1	0.083	0.0083	0.00077
Sand	9.6	1.6	1.1	1.0	1.0	0.67	0.11
Loam	2.0	1.7	1.7	1.7	1.7	1.6	1.4
Silt	0.96	0.94	0.93	0.93	0.93	0.93	0.91
Clay	0.11	0.10	0.10	0.10	0.10	0.10	0.10

In Figure 2 we have eliminated the soil pressure gradient (effect of permeability) and have plotted percent error versus wind speed for all five soils. Again, the wind speed has an effect similar to that in Figure 1. However, Figure 2 shows the effect of the product $n\alpha D$, porosity times diffusivity. The higher-$n\alpha D$ soils have higher percent errors at all wind speeds. Again, the higher-surface-flux (higher-$n\alpha D$) cases need a higher escape velocity to maintain a $c_a = 0$ boundary condition. (Note: $n\alpha D$ for no pressure gradient is $n\sqrt{(\lambda D)}$.)

In Table 3 we show the effect of a stagnant condition (no wind) on the percent error; this is the condition of maximum error for all soils. The high-permeability and high-$n\alpha D$ soils give the highest errors. This table indicates that for stagnant conditions, such as in a poorly ventilated house crawl space, the type-one boundary condition with $c_a = 0$ would be a source of significant error in most soils. One method of reducing the error in both type-one and type-three boundary conditions would be to use the actual $c_a > 0$ if it is known, since an unventilated system would build up the radon concentration.

EFFECTS OF WIND ON RADON CONCENTRATION PROFILES BELOW A TWO-DIMENSIONAL, SLAB-ON-GRADE DWELLING

To examine the effects of wind on subsurface radon concentrations around structures, we used the Rn3D code in two dimensions. This code is described in a previous paper (Holford et al., 1989) and in a companion paper in this volume (Holford et al., this volume). To simplify the approach even further than the two-dimensional limitation, we chose for the dwelling a simple, slab-on-grade rectangle, 3 m high and 15 m wide (Figure 3), laid on a homogeneous dry soil (the same dry soil as in the previous section). The radium activity was 110 Bq kg^{-1} with a radon emanation coefficient of 0.1. The following bulk dry soil densities (ρ_s, kg m^{-3}) were assumed: gravel, 1275; sand, 1934; loam, 1494; silt, 1583; and clay, 1290. Here, $\phi = 110\ \rho_s\ (0.1)$, as in equation (1).

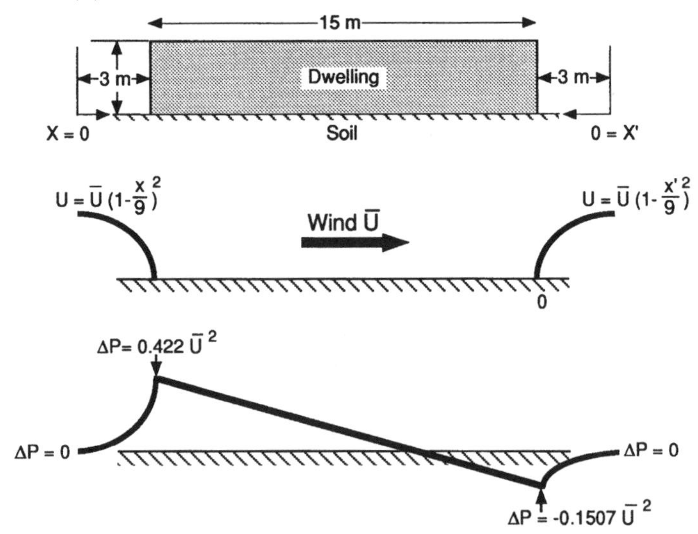

Figure 3. Schematic diagram of dwelling wind and pressure profiles. U = local wind velocity, m s^{-1}; $\bar{U}$ = wind velocity away from building or soil; ΔP = local pressure difference above atmosphere, Pa; x,x' = horizontal coordinates, m.

As noted in Figure 3, the wind was assigned to be perpendicular to the infinite third dimension of the dwelling. A simple representation of the wind horizontal velocity was assumed so that we could calculate a surface pressure profile near the dwelling. The upstream and downstream surface pressures at the two building-soil-air points were based on actual measurements of wind pressure coefficients (Simiu and Scanlon, 1978). A linear relationship for the horizontal pressure gradient between those points along the dwelling-soil interface was assumed. Pressures noted are in pascals for wind velocities in m s^{-1}. We used a type-one boundary condition with $c_a = 0$ at the air-soil interface and type two ($\frac{dc}{dt}\Big|_{z=0} = 0$) boundary condition under the slab.

Figure 4 is a plot of the radon concentrations at steady-state conditions (with $c_a = 0$ at the soil-air interface) for four wind speeds at the dwelling/dry-sand interface. At zero wind, the plot is symmetric, as one would expect. As the wind speed increases above zero, the profile becomes unsymmetric, as shown. At the highest wind speed (25 m s^{-1}), the upstream half of the building-soil interface concentration is significantly depressed; only at the downstream edge is it slightly elevated over the zero wind profile. The nonsmooth shape of the profiles is due to a coarse grid space. If we average the concentration profile along the slab at 25 m s^{-1}, we find that this average is about half the zero-wind average subslab concentration.

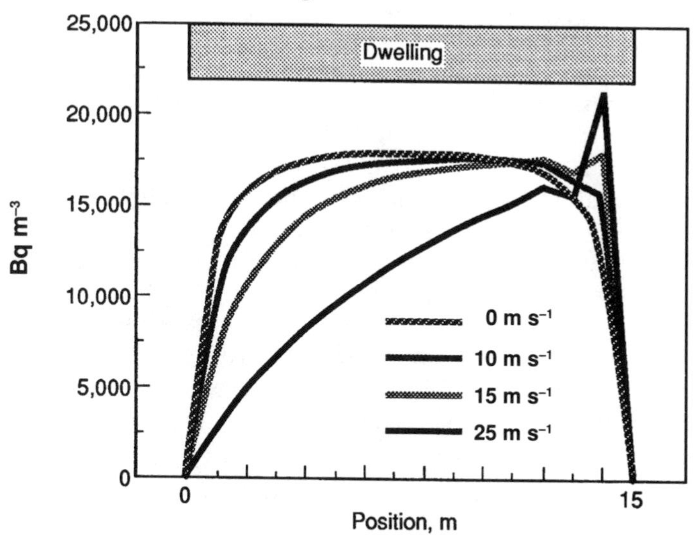

Figure 4. Steady-state radon concentration at dry-sand/dwelling interface.

Figure 5 is a plot of this subslab average concentration versus wind speed for each of the dry soil types. Only in the case of the highly permeable gravel is the reduction of the subslab concentration significant. These results immediately raise the question: Can buildings be laid over gravel to take advantage of its high permeability for passively reducing radon subslab concentrations using natural winds?

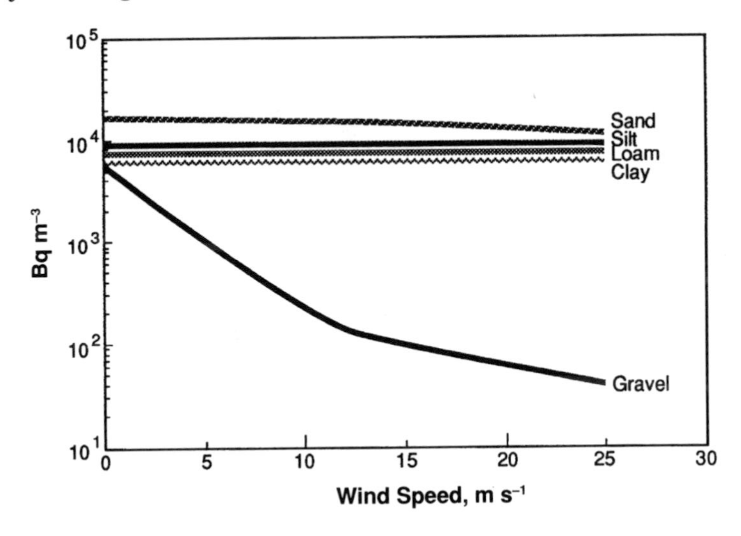

Figure 5. Average dry-soil/dwelling interface concentrations.

To partially answer this question we laid the study dwelling on a gravel bed on top of 47.5% saturated clay (emanation coefficient, 0.4; diffusivity, 8.6×10^{-9} m^2 s^{-1}; permeability, 2.8×10^{-16} m^2). The average subslab concentration of radon was then calculated at three wind speeds for gravel-bed thicknesses up to 1 m. The calculated data are plotted in Figure 6. The surprising result was that for all three wind speeds, the beneficial effects were realized at a bed depth of only 0.1 m (4 in.), with little benefit gained by increasing the bed depth. The gravel (dry) had the same radium activity as in Figure 5. For even a modest windspeed of 2.5 m s^{-1}, the reduction in subslab concentration was more than threefold for the 0.1-m gravel bed.

The next questions are: What will an annual average wind spectrum do to the subslab concentration? Can a large leak path into the dwelling through the slab seriously disturb the subslab concentration? The answers to these questions are forthcoming in future papers.

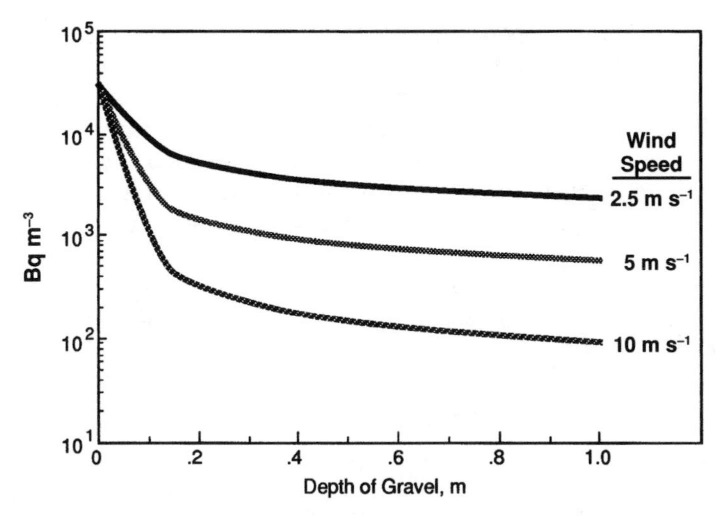

Figure 6. Average concentration under dwelling versus gravel depth and wind speed. Underlying soil consisted of 47.5% saturated clay under gravel.

We also addressed the following question since we ran Rn3D in a steady-state mode: How long does it take for a steady-state profile to be established after a sudden change in wind velocity? For example, do wind-induced profiles have any meaning in the steady state, since the wind is highly variable? To answer these questions, we now look at an advective diffusivity for each soil type and apply it to the study slab. This advective diffusivity (D_{Adv}) derives from the unsteady-state form of the flow equation and Darcy's Law:

$$D_{Adv} = P_0 K/2n\mu, \tag{12}$$

where P_o is the absolute pressure. A characteristic time, τ, also arises in the flow equation if we assign some characteristic length, ℓ, to the flow path:

$$\tau = \ell_2/D_{Adv}. \tag{13}$$

We assigned ℓ = 7.5 m, or half the dwelling width. For the soils in Table 2, we constructed Table 4, which shows that for gravel, τ = 0.0152 h, or 1 min. Therefore, one would expect the steady-state concentration profiles under the study dwelling laid over gravel to be established in a few minutes for each minute-scale variation in wind speed. This rapid response gives confidence that a steady-state analysis will suffice in analyzing wind effects on houses laid over gravel beds. With such a rapid response, any hour-to-hour changes in wind direction will result in a rapid adjustment in the gravel-bed air flow under a dwelling. Also, even with wind velocity and direction fluctuations, any wind will cause dilution to occur under the slab. Therefore, average subslab concentrations will be reduced.

Table 4. Values of soil advective diffusivities and characteristic time for the dwelling shown in Figure 3.

Soil type	D_{Adv}, $m^2\ s^{-1}$	τ, h
Gravel	1.030×10^{1}	0.0152
Sand	3.544×10^{-3}	4.41
Loam	1.291×10^{-4}	121
Silt	1.049×10^{-5}	1,490
Clay	5.485×10^{-7}	28,500

CONCLUSIONS

This paper has examined the effects of atmospheric variables, particularly wind, on radon fluxes from the soil subsurface. For two distinctly different influences of the wind, one in sweeping radon away from soil-air surfaces and the other in causing pressure gradients around dwellings, we have arrived at the following conclusions:

- Wind can affect radon fluxes from soils.
- Representation of the soil/air boundary with a type-one boundary condition is frequently adequate.
- The type-one boundary condition can be inadequate for stagnant conditions and for high-permeability soils that experience strong pressure gradients.
- The type-three boundary condition can provide a suitable alternative to the type-one boundary condition when the latter is unsuitable.
- The suitability of the type-one condition can be estimated with the dimensionless number $(-V_z + n\alpha D)/k$ if $c_a = 0$.
- Estimating the type-three mass transfer coefficient is not always straightforward.

- Wind pressures on dwellings can alter the radon concentration profiles beneath the dwelling, especially if highly permeable soils surround it.

REFERENCES

Bird, RB, WE Stewart, and EN Lightfoot. **1960**. *Transport Phenomena*. John Wiley & Sons, New York.

Clements, WE. **1974**. *The Effect of Atmospheric Pressure Variation on the Transport of ^{222}Radon from Soil to the Atmosphere*, Ph.D. Dissertation. New Mexico Institute of Mining and Technology, Socorro, NM.

Holford, DJ, GW Gee, PC Owczarski, and HD Freeman. **1988**. A finite element model of radon advection and diffusion in unsaturated cracked soils. EOS-Trans Am Geophys Union 69:1216 (abstract).

Holford, DJ, SD Schery, JL Wilson, and FM Phillips. **1989**. *Radon Transport in Dry, Cracked Soil: Two-Dimensional Finite Element Model*, PNL-7116. Pacific Northwest Laboratory, Richland, WA.

Owczarski, PC, DJ Holford, HD Freeman, and GW Gee. **1990**. Effects of changing water content and atmospheric pressure on radon flux from surfaces of five soil types. Geophys Res Lett 17:817-820.

Simiu, E and RH Scanlon. **1978**. *Wind Effects on Structures*. John Wiley & Sons, New York.

QUESTIONS AND ANSWERS

Q: Reimer. . . . certainly the silt and the effect of the clay cracking become an important modification to the soil. Can you incorporate these types of cracking into your three-dimensional model?

A: Owczarski. Well, I think if we had a way of getting some kind of bulk coefficient like Diana was talking about—that bulk permeability—if we knew how to measure that, that would be beautiful. That would also crank right in, I think, directly. In fact, I think that's the challenge that we're going to leave with the audience: How are we going to do this for real soil with cracks, rocks, etc. (the whole ballgame).

Q: Mosley. I came in a little late, and I may have missed something, but the dilution that you were measuring in the aggregate or subslab region. Did you have any indoor measurements to correlate with? Was that dilution occurring because it was going in the house, or was it occurring because it was being flushed out of the soil?

A: Owczarski. There was no transport in the house. In fact, in the model we assume there was zero flux across the subslab. In fact, even if you had major cracks it would not greatly affect the soil concentration under the slab. Usually what comes into the house only takes out a small fraction of the total amount of radon that's available.

Q: Mosely. Depends on how much you take out and how fast, right? Depends on how much you take *in* and how quickly as to whether you dilute the source or not.

A: Owczarski. Yes, but you'd have to have a pretty good hole in the floor to take out a substantial quantity.

C: Mosley. Only a few cubic feet per minute would do it.

A: Owczarski. That's a pretty good leak.

Q: Morley, Ministry of Health, British Columbia. As a mitigation or new construction strategy, would you suggest placing a layer of gravel totally underneath the house and underneath its foundation? What sort of contact does it need with the outside air beyond that?

A: Owczarski. Well, it needs to be totally in communication with the outside air at the edge of the building, and I'm suggesting that this has possibilities. We need to look at what the undersoil has to be in terms of this whole thing, too. Most of the radon is going to be supplied by the soil that's underneath the house. We need to look at its characteristics as well.

Q: Riddle, Cavalier Engineering, Spokane. Regarding the siting of various structures. From a practical standpoint, maybe you have another large building upwind, or maybe you have a tree or some other object that will create different flows. In your opinion, wouldn't we see much higher levels in extremely permeable soils or with reference to your gravel layer that has full continuity from one end of the house to the other?

A: Owczarski. Well, of course, obstructions that can interrupt this windflow and cause irregular patters around a house are certainly going to change things quite a bit. There are a lot of complications that can arise from this whole thing.

CHARACTERIZATION OF RADON ENTRY RATES AND INDOOR CONCENTRATIONS IN UNDERGROUND STRUCTURES[a]

T. B. Borak, F. W. Whicker, L. Fraley, M. S. Gadd, S. A. Ibrahim, F. A. Monette, R. Morris, and D. C. Ward

Department of Radiology and Radiation Biology, Colorado State University, Ft. Collins, Colorado

Key words: *Radon entry, underground soil structures*

ABSTRACT

An experimental facility has been designed to comprehensively determine the influence of soil and meteorological conditions on the transport of radon into underground structures. Two identical basements are equipped to continuously monitor pressure differentials, temperatures, soil moisture, precipitation, barometric pressure, wind speed, wind direction, natural ventilation rates, and radon concentrations. A computerized data acquisition system accumulates and processes data at the rate of 6000 points per day. The experimental design is based on performing experiments in one structure, with the other used as a control. Indoor radon concentrations have temporal variations ranging from 150 to 1400 Bq m^{-3}. The corresponding entry rate of radon ranges from 300 to 10,000 Bq h^{-1}. When the radon entry rate is high, the indoor radon concentration decreases, whereas elevated radon concentrations seem to be associated with slow but persistent radon entry rates. This inverse relationship is partially due to compensation from enhanced natural ventilation during periods when the radon entry rate is high. Correlations between measured variables in the soil and indoor-outdoor atmospheres are used to interpret these data. This laboratory has the capability to generate essential data required for developing and testing radon transport models.

INTRODUCTION

High concentrations of radon gas indoors have been implicated as a serious health hazard. These elevated levels of radon are the result of a complicated set of circumstances. In general, the most prominent source of radon is the soil surrounding the structure. This radon can be transported by several mechanisms toward the interface of the soil

(a) This work is funded through DOE Research Grant DE-FG02-87ER60581.

and a structure. Cracks, joints and other porous pathways in the structural barrier at this interface permit the entry of radon. The resulting concentration inside the structure depends on the radon entry rate and the residency time of indoor air.

There has been a considerable amount of effort in characterizing indoor radon concentrations, Perritt et al., 1990; Dudley et al., 1990; Borak et al., 1989. Also, many groups created models to predict indoor radon concentrations (Nazaroff, 1988; Revzan and Fisk, 1990; Arnold, 1990). Others have made measurements in occupied dwellings to develop predictive models or to estimate the apportionment of radon source terms (Holub et al., 1985; Harley, 1986; Turk et al. 1990; Hubbard et al., 1988; Nazaroff et al., 1985; Stoop et al., 1990).

Our study was initiated to verify theories for radon migration into underground structures. We have a scientific field facility with the equipment and personnel required to answer basic questions about radon transport and entry. It was designed to provide a compromise between a simple hole in the ground and a fully occupied family dwelling. Tests with appropriate controls can be designed and executed with a high expectation of obtaining results.

METHODOLOGY

Two virtually identical subterranean structures (chambers) were constructed; one serves as a test chamber, the other serves as a control. The concept is unique because these chambers use conventional construction techniques to simulate miniature basements but eliminate confounding factors that would be introduced by occupants' activities.

The facility is located on the campus of Colorado State University, in a flat, open area with no tall structures that would directly interfere with meteorological conditions. The soil type, which was carefully analyzed, is characteristic of this region (Table 1).

Figure 1 is a schematic diagram of the facility, showing the relative location of the two chambers, which are separated by about 25 m. A meteorology station is located between the chambers, and a data transmission line transfers information from one chamber to the other.

Figure 2 shows an elevation of one structure. Soil was excavated for 2 m on all sides of both chambers; it was then mixed in one large pile before being backfilled alternately around each chamber. This homogenizing of the soil was done in order to eliminate a single source or hot spot that could affect the radon concentration in either structure. During the backfilling operation a soil-testing firm monitored the moisture and bulk density of the compacted soil. The poured-concrete footings, floor, and walls are typical of construction in this region.

Table 1. Properties of soil at the Colorado State University radon project.

Composition %				
	Sand	Silt	Clay	Texture
Topsoil	46	26	28	Sandy Clay, loam
Subsoil	49	27	24	Sandy Clay, loam

Property	Value
Grain Density	2660 ± 60 kg m^{-3}
Dry Bulk Density	1622 ± 60 kg m^{-3}
Porosity	0.39 ± 0.3
Radium Content	52 ± 4 Bq kg^{-1}
Emanation Fraction (dry)	0.10 ± 0.02

Nutrients (ppm)			
	NO_{3-N}	P	K
Topsoil	88	139	1423
Subsoil	78	25	279

Depth (m)	Permeability (m^2) Backfilled soil	Undisturbed soil
1.0	5×10^{-14}	5×10^{-13}
1.8	2×10^{-13}	1×10^{-12}

Radon concentrations in soil gas were measured using grab samples in 100-ml scintillation flasks. Radon gas concentrations in the basements were continuously measured using a flow-through scintillation flask. The system has a volume of approximately 1 L and a sensitivity of 0.1 counts per minute (CPM) per Bq m^{-3} (4 CPM per pCi L^{-1}).

The natural ventilation rate was determined by using a constant regulated source of the trace gas, SF_6. A continuous-sampling infrared spectrophotometer monitored the concentration of this gas as a function of time. This system permits computation of ventilation rates without requiring steady-state conditions.

A data acquisition system based on an IBM AT microcomputer has been designed, tested, and installed. The system uses a 32-channel multiplexer in each chamber that can be connected to any combination of sensors. The system is interrogated every 2 s and averaged over 15-min intervals. The averaged data are stored on hard disk. The system can be accessed remotely so that the contents of the disk can be downloaded to another computer on campus.

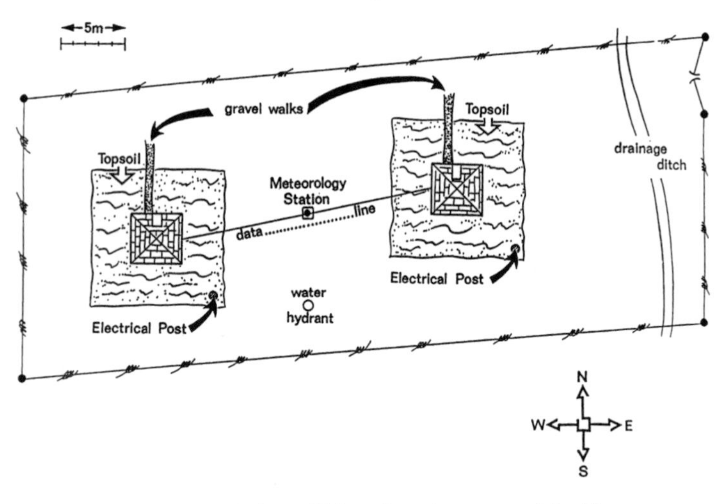

Figure 1. Diagram of Colorado State University radon research facility.

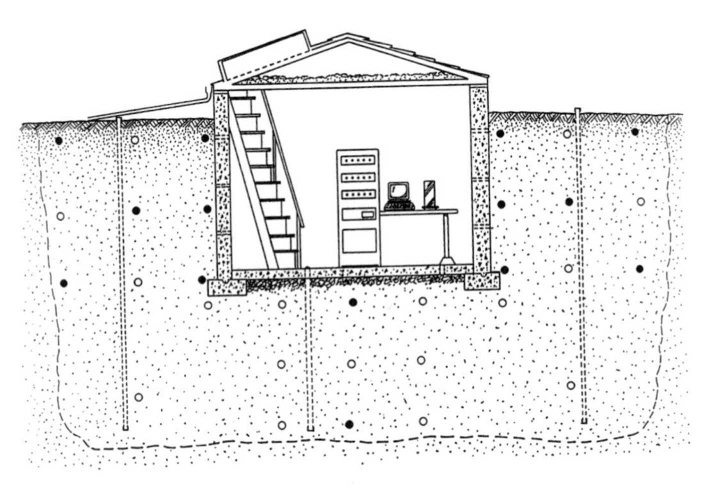

Figure 2. Diagram showing side elevation of Colorado State University radon research facility. Solid circles are sensors to measure temperature and pressure differentials between inside and outside of the chamber. Open circles represent pressure differential ports only. A total of 30 solid circles and 161 open circles surround each chamber. Dashed lines show locations of 3 of 5 access tubes that accommodate a neutron moisture probe.

RESULTS

Soil Data

Radon concentrations in soil gas were measured on a weekly schedule from November 1989 through August 1990. Samples were extracted through the tubes used for measuring pressure differentials at depths of 30, 100, 180 cm. Figure 3 shows the average radon gas concentration near the structures at each depth. The sample uncertainty based on counting statistics is approximately equal to the dimensions of the circular symbols.

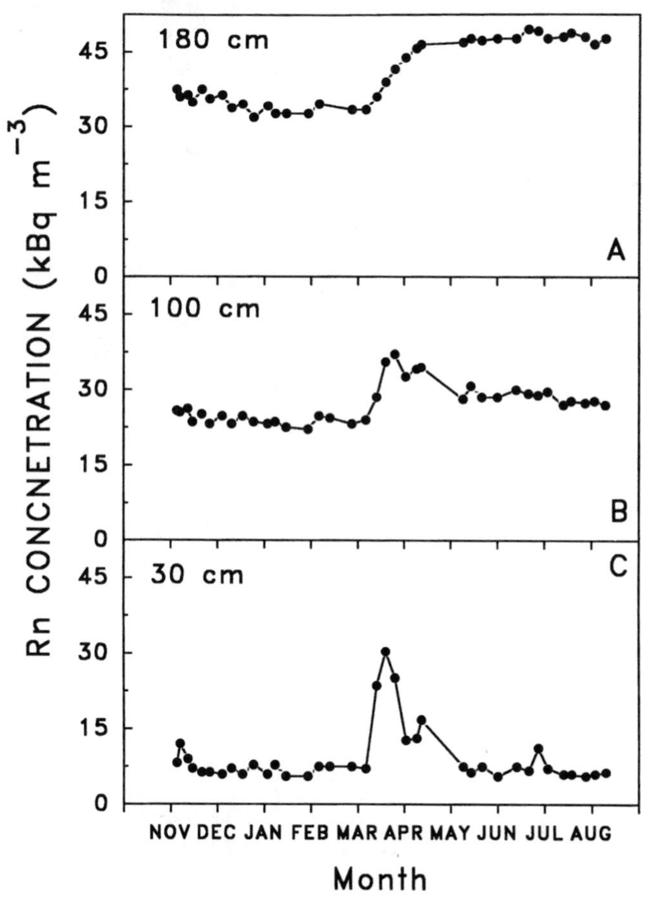

Figure 3. Soil gas radon concentrations near the facility at depths of 30 cm, 100 cm, and 180 cm, respectively, versus time.

From November to March, the concentrations increased with depth and were, for the most part, constant over time. The situation is similar from May through August, although there was a pronounced increase in the soil radon concentration in early March. This is followed by a decrease to pre-March levels at 30 cm. However, at 100 and 180 cm, the radon concentrations remained elevated.

In an attempt to explain these data, the average soil moisture content (measured with a neutron gauge) and a history of precipitation events are shown in Figure 4. The period from November to March was relatively dry; on March 6 a heavy rain deposited 3 cm of water, followed by a snowstorm that contained over 5 cm of moisture. Precipitation events during the summer did not seem to have much effect on the moisture profile because they were typically of short duration with considerable runoff.

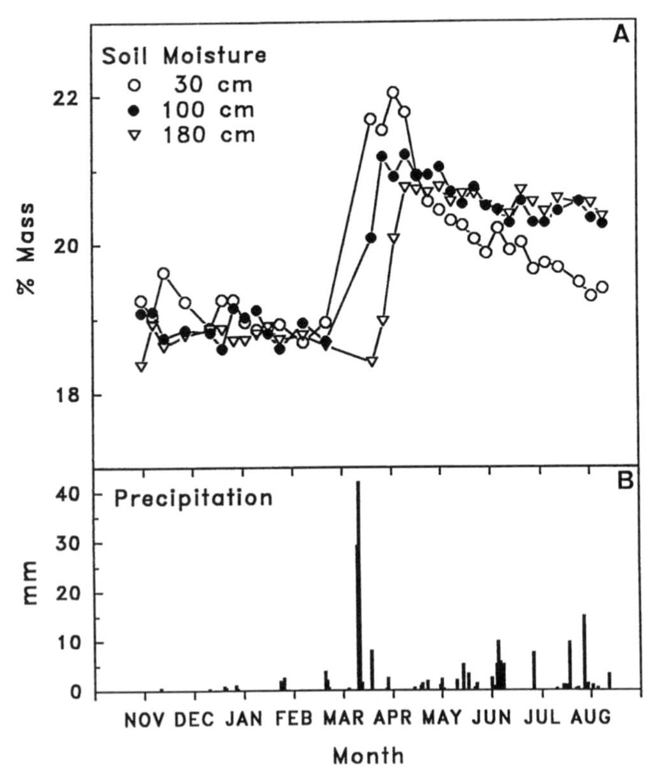

Figure 4. Soil moisture and precipitation at the radon research facility from November, 1989, through August, 1990.

Other soil parameters and meteorological data were analyzed to determine their influence on soil radon concentrations. Figure 5 includes a plot of soil temperature at 100 cm along with moisture and radon concentrations. Although the sample frequency was not sufficiently high to permit extensive quantitative analysis, a stepwise regression procedure identified linear correlations between these variables. At every depth, moisture was the dominating factor accounting for 60%, 80%, and 90% of the variance in radon concentrations at 30, 100 and 180 cm deep, respectively. Barometric pressure was not included in this analysis because results of sampling soil once a week could not be compared to a single reading of barometric pressure that can change from hour to hour.

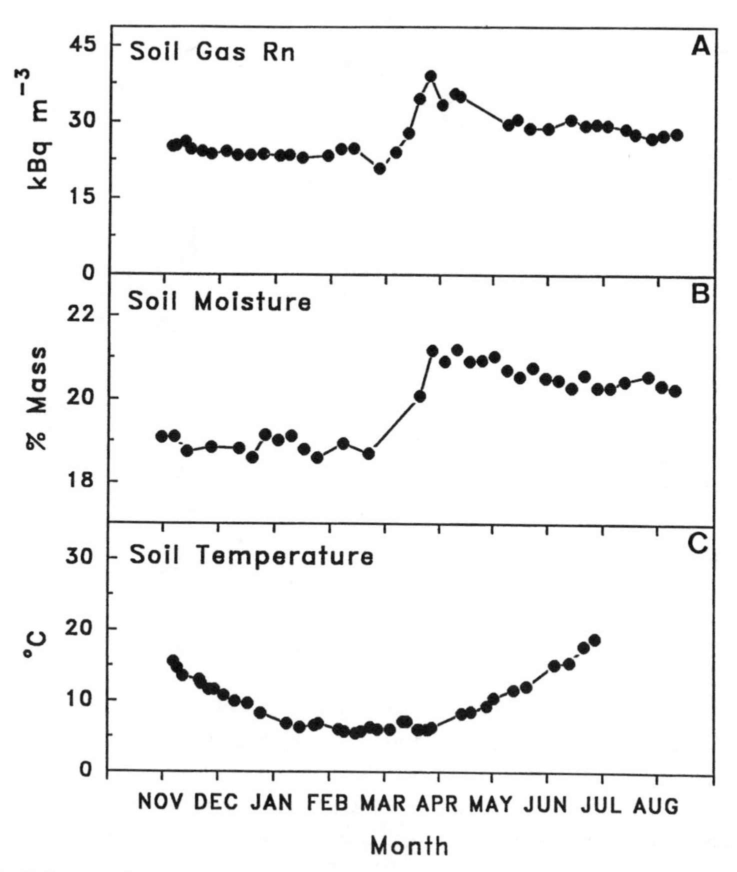

Figure 5. Soil-gas radon concentrations (A), soil moisture (B), and soil temperature (C) at the radon research facility, at 100 cm depth, from November, 1989, through August, 1990.

Indoor Data

Figure 6 is a plot of monthly arithmetic average radon concentrations in one chamber from January, 1989, to August, 1990. The monthly average radon concentrations ranged from 700 to 1200 Bq m^{-3} (20 to 30 pCi L^{-1}) (Figure 6) showing no dramatic temporal variations.

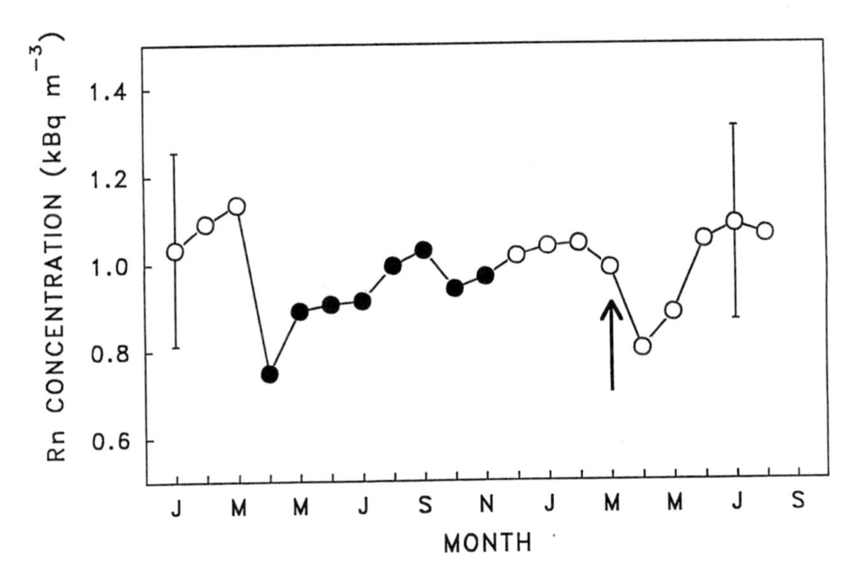

Figure 6. Monthly average indoor radon concentrations. Error bars represent one standard deviation of 15-min measurements used to compute the average. Arrow indicates time of major precipitation event. Sample size was always more than 2000, and uncertainty in estimate of the mean was within the dimensions of the circular symbol. Closed circles represent periods of time when all access tubes and ports were closed; open circles represent times when all 190 ports were open, creating a communication area of 24 cm^2 directly to soil.

The daily averaged radon concentration for the period just preceding and following the March precipitation event (Figure 7A) shows temporal variations, but no peculiarity on March 6 to 7. Figures 7B and 7C show plots of indoor-outdoor temperature differences and barometric pressure. A stepwise linear regression using meteorological data indicates that for this time period outdoor temperature was the most important factor in describing the variability of indoor radon concentrations. Similar results were obtained for other time periods, but the influence of temperature was not always as pronounced.

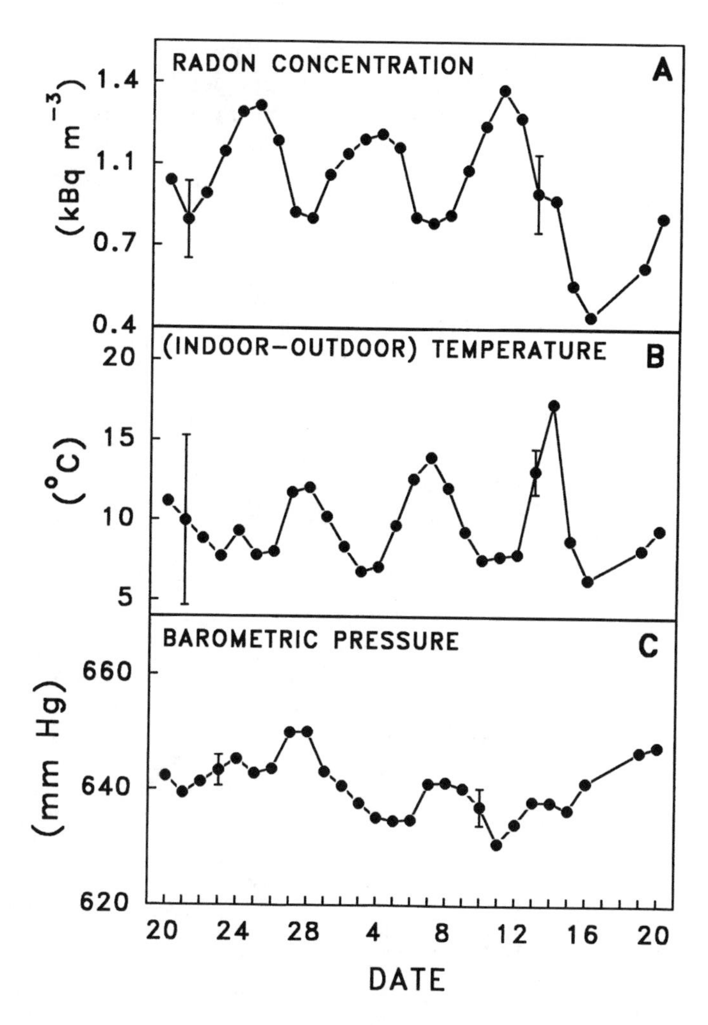

Figure 7. Daily average indoor radon concentration (A), indoor-outdoor temperature difference (B), and barometric pressure (C), for period surrounding precipitation event. Error bars represent one standard deviation of 15-min measurements.

Indoor radon concentrations depend on radon entry rate and mean residency time, which is controlled by the natural ventilation rate. The ventilation rate was determined continuously, using the trace gas method described earlier. Figure 8A is a plot of daily averaged ventilation rate, h^{-1}, as a function of time from February 27 to March 21, 1990. If the radon concentration and ventilation rate are known, it is possible to estimate the radon entry rate (Figure 8B). Figure 8C shows the daily averaged wind speed during this period. It appears that when wind speed is high, the radon entry rate increases. For high radon entry rates, the radon concentration (Figure 8D) actually decreased because of the corresponding increase in the ventilation rate during this period.

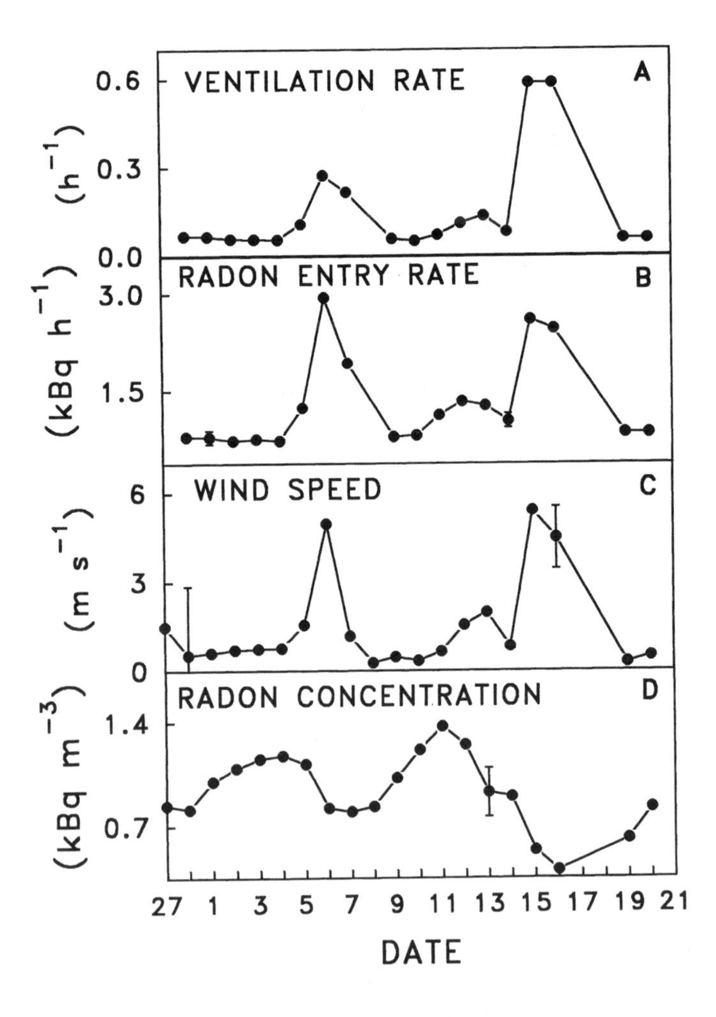

Figure 8. Daily average ventilation rate (A), radon entry rate (B), wind speed (C) and indoor radon concentration (D) for the period surrounding the precipitation event. Error bars represent one standard deviation of 15-min measurements.

To explore the dynamics of this situation, the data must be analyzed in detail greater than that permitted by monthly or daily averages. There is a clear dependence of both soil and outdoor pressure differences on wind speed (Figures 9A, 9B, and 9C). Figure 10A shows a quadratic relationship that indicates that pressure differences between the soil and indoor air are a function of wind speed squared. It can be seen (Figure 10B) that the pressure differences are resolved into two curves. The lower curve indicates a larger pressure difference when the wind is blowing from the south than when the wind is blowing with the same speed from the north (upper curve). Attempts to include other meteorological parameters as predictors of pressure differences were unsuccessful.

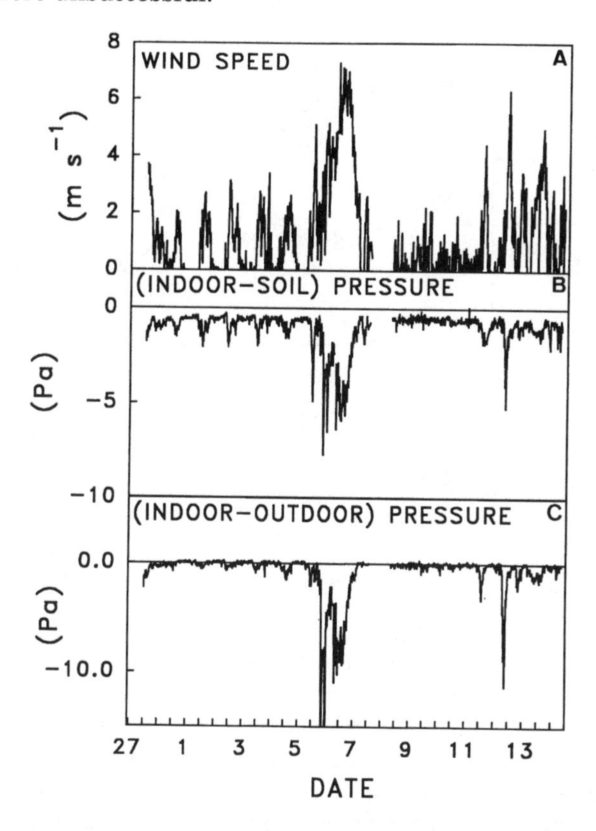

Figure 9. Fifteen-minute averaged wind speed versus time (A), indoor minus soil pressure differential (B), indoor minus outdoor pressure differential versus time (C), for period surrounding precipitation event. Pressure differences versus time between inside of facility and a point located in soil at 1 m depth and 0.3 m from the south wall.

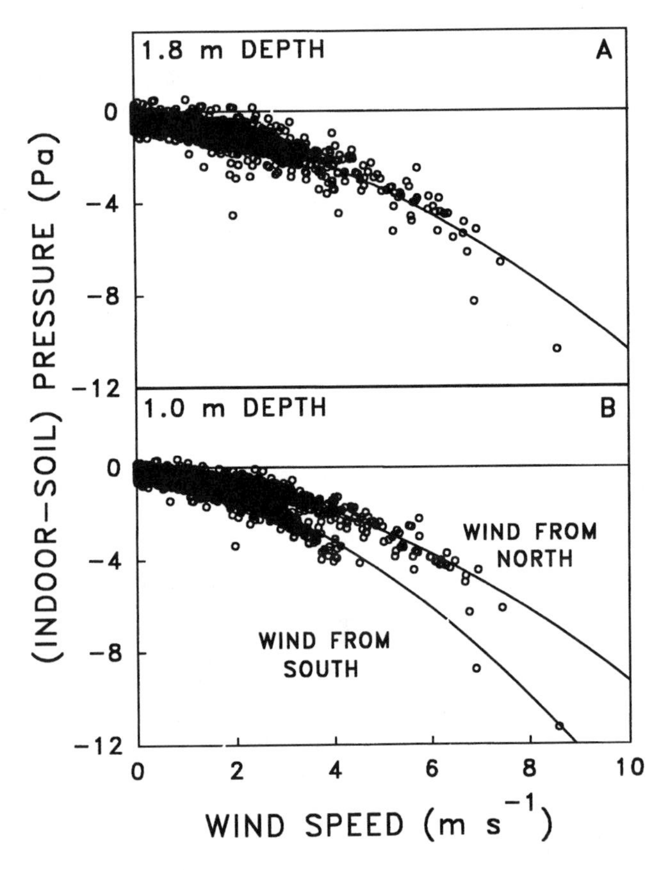

Figure 10. Indoor minus soil pressure differential versus wind speed for pressure ports located 0.3 m from the south wall and at 1.8 m (A) and 1.0 m (B) depth.

Figure 11 is a plot of indoor radon concentrations vs pressure differences. There is no obvious correlation between these two variables. However, in Figure 8 we observed that the radon entry rate increased with wind speed, and from Figure 9 we concluded that wind speed increased the pressure difference between the soil and indoors. Thus, the high values of radon entry are associated with wind-generated pressure differentials. There is also, however, a persistent but slower radon entry rate even when there is little or no wind. It is interesting that the highest indoor radon concentrations are observed when the radon entry rate is low.

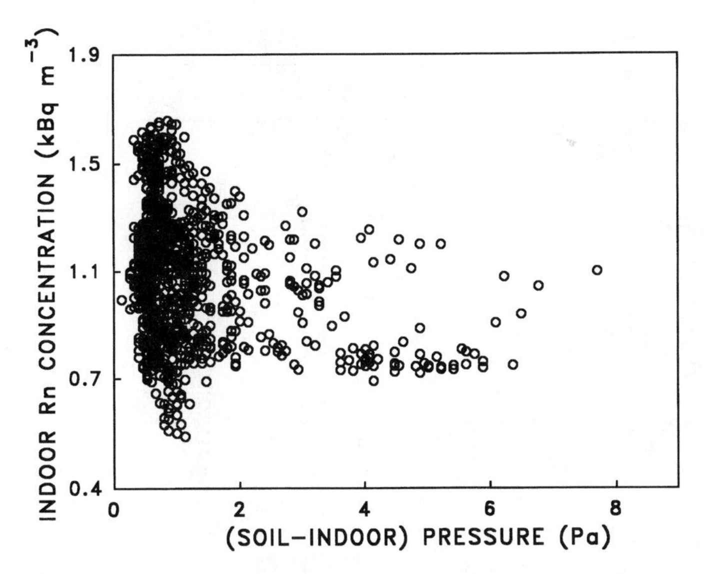

Figure 11. Scatter-plot of indoor radon concentration versus soil minus indoor pressure differential.

To understand the driving force responsible for the slow radon entry, the chamber was modified to include a variable-speed fan in the door. The fan speed was selected to intentionally pressurize the house. Figure 12A shows the pressure differentials between outdoors and two soil locations at different depths beneath the surface under normal conditions. Figure 12B shows the same pressure differentials when the fan was operational. The fan clearly pressurized the structure except for periods of high wind speeds where there were negative pressure excursions.

The radon entry rate was determined for these conditions. In order to examine the constant term, radon entry rate was computed and averaged only when the wind speed was less than 1 m s^{-1} (Table 2).

When the fan was off, the natural ventilation rate was less than 0.1 h^{-1} for periods of little or no wind. The indoor-outdoor pressure differentials were consistent with zero, and the structure was negative with respect to the soil at a depth of 1 m. The combined radon entry rate for these conditions was 0.9 ± 0.1 kBq h^{-1}. When the fan was operational, the ventilation rate increased to 0.3 h^{-1}. The structure was positive with respect to both outdoors and soil. The radon entry rate corresponded to 0.8 ± 0.1 kBq h^{-1}. These two numbers are indistinguishable.

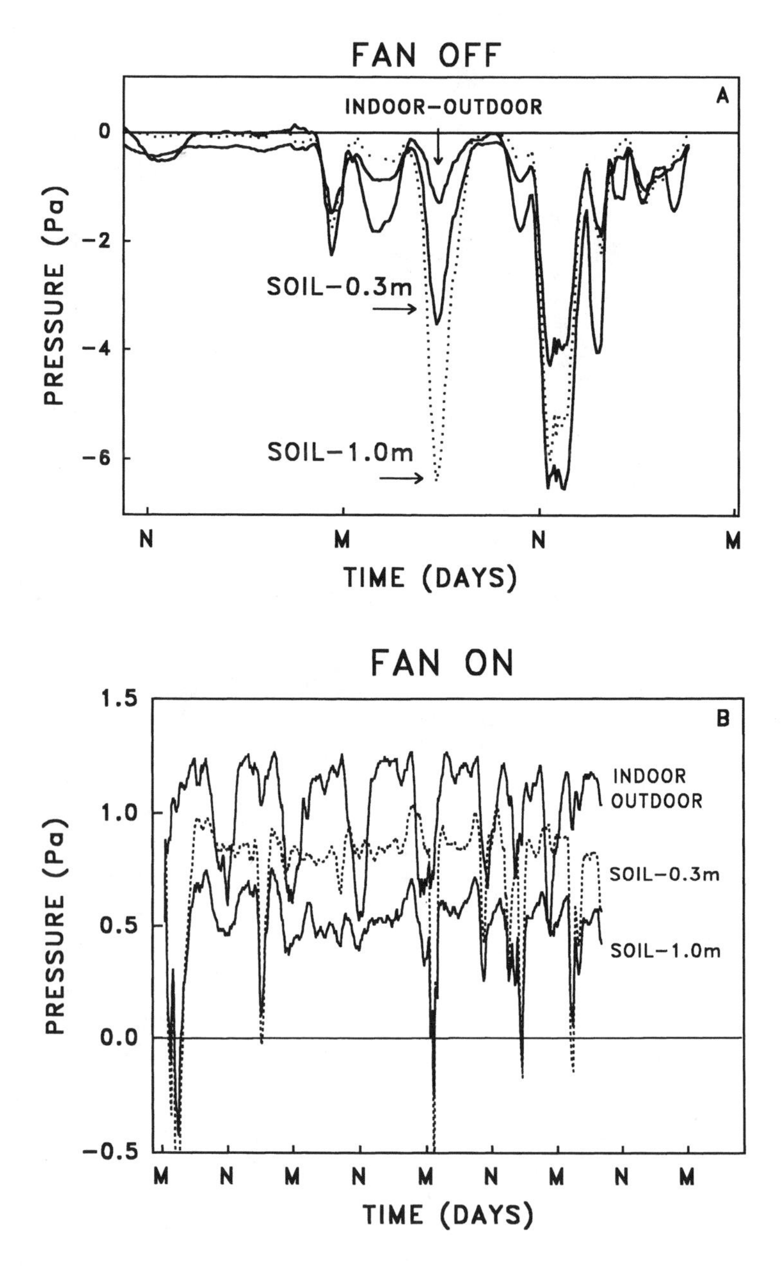

Figure 12. Pressure differentials versus time of day (N = noon, M = midnight), with blower fan off (A), and on (B).

Table 2. Comparisons of averaged ventilation rate pressure differentials, and radon entry rates with and without the external fan.[a]

Fan	Ventilation Rate, λv (h^{-1})	Pressure Differential, Δp (Pa)		Rn ($kBq\ h^{-1}$)
		Indoor-Outdoor	Indoor-Soil at 1 m	
Off	0.04 ± 0.01	—	—	1 ± 0.1
On	0.30 ± 0.01	0.8	0.3	0.9 ± 0.07
On	0.33 ± 0.01	1.2	0.6	0.7 ± 0.07
Off	0.02 ± 0.02	–0.02	–0.3	0.8 ± 0.04

[a] Uncertainties are 1 standard deviation of the sample used to compute the mean values.

The effective leakage area was determined according to the methods described by Sherman (1980). For the induced pressure differences indicated in Table 2, the computed leakage area was approximately 30 cm^2. This is considerably smaller than values quoted for houses. However, in the latter cases much higher pressure differences are used to determine leakage area.

CONCLUSIONS

Radon gas concentrations in the silty clay surrounding the chambers increased with depth. However, there are no apparent "seasonal" variations. The most important factor accounting for the variability of radon in soil is moisture, which was dominated by a single precipitation event. Moisture changes the soil conditions so that a simple, one-dimensional diffusion model cannot be used to describe long-term transport from soil gas to the atmosphere.

Monthly averaged radon concentrations in the underground chambers did not exhibit obvious seasonal variations, in contrast with data in published radon surveys. In the latter, lifestyle or demands of the occupants may be the controlling force on indoor concentrations rather than the physical forces responsible for radon transport and entry. A stepwise multiregression model indicates that when no attempt is made to control indoor temperature, outdoor temperature is the best predictor of daily average Rn concentration. However, this generally accounts for less than 50% of the day-to-day fluctuation.

Indoor radon concentrations were not correlated with pressure differences between the structure and surrounding soil.

For these chambers, the radon entry rate has two components, one constant and another that changes with time. The time-dependent component correlated with pressure differentials that were principally generated by wind speed and direction. Thus, as the wind speed increased, the radon entry rate increased. For periods of low wind speed the constant radon entry term dominated. This term remained constant even when the chamber was pressurized by 0.5 Pa with respect to the surrounding soil. It could be responsible for more than 50% of the total radon entering the chambers.

Wind speeds that increased the radon entry rate also increased the ventilation rate. For these chambers the increased ventilation rate compensated for the increased radon input. Thus, the periods of highest indoor concentration occurred when the radon entry rate was low.

A popular technique for mitigating radon concentration in houses is termed subslab depressurization. However, we have shown that there is an important radon entry term that is independent of pressure differentials. Thus the term depressurization is, perhaps, an oversimplification. Generally the pressure differences generated by these mitigation systems are many hundreds of pascals. In this regimen the system not only depressurizes but is also responsible for removal of the radon inventory surrounding the structure. Thus, subslab elimination or subslab interception would be more appropriate. It is possible, however, that there are more efficient and economical approaches to mitigation based on a better understanding of the two radon entry rates discussed here.

REFERENCES

Arnold, LJ. **1990**. A scale model of the effects of meteorological soil and house parameters in soil gas pressures. Health Phys 58:559-574.

Borak, TB, B Woodruff, and RE Toohey. **1989**. A survey of winter, summer and annual average ^{222}Rn concentrations in family dwellings. Health Phys 57:465-470.

Dudney, CS, AR Hawthorne, RG Wallage, and RP Reed. **1990**. Radon 222, ^{222}Rn progeny and ^{220}Rn progeny levels in 70 houses. Health Phys 58:297-311.

Harley, N. **1986**. Source term apportionment techniques for radon. In: *Indoor Radon*. Air Pollution Control Association, Pittsburgh, PA.

Holub, RF, R Droullard, T Borak, W Inkret, J Morse, and J. Baxter. **1985**. Radon-222 and ^{222}Rn progeny concentrations measured in an energy-efficient house equipped with a heat exchanger. Health Phys 49:267-277.

Hubbard, L, K Gadsby, D Bohac, A Lowell, D Harrje, R Scolow, T Matthews, and C Dudney. **1988**. Radon entry into detached dwellings: House dynamics and mitigation techniques. Radiat Protect Dosim 24:491-495.

Nazaroff, WW. **1988**. Predicting the rate of ^{222}Rn entering from soil into the basement of a dwelling due to pressure driven flow. Radiat Protect Dosim 24:199-202.

Nazaroff, WW, H Feustel, AV Nero, KL Revzan, DT Grimsrud, MA Essling, and RE Toohey. **1985**. Radon transport into a detached one-story house with a basement. Atmos Environ 19:31-46.

Perritt, RL, TD Hartwell, LS Sheldon, ML Smith, and JE Rizzuto. **1990**. Radon 222 levels in New York state homes. Health Phys 58:147-155.

Revzan, KL and WJ Fisk. **1990**. *Modeling Radon Entry into Houses with Basements: The Influence of Structural Factors*, LBL-28109. Lawrence Berkeley Laboratory, Berkeley, CA.

Sherman, MH. **1980**. *Air Infiltration in Buildings*, LBL 10712. Lawrence Berkeley Laboratory, Berkeley, CA.

Stoop, P, FJ Addenkamp, EJT Loos, RJ deMeijer, and LW Put. **1990**. Measurements and modeling of radon infiltration into a dwelling. In: *The 1990 International Symposium in Radon and Radon Reduction Technology*, Vol. III, USEPA Conf EPA/600 19-90 1005C. US Environmental Protection Agency, Research Triangle Park, NC.

Turk, BH, J Harrison, RJ Prill, and RG Sextro. **1990**. Developing soil gas and ^{222}Rn entry rate potentials for substructure surfaces and assessing ^{222}Rn control diagnostic techniques. Health Phys 59:405-419.

QUESTIONS AND ANSWERS

Q: Johnson, PNL. Tom, I think you're basically proving some of the mechanisms that I was trying to argue could be in place some years ago. The question I have is, did you say anything about the materials the house was built of? Could that long-term radon input come from the materials as opposed to outside?

A: Borak. That's a very good point. We have characterized the concrete with regard to the radium content and the radium content is similar to that of the soil. I think it is important, and we are going to try to investigate it. So we need to do all of this, and we cannot make any comments as to what fraction of that slow entry rate is coming from the building materials. But it isn't an extremely high radium content with regard to the soil.

Q: Harbottle, Brookhaven National Laboratory. By chance, one of the first samples of soil on which I ever measured the emanation coefficient of radon and thoron was a boroll from the vicinity of Ft. Collins. I assume that's the soil that your house is buried in.

A: Borak. We've got this almost uniform combination of silt sand and loam.

Q: Harbottle. No, I think the pedagalogical term for what you have is a particular type of boroll. In any case, it came from there, and that soil showed a very high sensitivity to moisture content in terms of the emanation coefficient. I think we got emanation coefficients of about 25 or 30% with the normal soil moisture. When we dried it out this went down to about 5%. There is a huge change in the ability of the soil to emanate with moisture, and that would correlate very nicely with your observation.

A: Borak. I think you're right, but to a certain extent we have measured it. We get a 10% emanation coefficient for dry soil, but even when we started out with soil that was already moist, quite moist, we believe that we are already over the hump such that this increase that we saw after a precipitation event can be explained as due to an increase in the emanation coefficient. We tried to explain why we are seeing all of this, and changes in the emanation coefficient from 10% moisture to 15% moisture isn't going to explain a factor of almost 2 in the radon concentration. But it certainly is very, very important when you have very dry soil, and then you start to give it a little bit of moisture.

Q: Harbottle. Was there any appreciable change in the water-table level during that time so that the rainfall could have produced a rise in the water table and exerted a pumping action forcing radon up from lower depths.

A: Borak. We did not see it, at least to the depths that we were characterizing the radon. We could probably go back and try to go down deep. I'm not exactly sure how much the water table was pumping.

Q: Haughey, Rutgers University. Tom, that was very good, but I suspect that there are major differences in different parts of the country. We just had a Ph.D thesis completed where a student took a house about 10 years old and studied the effect of a heating unit that took air from outside versus one that took air from the basement.

But with relation to your work, the most significant thing we found was the ground water. On one event, which involved rainfall for an entire week-end, where the ground water was literally flowing down the plain, the concentration in the basement went up by a factor of 8, and the student suggested an explanation for this. The piston action of the ground water forced the radon right into the basement of the house.

A: Borak. First of all, your very first comment is right. We have to be a little careful because the data, I believe, are correct, but it's representing the system we have, and it's a simplified system. Nevertheless, I think we can test the model. We do believe that the reason we have this buildup in the radon concentration in this soil was because of a capping action. As a matter of fact, we tried to do the one-dimensional diffusion equation that Peter Owczarski was talking about. In the dry soil we got a beautiful fit to that equation that gave us a diffusion length of 1.5 meters, which everybody is happy with. We then repeated it, and we got a diffusion length of about 4 meters, which is greater than air, which is saying that things are so complicated that you are not having homogeneous situations. So it is interesting that when you had the capping action, you saw an increase in the indoor radon, whereas we did not; but we were not intentionally heating. I think we can do these experiments though, because now we can take one of the houses and put a heater in it, and we can begin to look at that, whereas we still have the other house as a control. So the situation, I think, is going to present itself where we can start sorting this stuff out piece by piece.

C: Haughey. Right. Incidentally, you might be interested in knowing that the concentration in the basement with air taken from outside was higher than it was with air taken from inside. And the explanation we found for that was a very interesting feature of construction in the east. That is, around a chimney you have to have an air space. That's called the chase, and when you take air from inside the basement the air comes down the chase and dilutes the radon.

A: Borak. This is why we really were very careful about our ventilation, and we can do this in our place. We don't make any assumptions, as I said, of steady state. We are constantly measuring it, and we think we can get this time variation, and we can look at even higher frequencies rather than the daily averages that we showed.

Q: Rose, Pennsylvania State University. I'm wondering with regard to this radon increase when the soil moisture went up, whether you've looked at just the effect of the increased proportion of floor space that's filled with water. No doubt there is also some transport-related effect, particularly for the more shallow ones, but we found in eastern soils that just the changes in moisture saturation would explain the biggest part of our seasonal changes in radon and soil gas.

A: Borak. I believe you have a copy of the student's thesis on the way, Art, where we address this. And what we found out was that the decrease in the floor space due to the increased moisture wasn't sufficient, even taking into account that when you decrease the floor space with water, some of the radon is going to be inside of that water. So we couldn't quite explain all of it. But I'm wondering if we aren't doing the following: we're extracting the radon from the soil with a syringe. We're purging the tube with a syringe. What we might be doing there is, in the location where we're extracting the radon, depressurizing it momentarily, and we're also taking radon out of the moisture. Do you see what I'm saying? I don't know how to untangle that. But from our point of view, it looked like it was available, but the floor space alone did not explain this large difference in the measured radon concentration of soil gas.

RADON AND RADON PROGENY SOURCES

SOIL-GAS AND INDOOR RADON DISTRIBUTION RELATED TO GEOLOGY IN FREDERICK COUNTY, MARYLAND

S. L. Szarzi, G. M. Reimer, and J. M. Been

U.S. Geological Survey, Federal Center, Denver, Colorado

Key words: *Soil-gas radon, geology, Maryland, indoor radon, radon potential*

ABSTRACT

Soil-gas radon concentrations vary in response to geologic controls in Frederick County, Maryland, and the variation leads to different radon availabilities for potential indoor accumulations. Quartzites, which form the core of ridges and mountains of the southern and western part of the county, have a mean soil-gas radon concentration of 26 kBq m^{-3} (700 pCi L^{-1}). Phyllites, found in the Piedmont province in the eastern part of the county, have a mean soil-gas radon concentration of 59 kBq m^{-3} (1600 pCi L^{-1}). Many indoor radon measurements for homes in the southeast portion of the county, made by means of charcoal canisters, exceeded 1850 Bq m^{-3} (50 pCi L^{-1}). Homes built in areas where the soil-gas radon concentrations were greater than 75 kBq m^{-3} (2000 pCi L^{-1}) may have indoor radon concentrations that exceed 150 Bq m^{-3} (4 pCi L^{-1}), the current action level recommended by the U.S. Environmental Protection Agency. Data obtained in studies like ours throughout the United States are essential to identify "hot spots" which may produce elevated indoor radon levels of significant risk.

INTRODUCTION

Over the last few years, there has been growing recognition for the need to identify the radon potential of various areas within the United States. Not only does this information encourage homeowners to test their residences for radon, especially in higher-potential areas, but it also can assist land planners, developers, and local governments establish building standards to deal with the radon issue.

Also recognized recently is the fact that geology and related disciplines have a primary role in defining the regions of higher radon potential (Reimer, 1990a; Gundersen et al., 1988). A combination of geophysical gamma-ray measurements, pedological descriptions of surficial materials, and geochemical measurements of actual soil-gas radon concentrations form a powerful technological basis for estimating radon potential.

Current studies are identifying the regional differences that climate and physiography have in modifying the basis for estimating radon potential. For example, previous studies suggested that the annual seasonal radon soil-gas concentrations were high in winter and low in summer (Dyck, 1978; Schery and Gaeddert, 1982). However, in many regions of the country, particularly where soils have high moisture content in winter, as in central Pennsylvania, the radon soil-gas concentrations are high in the summer and low in the winter (Washington and Rose, 1990).

We report the results of a study designed to test the geologically based techniques to estimate the radon potential. Frederick County, Maryland (Figure 1) was chosen for study because it was representative of a diverse geological setting. Furthermore, it is adjacent to Montgomery County, Maryland, for which a radon-potential map has been prepared by the U.S. Geological Survey (USGS) (Gundersen et al., 1988). Comparison of the radon potential, derived from the Frederick County indoor concentrations, and that derived for adjacent Montgomery County provide a basis for evaluating the success of our measurement techniques.

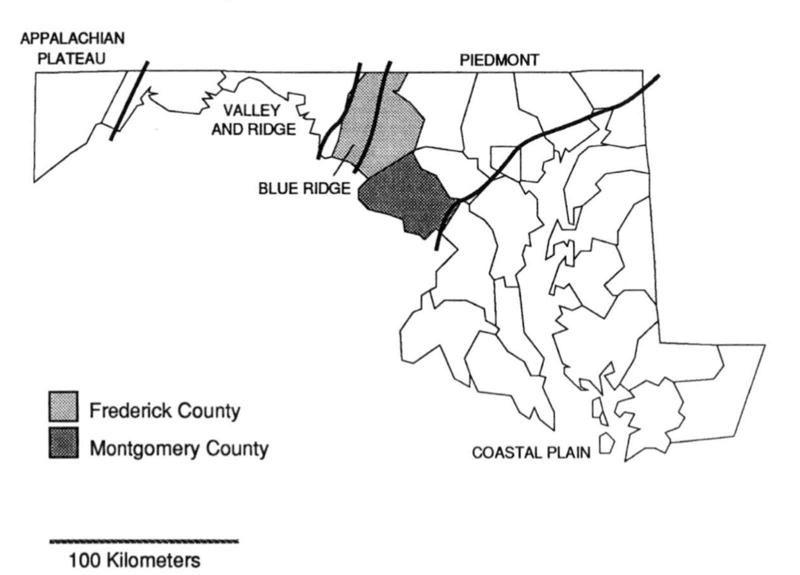

Figure 1. Map showing location of Frederick County and Montgomery County, Maryland, as well as location of major physiographic provinces in Maryland. Of the five physiographic provinces in the state, three (Valley and Ridge, Blue Ridge, and Piedmont) are in Frederick County.

STUDY AREA

Three physiographic provinces are represented in the county, from east to west (Figure 1):

1. The Piedmont is characterized by igneous and metamorphic rocks with the bedrock consisting of phyllite, schist, gneiss, gabbro, and metamorphosed sedimentary rocks, and igneous rocks of probable volcanic origin. Within the Piedmont, in the central part of the county, is the Frederick Valley, which is underlain by limestone and dolomite.
2. The Blue Ridge is an exposed anticline with quartzite ridges and valleys consisting of gneiss and volcanic rocks.
3. The Valley and Ridge is characterized by intensely folded and faulted sedimentary rocks. Limestone and dolomite are dominant in the eastern part of this province.

The ridges in the western section of the county are about 500 m above sea level. The terrain slopes unevenly eastward to a plain averaging about 150 m above sea level. The climate is defined as humid continental with a yearly mean temperature of about 13°C and 60 to 80 cm of precipitation per year. Farming is giving way to industrialization and urbanization. The county is centered on the Interstate 270 urban corridor, which extends northward from Washington, D.C.

METHODS

Soil-gas radon concentrations were determined using a probe method to obtain a grab sample. The probe (Figure 2) is a small-diameter, thick-walled, carbon steel tube. The probe length can vary, but, typically, 0.75 to 1.0 m is the depth from which samples are taken. This depth is below the major influences of meteorological variables and one that frequently encounters the lower B or upper C horizon for most soils (Reimer, 1990b; Hesselbom, 1985).[a] The probe is pounded into the ground, using a split, barbell-shaped deadweight as a sliding hammer. The probe tip has a small machine screw threaded into the lower end to prevent dirt from entering; five sets of holes are drilled through the side of the tube near the bottom. A wire inserted in the probe prevents dirt from entering the side holes. The sample is taken

(a) Soil development occurs in stages as the bedrock weathers. Discrete soil horizons form, representing differing maturity of development. Soil horizons are classified by letters, from the surface to the bedrock: A is the organic-rich layer; B is the zone of illuviation; C is the weathered parent material; and D is the underlying rock.

from the open end of the probe through a needle-guide fitting that contains a septum. The probe is purged, and controlled volumes of soil gas are collected using a hypodermic syringe. Samples can be stored in the syringes if the needle is capped to prevent gas loss or exchange with the atmosphere. Tests have shown that the syringes retain helium gas for several days before diffusion causes significant dilution (Reimer et al., 1979). The sample is then analyzed in the field using a portable alpha-scintillometer. Because the soil-gas radon concentrations are generally high, 15 kBq m^{-3} (400 pCi L^{-1}) and greater, and the range is large, 15 to 150 kBq m^{-3} (400 to 4000 pCi L^{-1}) and greater, precision of about 1.5 kBq m^{-3} (40 pCi L^{-1}) is adequate.

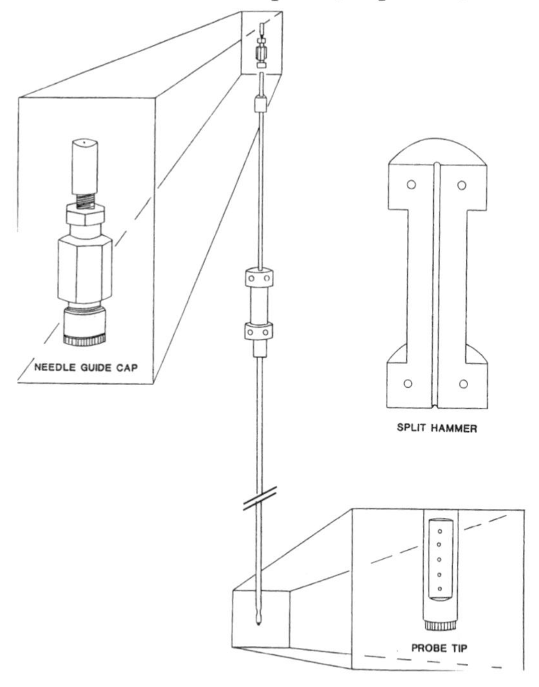

Figure 2. Schematic diagram of soil-gas sampling probe used in this study for collecting samples for radon analysis. Probe is small-diameter, carbon-alloy steel tube, in which base has been capped and side holes drilled for sample to enter. Probe is placed in ground by pounding barbell-shaped sliding hammer between two pounding collars. Samples are collected with hypodermic syringe through septum-containing cap placed on top of probe after pounding it into the ground.

Surface gamma-ray measurements were made using both a hand-held, non-energy-discriminating scintillometer and an energy-discriminating spectrometer. The energy-discriminating spectrometer identified potassium-40 (1.38-1.56 MeV gamma-ray range), equivalent thorium (2.44-2.77 MeV gamma-ray range for ^{208}Tl), and equivalent uranium

(1.66-1.90 MeV gamma-ray range for ^{214}Bi). The major geologic descriptions were obtained from a geologic map of the county (Jones and Stose, 1938) and are given in Table 1. Because this is the only geologic map available for the entire county at this scale, the unit names and locations shown on the map were used for sample grouping. Many of the names are of historical interest only and are no longer used. However, the unit and rock-type descriptions are illustrated sufficiently for the purposes of discussion in our paper.

Indoor radon concentrations were provided by the Geological Survey of Maryland from a proprietary list of measurements they sponsored (James Brooks, Maryland Geological Survey, personal communication, 1989).

Table 1. Description of geologic unit, or rock type, abbreviations (listed alphabetically) used in Figures 6 and 7. Units and ages, taken from geologic map of Jonas and Stose (1938), are used only for descriptive purposes. Much of the nomenclature and descriptions have changed as later studies provided additional geologic detail. Many of the Precambrian units have been revised to lower Paleozoic.

Unit Reference Number	Map Symbol	Geologic Unit or Rock Type and Description	Geologic Age
1	arh	Aporhyolite (blue and purplish-gray rhyolite)	Precambrian
2	arr	Aporhyolite (reddish-purple rhyolite)	Precambrian
3	Єa	Antietam Quartzite (gray to buff ferruginous quartzite)	Lower Cambrian
4	Єf	Frederick Limestone (blue limestone)	Upper Cambbrian
5	Єh	Harpers Phyllite (gray phyllite and slate)	Lower Cambrian
6	Єl	Loudoun Formation (sandy phyllite and arkosic quartzite)	Lower Cambrian
7	Єls	Loudoun Formation (blue and green slate)	Lower Cambrian
8	cmb	Catoctin Metabasalt (hornblende, albite, and epidote basalt)	Precambrian
9	Єt	Tomstown Dolomite (dark dolomite)	Lower Cambrian
10	Єw	Waynesboro Formation (red shale and sandstone)	Lower Cambrian
11	Єwq	Weverton Quartzite (vitreous white quartzite)	Lower Cambrian
12	ij	Ijamsville Phyllite (blue to green phyllitic slate)	Precambrian
13	ma	Catoctin Metabasalt (blue amygdular andesite)	Precambrian
14	mb	Metabasalt (green schistose basalt)	Precambrian
15	mdg	Mica schist and hornblende diorite with granitic intrusions	Precambrian
16	mg	Marburg Schist (muscovite-chlorite schist)	Precambrian
17	mrh	Metarhyolite and metaandesite (reddish-purple to blue schistose flows)	Precambrian
18	Og	Grove Limestone (fine-grained limestone)	Lower Ordovician
19	Ogq	Grove Limestone (dolomite and limestone containing quartz grains)	Lower Ordovician
20	qtz	Quartzite beds (white quartzites with sericite partings)	Precambrian
21	rt	Rhyolite tuff (sericite-quartz schist)	Precambrian
22	TRg	Gettysburg Shale (red shale and sandstone)	Upper Triassic
23	TRlc	New Oxford Formation (limestone conglomerate)	Upper Triassic
24	TRno	New Oxford Formation(red shale and gray to red arkose)	Upper Triassic
25	TRqc	New Oxford Formation (quartz in red sandy matrix)	Upper Triassic
26	ub	Urbana Phyllite (chlorite phyllite with slaty layers)	Precambrian
27	was	Wissahickon Formation (albite-chlorite schist facies)	Precambrian
28	wm	Wakefield Marble (white fine-grained marble)	Precambrian

RESULTS

The distribution of soil-gas radon varies throughout the county and seems to be strongly related to the underlying geologic unit or rock type, especially for high and low radon concentrations. The county-wide distribution of radon in soil gas appears as a log-normal distribution (Figure 3). In certain areas, such as the quartzite ridges in the western part of the county, the soil-gas radon concentrations are low. An example of this area is represented by unit 6 (arkosic quartzite) in Table 1. Other areas, such as the metamorphic region in the southeast part of the county near the community of Mt. Airy, have very high soil-gas radon concentrations. This is represented by units 12 (phyllitic slate) and 16 (schist) in Table 1. The soil-gas radon distribution in the southeast metamorphic area follows a more normal distribution (Figure 4).

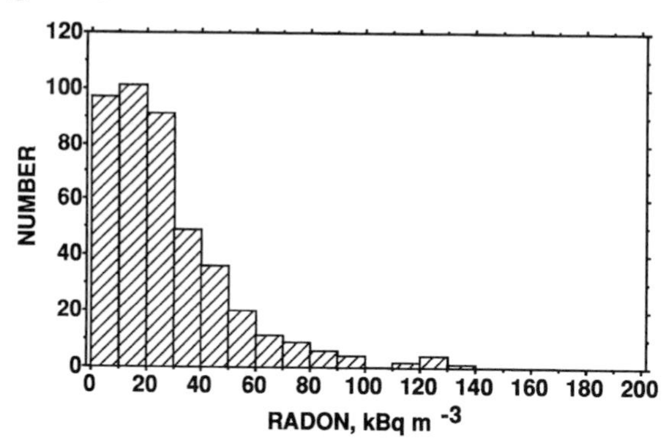

Figure 3. Histogram of radon soil-gas in Frederick County, Maryland. Log-normal distribution has mean concentration of 27 kBq m^{-3} (740 pCi L^{-1}); figure does not contain samples from Mt. Airy area.

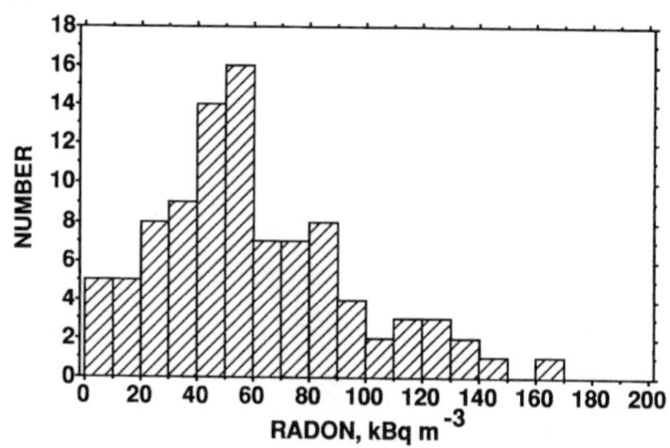

Figure 4. Histogram of radon in soil gas in Mt. Airy, Maryland. Normal distribution has mean soil-gas radon concentration of about 60 kBq m^{-3} (1600 pCi L^{-1}), distinctly higher than county-wide distribution.

The study indicates that the soil-gas radon concentrations were not correlated with the equivalent uranium (eU) concentrations (Figure 5). Because the field measurements were not made at the same locations as the indoor radon measurements, any correlations with indoor radon, other than when grouped within similar lithology where significant numbers of samples are available, would be meaningless. The soil-gas radon had characteristic concentrations when categorized by unit or rock type, but there was so much variance that the rock types could rarely be identified exclusively by the radon concentration (Figure 6). Only the high and low concentrations sufficiently characterized a unit. The distribution of indoor radon by geologic unit (Figure 7) shows that the metamorphic rocks have higher than average radon concentrations. These units and rock types are cross referenced to units 12, 14, 16, and 28 in Table 1.

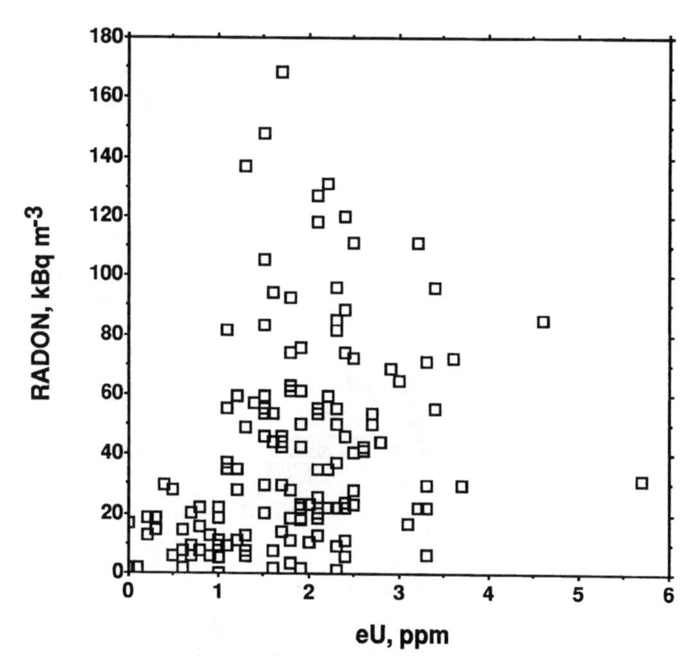

Figure 5. Correlation of equivalent uranium (eU) and radon soil gas for samples taken in same location in Frederick County, Maryland. Although there is no linear correlation between soil-gas radon and eU, there appears to be a regime which excludes radon concentrations greater than 40 kBq m^{-3} (1100 pCi L^{-1}) when eU concentrations are less than 1 ppm.

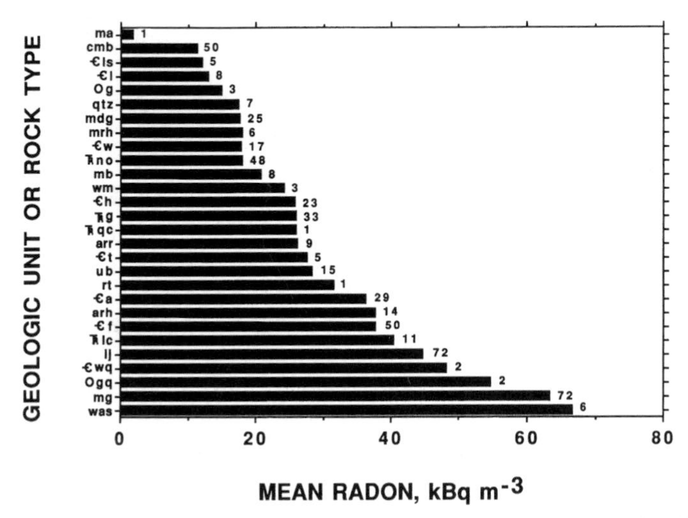

Figure 6. Means of soil-gas radon by geologic unit, or rock type, as described by Jonas and Stose (1938). Data from both Frederick County and Mt. Airy, Maryland, are included. Concentrations are grouped from low to high. Description of geologic units can be found in Table 1. Numbers following bars are the number of samples collected within each unit. Because variance from means can be large, only high and low concentrations could characterize the units with adequate sampling. See text for detailed explanation.

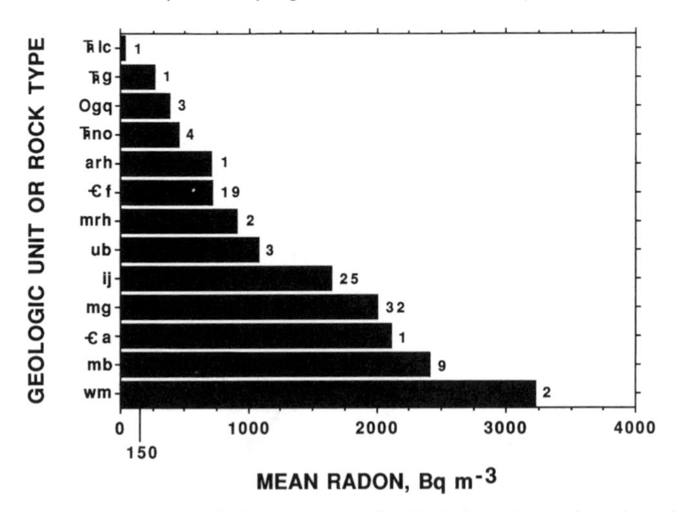

Figure 7. Indoor radon concentration (measured with 3-day charcoal canisters) by geologic unit, or rock type, as described by Jonas and Stose (1938). Data from both Frederick County and Mt. Airy, Maryland, are included. Description of geologic units can be found in Table 1. Concentrations are grouped from low to high. Numbers following bars are the number of samples collected within each unit. Different units, such as 4 (Cf), 12 (ij), and 16 (mg), with adequate numbers of sample pairs, show a qualitative correlation between soil-gas radon and indoor radon.

DISCUSSION

The data indicate that the radon potential can be estimated from soil-gas radon concentrations, geology, and gamma-ray signatures. Although no one technique is sufficient to make this estimate, when used together they provide an excellent combination to obtain data on which predictions can be based. The highest probability of accurate prediction is at the extremes of the ranges measured. That is to say, where the eU or soil-gas radon concentrations are high or low and where the geology indicates lithologies or structures, such as faults and joints, that provide conditions where uranium or radium are concentrated or depleted in the soil, radon potential can be estimated with great confidence.

The soil-gas radon measured by the probe technique is that available for transport at the time of sampling. Many of the variable parameters that control the availability, such as emanation, soil-moisture, and permeability, are coupled to some degree. Therefore, the soil-gas radon concentration is also variable, largely in response to meteorological conditions. Given a sufficiently large data base of controlling parameters, it may be possible to estimate the soil-gas radon concentration at any time from a measurement taken when a different set of conditions prevail. However, the large concentration range of radon concentrations permits several regimes to be identified without the need for concentration adjustment. An example is seen in the radon potential map derived for Montgomery County, Maryland. There, the boundary for radon potential was 75 kBq m^{-3} (2000 pCi L^{-1}) for high to moderate potential, and 20 kBq m^{-3} (500 pCi L^{-1}) for moderate to low potential (Gundersen et al., 1988)

In Frederick County, Maryland, the overall soil-gas radon distribution appears to be log normal (Figure 3), which is consistent with data from other studies (Reimer and Gundersen, 1989; Gundersen et al., 1988). The region of high concentrations near Mt. Airy (Figure 4) would contribute mostly to the tail of the county-wide distribution. The county-wide indoor radon concentrations (Figure 8) and the concentrations for Mt. Airy (Figure 9) are skewed toward the high concentrations. The skewness may be an artifact of the minimum detection limit of the charcoal measurement.

Because of different sampling conditions, a direct correlation between soil-gas radon and the eU concentration would be rare (Duval and Otton, 1990). The techniques sample different areas of the soil column.

The gamma-ray spectrometer is reading the total (but volume-integrated) ^{214}Bi, and the mobile soil-gas radon is provided only by the emanated fraction. Therefore, no correlation is expected for the overall analyses. Although there is no linear correlation between soil-gas radon and eU, there appears to be a regime which excludes radon concentrations greater than 40 kBq m^{-3} (1100 pCi L^{-1}) when the eU concentrations are less than 1 ppm.

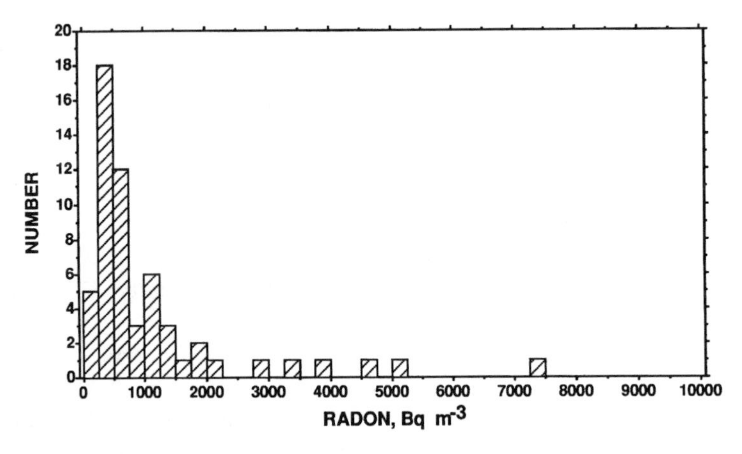

Figure 8. Histogram of indoor radon concentrations in Frederick County, Maryland. Mt. Airy data not included. Arithmetic mean is 1100 Bq m^{-3} (30 pCi L^{-1}).

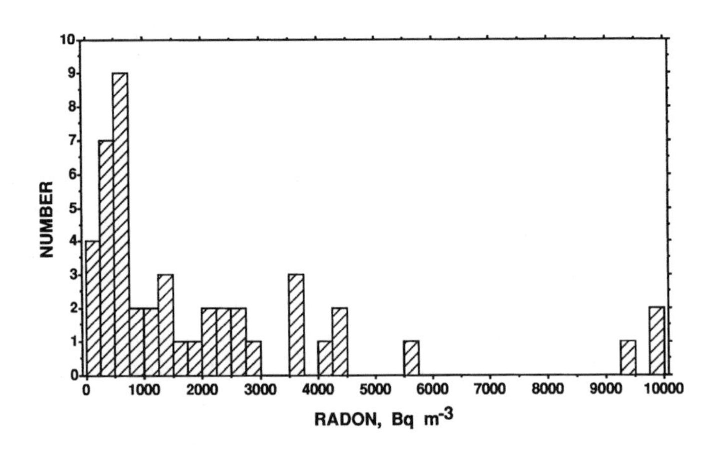

Figure 9. Histogram of indoor radon in Mt. Airy, Maryland. Arithmetic mean, 2100 Bq m^{-3} (56 pCi L^{-1}), is significantly higher than county-wide mean.

CONCLUSION

Geologically based techniques seem to provide an excellent approach to estimating the radon potential of an area. The methods are fast, efficient, and cost-effective and can be modified to accommodate a wide spectrum of soil conditions. The techniques have the advantage of being adaptable to a variety of scales, from a single building lot to county-wide distribution. The measurements can be compared to models being developed to describe the transport of gas across the soil/structure boundary. Other data, such as soil permeability, can be integrated into this basic methodology to improve site-specific interpretations.

REFERENCES

Duval, JS and JK Otton. **1990**. Radium distribution and indoor radon in the pacific northwest. Geophys Res Lett 17:801-804.

Dyck, W. **1978**. The mobility and concentration of uranium and its decay products in temperate surficial environments, pp. 57-100. In: *Uranium Deposits, Their Mineralogy and Origin*, MM Kimberly (ed.). University of Toronto Press, Toronto, Canada.

Gundersen, LCS, GM Reimer, CR Wiggs, and CA Rice. **1988**. Map Showing Radon Potential of Rocks and Soils in Montgomery County, Maryland, scale 1:62,500. US Geol Surv Miscell Field Studies Map MF-2043. US Geological Survey, Reston, VA.

Hesselbom, Å. **1985**. Radon in soil-gas; a study of methods and instruments for determining radon concentrations in the ground. Sver Geol Unders Arsb 803.

Jonas, AI and GW Stose. **1938**. Geologic Map of Frederick County, Maryland and Adjacent Parts of Carroll and Washington Counties, scale 1:62,500. Maryland Geologic Survey, Baltimore, MD.

Reimer, GM. **1990a**. The occurrence and transport of radon in the natural environment. Geophys Res Lett 17:799.

Reimer, GM. **1990b.** Reconnaissance techniques for determining soil-gas radon concentrations: An example from Prince Georges County, Maryland. Geophys Res Lett 17:809-812.

Reimer, GM and LCS Gundersen. **1989**. A direct correlation among indoor radon, soil gas radon and geology in the Reading Prong near Boyertown, Pennsylvania. Health Phys 57:155-160.

Reimer, GM, EH Denton, I. Friedman, and JK Otton. **1979**. Recent developments in uranium exploration using U.S. Geological Survey's mobile helium detector. J Geochem Explor 11:1-12.

Schery, SD and DH Gaeddert. **1982**. Measurements of the effect of cyclic atmospheric pressure variation on the flux of 222-radon from the soil. Geophys Res Lett 9:835-838.

Washington, JW and AW Rose. **1990**. Regional and temporal relations of radon in soil gas to soil temperature and moisture. Geophys Res Lett 17:829-832.

DETERMINATION OF RADON-222 IN GROUND WATER USING LIQUID SCINTILLATION COUNTING—SURVEY OF CAREFREE CAVE-CREEK WATER BASIN IN ARIZONA

J. M. Barnett[1], J. W. McKlveen[1], and W. K. Hood, III[2]

[1]Arizona State University, Tempe, Arizona

[2]Arizona Department of Environmental Quality, Phoenix, Arizona

Key words: *Radon, liquid scintillation, water*

ABSTRACT

Well water used in homes may contribute additional ^{222}Rn to the indoor radon concentration. Our research objectives are to establish a method to measure radon in ground water using liquid scintillation (LS) spectrometry, and to determine the lung dose from the radon released into the air.

The method involves collecting a nonaerated, slow, steady flow of water from a pumping well into a 437-ml (16-oz) glass bottle. A high meniscus assures no head space, and the sample is capped. In the laboratory, standard 22-ml LS glass vials are filled with 10 ml of a toluene-based, mineral oil, LS cocktail and two 5-ml sample aliquots. The vial is capped tightly, shaken vigorously, and placed in the LS counter.

Equilibrium was established in about 3.5 h, after which samples were counted for 100 min each. Only radon and daughters were measured. According to NUREG/CR-4007, the lower limit of detection is 1.9 Bq L^{-1} (51 pCi L^{-1}) in the window of interest. The radon progeny detection efficiency was between 320 and 330% per unit radon activity (accounting for the detection efficiency of each alpha particle and the beta continuum), and the average background was approximately 6 counts per minute.

We expect that wells containing radon concentrations between 100 and 1000 Bq L^{-1} would produce an effective dose equivalent to the lungs of 0.4 to 0.7 mSv y^{-1} (40-70 mrem y^{-1}). Our study of 28 wells in Carefree-Cave Creek indicates that in 25% of the wells, radon levels were over 100 Bq L^{-1} (2700 pCi L^{-1}). Twelve wells were chosen for monthly monitoring to ensure the efficiency of the methodology. This simple method allows us to count a large number of samples over a short time period.

INTRODUCTION

The first round of groundwater sampling for radon-222 began in August, 1989. Nine wells were for private use, two wells were for golf course irrigation use, and the remaining 17 wells were for public use and owned by a variety of water companies.

To determine ambient groundwater quality, a baseline of water-quality data was established by selecting those from representative wells from the Carefree/Cave-Creek area. Data included organic, inorganic, and radiochemical analyses. A primary focus of this study was to establish a method to measure radon in ground water using liquid scintillation (LS) spectrometry, and to estimate lung dose and mortality rate from radon released to indoor air. Repeated sampling for radon from selected wells verified our laboratory techniques and established the continuity and validity of the data, which included radon concentration, estimated dose rate, and mortality rate.

Radon-222, a noble gas, is odorless, colorless, water soluble, radioactive, and is the heaviest known gas. It differs from all other uranium decay products in that it is a gas and does not form chemical bonds. Its solubility in water is inversely proportional to water temperature. In the dissolved phase, radon moves freely through porous, permeable geologic materials, and from high-pressure to low-pressure areas. Radon-222 is the only isotope sufficiently abundant and long lived to be of health concern.

Early in the development of LS counters, environmental monitoring was mostly for weak beta-emitters (tritium and ^{14}C). However, over the past 15 to 20 y, LS counting has been applied to the analysis of ^{90}Sr, and to plutonium, thorium, uranium, radium, and radon (Schönhofer and Henrich, 1987). The main advantage of LS is its high efficiency for alpha particle detection, which is virtually 100%. Detection of radon in water using LS counting was demonstrated by Prichard and Gesell (1977). Improvements in LS counters and cocktails, and modification of the Prichard and Gesell method has made LS counting ideal for detecting radon in water.

Wells were selected for sampling in cooperation with water companies and private well owners in the Carefree/Cave-Creek water basin area. The basin, located near 111° 54' 00" longitude and 33° 50' 00" latitude, covers a land mass of approximately 64 km^2 (25 mi^2). It is approximately 56 km (35 mi) northeast of Phoenix, Arizona (Figure 1). Varied hydrogeologic conditions prevail in the basin. Basalt flows are interbedded with silt, sand, and gravel units. Depth to water ranges from 3 m to 30 m (10 ft to 100 ft) below the land surface, depending, to some extent, on the topography. Water was collected from 28 wells and analyzed for radon, using LS counting. The first set of results was used as a basis to select a smaller group of wells for a 6-mo study period.

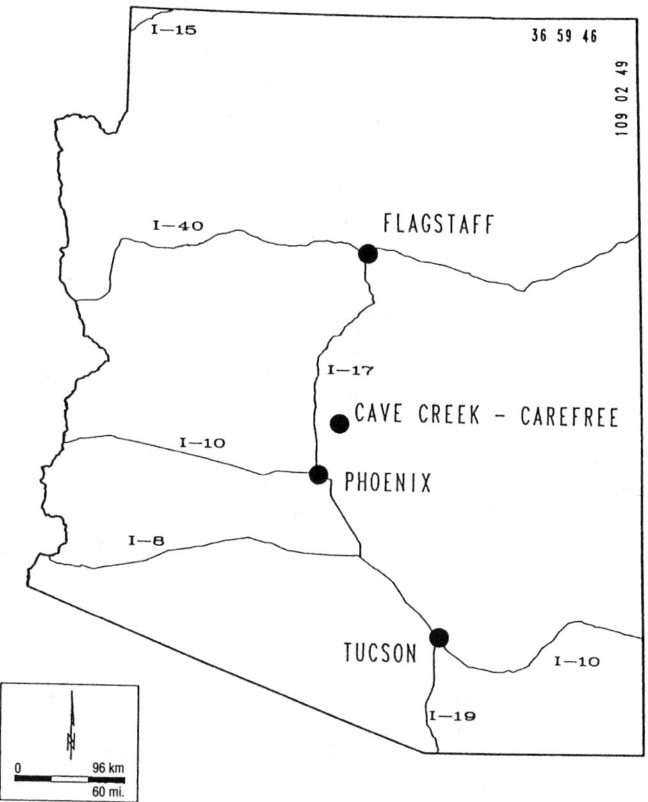

Figure 1. Relationship of Carefree/Cave-Creek water basin to major cities in the State of Arizona.

METHODS

Water collection for radon analysis took place in the Carefree/Cave-Creek area from standard operating wells equipped with pumps. If the pump had to be turned on manually to override the automatic timer system prior to sample collection, it was allowed to run long enough to evacuate stagnant water from the well casing (purging) so that a representative sample could be obtained. The sample was analyzed, using LS counting, at Arizona State University Nuclear Engineering Sciences Laboratory.

All sample containers had externally affixed water- and tamper-proof labels that bore, in indelible ink, sample name, well location, date and time of collection, and samplers' name(s). Radon samples were collected as near the well head as possible in 437-ml (16-oz) glass containers, with minimal sample aeration. A clear plastic hose was inserted into the spigot, and a slow, steady flow of water was obtained. The bottle was gently and completely filled so that zero head space was present; it was also overfilled to obtain a representative sample and to minimize any radon losses during collection. A high meniscus was obtained on the lip of the bottle before tightly sealing the bottle with a Teflon-coated cap. No preservation of the sample was necessary.

In the laboratory, standard 22-ml borosilicate, low-potassium glass LS vials were placed in an LS rack. The caps of the vials, which were also Teflon-coated, were marked with the sample name, laboratory identification number, and date and time of sample preparation. First, 10 ml of a high-efficiency, mineral-oil, toluene-based scintillation cocktail was added to a glass vial. Next, a sample of water was carefully opened, and two 5-ml aliquots were pipetted gently down the side into the vial. The water was sealed beneath the cocktail with minimal radon loss. The vial was quickly and tightly capped with the appropriately labeled cap, vigorously shaken, and placed in the rack, which was placed in the LS counter.

Samples were counted after reaching secular equilibrium, approximately 3.5 h after sample preparation, and counted for 100 min each. Additionally, a minimum of one background and one standard was counted with each set of samples for 100 min and 10 min, respectively. Repeat and duplicate samples were also counted to monitor method reliability as well as quality assurance and quality control.

Background standards were prepared using deionized water; standards were prepared from a radium solution of known activity. Standards were used after secular equilibrium was established (approximately 40 days).

RESULTS

The radon activity for various wells sampled beginning in August, 1989 (Table 1), has a lognormal distribution with a geometric mean of 46.5 Bq L^{-1} (1255 pCi L^{1}) and a geometric standard deviation of

2.7. Although the geographic distribution of radon in Arizona ground water is not known with certainty, Dixon and Lee (1988) and Longtin (1988) have shown mean radon concentrations ranging from 21 to 54 Bq L^{-1} (580 to 1450 pCi L^{-1}) for the entire state.

Table 1. Radon-222 concentration in well water from the Carefree/Cave-Creek area, sampled beginning in August, 1989.

Well number	Longitude	Latitude	Rn activity, Bq L^{-1}
1	111-50-05	33-53-03	301.9
2	111-50-01	33-49-26	66.7
3	111-52-30	33-48-17	25.6
4	111-52-50	33-48-47	40.0
5	111-52-55	33-49-23	47.8
6	111-53-09	33-49-50	68.5
7	111-53-24	33-48-47	28.5
8	111-53-24	33-49-15	60.7
9	111-53-59	33-49-22	72.2
10	111-54-30	33-49-32	34.8
11	111-54-48	33-50-47	36.3
12	111-55-01	33-49-51	61.1
13	111-55-13	33-52-20	135.9
14	111-55-13	33-50-47	36.3
15	111-55-52	33-49-55	189.6
16	111-55-58	33-51-38	10.4
17	111-56-34	33-51-15	15.9
18	111-56-39	33-50-45	39.6
19	111-56-42	33-49-48	150.4
20	111-56-48	33-49-56	214.8
21	111-56-56	33-49-50	175.2
22	111-57-14	33-51-03	56.7
23	111-57-17	33-53-03	103.0
24	111-57-19	33-50-12	9.6
25	111-57-48	33-50-16	14.4
26	111-57-54	33-49-48	9.6
27	111-58-06	33-52-34	19.5
28	111-58-10	33-49-10	14.1

Twelve wells were selected for the approximately 6-mo monitoring period from January, 1990, to July, 1990. Samples were taken first from wells with activity that exceeded 100 Bq L^{-1} (2700 pCi L^{-1}), then from wells with more than 50 Bq L^{-1} activity. Monthly radon activity and averages (Table 2) ranged from 52 to 300 Bq L^{-1} (1400 to 8150 pCi L^{-1}). Most wells showed little variation in radon concentration; however, wells 1, 15, 20, and 21 showed variations of over 55 Bq L^{-1} (1500 pCi L^{-1}) from the lowest to highest radon level measured.

Counting efficiency for the LS counter's total window (710-845) is 325% where the radon-222, polonium-218, and polonium-214 alphas are each counted with nearly 100% efficiency. The average background in the total window is about 6.2 counts per minute. The lower limit of detection is then 1.9 Bq L^{-1} (51 pCi L^{-1}) for a 1-mo-old background sample; otherwise, for background samples prepared the same day as the water samples, the lower limit of detection is just below 0.7 Bq L^{-1} (20 pCi L^{-1}).

Table 2. Monthly ^{222}Rn concentrations and averages of 12 wells sampled from January, 1990, to July, 1990.

Well number	Monthly radon activities, Bq L^{-1}							
	Jan	Feb	Mar	Apr	May	Jun	Jul	Avg
1	289.6	279.7	259.7	316.3	336.4	321.7	300.3	301.8
2	52.9	56.9	61.5	—[a]	59.9	67.9	70.7	61.6
6	—[a]	77.7	72.6	65.8	73.1	77.6	55.2	70.3
7	—[a]	48.7	—[a]	42.9	56.3	52.0	58.7	51.7
9	66.1	65.7	66.3	64.1	65.9	70.7	74.2	67.6
12	78.0	83.0	—[a]	77.0	74.4	76.2	78.3	77.9
13	99.1	115.5	110.7	101.1	115.3	129.9	102.5	110.6
15	172.3	174.2	184.0	—[a]	184.4	169.1	134.8	169.8
19	131.8	146.4	130.0	—[a]	156.6	163.8	—[a]	145.7
20	163.9	221.9	214.1	209.8	162.6	146.2	206.4	189.3
21	72.7	101.6	151.9	—[a]	106.4	146.8	51.2	105.1
23	105.4	92.0	109.8	118.9	121.9	112.4	—[a]	110.1

[a] Sample not obtained.

CONCLUSIONS

Variability in results of analyses for particular water-quality parameters, such as radon activity, may occur spatially with location and depth. Activity may also vary temporally in replicate groundwater samples. Variability may be due to natural effects such as contaminant source fluctuations and hydrogeologic interactions. Other factors may include well design, sampling devices, and sampling protocols.

The short-term sample collection effort in this study should be viewed only as a "snapshot" of the actual conditions prevalent during the sampling period. From the replicate study, it appears that it is necessary to monitor the radon concentration of well water over a longer period of time because of sample variability. A 1-y study would allow a reasonable average of radon concentration in samples from each well.

Liquid scintillation counting is quick and reliable. Sample preparation takes 1 to 2 min per sample, and the method allows for measuring a wide range of concentrations. The average total error of a sample concentration is approximately 6%. New scintillation cocktails, currently under development, which are both more efficient and biodegradable, will continue to improve the LS method.

ACKNOWLEDGMENT

The authors thank Ms. Melissa Chu, Ms. Jeanmarie Haney, and Mr. David Totman for their assistance during the project.

REFERENCES

Crawford-Brown, DJ and CR Cothern. **1987**. A bayesian analysis of scientific judgment of uncertainties in estimating risk due to ^{222}Rn in U.S. public drinking water supplies. Health Phys 53:11-21.

Dixon, KL and RG Lee. **1988**. Occurrence of radon in well supplies. J Am Water Works Assoc 80:65-70.

Kearfott, KJ. **1989**. Preliminary experiences with ^{222}Rn gas in Arizona homes. Health Phys 56:169-179.

Longtin, JP. **1988**. Occurrence of radon, radium, and uranium in groundwater. J Am Water Works Assoc 80:84-93.

Prichard, HM and TF Gesell. **1977**. Rapid measurements of ^{222}Rn concentrations in water with a commercial liquid scintillation counter. Health Phys 33:577-581.

Schönhofer, F. and E. Henrich. **1987**. Recent progress and applications of low level liquid scintillation counting. J Radioanal Nucl Chem 115:317-333.

QUESTIONS AND ANSWERS

Q: Reimer, USGS. On your sampling method, do you have any feel for how much radon would be lost by sampling within an open container from the turbulence of the water flowing through the hose into the bottle?

A: Barnett, Arizona State University. We did in-field sampling and the bottle method sampling, and we found that the loss was minimal. In fact, there were some cases where the sample that we took in the field had less of a radon concentration than when we actually brought it back to the laboratory and analyzed it. So realistically, within the error of the deviations (by the way, total average error of any sample is about 6% as a maximum) there was no difference between measuring in the field or bringing the water back to the lab.

RADON EXHALATION RATE FROM COAL ASHES AND BUILDING MATERIALS IN ITALY

A. Battaglia,[1] D. Capra,[1] G. Queirazza,[2] and A. Sampaolo[3]

[1]CISE SpA, Segrate, Italy

[2]ENEL/CRTN, Milano, Italy

[3]ENEL/CRC, Brindisi, Italy

Key words: *Building materials, exhalation rates, fly ash, natural radioactivity, radon*

ABSTRACT

The Italian National Electricity Board, in cooperation with Centro Informazioni Stubi Esperienze (CISE) has a program to assess the hazards connected with using fly ash in civil applications as partial substitutes for cement and other building materials.

We investigated the natural radioactivity levels of more than 200 building materials. The survey involved materials available in Italy, categorized by geographical location and type of production. We also examined approximately 100 samples of fly ash from United States and South African coal, obtained from Italian power plants.

Exhalation rates from about 50 powdered materials were determined by continuously measuring radon concentration growth in closed containers. Measurements were also performed on whole bricks, slabs, and tiles. Details about the high-sensitivity measuring devices are presented.

The influence of fly ash on exhalation rates was investigated by accurately measuring radon emanation from slabs with various ash/cement ratios and with slabs of inert materials having various radium concentrations.

We will discuss results of forecasting indoor radon concentrations under different ventilation conditions. Two identical test rooms are being built, one with conventional and one with fly-ash building materials, to compare theoretical calculations with experimental data. Specifications for instruments to control and to measure the most important parameters are also discussed.

INTRODUCTION

The use of coal for producing electricity in Italian power plants has brought up problems connected with the disposal or reclaiming of coal ash. The Italian National Electricity Board (ENEL) has a program to assess the hazards connected with using coal ash in civil applications as a partial substitute for cement and other building materials. This program, carried out in collaboration with CISE, refers specifically to radon emanation characteristics, because radon is the major radiological hazard in coal ash.

Our investigation, which extended over all of the Italian national territory, covered the following points: (1) the naturally occurring radionuclide content of different building materials selected by manufacturer's location and capacity and from a large number of ash samples taken from Italian power plants; (2) the radon emanation characteristics of those materials; (3) the radon emanation from materials using ash as a partial substitute for cement in concrete.

Laboratory test results will be used to predict radon concentrations in houses built of these materials, using a mathematical model currently being developed. The validity of this model will be tested by measuring indoor radon concentrations in two identical rooms that simulate those in a typical Italian house. One room is made of conventional building materials, the other of prefabricated panels containing known amounts of coal ash.

MATERIALS AND METHODS

Naturally Occurring Radionuclide Analysis

Germanium-detector gamma-ray spectrometry was used to determine ^{226}Ra, ^{228}Ra, and ^{40}K concentrations. Measurements were made in sealed containers 15 days after preparation of the samples to allow ^{222}Rn to attain secular equilibrium with ^{226}Ra. Radium-226 was assayed by assessing the activity of the ^{222}Rn daughters ^{214}Pb and ^{214}Bi. Analytical results (mean and standard deviation) are summarized in Table 1.

Emanating Power and Exhalation Rate

Emanating power is the ratio between the amount of radon given up to the environment and the amount of radon produced in the sample. Exhalation rate (Bq kg^{-1} s^{-1}) is the activity yielded by the sample per unit of mass and time.

Table 1. Concrete slab components in tests A and B; values expressed for dry materials.

	Inert materials		Cement		Water		Fly ash	
	A White Limestone	B Sand and gravel	A Portland	B Portland	A	B	U.S.	South Africa
% in weight	81	76	12.0	16.5	7.0	7.5	—	—
^{226}Ra, Bq kg^{-1}	1.6	17	12.8	15.4	—	—	175	150
Exhalation rate, Bq kg^{-1} s^{-1}	0.3	2.4	2.1	4.0	—	—	3.7	3.5
Emanating power, %	10	6.5	8.0	12.4	—	—	1.0	1.1

These two quantities were measured in (1) powder samples, analyzed without grinding, in cylindrical containers 11 cm dia and 8 cm high; (2) whole bricks and concrete blocks (crushed in pieces of 4 to 10 mm dia) analyzed in both their original form and ground up; (3) slabs (20 x 20 x 10 cm).

The effect of grinding on emanating power was assessed by measuring exhalation rate in samples with different particle sizes. Tests showed a slight increase in emanation for particle sizes of less than 0.5 mm, which agrees with other reports in the literature (Morawska, 1989).

Radioactive gas emanation was assayed by continuously measuring radon concentration in a box sized to ensure easily measurable concentrations and minimize back-diffusion within acceptable limits. Continuous measurements were made using CISE-designed flask monitors (Thomas and Countess, 1979) with an efficiency of 0.27 cpm Bq^{-1} m^{-3} and a background of 1.2 cpm. The free volume of the box/flask-monitor system (Figure 1) was 6.5 L for powdered samples and 15 L for larger samples.

Exhalation rate was calculated by evaluating the initial slope of the growth of the radon activity in a sealed container (Jonassen, 1983). Figure 2 shows radon concentration growth in a coal-ash-charged concrete sample, determined by making repeated measurements on the same sample.

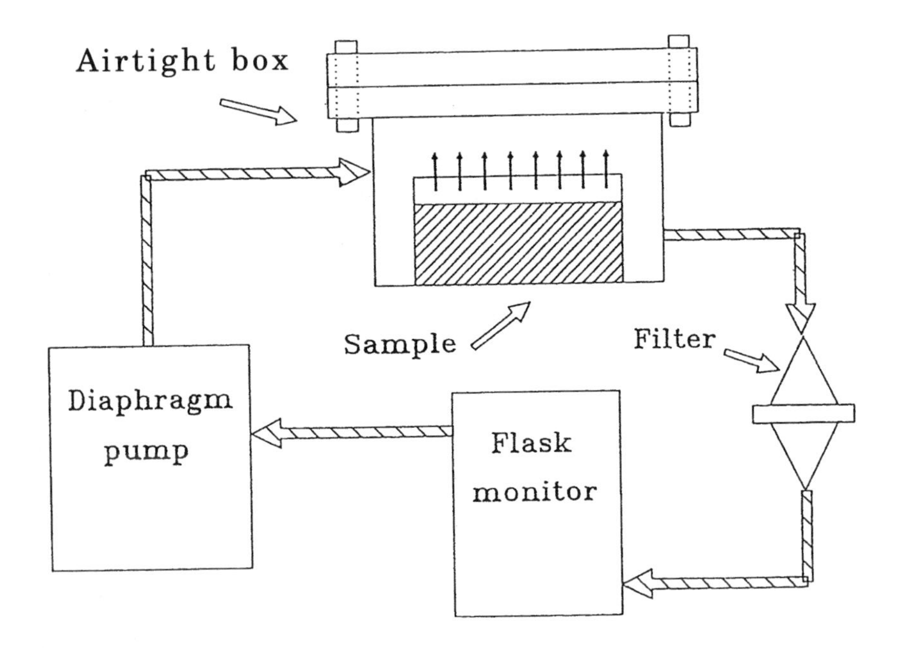

Figure 1. System for measuring exhalation rate of radon from cement.

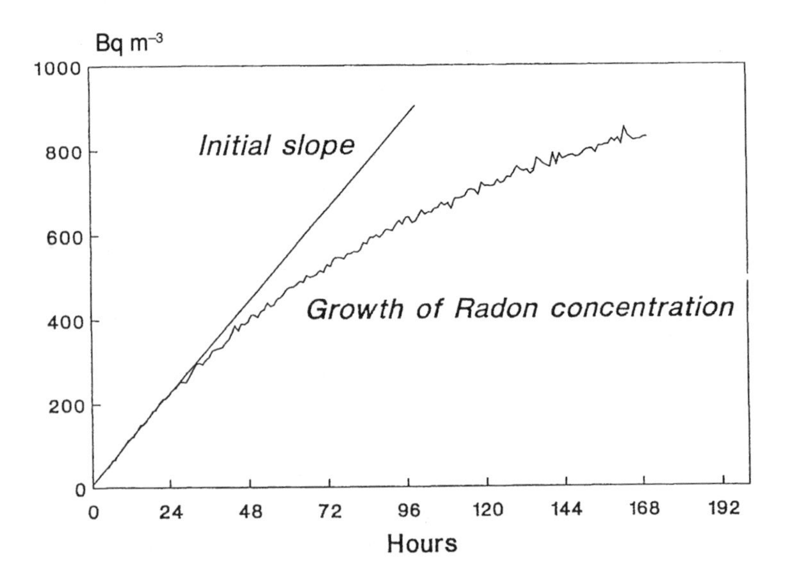

Figure 2. Growth of radon concentration from a 5% fly-ash cement slab.

Flask monitors were calibrated with an emanating source (PYLON Electronic Development mod. RN 1025-20) during the 3rd CEC Intercomparison at the National Radiation Protection Board (UK) (Miles and Sinnaeve, 1988); precision was better than 10%. Analysis results (mean and standard deviation) are summarized in Table 2.

Table 2. Natural radioactivity in fly-ash samples and building materials.

	^{226}Ra Bq kg^{-1}	^{228}Ra Bq kg^{-1}	^{40}K Bq kg^{-1}
Fly ash U.S. (71 samples)	170 ± 40	128 ± 26	471 ± 102
Fly ash South Africa (30 samples)	173 ± 25	176 ± 24	185 ± 53
Portland cement (27 samples)	27 ± 16	18 ± 13	211 ± 41
Pozzolano cement (21 samples)	49 ± 26	45 ± 39	394 ± 203
Sand and gravel (55 samples)	16 ± 6	18 ± 10	397 ± 217
Bricks (94 samples)	42 ± 16	45 ± 12	702 ± 196
Hydrated limes (10 samples)	10 ± 6	0.8 ± 0.5	8 ± 7
Tuffs (26 samples)	165 ± 119	196 ± 143	1640 ± 521
Pozzolano (15 samples)	208 ± 138	248 ± 160	1662 ± 445
Peperino (6 samples)	155 ± 57	187 ± 80	1424 ± 137

Radon Emanation from Concrete with Fly Ash

The effect of fly ash on radon emanation from concrete was investigated in two tests carried out on concrete in which cement was replaced with various percentages of fly ash.

Concrete samples (20 x 20 x 10 cm) were prepared according to a standard procedure for the study of mechanical properties and stored for 28 days in a climatic chamber maintained at 20 ± 2°C temperature and >90% relative humidity. Standard deviation of measurements on 10 samples with the same composition was better than 5%.

Two sets of samples were made of inert materials with different radioactivity levels (see Table 1). In the first set, Portland cement was replaced with 5%, 15%, and 30%, respectively, by weight, of either American or South African coal ash. In the second set, cement was replaced with 5%, 15%, 26%, and 40%, respectively, of the same kind of ash. Calculated exhalation rates for different mixtures of dry components and experimental data obtained from concrete slabs are given in Table 3.

Table 3. Exhalation rates and emanating power of fly ash and building materials.

		Exhalation rate, μBq kg^{-1} s^{-1}			
		A		B	
		Calculated	Measured	Calculated	Measured
	Without fly ash:	0.50	2.94 ± 0.14	2.48	5.49 ± 0.06
5% fly ash	U.S.	0.50	2.27 ± 0.11	2.64	5.77 ± 0.24
	South Africa	0.50	2.58 ± 0.06	2.63	5.63 ± 0.97
15% fly ash	U.S.	0.52	2.95 ± 0.03	2.94	5.74 ± 0.01
	South Africa	0.52	3.30 ± 0.15	2.91	6.01 ± 0.10
26–30% fly ash	U.S.	0.55	2.91 ± 0.14	3.27	6.65 ± 0.21
	South Africa	0.55	3.66 ± 0.13	3.22	6.63 ± 0.24
40% fly ash	U.S.	—	—	3.70	6.81 ± 0.01
	South Africa	—	—	3.62	7.20 ± 0.18

Physical Model

To assess the effect of coal ash on indoor radon concentrations, two identical rooms are being set up. One, already completed, is made of conventional building materials; the other, under construction, is made of prefabricated panels containing coal ash. The dimensions of these rooms are 3.4 x 3.8 x 2.15 m, and they are located on top of a 40-cm-high crawl space. Facilities are provided to produce any desired temperature, humidity and air-change conditions inside these rooms. The air-change rate is controlled by an SF_6 measurement system.

We continuously monitor radon concentrations inside the rooms, in the open air, and in the crawl space, as well as meteorological parameters (outdoor and indoor temperatures, differential atmospheric pressure, wind speed and direction, humidity).

RESULTS AND DISCUSSION

Naturally Occurring Radionuclides

Table 2 shows that significant differences between American and South African coal ash were detected only for ^{228}Ra and ^{40}K concentrations. The ^{226}Ra level in these samples is comparable with that found in tuff, pozzolana and peperite samples, while concentrations in cements and other building materials are 4 to 10 times lower.

Exhalation Rates

Exhalation rates in all groups of materials exhibit a marked variability, as shown by the high standard deviations involved (Table 4). Insofar as mean values are concerned, fly-ash exhalation rates are intermediate between those of Portland-pozzolano cements and the other materials tested. Fly ash has very low emanating values compared with those of brick.

Table 4. Exhalation rates and calculated values for dry components for slabs with different fly-ash composition.

	Exhalation rate, μBq k^{-1} s^{-1}	Emanating power, %
Fly ash (15 samples)	4.4 ± 3.4	1.2 ± 0.7
Portland cement (17 samples)	2.1 ± 1.3	5.2 ± 4.4
Pozzolano cement (15 samples)	9.3 ± 7.0	8.8 ± 4.4
Sand and gravel (27 samples)	2.3 ± 1.5	8.3 ± 3.9
Bricks (30 samples)	1.3 ± 1.0	1.7 ± 1.5
Hydrated lime (6 samples)	1.2 ± 0.5	8.6 ± 5.8

Effect of Adding Coal Ash to Concrete on Radon Emanation

In the test summarized in Table 4, American (^{228}Ra = 175 Bq kg^{-1}; exhalation rate, 3.7 μBq kg^{-1} s^{-1}) and South African (^{226}Ra = 150 Bq kg^{-1}; exhalation rate, 3.5 μBq kg^{-1} s^{-1}) coal ash was used.

Two types of inert material with different ^{226}Ra content and different exhalation rates were tested. Type A is a mixture of crushed white limestone of various sizes (^{226}Ra = 1.6 Bq kg^{-2}; exhalation rate, 0.3 μBq kg^{-1} s^{-1}). Type B is a mixture of sand and gravel from alluvial deposits in the Po Valley (^{226}Ra = 17 Bq kg^{-1}; exhalation rate, 2.4 μBq kg^{-1} s^{-1}). The Portland cement used in the tests showed rather high exhalation rates (2.1 and 4.0 μBq kg^{-1} s^{-1}, respectively).

Results clearly show that the addition of fly ash caused an increase in radon emanation. However, this increase is small compared with changes caused by using different kinds of inert materials and cement stocks. Samples with South African coal ash exhibited slightly higher radon emanation than those with American coal ash.

In any case, the radiological hazards connected with the use of fly ash in the building industry seem to be negligible, generally in the 10-20% range (Quinolos et al., 1989).

Results also show that the exhalation rate of the concrete samples is significantly higher than the sum of the exhalation rates of the single components. This is presumably because the radium atoms have a different location in the concrete structure and, to some extent, to the presence of water (Van der Lugt and Scholten, 1985).

Assuming that the inert materials in concrete do not change their emanation properties, the radon emanation of cement alone may be estimated in the concrete tested. In test A, the cement exhalation rate increased from 0.26 to 2.7 μBq kg^{-1} s^{-1}; in test B, from 0.66 to 3.7 μBq kg^{-1} s^{-1}. Concrete samples with fly ash exhibited increases on the same order.

CONCLUSION

The investigation carried out on more than 100 Italian inert materials showed a substantial homogeneity in natural radioactive levels, except for volcanic materials (tuff, pozzolana, etc.). The technique used for determining radon emanation proved highly reliable. Tests on concrete made with coal ash showed that the radiological hazard from radon was relatively negligible in comparison with that connected with using other building materials.

Results presented in this paper are preliminary, and further tests will be carried out using other building material mixtures having different radium content. Tests to be carried out in the simulated rooms will provide a quantitative evaluation of the fly-ash radiological impact under different environmental conditions.

ACKNOWLEDGMENT

We thank Mr. Guido Viglieno of I.S.M.E.S., Bergamo, for the preparation of the concrete samples and for our useful and interesting talks.

REFERENCES

Jonassen, N. **1983**. The determination of radon exhalation rates. Health Phys 45:369-376.

Miles, JCH and J Sinnaeve. **1988**. *Results of the Third CEC Intercomparison of Active and Passive Detectors for the Measurement of Radon and Radon Decay Products*, Report EUR 11882. Commission of European Communities, Lyons, France.

Morawska, L. **1989**. Investigation of a specific surface area of a material on the basis of Rn-222 emanation coefficient measurements. Health Phys 57:23-27.

Quinolos, LS, GJ Newton and MH Wilkening. **1989**. Estimation of indoor Rn-222 from concrete. Health Phys 56:107-109.

Thomas, JW and RJ Countess. **1979**. Continuous radon monitor. Health Phys 36:734-738.

Van der Lugt and LC Scholten. **1985**. Radon emanation from concrete and the influence of using fly ash in cement. Sci Tot Environ 45:143-150.

ORPHAN RADON DAUGHTERS AT DENVER RADIUM SITE

R. F. Holub, R. F. Droullard, and T. H. Davis

Bureau of Mines, Denver, Colorado

Key words: *Radon, thoron and their progeny, outdoors*

ABSTRACT

During 18 mo of sampling airborne radioactivity at a National Priority List ("Superfund") site in metropolitan Denver, Bureau of Mines personnel discovered radon daughters that are not supported by the parent radon gas. We refer to them as "orphan" daughters because the parent, radon, is not present in sufficient concentration to support the measured daughter products. Measurements of the "orphan" daughters were made continuously, using the Bureau-developed radon and working-level (radon-daughter) monitors. The data showed high equilibrium ratios, ranging from 0.7 to 3.5, for long periods of time. Repeated, high-volume, 15-min grab samples were made, using the modified Tsivoglou method, to measure radon daughters, to which thoron daughters contributed 26 ± 12%. On average, 28 ± 6% of the particulate activity was contributed by thoron daughters. Most samples were mixtures in which the ^{218}Po concentration was lower than that of ^{214}Pb and ^{214}Bi, in agreement with the high-equilibrium factors obtained from the continuous sampling data. In view of the short half-life of radon progeny, we conclude that the source of the orphan daughters is not far from the Superfund site. The mechanism of this phenomenon is not understood at this time, but we will discuss its possible significance in evaluating population doses.

INTRODUCTION

In 1988, the U.S. Bureau of Mines initiated an on-site monitoring effort for airborne radioactivity at the Robinson Brickyard (Robco), in downtown Denver, Colorado. The property is a major part of the Denver Radium National Priorities List (NPL) Site. The site is identified in the NPL because of radium, uranium, and associated radioactive materials found on the property. The Bureau of Mines has been identified as the "responsible" party.

The goal of the monitoring effort was to document the radiation emanating from the site, to establish background radiation levels, and to establish the effectiveness of remediating the site. A secondary purpose was to test a newly developed airborne radiation monitor for future use in environmental radiation monitoring.

EQUIPMENT USED

The monitoring site at the Robco property consists of the newly developed Environmental Radiation Monitor (ERM), based on Whittlestone's 1985 concept (Whittlestone, 1985), supporting instrumentation and recording devices, and a separate meteorological station. The monitoring site, west of the Robco property, was selected because of its close proximity to excavation activity on the site, the availability of electrical power, and the relative security of the trailer and equipment. Because the ERM must be operated at temperatures above freezing and within the operational range of electronics equipment, it is in a heated and air-conditioned trailer. Details about the operation of ERM will be given elsewhere.

The trailer (Figure 1) houses the ERM and its supporting electronic and recording equipment. Outside, the box contains the radon- and thoron-daughter detector and input for the radon detector, all at 1 m above ground. The weather station, mounted on the trailer roof, consists of the mast, which contains the sensors that measure wind speed and direction, air temperature, relative humidity, and barometric pressure. The necessary transducers and data acquisition system are located inside the trailer.

Figure 1. Trailer, weather station, and outside housing for the continuous WL detector and inlet for the radon detector. The Robco site is to the right.

Figure 2 shows the ERM inside the trailer. The ERM (Figure 3) is a version of a well-known, two-filter method. Its main functional parts are an aerosol generator, a measurement chamber, a Geiger-Müller detector, and supporting filters, pumps, and a flow meter. (Reference to specific products does not imply endorsement by the Bureau of Mines.) The total volume of the chamber is 400 L. The first filter is a 10.5-cm, glass-fiber filter; the second is a 2.5-cm filter faced with an end-window Geiger-Müller detector.

Figure 2. Environmental Radon Monitor (ERM), located inside trailer. At lower right is aerosol generator.

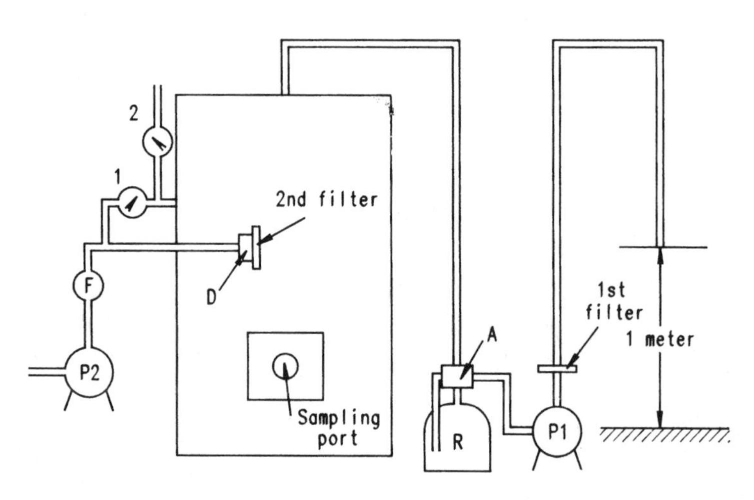

Figure 3. Environmental Radon Monitor (ERM); total volume of chamber is 400 L. Key: A, aerosol generator; R, distilled water reservoir; P, Pumps, 1 and 2; F, flow meter/regulator; D, Geiger-Müller detector; [gage symbol], Gages 1 and 2.

The critical feature, the aerosol generator, or atomizer head, has been described previously (Droullard et al., 1984). Typical diffusion battery/condensation nuclei counter measurement distributions of aerosol sizes are about 0.04 μm. Typically, the concentration of aerosol that fills the chamber has a particle number concentration that varies from 70 x 10^3 cm^{-3} to 110 x 10^3 cm^{-3}, with little change over extended periods of time.

Surprisingly, we use "distilled" water bought in grocery stores for aerosol generation. Pure, 18-MΩ, demineralized water, when used for the same purpose and filtered through a 0.8-μm filter, gave similar results. A preliminary conclusion is that water and, according to our findings, even air, contain microorganisms (and their debris after passing through the atomizer head with a pressure drop about 1.5 atm) that are responsible for the aerosols shown in Figure 3. We have also observed that the longer the water stands, the greater the number of these aerosols that fill the monitor chamber, to levels as high as 140 x 10^3 m^{-3}. This increase is presumably the result of microbial growth in the reservoir. Four liters of distilled water last for several weeks (depending on the relative humidity outside the chamber), providing sufficient time for the bacterial growth. We are developing a method to ensure that the aerosol level in the chamber is constant.

Introducing condensation nuclei at these concentrations greatly diminishes plate out of the radioactive aerosols on the walls of the chamber (Whittlestone, 1985). This process is the critical step in ensuring high aerosol sensitivity for radon collection. The radioactive aerosols are then collected and detected on the second, or output, filter.

The ERM can measure about 200 counts per minute per picocurie per liter (Holub and Droullard, 1989) and can provide instantaneous radiation readings of sufficient sensitivity to enable detailed analysis of sources, trends, and concentration variations. Most environmental radiation measurements have been confined to long-term integrating devices, usually for 60 to 90 days. (Gesell, 1983)

Radon and thoron daughters were continuously measured, in aggregate (Droullard and Holub, 1985) and individually, using grab-sampling methods (modified Tsivoglou method for radon), and counted on separate systems located in the trailer. Thoron daughters were measured using the standard Kusnetz method.

RESULTS AND DISCUSSION

Over 180,000 data points, collected continuously since October, 1988, have been amassed. The entire record now spans 22 mo for the parameters shown; namely, radon, radon daughter, working level ratio, air temperature, wind direction, wind speed, relative humidity, and barometric pressure. While certain correlations and patterns may be seen in this short record, some confounding factors are also present. The entire record will require extensive statistical analysis to extract the underlying relationships.

One confounding factor will be discussed in this report: The discovery of working level (disequilibrium) ratios greater than unity. Figure 4 presents these ratios for 5 days in November, 1988. The “orphan daughters” were observed in the morning hours each day except where there were strong winds.

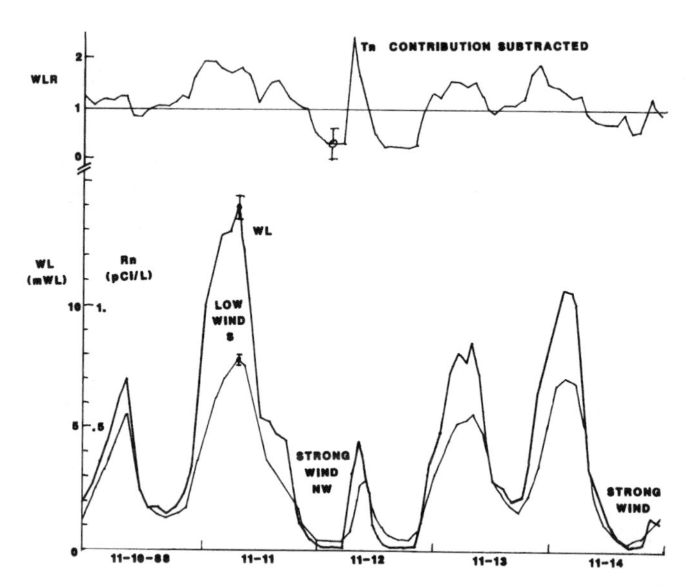

Figure 4. Plot of WL, radon, and WL ratio for typical 5-day period during November, 1988. Thoron contribution was subtracted. Error bars at very low WL are estimated as relatively large because of uncertainties in the ^{212}Pb contribution. WLR = working level ratio.

The values of WL, radon, and equilibrium ratio have had the thoron component (primarily ^{212}Pb, thorium B) subtracted. Even though we have not yet installed a solid-state detector and multichannel analyzer to measure radon and thoron components continuously, we have determined the ratio of these components for both continuous and grab-sampling measurements. For the former, based on eight measurements, it is 3.2 ± 0.8; for the latter, based on about 60 grab-sample measurements, it is 3.6 ± 1.7. Assuming that radon and thoron follow each other (Grum, D, and S Schery, 1990; private communication, New Mexico Institute of Mining Technology, Soccorro, NM), and because the daily variations from the thoron-daughter components are damped about 15 times more than those for the radon component, a constant value of 0.0019 WL and of 0.017 pCi L^{-1} for WL and radon, respectively, was subtracted from the combined (aggregate) data. The working level ratio (WLR), or disequilibrium factor curve (top, Figure 4) was obtained from these two subtracted components. To make sure the high WLR is not caused by miscalibration, both the WL monitor and EMR were recalibrated several times. The calibration factors have always been within ±5% of the assumed values. A separate calibration factor for thoron daughters has also been determined; it was about 10% lower than for radon daughters.

To further check our unusual findings we have taken about 90 grab-sample measurements (using the modified Tsivoglou method), usually in the morning hours when the activity was still quite high. Most of the measurements were duplicated by RT Beckman and his collaborators from the Mining Safety and Health Administration (MSHA). Their counting system was independent, and the results (Figure 5) have confirmed ours several times before (Cooper and Holub, 1990). The plot is adapted from triangular plots, which we have used (Holub and Droullard, 1977) to show the mixtures of radon daughters in mines, by enlarging the area of interest, which is low in ^{218}Po component. Unlike our data, most of the mine mixtures were higher in ^{218}Po. The first daughter, ^{218}Po, then decays with a 3.05-min half-life, then changes into ^{214}Pb, etc., until it reaches a position where the ^{218}Po concentration equals zero. The dashed curve is the natural aging curve of radon daughters from originally pure ^{218}Po to the equilibrium point, where $^{218}Po = {}^{214}Pb = {}^{214}Bi$.

Two features of the measurements are striking: (1) most mixtures are very low in ^{218}Po; (2) simultaneous measurements have large errors, especially in ^{218}Po concentrations. The first feature, we believe, is confirmation of the fact that radon daughters are not supported by sufficient quantities of a radon parent, even though radon has a much longer half-life than its daughters. Because the daughters exist in the atmosphere for short periods of time without their parent, we coined the name "orphan daughters," which succinctly describes the phenomenon.

This "orphan daughter" situation is important because our monitoring data suggest that measuring only radon does not give an upper limit to determine the exposure. Unfortunately, the magnitude of the error in ^{218}Po measurements makes it very difficult to pinpoint the distance of the source. Strong winds can reduce the airborne radioactivity, both gaseous and particulate, to negligible amounts. However, on less-windy days, outdoor exposures, especially to cancer-causing particulate radioactivities, may be several times higher than generally assumed.

It is premature to speculate on the mechanism of orphan daughter formation. However, it is conceivable that electrical or gravitational forces may provide mechanisms to separate gaseous and neutral radon from its solid phase and, possibly, from its charged daughters.

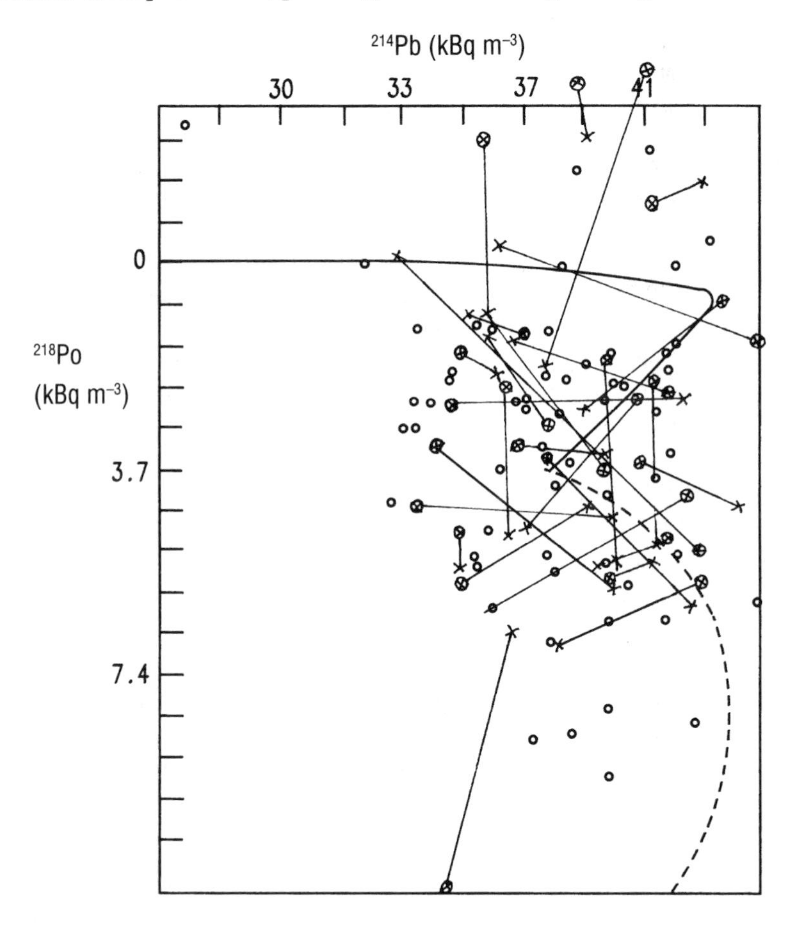

Figure 5. Plot of approximately 100 measurements of radon-daughter mixtures, made using modified Tsivoglou method. Open circles are earlier Bureau of Mines measurements; X, Mining Safety and Health Administration (MSHA); ⊗, measurements made simultaneously by Bureau of Mines and MSHA personnel. Length of straight lines joining simultaneous measurements gives indication of errors. Dashed line is natural aging curve marking mixtures prior to equilibrium. All data points normalized to 1 WL.

CONCLUSIONS

The ERM has provided a new level of sensitivity in airborne radioactivity measurements, similar to Whittlestone's but with different aerosol generation. The existence of "orphan daughters" is a phenomenon that is not yet understood; one that we hope will be confirmed when these sensitive measurements are repeated by others.

REFERENCES

Cooper, WE and RF Holub. **1990**. *International Intercalibration and Intercomparison Program, Radon Daughter Measurements*, IC 9257. US Bureau of Mines, Washington, DC.

Droullard, RF, TH Davis, EE Smith, and RF Holub. **1984**. *Radiation Hazards Test Facilities at the Denver Research Center*, IC 8965. US Bureau of Mines, Washington, DC.

Droullard, RF and RF Holub. **1985.** *Continuous Working Level Detectors*, IC 9029. US Bureau of Mines, Washington, DC.

Gesell, TF. **1983**. Background atmospheric ^{222}Rn concentrations outdoors and indoors: A Review. Health Phys 45:289-302.

Holub, RF and RF Droullard. **1977**. Evaluation of various radon daughter measurement methods, pp. 197-219. In: *Workshop on Methods for Measuring Radiation in and Around Uranium Mills*. Atomic Industrial Forum, Washington, DC.

Holub, RF and RF Droullard. **1989**. Further Tests of Modified Crump and Seinfeld Wall Loss Theory in Aerosol Chambers used in Two Filter Methods. Presented at American Association for Aerosol Research Annual Meeting, Reno, NV. October 9-13, 1989, Abstract 6C.4.

Whittlestone, S. **1985**. High-sensitivity radon detector incorporating a particle generator. Health Phys 49:847-852.

METHODS TO CONTROL RADON AND RADON PROGENY EXPOSURE

AIR CLEANING AND RADON DECAY PRODUCT MITIGATION

P. K. Hopke,[1] C.-S. Li,[2] and M. Ramamurthi[3]

[1]Clarkson University, Potsdam, New York

[2]John B. Pierce Foundation Laboratory, Yale University School of Medicine, New Haven, Connecticut [(a)]

[3]Battelle Columbus Laboratories, Columbus, Ohio

Key words: *Air cleaners, radon mitigation, unattached fraction, size distribution, room model*

ABSTRACT

We evaluated air cleaning as a means to mitigate risks arising from exposure to indoor radon progeny in several single-family houses in the northeastern United States, using a new, automated, semi-continuous activity-weighted size distribution measurement system. Measurements included radon concentration, condensation nuclei count, and activity-weighted size distribution of radon decay products. Measurements were made with and without the air cleaning system operating. The influence of particles generated by various sources common to normal indoor activities on radon progeny behavior was evaluated. Aerosols were generated by running water in a shower, burning candles, smoking cigarettes, vacuuming, opening doors, and cooking. Both a filtration unit and an electrostatic precipitator were evaluated. Using a room model, the changes in attachment rates, average attachment diameters, and deposition rates of the "unattached" fraction with and without the air cleaning systems were calculated. The air cleaner typically reduced the radon progeny concentrations by 50 to 60%.

INTRODUCTION

Recently, the U.S. Environmental Protection Agency (USEPA) estimated 20,000 deaths annually from radon-induced lung cancer (Puskin and Nelson, 1989). Because inhalation and lung deposition of radon progeny produce adverse health effects, it is necessary to control exposure to radon progeny in order to lower the risk from indoor radon. Three ways of controlling radon progeny in the air space of houses are: (1) prevent radon infiltration (basement pressurization, basement sealing, subslab depressurization, choice of building materials); (2) dilute the radon and radon progeny inside the house (air-to-air heat exchanger, ventilation); (3) use air-cleaning systems for direct

(a) Current Address: Institute of Public Health, College of Medicine, National Taiwan University, Taipei, Taiwan, ROC.

removal (filtration, mixing fan, electric field methods, radon adsorption, and ion generator). Various methods to reduce infiltration or to lower concentrations through increased ventilation have been under study for several years and can often be highly effective. However, we need interim solutions, like air cleaning, which lower exposure while long-term solutions are pursued. Because Rudnick et al. (1983) suggested that air-cleaning systems may increase risk in some situations, the USEPA (1986) has not endorsed the use of air cleaners as a method of reducing the risk from radon decay products in indoor air.

Previous evaluations of air-cleaning systems undertaken in test chambers or in indoor environments are reviewed in Hopke et al. (1990). The studies demonstrated that air-cleaning systems can effectively remove radon decay products from indoor air, although the reduction of ^{218}Po is smaller than that of ^{214}Pb and ^{214}Bi. At the same time, the particles are removed from the air and, as a result, the "unattached" fraction increases, especially for ^{218}Po. Because the "unattached" fractions of the radon progeny may more effectively deposit their radiation dose in lung tissue than attached fractions, the efficacy of air cleaning as a means of mitigating hazards from indoor radon has been questioned.

The main problem in previous studies is that systems used to measure radon progeny could not determine their full size distribution, especially when particle diameters were smaller than 10 nm. Size measurement methods and results are not clearly stated. Therefore, research is needed to measure the concentration and size distribution of radon progeny activities when air-cleaning devices are employed.

MATERIALS AND METHODS

We evaluated an air filtration system in a single-family house in Princeton, New Jersey. We measured the concentration and size distribution of radon progeny with and without the air cleaner and with various methods of generating particles. Aerosols were generated by burning candles, allowing cigarettes to smolder, vacuuming, opening doors, drying clothes, and cooking. Using a simple room model (Jacobi, 1972; Porstendörfer et al., 1978), we calculated changes in attachment rate, average attachment diameter, and deposition rate of the "unattached" fraction with and without the air filtration system.

AIR FILTRATION SYSTEM

We used a Pureflow Air Treatment System (Amway Corporation, Ada, MI). It is a multistage filtering system containing a total of five separate filters: a flexible, foam prefilter which traps hair, lint, and large dust particles; two activated-carbon filters that contain 2 lb each of proprietary blends of various specially treated activated carbons, supported uniformly in honeycomb beds, covered with white filter media, and encased in a high-quality aluminum frame; a final carbon filter composed of activated carbon bonded to a nonwoven substrate; and a high-efficiency particulate air (HEPA) filter that is designed to maximize filter efficiency, eliminate leaks, and reduce the possibility of breakage. This HEPA filter contains the same filtering material used in many hospital surgical-suite air filtration systems and industry "clean" rooms. The prefilter and activated-carbon filters are installed at the back of the system; the third carbon filter and the HEPA filter are in the front. The system's microcomputer allows programming of up to four different sets of ON-OFF times and fan speeds for automatic operation; there are four fan speeds: 1.13, 2.26, 3.40, and 4.25 $m^3\ min^{-1}$. The system gives maximum particle-removal efficiency when operating at the highest speed.

MEASUREMENT SYSTEM FOR RADON-PROGENY SIZE MEASUREMENTS

The measurement system uses sampler/detector units incorporating wire-screen and alpha surface-barrier detectors to determine the activity size distributions (Ramamurthi and Hopke, 1991) on a semi-continuous basis. Screens are wrapped around the sampler, covering an entrance slit except for one sampler, which is left open to measure the total progeny concentration. The system draws air simultaneously through all six units at independent flow rates (parallel operation). Each sampler-detector unit is thus an independent stage that separates the airborne activity based on its diffusion coefficient. Detectors in the sampler/detector units are connected to a multichannel analyzer through a multiplexer. The personal-computer-based analysis system allows us to determine the concentrations of ^{218}Po, ^{214}Pb and ^{214}Bi (or ^{214}Po) that penetrate to the filter in each sampler-detector unit using alpha-spectroscopy with counting while sampling. Given the penetration characteristics and the measured detection efficiency of each sampler, the activity size distributions can then be calculated from each of the observed radon-progeny activities and for the potential alpha energy concentration (PAEC) (Ramamurthi and Hopke, 1990).

FIELD SAMPLING IN PRINCETON, NEW JERSEY

Our measurements were made in collaboration with the Center for Energy and Environmental Studies (CEES) at Princeton University, which has identified and instrumented houses for continuous measurement of radon and certain physical parameters. These include basement, bedroom, and subslab radon concentrations; pressure differences across the basement/outside south interfaces, basement/subslab, basement/upstairs, and basement/outside east interface; temperatures in the basement, upstairs, outside, and in soil. House 21 was a one-story residence with a living room, a dining room, a kitchen, two bedrooms, a study, two bathrooms, and basement (Figure 1). Activity size distributions were also measured in the living room (10 measurements) and the master bedroom (more than 100) over a 2-wk period. Particle concentrations were measured by using a Gardner (Gardner Associates, Schenectady, NY) manual condensation nucleus counter. The concentrations and size distributions of radon progeny were determined using a 75-min total interval to maximize the number of distributions measured on the evolving aerosol. The air filtration system was evaluated by making measurements using different types of particle generation with and without the air cleaner operating. Particles were produced by a candle burning, a cigarette smoldering, vacuuming (electric motor), and cooking. The particle concentration in the kitchen was always 80,000 to 100,000 cm^{-3} because of the gas-stove pilot lights. In the living room and dining room, particle concentration was in the range of 9,000 to 13,000 cm^{-3}, and in the two bedrooms, it was 5,500 to 10,000 cm^{-3} with the doors open.

The air-cleaning studies were performed in the master bedroom, which adjoined one of the bathrooms. The infiltration rate measured by CEES was approximately 0.3 h^{-1} with the interior doors open and all windows closed. Therefore, the ventilation rate was assumed to be 0.2 h^{-1} with the bedroom door closed. Figure 2 shows typical activity size distributions without air filtration and with the air-filtration system running for 2 h.

With the bedroom door closed and the air-filtration system off, conditions in the bedroom were: (1) particle concentration, about 2,500 - 3,000 cm^{-3}; (2) unattached fraction of ^{218}Po, between 0.5 and 0.7;

(3) unattached fraction of PAEC, between 0.2 and 0.4; (4) equilibrium factor, in the range of 0.3 and 0.5; (5) attachment rate, between 6 and 14 h^{-1}; (6) average attachment diameter, about 50 - 90 nm; and (7) deposition rate of unattached fraction, around 4 - 12 h^{-1}. With the bedroom door closed and the air filtration operating, more than 95% of ^{218}Po was in the 0.5- to 1.6-nm size range. The fractions of ^{214}Pb and ^{214}Bi in the 0.9-nm size range were about 80% and 50%, respectively. These results show that the air-filtration system effectively removed most of the particles in the bedroom. There was an insignificant fraction of the three distributions in the 1.6- to 16-nm size range.

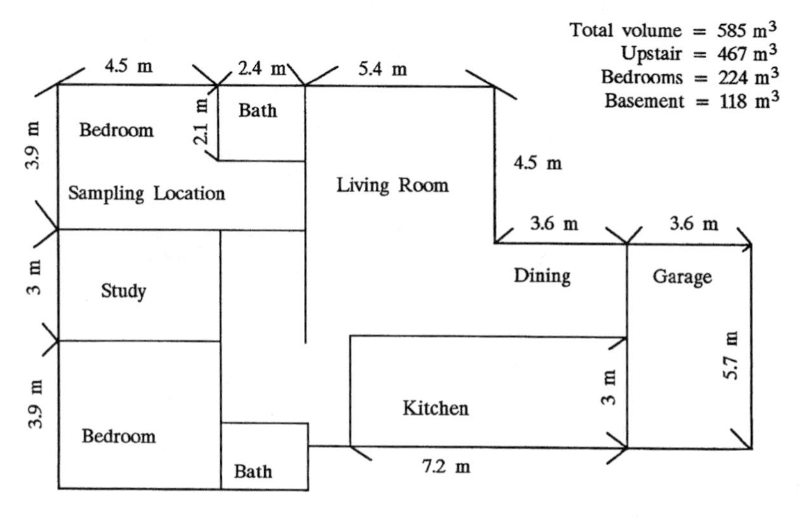

Figure 1. Princeton House No. 21 (first floor plan), in which radon decay products were measured.

With the bedroom door closed and the filtration system on, conditions were: (1) particle concentration, about 1,500 - 2,000 cm^{-3}; (2) unattached fraction of ^{218}Po, between 0.93 and 0.97; (3) unattached fraction of PAEC, between 0.67 and 0.83; (4) equilibrium factor, in the range of 0.10 - 0.16; (5) attachment rate, 0.4 - 2.0 h^{-1}; (6) average attachment diameter, about 20 - 30 nm; and (7) deposition rate of unattached fraction, between 8 and 16 h^{-1}. The average working level reduction was about 55%.

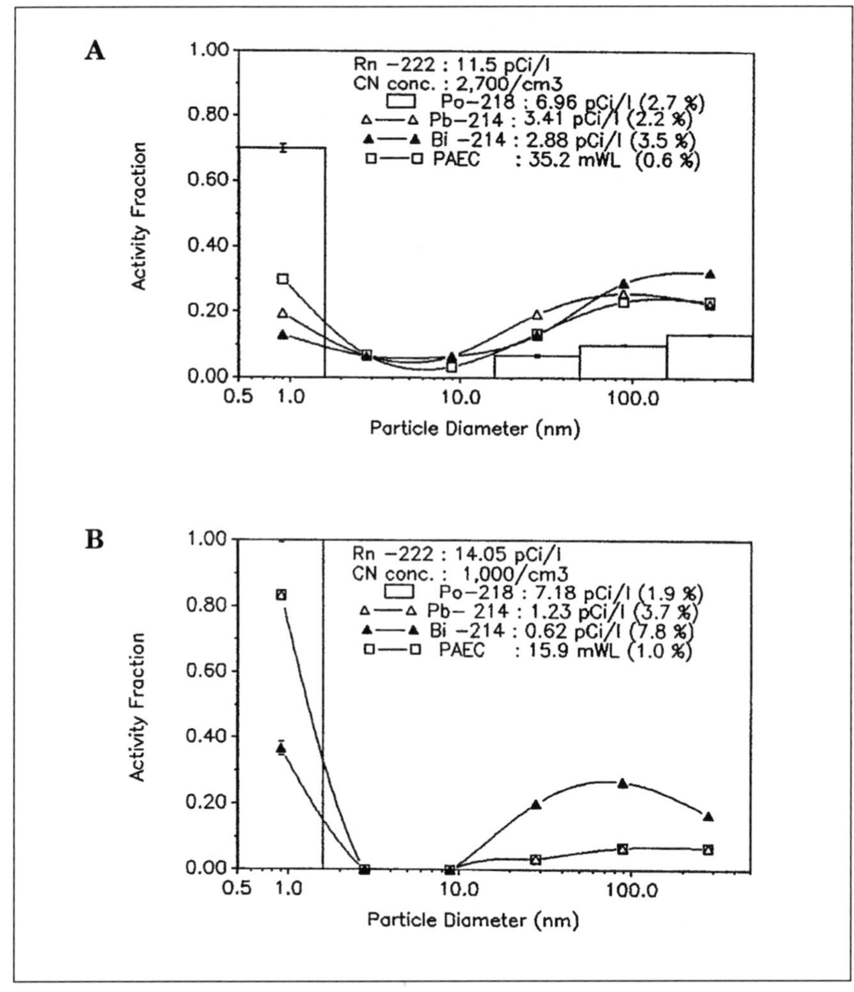

Figure 2. Radon progeny-activity size distributions in closed bedroom of house described before, A, and after, B, starting air cleaner. PAEC = potential alpha energy concentration.

The influence of cigarette smoke on radon progeny was evaluated in the closed bedroom (Figures 3 and 4). A cigarette was allowed to burn for 20 min, producing only sidestream smoke. Measurements were made 5 min after lighting the cigarette (5-20 min), 60 min after the end of the burn interval (80-95 min from the experiment start), and 135 min later (155-170 min).

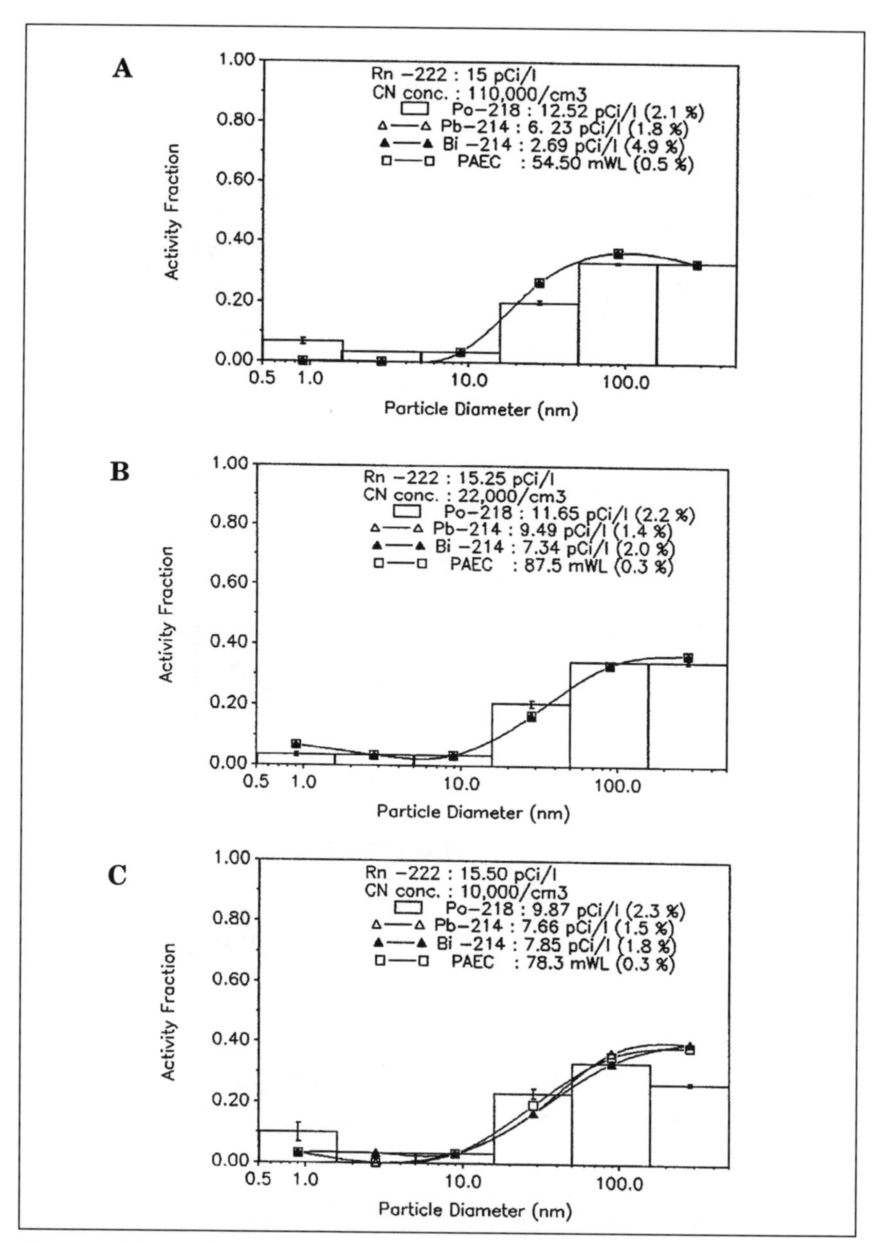

Figure 3. Changes in radon progeny-activity size distribution produced in Figure 1, while burning a cigarette in a closed room. A, without air filtration; B, 60 min after cigarette-burning ceased; C, 135 min after cigarette-burning ceased. PAEC = potential alpha energy concentration.

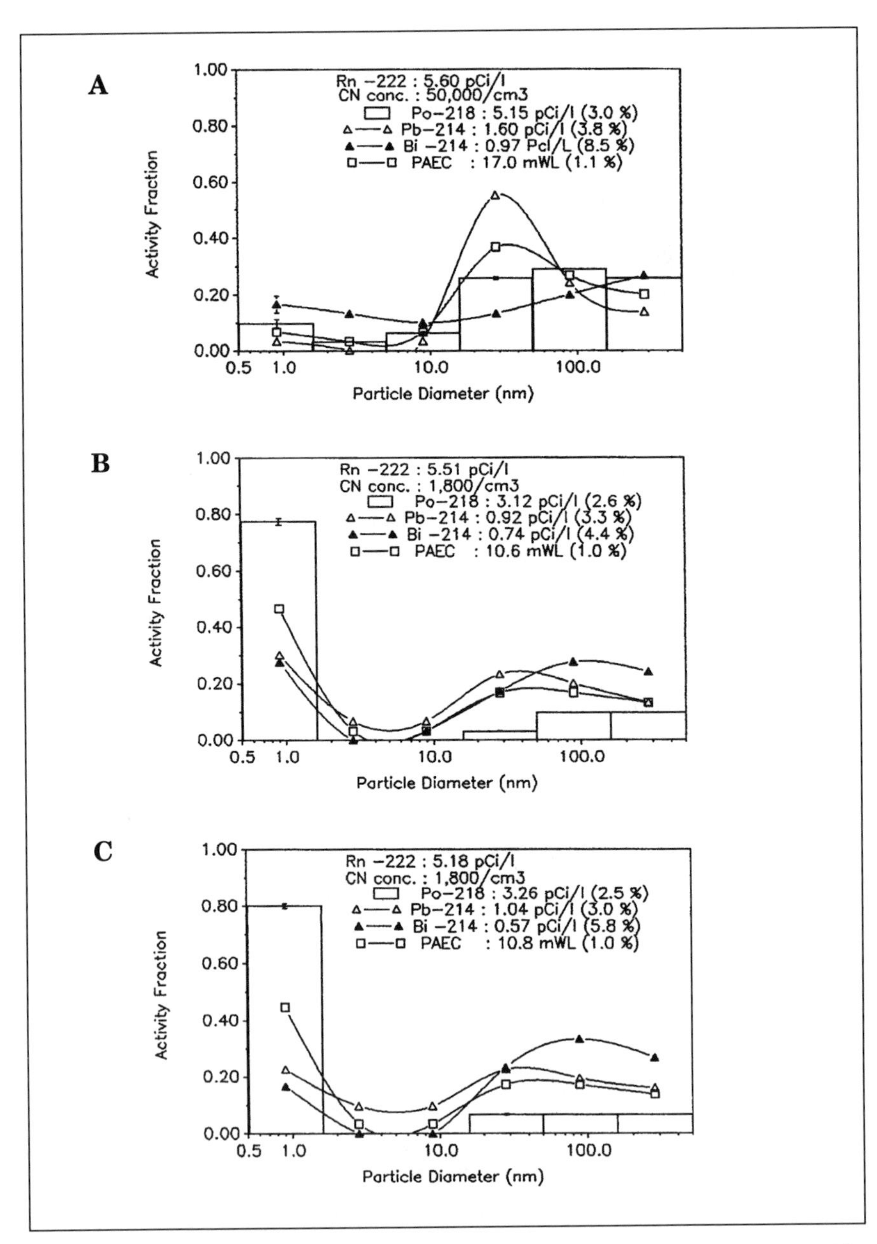

Figure 4. Changes in radon progeny-activity size distribution produced in Figure 1, while burning a cigarette in a closed room. A, with air filtration continuously operating; B, 60 min after cigarette-burning ceased; C, 135 min after cigarette-burning ceased. PAEC = potential alpha energy concentration.

For measurements without the air cleaner, the fraction of ^{218}Po in the 0.5- to 1.6-nm size range changed from 60% to 8%. The fraction of ^{214}Pb and ^{214}Bi in this smallest size range was about 10% before the cigarette was lit and decreased to essentially zero. The fraction of three distributions in the 1.5- to 15-nm size range stayed the same. A large increase (from 40% to 80%) in the 50- to 500-nm size range was observed for ^{218}Po, but changes for ^{214}Pb and ^{214}Bi were small (from 35% to 40%). Without the filtration system, the changes produced by cigarette smoke on the behavior of radon progeny were: (1) particle concentration increased significantly (100,000 cm^{-3}) while the cigarette burned, then declined to a steady-state concentration of around 8,000 cm^{-3}; (2) unattached fraction of ^{218}Po and PAEC decreased from 0.67 to 0.13 and from 0.30 to 0.07 (after 5 min); to 0.10 and to 0.10 (60 minutes after the end of the burn); to 0.17 and to 0.07 (135 minutes after the burn); (3) equilibrium factor increased from 0.31 to 0.40 (after 5 min); to 0.60 (60 min); to 0.55 (135 min); (4) attachment rate increased from 8 h^{-1} to 110 h^{-1} after 5 min; to 120 h^{-1} at 60 min later; to 60 h^{-1} at 135 min after the end of the burn; (5) average diameter decreased from 60 nm to 50 nm after 5 min, increased to 120 nm at 60 min, and remained constant to 135 min after the end of active aerosol generation.

The influence of the air cleaner was evaluated in the closed bedroom with the air-filtration system running continuously. The specifications for burning the cigarette and the sampling times were the same as in the previous experiment. The fraction of ^{218}Po changed from 95% to 10% in the smallest size range after 5 min, then increased to 80% in the later distributions. The fraction of ^{214}Pb and ^{214}Bi in the 1.6- to 5.0-nm range was ~8% after 5 min, but returned to ~25% by the 60-min measurement. Very little activity appeared in the 1.6- to 15-nm size range for samples at 60 or 135 min, but 15% was present at 5 min. The increase (from 20% to 80%) in the "attached" mode peaked in the 15- to 500-nm size range after 5 min, then returned to the background condition by the second measurement. With the air filtration system, the effect on the radon decay products by cigarette smoke were: (1) particle concentration increased significantly (60,000 cm^{-3}) after 5 min, then reached steady state around 1,800 cm^{-3}; (2) unattached fraction of ^{218}Po and PAEC decreased from 0.93 to 0.13 and from 0.66 to 0.11; (3) equilibrium factor increased from 0.13 to 0.3; (4) attachment rate increased from 1 h^{-1} to 100 h^{-1} after 5 min, to 4 h^{-1} by the 60-min sample; (5) average diameter increased from 40 nm to 60 nm after 5 min, and to 60 nm after 60 min; (6) deposition rate of unattached fraction increased from 8 h^{-1} to 12 h^{-1} after 5 min. With the air filtration system, the working level decreased 25% after 5 min, 68% after 60 min, and 60% after 135 min, compared with the working level observed with the unfiltered cigarette smoke.

Particle concentration reached 40,000 cm^{-3} during vacuuming. By the second measurement, particle concentration declined to a steady state of 1,800 cm^{-3} (Figures 5 and 6). Activity in the smallest size range was 70% for ^{218}Po and 30% for ^{214}Pb and ^{214}Bi. The unattached fraction of ^{218}Po decreased from 0.93 to 0.80, and the unattached fraction of PAEC decreased from 0.74 to 0.62. The equilibrium factor did not change significantly. A moderate decrease in the unattached fraction during vacuuming was observed both with and without the air filtration system. A 25% increase in the 1.5- to 50-nm size range was observed for ^{218}Po because decay products attached to the small particles generated from vacuuming. A 30% increase in the 1.5- to 150-nm size range was also observed for all three size distributions. The size distributions of the 60- and 130-min measurements were very similar to those during background measurement. This might be because larger particles were removed in the bedroom and because of a lower attachment rate of aerosols during vacuuming. The attachment rate increased from 1 h^{-1} to 3.5 h^{-1} during the active particle period, and decreased to 0.8 h^{-1} after 60 and 135 min, respectively. The average diameter decreased from 30 nm to 15 nm during the active particle period, and to 26 nm after 60 and 135 min, respectively. The average reduction in PAEC per Bq m^{-3} radon was 58% during the active particle period, 59% after 60 min, and 50% after 135 min, compared with the PAEC observed with vacuuming alone.

CONCLUSIONS

Under normal conditions in this domestic environment, the unattached fraction of PAEC varied over a wide range (7% to 40%), and the equilibrium factor was between 0.13 and 0.50. However, these two terms are strongly dependent on when measurements were made in the house. Because a large number of particles were generated from normal household activities, the working level increased during that time, and the unattached fraction decreased. Particles generated from cigarette-smoking and cooking dramatically shifted most radon progeny to the attached mode, where they remained for a long time after aerosol generation. Particles produced by candle-burning and vacuuming were much smaller (average attachment diameters, ~15 nm) and decreased the unattached fraction. However, size distributions had returned to background conditions by 150 min later.

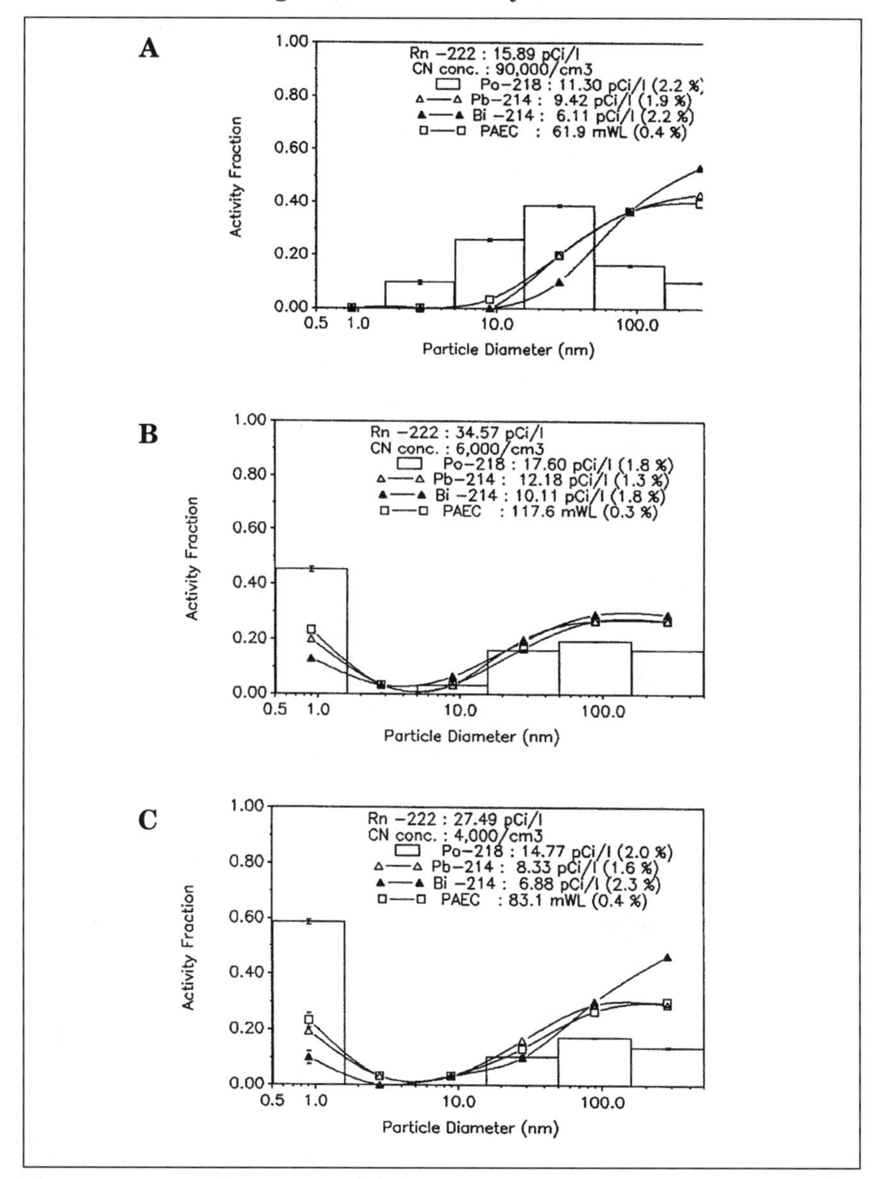

Figure 5. Changes in radon progeny-activity size distribution produced in Figure 1, while vacuuming in a closed room. A, without air filtration; B, 60 min after vacuuming ceased; C, 135 min after vacuuming ceased. PAEC = potential alpha energy concentration.

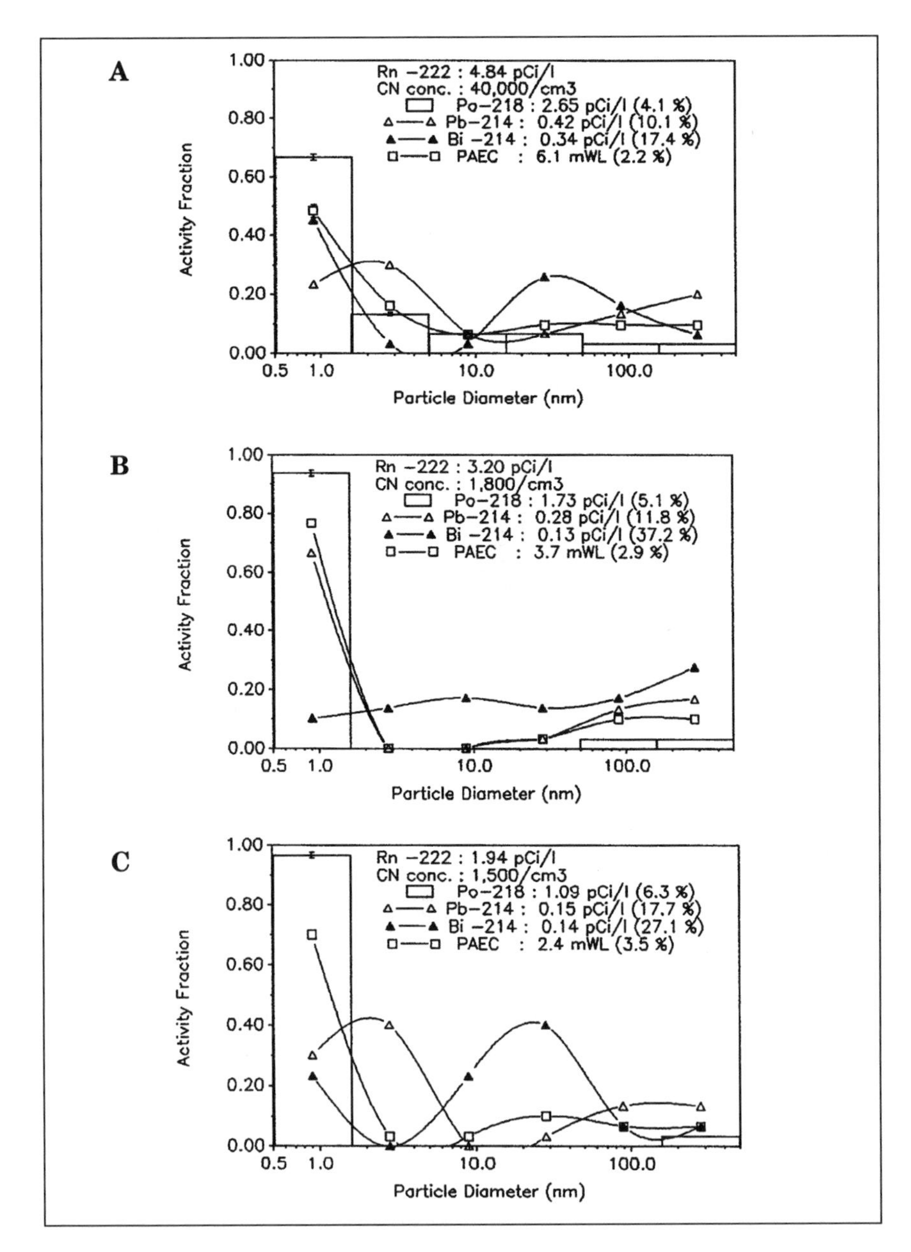

Figure 6. Changes in radon progeny-activity size distribution produced in Figure 1, while vacuuming in a closed room. A, with air filtration continuously operating; B, 60 min after vacuuming ceased; C, 135 min after vacuuming ceased. PAEC = potential alpha energy concentration.

With the air filtration system operating, the unattached fraction of PAEC substantially increased to 75% to 80%, and the equilibrium factor was about 0.14. The working level reduction was about 45% to 55%. During candle-burning, cigarette-smoking, and vacuuming in the bedroom, working levels were reduced by air filtration about 60%. During cooking, working levels in the bedroom were reduced by air filtration about 40%.

ACKNOWLEDGMENTS

This work was supported by the New Jersey Department of Environmental Protection through contract J89-62 and by the U.S. Department of Energy through Grant No DE FG02 89ER60876. The collaboration and assistance of K. Gadsby, A. Cavallo, and R. Socolow of the Center for Energy and Environmental Studies at Princeton University are gratefully acknowledged.

REFERENCES

Hopke, PK, CS Li, and M Ramamurthi. **1990**. *Review of the Scientific Studies of Air Cleaning as a Method of Mitigating the Health Risks from Radon and its Decay Products in Indoor Air*, Report to New Jersey Department of Environmental Protection, Trenton, NJ.

Porstendörfer, J, A Wicke, and A Schraub. **1978**. The influence of exhalation, ventilation and depositional processes upon the concentration of radon (^{222}Rn), thoron (^{220}Rn), and the decay products in room air. Health Phys 34:465-473.

Puskin, JS and CB Nelson. **1989**. EPA's perspective on risks from residential radon exposures. J Air Waste Management Assoc 39:915-920.

Ramamurthi, M and PK Hopke. **1989**. On improving the validity of wire screen "unattached" fraction Rn daughter measurements. Health Phys 56:189-194.

Ramamurthi, M and PK Hopke. **1990**. Simulation studies of reconstruction algorithms for the determination of optimum operating parameters and resolution of graded screen array systems (nonconventional diffusion batteries). Aerosol Sci Technol 12:700-710.

Ramamurthi, M and PK Hopke. **1991** An automated, semi-continuous system for measuring indoor radon progeny activity-weighted size distributions, d_p: 0.5-500 nm. Aerosol Sci Technol 14:82-92.

Rudnick, SN, WC Hinds, EF Maher, and MW First. **1983**. Effect of plateout, air motion and dust removal on radon decay product concentration in a simulated residence. Health Phys 45:463-470.

USEPA. **1986**. *A Citizen's Guide to Radon*, ODA-86-004, U.S. Environmental Protection Agency, Washington, DC.

DEVELOPMENT OF APPARATUS AND PROCEDURES FOR EVALUATING RADON-RESISTANT CONSTRUCTION MATERIALS

T. D. Pugh,[1] M. B. Greenfield,[1] J. MacKenzie,[1] and R. J. de Meijer[2]

[1]Florida A & M University, Tallahassee, Florida

[2]Kernfysisch Versneller Instituut, Groningen, The Netherlands

Key words: *Materials, construction, radon resistance, evaluation apparatus*

ABSTRACT

Laboratory facilities and apparatus have been constructed to measure radon exhalation from, and radon permeability through, various construction materials. This phase of the project has focused on development of test apparatus and evaluation of instrumentation. Results indicate significant spatial variability in the radon permeability of polyethylene, even when all test samples were selected from the same roll of material, and when no visible differentiation could be made regarding sample quality. Implications for code enforcement are described, and recommendations are offered for refinement of equipment and the measurement process, prioritization of future materials testing, and specific building code provisions, based on our results.

INTRODUCTION

The distribution of radon concentrations in dwellings is usually well described by a lognormal probability function. This indicates that many factors determine the concentration in a specific structure. In general, the contribution of the underlying soil is considerably larger than that of the building materials. The physics of radon transport is described either by diffusion or pressure-driven flow. The fact that concentrations are nevertheless lognormally distributed indicates the complexities underlying indoor radon concentrations.

BACKGROUND

Previous studies by others who measured radon concentrations both in the air inside houses and in the surrounding soil indicate that some residences tend to have low indoor radon concentrations, despite being built without special consideration for radon resistance (Nagda et al.).

It is commonly assumed that reduction of convective radon entry into a building yields the most significant results. The work of Scott and Findlay (1987), in which test houses were built on reclaimed phosphate mining land with various passive barrier techniques incorporated in the floor and substructure illustrates the point. These houses were shown to be resistant to radon entry and accumulation; e.g., soil gas ≈ 148 kBq m^{-3} (≈ 4,000 pCi $^{-1}$), and indoor air ≈ 55 Bq m^{-3} (≈ 1.5 pCi L^{-1}), supporting this assumption. These studies, however, were not designed to evaluate the contribution of convection and diffusion through each component in the barrier system. Nor can their methodology be used to evaluate new materials or techniques, except by construction of another test structure. Therefore, they do not provide information which may be used to predict the reduction factor of the barrier or to indicate the radon level after completion of the structure.

To predict the reduction factor of structures with any certainty, one must know the convective and diffusive radon transport characteristics of each component of the barrier. Moreover, as states begin to regulate the construction industry with regard to radon, it is logical to anticipate that many products will be marketed that will purport to reduce or eliminate radon transmission. A standard method for quantifying the effectiveness of these products is imperative.

Radon entry via diffusion is often slight relative to convective transport in structures with elevated indoor radon levels. However, the national goal of achieving indoor radon concentrations that approach outdoor ambient levels makes radon entry via diffusion relatively more important. Clearly, as convective transport is eliminated, diffusive entry will become dominant. The radon reduction rate, which is determined by blocking various routes of entry, is limited by radon diffusion through the barrier materials. It is important to evaluate building assemblies rather than only single pieces of material. Others have evaluated the transmission of radon through various substances such as concrete (Zapalac, 1983) and polyethylene (Hammon et al., 1977). However, the building substructure is composed of many separate materials. To use an analogy, glass is airtight, while windows leak.

The complexity of several materials that are assembled in complicated geometries makes analysis of these effects impractical. In the field of thermal resistance analysis, where multiple transport mechanisms are active, physical testing of entire assemblies has long been the norm. It will ultimately be necessary to measure, in their final assembled state, the radon resistance of various combinations of materials commonly required for building construction to properly evaluate their resistance to radon diffusion.

The aim of our investigation was to understand the modes of reducing indoor radon concentrations and to indicate the level to which the reduction will lead. We hope to achieve this by measuring the radon transparency of building materials and full-scale assemblies in a radon-diffusion test chamber. As a first step, we investigated the possibilities and limitations of the test facility and the development of a model to describe our observations.

EXPERIMENTAL TECHNIQUE

The concept of the present setup is similar to that of previous radon diffusion chambers: two gas-tight volumes, separated by the assembly or material being tested. One volume is charged from a radon source, and the other is connected to a detection device.

The dimensions of the setup are a compromise between having the minimum acceptable size of the assembly test sample and being practical to handle without large, motor-driven equipment. An additional boundary condition is the sensitivity of radon detection, which is related to the surface area of the sample, with "off-the-shelf" equipment.

As a result (see Figures 1 and 2), a rectangular test chamber was designed, with an inner surface of 43 x 64 cm (17 in. x 25 in.). This accommodates a sample size of 41 x 61 cm (16 in. x 24 in.), while providing a 1.2 cm (0.5-in.)-wide sealant space around the edge of the sample. The chambers are fabricated from A36 steel. Both the upper and lower chambers are about 10 cm (4 in.) deep, providing adequate clearance for capping pipes that pass through the test sample, while maintaining a reasonable volume for the source material. The two halves of the chamber are sealed by a 5-cm (2.5-in.)-wide, 5-mm (0.125-in.)-thick neoprene gasket.

For detecting high radon concentrations in the source chamber, 0.3-L Lucas cells were used; for the concentration in the detection chamber, Pylon Trace Environmental Level (TEL) monitors were used. Both detection systems were used in combination with Pylon AB-5 counters. A standard measurement protocol was established: Chambers were ventilated with nitrogen to outdoor air prior to disassembly. The test material or assembly was mounted in the test chamber, and nitrogen was introduced to raise chamber pressure 5 kPa, monitored with a digital micromanometer. A background alpha count was then performed for at least four nitrogen cycles (24 h), then radon was introduced in the source chamber. Alpha counts, recorded internally by the counters, were downloaded into the computer at the end of each run. Pressures were recorded continually and stored in hourly increments, and flows were monitored and recorded manually.

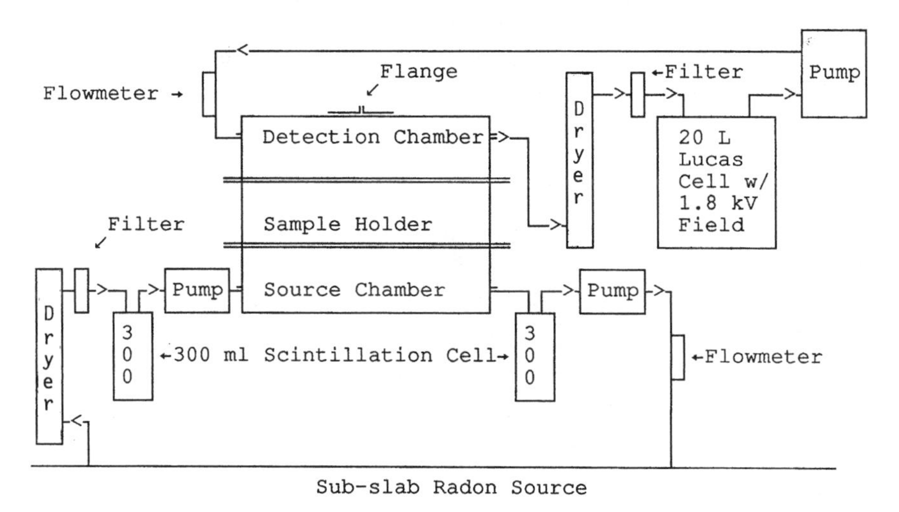

Figure 1. Diagram of the typical experimental setup for evaluating radon-resistant materials.

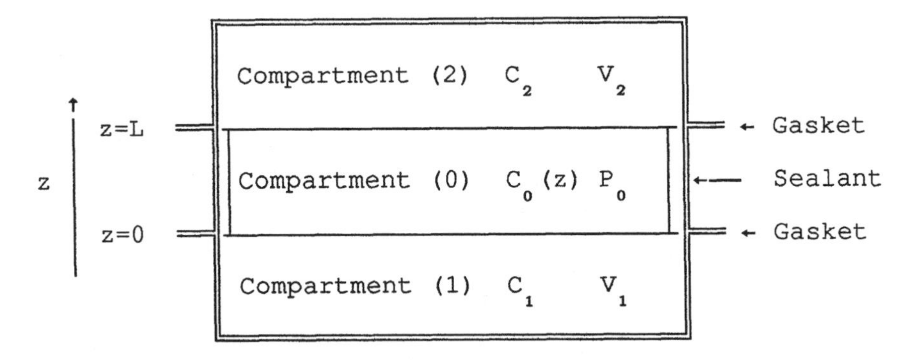

Figure 2. Schematic representation of a material which produces P_0 Bq m^{-3} s^{-1} placed between two compartments with volumes V_1 and V_2, and radon concentrations C_2 and C_2

This system may be adequate for materials with a high permeability to radon, but it has deficiencies when testing low-permeability materials. Therefore, for very low concentrations, a flange with a surface barrier detector was recently installed in the flat surface of one detection chamber. By applying a high voltage between the setup and a foil just in front of the detector, radon decay products are electrostatically collected on the foil and are detected by counting the alpha-particles emitted in their decay. The system has recently been optimized by Aldenkamp et al. (1987).

With the TEL-based detection systems, the radon and nitrogen mixture in the detection chamber is circulated for an hour via a filter, a desiccant, a flowmeter, the TFL and the AB-5 pump. The hour of pumping is followed by five intervals (1 h each) of counting. The results of only the last two intervals are used because equilibrium between radon and its decay products is reached after about 3 h, and the sensitivity of the detection system is known only for equilibrium situations. We found that the circulation was prone to leaks, therefore data at low concentrations could be misinterpreted when using Lucas cell or TEL techniques because radon leaked into the detection chamber from the room air.

Background measurements were made before and after each sample test after flushing the setup with dry nitrogen. For both the TEL-based and semiconductor detection systems, radon for the source chamber was obtained by circulating air from the building's subslab soil via a drier, a filter, a flowmeter and a Lucas cell, into the source compartment and back into the subslab soil. No matter what outside conditions were, the radon concentration in this source was surprisingly constant over a period of a month or longer. Concentrations of 18 kBq m^{-3} (500 pCi L^{-1}) were obtained in the source chamber in this way.

THEORETICAL DESCRIPTION

An expression for the one-dimensional diffusive transport of radon through materials was derived. Figure 2 presents the situation schematically: Compartment 1 has a constant radon concentration, C_1 Bq m^{-3}. It is assumed that the material in compartment 0 produces P_o Bq m^{-3} s^{-1} and has a diffusion constant of D_o.

The time-dependent equation describing the concentration, C_o, is given by Fick's Law:

$$\frac{dC_o}{dt} = D_o \frac{d^2C_o}{dz^2} + P_o - \lambda C_o, \tag{1}$$

where λ = the decay constant for radon.

In the steady state, $\frac{dC_0}{dt} = 0$, and equation 1 transforms into:

$$\frac{d^2C_0}{dz^2} - \lambda C_0/D_0 = P_0/D_0. \tag{2}$$

The boundary conditions for the equilibrium condition are:

$$C_0\ (z = 0) = C_1 \text{ and } C_0\ (z = L) = C_2. \tag{3}$$

For convenience, we introduce the diffusion length, ℓ_o, defined as:

$$l_0^2 = D_0/\lambda. \tag{4}$$

As can be deduced from equation 1, D_o has the dimension [$M^2\ s^{-1}$], and since λ has the dimension [s^{-1}], the dimension of ℓ_o is [m]. Substituting equation 4 in equation 2 yields:

$$\frac{d^2C_0}{dz^2} - C_0/l_0^2 = -P_0/D_0. \tag{5}$$

The solution of equation 5 is:

$$C_o(z) = A \sin h\ (z/\ell_o) + B \cos h\ (z/\ell_o) + P_o/\lambda. \tag{6}$$

From the boundary conditions in equation 3 it follows that:

$$A = \frac{C_2 - C_1 \cos h\ ß - (P_o/\lambda)\{1 - \cos h\ ß\}}{\sin h\ ß}, \tag{7a}$$

and $B = C_1 - P_o/\lambda$, with (7b)

$$ß = L/\ell_o. \tag{8}$$

At equilibrium, the concentration C_2 is given by:

$$E_o \cdot A_o = \lambda C_2 \cdot V_2, \tag{9}$$

where E_o is the exhalation from the area A_o, given as

$$E_o = -D_o \left[\frac{dC_o}{dz} \right]_{z=L} . \tag{10}$$

Substituting equations 6 – 8 and equation 10 in equation 9 yields

$$C_2 = \frac{A_o l_o [C_1 - (P_o/\lambda)\{1 - \cos h\, \beta\}]}{V_2 \sin h\, \beta + A_o l_o \cos h\, \beta} \tag{11}$$

or

$$C_2 = \frac{\gamma \cdot \delta}{\sin h\, \beta + \gamma \cos h\, \beta} , \tag{12a}$$

with

$$\delta = \frac{A_o l_o}{V_2} , \tag{12b}$$

and $\gamma = C_1 - (P_o/\lambda \{1 - \cos h\, \beta\}.$ (12c)

Substituting in equations 6 and 7 leads to

$$C_o(z) = \left[\frac{-\delta (\lambda \sin h\, \beta + \cos h\, \beta}{\sin h\, \beta + \gamma \cos h\, \beta} + \frac{P_o}{\lambda} \cdot \sin h\, \beta \right] \sin h\, (z/l_o) \tag{13}$$

$$+ [\delta - (P_o/\lambda) \cos h\, \beta] \cos h\, (z/\lambda_o) + P_o/\lambda.$$

PRELIMINARY RESULTS

At this stage of the investigation, three types of materials were tested: a 1-cm (0.5-in.)-thick gypsum wallboard sheet; a 0.3-cm (0.125-in.)-thick steel plate; and a 0.1-mm (0.004-in.)-thick polyethylene film. The first two samples were chosen as extremes of radon-transparent and radon-opaque materials, respectively; the third is a material often applied in residential construction as a water-vapor retarder.

As anticipated, experiments conducted with gypsum wallboard showed that $C_1 = C_2$ within approximately 10 h. Surprisingly, early data from the steel experiment indicated that radon was entering the detection chamber. A test vessel fabricated of galvanized pipe fittings and tenon tape was tested for background alpha-emissions, then used to determine if the grease used to seal the neoprene gaskets on the large test chamber was a radium source. The test was negative.

One chamber was then fitted with the surface barrier detector arrangement, and the steel experiment repeated with both the TEL-type detector and the surface barrier detector activated. It became clear that radon was, indeed, being detected in the upper compartment of the chamber. It was equally clear that the tubing, filters, desiccants, flowmeters, and pumps associated with the use of the TEL were prone to leaks that could introduce radon from the laboratory air at about 10 to 20 Bq m^{-3} ~(0.5 to 0.7 pCi L^{-1}). Thus, an effective minimum sensitivity was set by the ratio of the concentration in the detection compartment to that in ambient air. For our source, this translated to a maximum concentration gradient (C_1/C_2) of approximately 500:1.

Figure 3 shows results of the same experiment with only the surface barrier detector activated. The levels are lower than in previous experiments, confirming the possibility of pumped Lucas cells admitting radon from laboratory air into the detection circuit. It is still not certain that these are the only other entry routes, as this concentration is below that of the ambient air.

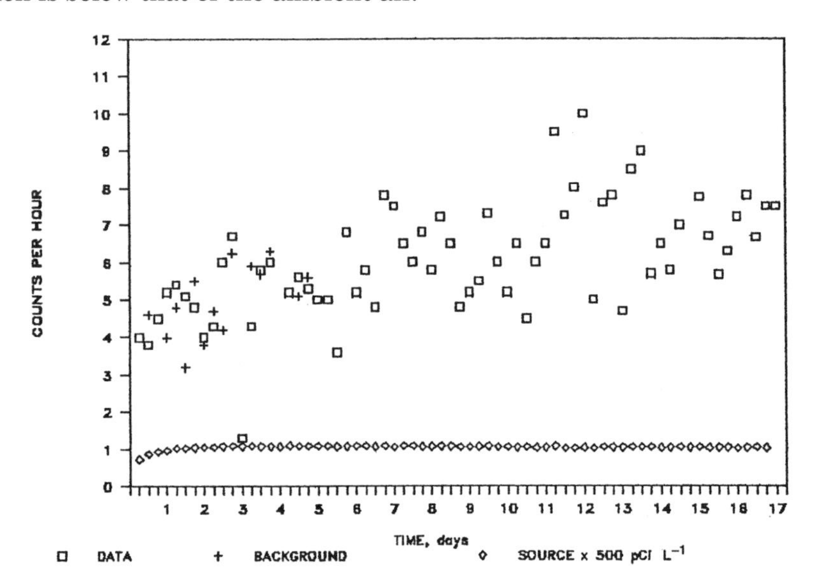

Figure 3. Radon concentration plotted against time for steel-plate experiment with a surface barrier detector.

DISCUSSION

Significant shortcomings encountered with this arrangement include leakage through associated tubes and fittings and difficulty in controlling the dynamic pressure head during pumping cycles. Efforts were made to maintain constant pressure gradients across the assembly by adjusting the flows of the pumps on the upper and lower chambers. However, since experiments with materials or assemblies sometimes took more than 2 wk each, the accuracy of the internal clocks in the Pylon AB-5 was a problem, as we experienced several minutes drift over the course of an experiment. Also, we found that using an open source system, such as we have with the circulation of soil gas, makes long, constant-pressure experiments extremely difficult. Since the soil gas is subject to fluctuations in barometric pressure, and the sealed detection chamber is not, significant pressure changes were observed in connection with weather changes. Pressure gradients observed in the chambers were comparable to those found in houses. However, interest in conducting a standardized radon transport experiment and increased interest in separating true diffusion from pressure-driven flow have made the shortcomings of this system unacceptable.

CONCLUSIONS

A need exists for a standardized test of full-scale building assemblies to establish the average radon transmission rate for each component. It is logical and important that radon be the test gas used to perform these tests.

The test chambers have proved suitable in design and can be scaled to larger sizes if necessary. For reasonable control of the pressure differential across the test surface, a constant volume and a sealed radon source would be preferable. Alternatively, the chambers should be fitted with a pressure- compensating, variable-volume device. For very low radon concentrations, a surface barrier detector in the test chamber would provide superior detection.

Separating the diffusive component of radon transport through materials will require some additional pressure-control apparatus to compensate for pressure-driven flow through the material.

REFERENCES

Aldenkamp, FJ, LW Put, and RJ de Meijer. **1987**. *Aspects of An Instrument for In Situ Measurements of Radon Exhalation Rates*, Technical Report 690. Kernfysisch Versneller Instituut, Groningen, The Netherlands.

Hammon, HG, K Ernst, and JC Newton. **1977**. Noble gas permeabilities of polymer films and coatings. J Appl Polymer Sci 21:1989-1997.

Nagda, NL, MD Koontz, RC Formann, WA Schoenborn, and LL Mehegan. *Florida Statewide Radiation Study*, Report Number IE-1808. GEOMET; Germantown, MD.

Scott, AG and WO Findlay. **1987**. *Production of Radon-Resistant Foundations*, Reports 84-05-021 and 84-05-021s. Florida Institute of Phosphate Research, Bartow, FL.

Zapalac, GH. **1983**. A time-dependent method for characterizing the diffusion of ^{222}Rn in concrete. Health Phys 45:377-383.

QUESTIONS AND ANSWERS

Q: McLaughlin, Dublin. . . . radon permeability for sealing material you use for your samples. Because I remember years ago trying to do something similar, and we got leaks through the sealing material.

A: Pugh, Florida A&M University. We basically used materials that we could locate in the flanges so that the sealant was not on the pathway. In other words, for radon to move from one chamber to the other, it would have to go out into the lab and back through another route and back in. So, in that sense I think we've compensated for it, but the direct answer is no. We wanted first to establish that the equipment was appropriate for that. In the literature, if we look at various sealing materials, it's our feeling that this particular setup probably would make them all look very good, but it wouldn't be of fine enough resolution to differentiate.

Q: Harbottle, Brookhaven National Laboratory. I was very interested to hear about the diffusion of radon in polyethylene. We've noticed the same thing. We measured it with a much more simple, less sophisticated apparatus than yours. We just took two desiccators and put the sheet of polyethylene between them, and then simply clamped them together so there was no possibility of an external route. The odd thing is that in some cases we found diffusion of radon through, in other cases we did not. From what you say there are apparently differences in different batches of sheeting. Is that correct?

A: Pugh. This is the exact same batch. It was the exact same roll. We unrolled approximately 20 feet of it very carefully in the lab on clean tabletops to make sure that we didn't puncture it with sand grains or anything of that sort, and then we cut out successive pieces that were approximately 30 x 36 inches. We fitted them over the flanges so that, again, there was no possible way for a direct short circuit. And we found that the equilibrium ratios we got were roughly similar. But in some cases we arrived at them very much sooner than in others, and there was some noticeable difference in the ultimate equilibrium ratios. So clearly, the flux through the material was different.

Q: Harbottle. We ran into this in a very odd way. We had ordered up some Marinelli cells for counting soil samples on a germanium counter, and we noticed with some of them that it was impossible to get to radon equilibrium. They seemed to give final values which were much too low, so we began looking into this. You might be interested to know that there are about two or three Russian papers dating back, maybe, 20 years, in which a chap measured very precisely the diffusion of radon and polyethylene and got the activation energies, frequency factors, did the whole job on it, but this is in a rather obscure Russian journal.

A: Pugh. I appreciate the information.

TECHNICAL AND PUBLIC POLICY CONSIDERATIONS AND DEVELOPMENT OF A CODE FOR CONTROL OF RADON IN RESIDENCES

M. Nuess and S. Price

Washington State Energy Office, Spokane, Washington

Key words: *Radon, indoor air quality, building codes, building design, building construction*

ABSTRACT

Building codes that address radon control in residential buildings are a relatively new development in the larger trend toward increased efforts to understand and control indoor air quality. A residential radon construction standard has been developed in the Pacific Northwest region of the United States. The Northwest Residential Radon Standard (NRRS) seeks to provide a measured public policy response that is commensurate with current knowledge of both the health risk and the state of building science. This paper reviews the range of potential public policy responses available to deal with radon as a public health problem, describes the policy framework on which the NRRS is structured, and explains the development process.

Time and budget constraints limited the scope of the NRRS to identifying the minimum set of measures necessary to reliably achieve radon reductions without impairing structural integrity, capability to control other indoor air pollutants, occupant comfort, or energy efficiency. Though it encourages measures that enhance the linkages among durability, indoor air quality, and comfort, it does not require them unless they are part of the minimum requirements necessary for radon control. As a result, the potential for optimizing indoor air quality and other factors is not impaired, but it is also not realized by the NRRS. Better control of radon is possible, but it requires broader dispersion of already available information, further development of technical support, supportive changes in other building codes, and the different emphasis of a whole-systems approach. The NRRS provides a useful interim step toward the larger goal of a more systemic approach.

INTRODUCTION

The Northwest Residential Radon Standard (NRRS) was developed under the auspices of the Washington State Energy Office with support from the Bonneville Power Administration (BPA), a federal regional power-marketing agency. Radon is a pollutant requiring a different policy response than some other indoor contaminants.

Because it originates from an external source, radon is dependent on certain aspects of building science for control. There is a need for governmental intervention to increase public awareness of the issue, encourage voluntary action by individuals, and create the opportunity for individuals to live free of high radon exposures.

THE REGIONAL CONTEXT

In 1980, the U.S Congress established the Northwest Power Planning Council (NWPPC), a regional body mandated to develop a plan for ensuring adequate supplies of electrical energy. The initial plan (and subsequent revisions) has emphasized conservation as the most cost-effective resource in the region. The Power Council's plan encourages the BPA to pursue the conservation resource aggressively. It is estimated that BPA has spent $1 billion on conservation programs, achieving electrical energy savings averaging $.02-.03 per kWh (NWPPC, 1986).

During indoor air quality research, an early component of BPA's programs to conserve energy in buildings, radon testing was initiated. Participating electric utilities tested residences throughout their service territories for radon levels during the heating season by means of alpha track monitors for a minimum of 3 mo. The result is one of the largest data sets ever collected on radon levels in residential buildings. Over 32,800 residential sites in approximately 400 townships were measured in Oregon, Washington, Idaho, Montana, and Wyoming (Figure 1).

The average radon concentration measured in roughly one-half of the 400 townships was greater than 37 Bq m^{-3} (1 pCi L^{-1}). None of the townships had an average radon concentration at or below 7.4 Bq m^{-3} (0.2 pCi L^{-1}), the new long-term national goal specified by Congress. A few areas had a large number of homes with elevated radon levels, notably the Spokane River valley on the border of Washington and Idaho. In the City of Spokane, Washington, nearly half the homes had levels above 150 Bq m^{-3} (4 pCi L^{-1}).

As part of the Pacific Northwest's aggressive pursuit of energy efficiency, the NWPPC developed Model Conservation Standards (MCS) for construction of new buildings. These standards require higher insulation levels in the building envelope, tighter building construction to reduce air leakage, ventilation provided by mechanical systems, and certain indoor air pollutant control measures. Roughly 100 building-code jurisdictions in the region have adopted the MCS.

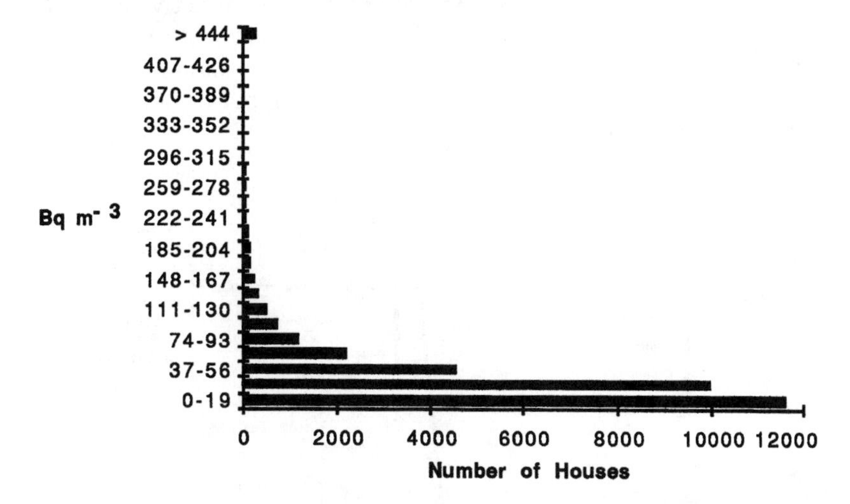

Figure 1. Distribution of 32,885 radon measurements in the Pacific Northwest. Alpha-track measurements from a minimum of 3 winter mo to 1 y. Data from Bonneville Power Administration. Distribution of test sites weighted toward large participating utilities in certain areas.

The MCS are the first adopted and enforced standards in the United States that begin to address radon. In addition to specific requirements for subslab gravel and crawl-space ventilation, the MCS contains an appendix which specifies technical measures to be incorporated in certain residential buildings.

In the summer of 1988, BPA contracted with the Washington State Energy Office's Energy Extension Service (WEES) to research and develop a model radon code for new residential construction. The WEES has had an active public education program on indoor air quality for the past decade. When radon became an issue of public concern, WEES was able to respond quickly with educational services. In a 1-y effort, WEES developed the Northwest Residential Radon Standard (NRRS; Figure 2).

A REGULATORY APPROACH—LOOKING FOR PRECEDENT

The task of determining an appropriate public policy response to the public health issue of radon presents interesting challenges. As a naturally occurring indoor air pollutant that originates largely from outside buildings, radon is categorically different from many other indoor contaminants. It is not generated by occupant activity, and it is not responsive, in large part, to behavioral adjustments by the occupant. In this light, radon appears appropriate for some level of regulation. The range of policy options for addressing public health threats is diverse (Figure 3).

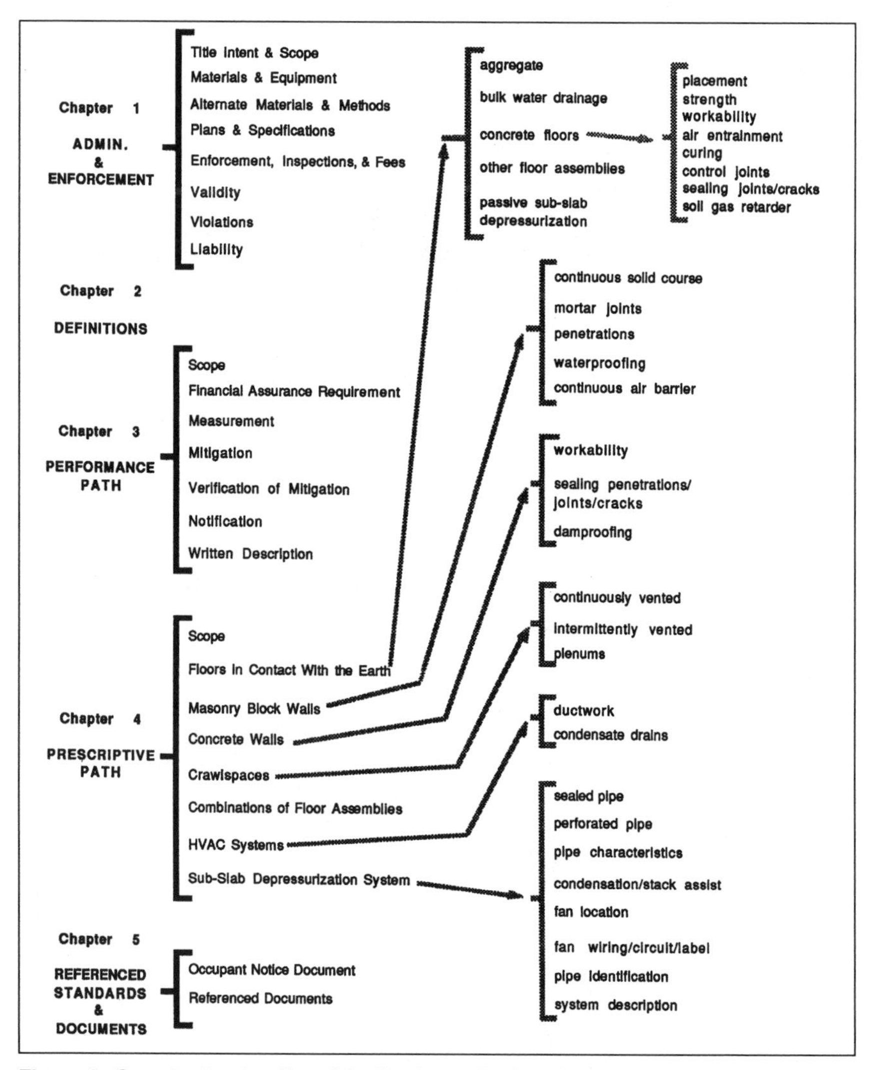

Figure 2. Organizational outline of the Northwest Residential Radon Standard.

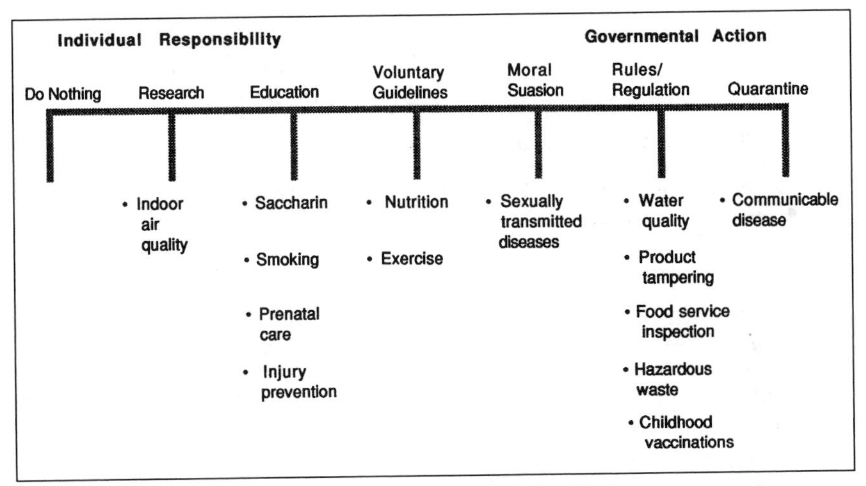

Figure 3. Range of public health policy options in the United States.

There are multiple levels of governmental response to health issues, and policy responses vary in the United States for different public health issues. It is illustrative to look at other health issues in light of the governmental response. The response to saccharin use and tobacco smoking relies almost exclusively on personal choice and public education (though in the case of tobacco smoke, local communities are becoming more aggressive in regulating where the activity can take place). Childhood vaccinations and AIDS are two well-publicized health issues that have received a stronger regulatory response.

THE SHARED RESPONSIBILITY VALUE

There has been very little regulatory control of radon in buildings in the United States. At the outset the project required many value choices about both the technical structure and policy framework of the code. The fact that a code would be developed at all assumed the problem warranted intervention by government, but at what point in the regulatory continuum?

The WEES assumed that in the area of radon, public health is a shared responsibility between individuals and government. Unlike outdoor air quality (where the costs and benefits of clean air cannot be rationally apportioned to an individual and attained through voluntary individual action), the benefits that accrue to the individual from voluntary actions to maintain healthy indoor air are clear (Spengler and Sexton, 1983). The WEES assumed that it was the role of government to empower individual choice by providing:

- education about radon health effects, measurement, control, etc;
- access to necessary resources by nurturing the development of necessary technology;
- quality control through industry coordination and regulation;
- regulation necessary to provide the healthier indoor air (including construction standards).

The WEES also assumed that the individual's freedom of choice should be preserved to the extent possible, and that it was the individual's responsibility to recognize the value of healthy indoor air and to choose whether or not to live in it.

Therefore, the NRRS was structured as governmental intervention that enables voluntary action by regulating the building in order to control radon and preserving the individual's option to live in a healthier indoor environment. It stops short of requiring an individual to test or mitigate in order to continue to live in that environment. Because it is a construction standard, its scope, which is very focused, addresses but one of several important regional and national issues with regard to radon and health.

SCIENTIFIC UNCERTAINTIES AND POLICY MOMENTUM

The issue of radon as an indoor air contaminant is relatively brief. In the United States, an ever-increasing understanding of radon as a threat to public health has generated governmental activity at federal, state, and local levels.

At the federal level, the U.S. Environmental Protection Agency (EPA) issued action guidelines to the public for mitigation activity based on radon test results. As research more firmly established radon as a public health threat and as the public's awareness of the radon issue increased, governmental response also increased. The EPA recommended the testing of all homes. The U.S. Surgeon General issued a report on the health threat presented by radon and encouraged all Americans to test their homes. Congress recently passed legislation that provided funds to the states for radon programs, established a long-term national goal to lower radon levels in buildings to outdoor levels (7.4 Bq m^{-3}), and mandated the development of National Model Construction Standards by June, 1990.

Despite increased levels of governmental activity, some uncertainty concerning radon remains:

- Estimates of the level of risk to human health at various exposure levels still vary.
- Measurement protocols need improvement.
- We do not have long-term experience with the techniques of radon control.
- Several technical questions remain unanswered (and probably unasked).

It is within this environment of scientific uncertainty and governmental desire to respond to the perceived threat that the NRRS had to be developed.

The WEES is confident that techniques required by the NRRS represent a reasonable and appropriate "good practice" standard at this time. It is evident that the radon control approaches required by the NRRS are very effective. Several radon mitigators utilize these techniques in the mitigation of existing residential buildings and guarantee their performance. However, it should be clearly understood that new information will likely emerge that will require changes in these measures.

THE VALUE OF A SYSTEMS APPROACH

The control measures required by the NRRS are intended to represent the minimal measures necessary to reliably reduce indoor radon levels without impairing structural integrity, capability to control other indoor air pollutants, occupant comfort, or energy efficiency. These measures are designed to mesh with current building practices, materials, and building codes. Hence, the NRRS requires:

- practical techniques that reduce the number of openings available for soil gas transport to indoor air, and
- a pressure-difference control system designed to override house/soil pressure-differences contributing to soil gas transport, such as stack, wind, and mechanical appliance effects.

The NRRS does not require:

- as-tight-as-possible building envelope construction;
- mechanical ventilation in all residences;
- decoupling of all combustion appliances from indoor air;
- attention to pressure-difference control in designing heating, ventilating, and air-conditioning systems.

These additional measures serve multiple purposes and cannot always be justified for one purpose alone. For example, mechanical ventilation (properly installed) would contribute to further reduction of indoor radon, but its contribution is more than an order of magnitude less than that of the subslab depressurization system capability required by the NRRS. Yet mechanical ventilation would enhance the control of other indoor air pollutants and increase occupant comfort, if installed in an air-tight house.

Envelope tightness could reduce the volume rate of soil gas transport by enhancing pressure-difference control capability at minimal energy cost. It would also enhance mechanical ventilation effectiveness, moisture control, comfort, and energy efficiency.

These and other measures could contribute significantly to further radon reductions. However, they would serve multiple purposes, and the costs should be appropriately proportioned. A reciprocal effect is that part of the cost of the required radon control measures, such as substructure/crawl-space sealing and subslab depressurization, could be charged to comfort, control of other indoor air pollutants, control of moisture (several tons/heating season removed from the soil), and control of other soil gas pollutants. As an example, Jim White, of Canada Mortgage and Housing Corporation, reported that garbage gases have been measured several kilometers away from landfill sites. Also, some bacteria, fungi, and viruses found in soils can produce serious health problems (White, 1988).

A whole-systems approach which attempted to optimize residential buildings for durability, health and safety, comfort, and energy efficiency would include at least the additional measures listed above. Such an approach would further rationalize the cost of radon control. The increased durability, safety, comfort, and energy efficiency could increase the net value of residential buildings.

Because of these limitations, the proposed NRRS is not an optimal standard. Better control of radon is possible, but it requires broader dispersion of already available information, further development of technical support, supportive changes in other building codes, and the different emphasis of a whole-systems approach. The WEES is encouraged to think that the NRRS serves to provide a useful step toward the larger goal of a systemic approach.

INTRAREGIONAL VARIABILITY

Radon exposures in some areas of the Pacific Northwest are relatively low, in some areas relatively high, and in some areas are unknown. It was originally intended that the NRRS would be offered to the region for optional adoption by local jurisdictions.

FLEXIBILITY—THE ROLE OF A DUAL-PATH STANDARD

The national model codes of the United States, e.g., the Uniform Building Code, are performance codes, which specify levels of performance rather than specific materials or procedures. You must attain the end goal but are free to choose the means of attainment. Performance codes allow flexibility, cost optimization, and the development of new and improved materials and systems.

On the other hand, a prescriptive standard requires installation of certain materials and systems and specifies a path that must be taken. Prescriptive standards/codes are simpler and easier to follow but lack the flexibility of performance codes, as well as the potentials for innovation and cost reduction.

Jim Gross, Deputy Director of the Center for Building Research, of the National Institute of Standards and Technology, has encouraged a dual-path standard: a performance standard with the option of specified measures "deemed to satisfy" the standard (Gross, personal communication, 1989). This seems the most practical approach. The proposed NRRS follows this dual-path pattern.

The NRRS seeks to provide increased protection for all new and significantly remodeled residential occupancies in any jurisdiction of the Pacific Northwest that chooses to adopt it. It seeks to limit exposure to indoor radon for occupants by requiring for every such occupancy either:

- demonstration of postconstruction tested indoor radon levels at or below 150 Bq m^{-3} (4 pCi L^{-1}), or
- installation of certain specified materials and systems during construction that reduce the potential for elevated indoor radon and establish the capability to further reduce radon levels should the owner desire.

Option 1 (Chapter 3 of the NRRS) is a performance requirement. If the building does not meet the performance specification it must be modified until it does. There are no specified control requirements to be met during construction. It allows both flexibility and the demonstration and use of new and different approaches to controlling radon.

Option 2 (Chapter 4 of the NRRS) specifies certain prescriptive requirements, primarily substructure and crawl-space sealing, and the rough-in of a subslab depressurization system. If the prescribed measures are correctly installed, there are no future responsibilities for radon control.

NEED FOR A LONG-TERM MEASUREMENT TEST

The performance path of the NRRS requires verification that the performance goal has been reached. The intent is to ensure, within a reasonable level of certainty, that a building will perform as required.

The EPA's Interim Protocols for Screening and Follow-up Radon and Radon Decay Product Measurements state that, "The EPA does not recommend taking any significant remedial action on the basis of a single screening measurement" (USEPA, 1987).

The screening measurement, a short-term test, can be a reasonably accurate measure of the radon levels during the actual test period, but the range and period of variation are too great to enable a reasonably accurate measure of the long-term average radon levels. Arthur Scott of American Atcon Inc. has suggested that the decision level of a short-term (3-day charcoal, for example) radon test is really very different from that of a long-term test (6- to 12-mo alpha track), and that short-term tests are not being interpreted correctly. Short-term tests cannot predict long-term averages. He indicated that if a long-term average radon exposure is really 185 Bq m^{-3} (5 pCi L^{-1}), then the probability of a short-term test result of 37 Bq m^{-3} (1 pCi L^{-1}) is the same as the probability of a short-term test result of 750 Bq m^{-3} (20 pCi L^{-1}) (Scott, personal communication, 1989).

Currently, the short-term test is being misinterpreted in many sectors. William Ethier, an attorney for the National Association of Homebuilders (NAHB), recently suggested at the National Radon Conference (Ethier, 1989), that utilization of a short-term test to imply that radon levels are below 150 Bq m^{-3} and therefore acceptable could provide reasonable grounds for a claim of fraud or misrepresentation if a long-term test later showed levels over 150 Bq m^{-3}. According to Ethier, NAHB takes the position that short-term tests should not be part of a real-estate transfer contract (Ethier, 1989).

In its report to the U.S. House of Representatives, the U.S. Committee on Energy and Commerce noted concern "...about people making decisions not to mitigate based on low readings from short-term radon tests. Accordingly, the Committee expects EPA to evaluate the appropriate use of results from short- and long-term tests by the public. In particular, the Committee expects EPA to consider whether the Agency should recommend that only results from long-term tests should be used" (Committee on Energy and Commerce, 1988).

The EPA screening protocol would be inappropriate for the NRRS because of its reliance on short-term measurements. One NRRS reviewer suggested that a separate measurement protocol be developed rather than relying on the EPA screening protocol. Another reviewer cautioned that developing a protocol in addition to that of the EPA might make it difficult to compare the results to measurements made elsewhere.

A longer-term measurement is necessary to attain a reasonable estimate of the building's actual performance and to avoid cheating by "smart" testers, who could affect results by coordinating test periods with rainfall, weather systems, and other factors. This requires addressing the additional difficulty of testing after occupancy. However, the positive side is that the occupant has the least incentive for fraud (unless he or she is preparing for resale).

The NRRS requires a long-term test by specifying adherence to certain EPA follow-up measurement protocols. According to the EPA's Interim Protocols for Screening and Follow-up Radon and Radon Decay Product Measurements, "The purpose of the follow-up measurement is to estimate the long-term average radon or radon decay product concentrations in general living areas with sufficient confidence to allow an informed decision to be made about risk and the need for remedial action" (USEPA, 1987).

SHOULD WE REQUIRE MONITORING FOR ALL RESIDENTIAL BUILDINGS?

Currently the NRRS requires monitoring only for the performance path because it is the responsibility of the builder to meet the performance standard. It does not require monitoring for the prescriptive path because the builder completes his or her responsibility upon complying with specifications which are "deemed to satisfy" the standard. At this point the responsibility for addressing indoor radon is passed to the owner. The proposed NRRS stops short of governing the owner or occupant.

Neither compliance path guarantees that, for any given residential building, future indoor radon levels will be below 150 Bq m^{-3}. If a building has conformed to the prescriptive path, the owner or occupant will not know the radon level until he or she tests. If a building has conformed to the performance path, there is no certainty that future events will not alter long-term average radon levels. Periodic measurements over the course of the useful life of any building built to this standard will be necessary if knowledge of radon levels is desired. For all governed buildings, the NRRS requires measures to:

- inform all future occupants of the radon control measures taken,
- strongly encourage them to test for radon,
- provide them access to further information about health effects, testing, and mitigation.

Some NRRS reviewers recommended monitoring ali new residential buildings. Other policy approaches were offered. For example, a member of EPA's National Radon Standards and Codes Work Group who has been involved in several mitigation demonstration projects, expressed a concern that the only workable way to reduce radon exposure in buildings is to have a standard that is at once both a performance and a prescriptive standard. Buildings would be built to specifications, tested, then mitigated if necessary. He felt that quality control was so essential, yet so lacking, that this approach might be necessary.

Testing following construction or remedial action and continued testing for several years afterward is warranted by the lack of knowledge of the short-term effects of specific measures in specific houses and the longevity of the effects of those measures. The WEES concluded that such follow-up testing should be encouraged (perhaps funded for research purposes) but not required.

NEED FOR EDUCATIONAL SERVICES

There is little system-wide coordination within the building industry. Many builders receive training on the job and must make do with what they have learned from this rather local sphere of influence. There is significant variation in construction methods by both geographical area and climate.

In addition, builders must survive in an economic milieu in which emphasis on first costs forces builders to resist any increase in housing costs. Builders face a forest of regulation and will, in many cases, be less than eager to comply with additional regulations.

Educational and technical support services will be of significant value. While no radically new construction techniques are required, many are new to large portions of the residential construction sector. An example is the soil gas retarder membrane required by the NRRS. Many reviewers supported its inclusion, considering it feasible and reasonable. Others were concerned about both the difficulty and cost incurred by having this technique required. It has become clear from several discussions that perceptions about this issue vary widely within the building trades.

Successful (and unsuccessful) experiences with the subslab membrane are closely linked to perceptions about correct concrete practices, and these perceptions also seem to vary widely. More stringent aggregate specifications and sealing techniques may also require educational services in the residential sector.

THE NRRS DEVELOPMENT PROCESS

Decisions about public health risks (in this case, radon) can be extremely complicated. They involve elements of risk assessment, risk management, and risk communication. All too often, difficult decisions about risk assessment and risk management are made remotely by experts, then poorly communicated to the public. Often the result is conflict, with experts feeling misunderstood, and the public feeling misused; often both are right. Conflicts about health risk issues usually contain the underlying issues of equity and control. "Public participation" is usually too late and does not involve the kind of information and power-sharing necessary for the realization of enduring policies. Risk communication, with the goal of an actively concerned public, and within a context of real openness to public input, is vitally important. It may be difficult, but it is both possible and necessary.

A good communications process can serve to align public perceptions with the perceptions of the scientific community. It can serve to eliminate the inappropriate extremes of either panic or apathy. It can empower a community with the sense that it can take charge and address the issues that confront it.

The process for developing the NRRS was very participatory. Input was solicited from a diversity of economic sectors, including realtors, builders and builder associations; technical specialists and generalists in the fields of building science, radon, and ventilation; consumer protection organizations; energy utilities; state and federal agencies; building code organizations; and research organizations.

While broadly solicited, the input was sequenced: technical input was solicited first, and the range of known technical solutions was identified. Technical specifications had to meet criteria for control effectiveness, ease of implementation by typical tradesmen, availability of materials, cost, compatibility with comfort, and compatibility with other indoor air pollutant control techniques. Legal and policy-related input followed. The effort began in June, 1988. A literature search was conducted. Researchers, mitigators, and policymakers who were known to have radon-related experience were contacted until a sense of closure with regard to available national resources had developed.

On October 1, 1988, an initial draft of the NRRS was completed and circulated for technical review. Circulation for legal review followed. More than 35 technical reviewers contributed comments about the initial draft. They included persons from the EPA, national research laboratories, university researchers, private-sector builders, contractors, radon mitigators, tradesmen, engineers, architects and product suppliers, building code officials and organizations, builder associations, state energy offices, BPA, and the NWPPC. The time allowed for the technical and legal review comment period had to be extended more than originally anticipated in order to obtain important and valuable review comments. The need for a longer review period may be in part due to the unanticipated intensity of activity in the radon industry in 1988, which included a national symposium, and the passage by the Congress on October 28, 1988, of the Indoor Radon Abatement Act, which set a new national goal that indoor radon levels should be no higher than outdoors.

In January, 1989, the EPA asked WEES to contribute to their effort to develop model construction standards by June, 1990, and to partake in a National Radon Standards and Codes Work Group. The group included persons representing the national Model Code Organizations [International Conference of Building Officials (ICBO), Southern Building Code Congress International, Building Officials and Code Administrators International, and Council of American Building Officials], U.S Housing and Urban Development, National Institute of Building Sciences, National Institute of Standards and Technology, National Association of Home Builders, Canada Mortgage and Housing Corporation, members of an American Society for Testing and Materials committee on radon, and representatives from states actively working on radon codes. In February, WEES presented an introduction to the first draft of the NRRS to that group and received several constructive comments.

The second draft of the NRRS was distributed March 30, 1989. It was circulated to a Policy Review Committee consisting of state and local officials in the general government, building code, and public health areas; policy level representatives from BPA, the NWPPC, utilities, and the shelter industry; the EPA National Radon Standards and Codes Work Group; the National Institute of Standards and Technology; and the Canada Mortgage and Housing Corporation. The second draft was also recirculated to technical and legal reviewers as a courtesy.

The first draft of a generic Implementation Plan was completed in May and circulated for review by a local advisory committee. The Implementation Plan is a guidance document intended to assist local jurisdictions with considering, adopting, and implementing the NRRS. The plan seeks to provide the conceptual framework for a reasonable, equitable, and informed process for consideration of the NRRS. It is not meant to encourage adoption of the NRRS; the intent is to encourage and enable a good choice.

The final Implementation Plan and final draft of the NRRS were completed in June, 1989.

In January, 1990, the ICBO Indoor Air Quality Committee unanimously recommended that the NRRS be considered for introduction as an Appendix to the Uniform Building Code and submitted as a code change during the 1991 cycle.

ACKNOWLEDGMENT

The authors deeply appreciate the willingness of the many capable (and therefore very busy) people who generously provided time, effort, encouragement and support to the development of the NRRS.

REFERENCES

Committee on Energy and Commerce. 1988. *Report (to accompany H.R.2837) to the U.S. House of Representatives*, Report 100-1047, October 4. Available from U.S. EPA, Washington, DC.

Ethier, W. **1989**. Comments at the March National Radon Conference, Cincinnati, OH. Available from U.S. EPA, Washington, DC.

ICBO. 1990. Building Standards: Preliminary Report of the Indoor Air Quality Code Development Committee. *Requirements for Radon Control for Group R. Division 3 Occupancies*. International Conference of Building Officials, 5360 South Workman Mill Road, Whittier, CA 90601.

NWPPC. 1986. *Northwest Conservation and Electric Power Plan*, Volume 1, p. 6-1. Northwest Power Planning Council, Portland, OR.

Spengler, JD and K Sexton. **1983.** Indoor air pollution; A public health perspective. Science 221:4605.

USEPA. 1987. *Interim Protocols for Screening and Followup Radon and Radon Decay Product Measurement*, No. 520/1/86-014. U.S. Environmental Protection Agency, Washington, DC.

White, JH. **1988.** *Radon—Just Another Soil Gas Pollutant?* Presented at 81st Annual Meeting of the Air Pollution Control Authority, Dallas TX, USA. Copies available from Canada Mortgage and Housing, 682 Montreal Rd., Ottawa, Ontario, Canada K1A 0P7.

QUESTIONS AND ANSWERS

C: Nuess. We were asked 2-1/2 years ago by Bonneville Power Administration to take the current level of awareness about radon, its health impacts, what we knew about ways to control radon through building science, and come up with a code that would reflect and be commensurate with what we knew about this issue. It took a 1-year effort to do that.

Q: Suess, WHO. A concentration standard?

A: Nuess. Thank you; yes. I didn't tell you what the actual performance target was. The standard was designed to be ratcheted down as we moved closer toward our national goal of ambient exposure indoors as well as outdoors. However, in the standard as drafted now we are saying that the performance level (action level) is 4 pCi per liter with the intent stated that as building science got better at this and as the technology developed, we would ratchet that down.

Q: Cohen, University of Pittsburgh. I must say, this thing about polyethylene doesn't square with my experience at all. I've tested lots of pieces of polyethylene, including a lot thinner than 4 mils and I've never seen any radon go through polyethylene. It is very hard to seal around the edges, but it seems to me that in the applications you are talking about, the sealing would be the fatal problem, rather than the polyethylene leaking. I wouldn't have any idea how you'd seal it.

Incidentally, I was going to say this after the last paper; we have a very easy way of measuring efficiency for membranes. We just put them over a charcoal canister and stick it in our radon chamber, and let it sit there for a week, and take it out and just count it. You get a limited detection of 2% very easily that way, and it's no trouble, takes no time, or anything like that. Although with polyethylene, I must say, we did find that we had to seal it between O-rings. Not only is polyethylene good—almost anything that looks like a solid plastic doesn't let radon through. There are exceptions, but they are really exceptions rather than the rule.

A: Nuess. OK, thank you. Our purpose, too, was not as a diffusion barrier, but as an air barrier for the poly. But your point is very appropriate for the diffusion issue.

Q: Riddle, Cavalier Corporation, Spokane. As you know, we have stated before that we don't see any success with the passive methods. We have been hired numerous times to rescue builders who have tried the passive methods. They had a home buyer who didn't care whether it was passive or active, he just wanted it to work, and it didn't. But I have a question regarding the notice that you have put in there to the home buyer. That's the first time I've seen it in several versions of your code, and I commend you that it's finally in there. But I'm not sure, is it only a notice when the builder has chosen the prescriptive path? It seems to me that whether the builder has chosen the performance path or the prescriptive path, the homeowner should be encouraged to spend $10 every year with Bernie Cohen's lab, or somebody, to test that house. So, which way is it?

A: Nuess. The notice advises the person to test. The notice says that under the prescriptive path the notice tells the occupant that this building may not have been tested at all, and that he is strongly advised to test. Under the performance path the standard says, although this building may have been tested, conditions can alter the way the building performs and that you should continue to test. This has been in the standard from the beginning.

However, I understand where your confusion comes from. It has not been with the Washington State proposed radon code, in which, based on your testimony a few weeks ago, I believe there are some changes.

Q: Morley, Radiation Protection, British Columbia. Has that standard or guideline been incorporated in any of the Washington State building codes, or is it left up to the areas to accept them, or what?

A: Nuess. No, this standard has not been adopted anywhere and is not being enforced anywhere. It was written as a model standard to provide people with a good baseline for further consideration. What has happened in Washington State was that the state legislature has mandated, as part of the newly adopted state energy code, that the energy code adopt certain provisions for the control of radon as well. And this standard, as well as any other standards that were currently in draft form or being developed in the United States and in Canada (the Canadian standard as well), were included as part of a study effort in developing the Washington

State standard. But what Washington has now got, at least, at the draft level, which is not yet complete, is not this standard but a different approach.

Q: Morley. How do the building codes work in the United States? Do you have a national code and then a state code?

A: Nuess. The United States has three: SBCCI (Southern Building Code Congress International), ICBO (International Congress of Building Officials) and BOCA. SBCCI does the southern and eastern United States generally; BOCA does the northern, central, and eastern United States; and ICBO does the western half of the United States. So there are three national model code organizations in the United States. Those three code organizations are "umbrellaed" by one coordinating body, CABO (Council of American Building Officials). Those bodies all write model codes, and states differ widely in the way they decide to have building codes. Some states, for example, have energy codes that are statewide and mandated, some states have none. In some states local jurisdictions have the primary authority for the adoption of codes, and in other areas it differs. So it's alphabet soup.

Q: Suess, WHO. One more question. I take it that all of the codes, standards, or whatever, are for private homes? Single family homes or dual family homes? What about apartment buildings? You have basements, you have construction material that may—

A: Nuess. This is very limited in scope in that we addressed it only as residential buildings of three stories or less. We did not address large multifamily apartment buildings, we did not address nonresidential buildings such as schools, which obviously need to be addressed. So it is a very small subset of the buildings that need to be addressed.

PREVENTING EXCESSIVE RADON EXPOSURE IN U.K. HOUSING

J. C. H. Miles, K. D. Cliff, B. M. R. Green, and D. W. Dixon

National Radiological Protection Board, Chilton, Didcot, Oxon, United Kingdom

Key words: *Radon prevention policy, homes*

ABSTRACT

In the United Kingdom (UK) it has been recognized for some years that some members of the population received excessive radiation exposures in their homes from radon and its decay products. To prevent such exposures, an Action Level of 400 Bq m^{-3} was adopted in 1987. In January, 1990, the National Radiological Protection Board (NRPB) advised that the Action Level should be reduced to 200 Bq m^{-3}, and this advice was accepted by the Government. It is estimated that exposures in up to 100,000 UK homes exceed this Action Level; this amounts to about 0.5% of the available housing. The UK authorities have developed a strategy for preventing such exposures: (1) Areas in which it is estimated that >1% of homes exceed the Action Level for radon are being designated as Affected Areas, and a program to map such areas is under way. Households in these areas are advised to have radon measurements made by NRPB under a "free" (Government-funded) scheme. (2) Householders found to have whole-house, whole-year average radon concentrations >200 Bq m^{-3} are advised to take remedial action and are provided with information on how this can be done. Partial grants toward remedial work are available in cases of financial need. So far, around 3000 such households have been identified. (3) Within Affected Areas, localities are being defined where new homes must incorporate precautions against radon exposure.

In addition to this strategy, a joint case-control study of the risks of radon in homes is being undertaken by the Imperial Cancer Research Fund and NRPB, supported by the UK Government and the Commission of the European Communities.

INTRODUCTION

The National Radiological Protection Board (NRPB) has a statutory duty to advise the UK Government on matters of radiological protection. In January, 1987, it issued formal advice that an Action Level of 20 mSv a year, or 400 Bq m^{-3} of radon, should be adopted (NRPB, 1987). In January, 1990, NRPB issued revised and strengthened advice, reducing the Action Level for radon to 200 Bq m^{-3} (NRPB, 1990). The advice was summarized in six recommendations, the foundation and implementation of which are discussed below.

NRPB RECOMMENDATIONS ON RADON

1. A new Action Level should be set for present homes. Its value, expressed as the annual average of the radon gas concentration in the home, should be 200 Bq m^{-3}.

Since NRPB introduced the original Action Level of 400 Bq m^{-3}, evidence has emerged that the risk of contracting lung cancer is greater than previously estimated (ICRP, 1987; BEIR, 1988). The evidence suggested that the risk from radon was about twice as high as previously thought, implying that a reduction in the Action Level by a factor of two or so would be appropriate. Such a reduction was introduced in 1990. On the basis of the new risk coefficient, it is calculated that radon is responsible for about 2500 deaths a year in the UK, and that the lifetime risk for the average person living in a home at the new Action Level is about 3%.

The Action Level was expressed as the annual average of the radon gas concentration for several reasons. The risk from radon is accumulated slowly and expresses itself over a lifetime, so it is important to base decisions about remedial actions on the long-term average, not on peak levels. This requirement for an estimate of the annual average level has implications for measurement strategies. Because radon concentrations vary widely from day to day and from season to season, it is essential that measurements be carried out over a long period. At present, the only reliable and inexpensive long-term monitors measure radon gas, not radon decay products. Dosimetric work (James, 1987) has shown that, in homes, the radiation dose received by lung tissue is more closely proportional to exposure to radon gas than to exposure to radon decay products; hence for both practical and dosimetric reasons, it is preferable to express the level as the long-term average of radon gas concentration.

The new Action Level has been adopted by the Government. It is an advisory level: No one is compelled to measure the radon in his home nor to reduce it if he does not want to do so.

2. Parts of the UK with a 1% probability or more of present or future homes that are above the Action Level should be regarded as Affected Areas. Such areas should be identified from radiological evidence and periodically reviewed.

The concept of radon Affected Areas was developed so as to concentrate attention and effort where it was most needed. It was recognized that any criterion for identifying such areas was to some extent arbitrary, but it was felt that it would be sensible to designate areas where 1% or more of homes were above the Action Level. Such areas were to be identified by NRPB from radon measurements in homes rather than by geological criteria or by using administrative boundaries. Although this approach is slower than the alternatives, it ensures that the boundaries chosen correspond to the actual risk of high radon levels.

National and regional surveys of radon in UK homes have shown that the majority of homes above the Action Level are likely to be found in the counties of Cornwall and Devon (Wrixon et al., 1988). More than 8000 homes in these counties have been surveyed by NRPB, using passive etched-track radon monitors, for 3, 6, or 12 mo. Techniques have been developed to use these data to estimate the proportion of homes in a given area likely to exceed a threshold (Miles et al., 1990). If Cornwall and Devon are divided into 5-km squares, all but a few squares have results of measurements of radon in homes; the squares without measurements have little or no population. For each grid square, the geometric mean radon concentration was calculated from the data available. Because of the wide spread of radon levels even within small areas, and the small numbers of measurements in some squares, these raw values are often misleading. For this reason, geometric infilling and smoothing routines were used to obtain a corrected estimate of the geometric mean in each case. From these results, the proportion of the housing exceeding the Action Level in each square was estimated, and the results mapped. Figure 1 shows the smoothed results of this exercise. On the basis of this work, the whole of Cornwall and Devon have been declared an Affected Area (Miles et al., 1990).

Surveys of radon in homes in the counties of Derbyshire, Northamptonshire, and Somerset will be made in 1991 to facilitate designation, if necessary, of further Affected Areas.

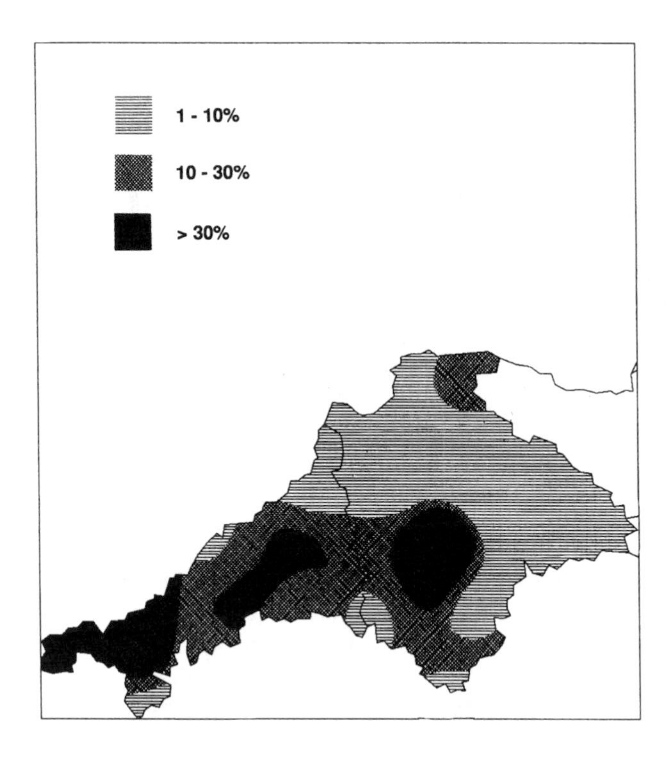

Figure 1. Estimated proportion of homes with radon measurement exceeding the Action Level (>200 Bq m^{-3}) in Cornwall and Devon, UK. Data smoothed.

3. Present homes in Affected Areas should have radon measurements. Homeowners with radon concentrations above 200 Bq m^{-3} should reduce them before the occupants receive a further time-integrated concentration of 1500 Bq m^{-3} y. The annual average of the concentration should then become as low as reasonably practicable.

Even within areas that are geologically homogeneous, a wide spread of radon levels is found in homes. This is due to local variations in the nature of the ground, details of the design and construction of the home, and the habits of the occupants. In order to discover which homes have high radon concentrations, it is therefore necessary to make measurements in each home that might be affected. The Department of the Environment (DOE) has agreed to fund radon measurements in such homes throughout England, and similar arrangements have been made with other Government departments for homes in Wales, Scotland, and Northern Ireland.

Householders who wish to have a radon measurement apply to NRPB, and if they are living in an area where radon may be a problem, they are sent two passive radon detectors. One detector is placed in the living area and one in the bedroom for a 3-mo measurement. The results are averaged and corrected for the typical seasonal variations in radon levels. Because there is some uncertainty in the seasonal correction, further measurements are sometimes made. If the corrected result is between 130 and 300 Bq m^{-3}, a second set of 3-mo measurements is undertaken, starting approximately 3 mo later, and a final result derived from the two sets. If the initial result is below 130 Bq m^{-3}, or the final result is below 200 Bq m^{-3}, the householder is advised that the concentration is below the Action Level, and remedial measures are not necessary. If the initial result is above 300 Bq m^{-3}, or the final result is above 200 Bq m^{-3}, the owner is advised to take action to reduce the radon concentration. Householders are given a booklet, *The Householders' Guide to Radon* (DOE, 1990), which offers advice on remedial measures. Various remedial techniques are described, and their application to different situations is discussed. The NRPB has devised a program to examine householders' responses to this advice and the effectiveness of the remedial measures taken.

The advice given by NRPB on the timescale for remedial action was introduced to ensure that greater urgency would be given to reducing the highest radon concentrations. The radon exposure specified, 1500 Bq m^{-3} y^{-1}, is the same as the lifetime exposure of an average member of the UK population.

4. Within Affected Areas, localities should be delimited for precautions against radon in future homes. Such localities should be established by the appropriate Government authorities and periodically reviewed.

It is clearly undesirable that new homes should have high radon levels; the technology to prevent this and anti-radon measures are more easily implemented during construction or before the home is occupied. In recognition, the UK Government issued Interim Guidance on the Construction of New Dwellings (DOE, 1988), which defined an area within the counties of Cornwall and Devon where precautions should be taken against radon. At the time, radon measurements were insufficient to outline the areas with the greatest radon problems, so the area was defined mainly on the basis of the underlying geology. The Government intends to issue, in 1991, a new Approved Document for the Building Regulations 1985, which will take into account the latest NRPB advice on Radon-Affected Areas.

5. For such localities, Government authorities should decide whether all homes should be constructed with precautions against radon or constructed in the ordinary way, tested for high radon levels, and remedied if necessary.

Two possible strategies for avoiding high radon levels in new homes are described in this recommendation. The choice of strategy is a matter for the Government rather than for NRPB. So far, the Government seems inclined to adopt the strategy of requiring full precautions within Affected Areas, where most of the homes with high radon levels are likely to be found, with a lower level of protection where fewer are to be found.

6. Homes with precautions against radon should be constructed in accordance with approved guidance issued by the appropriate Government authorities. Compliance with the guidance should offer reasonable assurance that concentrations are as low as reasonably practicable and at least below the Action Level.

Construction techniques for use in defined areas were described in the Interim Guidance (DOE, 1988). Provision for subfloor ventilation was advised in all cases, either by providing vents underneath suspended timber or concrete floors, or by providing a radon sump underneath solid floors. Mechanical air extraction through the vents was not required, but it was seen as a simple matter to add this later if necessary. For the areas where the most severe radon problems were expected, it was advised that a continuous membrane should also be installed across the whole area of the building, with particular attention paid to lapping of sheets and to seals around ducts penetrating the membrane. Testing of the efficacy of such preventive measures by the Government and NRPB is continuing. In common with various other UK building regulations, it is not mandatory to use the techniques suggested: other preventive techniques are also acceptable, provided they are effective.

RADON EPIDEMIOLOGY

In addition to implementing the strategy discussed above for preventing excessive exposure of the public to radon, Government departments and NRPB are supporting epidemiological studies to clarify the risk from radon in homes. Of particular note is a case-control study by the Imperial Cancer Research Fund and NRPB, which is also partly funded by the Commission of the European Communities. This study is designed to examine the current and past radon exposure of 600 lung-cancer cases in Cornwall and Devon and twice as many controls. Other epidemiological studies include cohort studies of radon and cancers among the residents of Cornwall and Devon, among children in England and Wales and among coal miners, and case control studies of radon and leukaemias—in adults and children—throughout the two countries.

CONCLUSIONS

In the UK, radon has been recognized as a serious environmental hazard in some areas. A sensible strategy has been developed to prevent problems occurring in new homes, to identify existing homes with high radon levels, and to reduce these to acceptable levels. Supporting health studies are also being conducted to refine risk estimates, with the intent to resolve the matter by the end of the decade.

REFERENCES

BEIR (Committee on the Biological Effects of Ionizing Radiation). **1988**. *BEIR-IV: Health Risks of Radon and Other Internally Deposited Alpha-Emitters*. NAS/NRC, Washington DC.

DOE. **1988**. *Building Regulations, 1985. Part C: Radon. Interim Guidance on Construction of new Dwellings*. Department of the Environment, London.

DOE. **1990**. *The Householders' Guide to Radon* (Second Edition), Department of the Environment, HMSO, London.

ICRP (International Commission on Radiological Protection). **1987**. *Lung Cancer Risk from Indoor Exposure to Radon Daughters*, ICRP Publication 50. Pergamon Press, Oxford.

James, AC. **1987**. A reconsideration of cells at risk and other key factors in radon daughter dosimetry, pp 400-418. In: *Radon and its Decay Products: Occurrence, Properties and Health Effects*. American Chemical Society, Washington, DC.

Miles, JCH, BMR Green, PR Lomas, and KD Cliff. **1990**. *Radon Affected Areas: Cornwall and Devon*. Documents of the National Radiological Protection Board, 1, 4, 37-43. HMSO, London.

NRPB (National Radiological Protection Board). **1987**. *Exposure to Radon Daughters in Dwellings*, NRPB ASP 10. HMSO, London.

NRPB. **1990**. *Limitation of Human Exposure to Radon in the Home,* Documents of the National Radiological Protection Board, No. 1. HMSO, London.

Wrixon, AD, BMR Green, PR Lomas, JCH Miles, KD Cliff, EA Francis, CMH Driscoll, AC James, and MC O'Riordan. **1988**. *Natural Radiation Exposure in UK Dwellings*, NRPB R-190. HMSO, London.

QUESTIONS AND ANSWERS

Q: Schlesinger, Israel. My question is the urgency limit of 1500 Bq m^{-3} in your earlier recommendation. Should you now go down to 750 because you just went down from 400 to 200 Bq m^{-3} for the new reaction level?

A: Miles. The earlier recommendations didn't contain 1500 explicitly. They had a recommendation that, at higher radon levels, particularly about 1000 Bq m^{-3}, action should be taken more urgently.

Q: Osborne, AECL Research, Canada. I think you were using your words very carefully so I was intrigued that in recommendation 3 you used the expression "as low as practicable," and in recommendation 6, or later on, it goes "as low as reasonably achievable." Was that intentional?

A: Miles. It should have been practicable in both cases. The reason that it is practicable rather than achievable is simply that practicable is a word commonly used in English case law and reasonably achievable is not, and therefore there are legal precedents for saying what is reasonably practicable.

Q: Johnson, PNL. I was intrigued by your last slide. Unfortunately, my attention was distracted, and I didn't quite hear. Was that all of the ongoing radiation-related epidemiological studies?

A: Miles. Those are all *radon* studies. There are other studies of other types of radiation, in particular, associated with Sellafield.

Q: Johnson, PNL. Yes, I know that. That's why I was looking at leukemia. Was that childhood leukemia specifically, the last two?

A: Miles. Could we have the last slide back, please?

Q: Johnson, PNL. I'm sure you are aware there's a presentation tomorrow about potential childhood leukemia from radon.

A: Miles. Right. Yes, there are several proposed studies; in fact, only the last one is a proposed study, the others are either ongoing or have got funding and are just about to get going. There is a correlation study of childhood cancer, all childhood cancer, a correlation study of leukemia, and the last one is a proposed study, a case control study, of leukemia.

Q: Johnson, PNL. That would be 1000 new cases per year that you introduce into that study?

A: Miles. Yes, 1000—not cases—1000 subjects. The size is subjects, which includes cases and controls.

Q: Suess, WHO. You have reached now, with your presentation of the present recommendations, what was proposed by WHO in 1985. Now my question obviously is, now that you have managed to make the first step when are you going to take the second step and reduce the action level again by half?

A: Miles. There are practical difficulties, as I am sure you know, in reducing the action level. I know there are proposals in the United States to reduce the action level, and some people, I understand, are quite alarmed by that proposal because of the practical difficulties.

INDEX TO AUTHORS